Cadmium Toxicity Mitigation

Amrit Kumar Jha • Nitish Kumar
Editors

Cadmium Toxicity Mitigation

Springer

Editors
Amrit Kumar Jha
Krishi Vigyan Kendra, Sahibganj
Birsa Agricultural University
Sahibganj, Jharkhand, India

Nitish Kumar
Department of Biotechnology
Central University of South Bihar
Gaya, Bihar, India

ISBN 978-3-031-47389-0 ISBN 978-3-031-47390-6 (eBook)
https://doi.org/10.1007/978-3-031-47390-6

This Springer imprint is published by the registered company Springer Nature Switzerland AG
The registered company address is: Gewerbestrasse 11, 6330 Cham, Switzerland

Paper in this product is recyclable.

Preface

Cadmium, a transition metal, has the chemical symbol of Cd. Cadmium was discovered in Germany in 1817, by a German scientist, Friedrich Strohmeyer. It is regarded as a severe environmental pollutant due to its extensive industrial use. Cd contamination of soil and water has drawn attention in the past decades, due to Cd contamination of groundwater and food sources, it still poses a serious risk to human health on a global scale. The speciation of Cd largely determines its toxicity, mobility, and bioavailability. Cd behavior in soil, soil-plant transfer, and accumulation in various plant parts depend on the chemical form, type of plant, and physico-chemical characteristics of the soil. Cd speciation and behavior in soil are significantly influenced by the soil microbial population. Plants can prevent Cd from entering the cell through external rejection of organic acids, amino acids, and proteins that combine with heavy metals. Accumulation of Cd in plants inevitably disrupts homeostasis, damages cell structure, and affects the dynamic balance of the antioxidant enzyme system.

This book "Cadmium Toxicity Mitigation" presents the most recent research on cadmium's chemistry, sources, acute and long-term impacts on the environment, human health, and mitigation measures. This book offers thorough examination of cadmium exposure, toxicity, and toxicity prevention so that readers can evaluate the hazards associated with it efficiently. The book compiles the most recent research on how cadmium affects the environment, human health, and remediation. Recent methods in cadmium detoxification, speciation, and molecular mechanisms are included, and it offers all the knowledge required for efficient risk assessment, prevention, and countermeasure. This book is a valuable resource to students, academics, researchers, and environmental professionals doing field work on cadmium contamination throughout the world.

Sahibganj, Jharkhand, India — Amrit Kumar Jha
Gaya, Bihar, India — Nitish Kumar

Preface

Acknowledgment

First and foremost, we would like to praise and thank God, the almighty, who has granted countless blessing, knowledge, and opportunity to accomplish this work. Thanks to all the authors of the various chapters for their contributions. It had been a bit of a long process from the initial outlines to developing the full chapters and then revising them in the light of reviewer's comments. We sincerely acknowledge the author's willingness to go through this process. We also acknowledge the work and knowledge of the members of our review panels, many of which had to be done at short notice. Thanks to all the people at Springer Nature especially Dr. Sofia Costa and Mr. Srinivasan Manavalan with whom we corresponded for their advice and facilitation in the production of this book.

Acknowledgments

Contents

Editors and Contributors

Editors

Amrit Kumar Jha is Scientist (Soil Science) and Head at Krishi Vigyan Kendra, Sahibganj, Birsa Agricultural University, Ranchi, Jharkhand, India. Dr. Jha completed his doctoral research in subject Soil Science and Agricultural Chemistry from Birsa Agricultural University, Ranchi, Jharkhand, India. He has more than 18 years of research experience in the field of soil chemistry, soil fertility, and soil pollution. He has published more than 40 research articles on cadmium in international and national journals, more than 11 technical bulletins and 2 books. He has received many awards such as Young Scientist Award conferred by Society of Krishi Vigyan in 2018, Best KVK Scientist Award conferred by Indian Society of Extension Education in 2018, and Best KVK Scientist Award conferred by Society of Krishi Vigyan in 2020.

Nitish Kumar is Senior Assistant Professor at the Department of Biotechnology, Central University of South Bihar, Gaya, Bihar, India. Dr. Kumar completed his doctoral research at the Council of Scientific & Industrial Research–Central Salt & Marine Chemicals Research Institute, Bhavnagar, Gujarat, India. He has published more than 70 research articles on cadmium in international and national journals, more than 20 book chapters and 10 books with Springer and Taylor & Francis. He has a wide area of research experience in the field of Agriculture & Crop improvement and Microbial & Environmental Biotechnology. Dr. Kumar is a recipient of the Young Scientist Award from the Science and Engineering Research Board (SERB) in 2014. He has received many awards/fellowships/projects from various prestigious government organizations like CSIR, DBT, ICAR and SERB-DST, BRNS-BARC, among others. He is a reviewer for various international journals and serves as an associate editor of the journal Gene (Elsevier).

Contributors

N. Acelas Grupo de Investigación Materiales con Impacto, MAT&MPAC, Facultad de Ciencias Básicas, Universidad de Medellín, Medellín, Colombia

Awais Ahmad Department of Plant Production, College of Food and Agriculture Sciences, King Saud University, Riyadh, Saudi Arabia

Naheed Akhtar Department of Biochemistry, Government College University Faisalabad, Faisalabad, Pakistan

Tayyaba Akhtar Institute of Soil and Environmental Sciences, University of Agriculture Faisalabad, Faisalabad, Pakistan

Arwa Abdulkreem AL-Huqail Department of Biology, College of Science, Princess Nourah bint Abdulrahman University, Riyadh, Saudi Arabia

Maha Aljabri Department of Crop Sciences, Faculty of Agriculture, Menoufia University, Shebeen El-Kom, Egypt

Majed Alotaibi Department of Plant Production, College of Food and Agriculture Sciences, King Saud University, Riyadh, Saudi Arabia

Jackeline Lizzeta Arvizu-Gómez Secretaría de Investigación y Posgrado, Centro Nayarita de Innovación y Transferencia de Tecnología (CENITT), Universidad Autónoma de Nayarit, Tepic, Nayarit, Mexico

Muhammad Asad Institute of Soil and Environmental Sciences, University of Agriculture Faisalabad, Faisalabad, Pakistan

Martin L. Battaglia Corporate Engagement, The Nature Conservancy, Arlington, VA, USA

Renata Bešta-Gajević Department of Biology, Faculty of Science, University of Sarajevo, Sarajevo, Bosnia and Herzegovina

P. Chaitra Department of Soil Science, PAU, Ludhiana, Punjab, India

Shahzad Ali Shahid Chatha Department of Chemistry, Government College University Faisalabad, Faisalabad, Pakistan

Vivek Chintada Department of Zoology, Sri Venkateswara University, Tirupati, India

Rohan J. D'Souza Department of Botany, St. John's College, Agra, Agra, India

Sabina Dahija Department of Biology, Faculty of Science, University of Sarajevo, Sarajevo, Bosnia and Herzegovina

Sridhar Dumpala Department of Aquaculture, Adikavi Nannaya University, Rajamahendravaram, Andhra Pradesh, India

Paulo J. C. Favas School of Life Sciences and the Environment, University of Tras-os-Montes e Alto Douro, Vila Real, Portugal
Faculty of Sciences and Technology, MARE – Marine and Environmental Sciences Centre, University of Coimbra, Coimbra, Portugal

E. Flórez Grupo de Investigación Materiales con Impacto, MAT&MPAC, Facultad de Ciencias Básicas, Universidad de Medellín, Medellín, Colombia

A. Forgionny Grupo de Investigación Materiales con Impacto, MAT&MPAC, Facultad de Ciencias Básicas, Universidad de Medellín, Medellín, Colombia

Iqra Ghafoor Institute of Soil and Environmental Sciences, University of Agriculture Faisalabad, Faisalabad, Pakistan

Alejandro Hernández-Morales Facultad de Ciencias Químicas, Universidad Autónoma de San Luis Potosí, San Luis Potosí, San Luis Potosí, Mexico
Facultad de Estudios Profesionales Zona Huasteca, Universidad Autónoma de San Luis Potosí, Ciudad Valles, San Luis Potosí, Mexico

Shumailah Ishtiyaq Department of Botany, St. John's College, Agra, Agra, India

Sadia Javed Department of Biochemistry, Government College University Faisalabad, Faisalabad, Pakistan

C. Jimenez-Orozco Grupo de Investigación Materiales con Impacto, MAT&MPAC, Facultad de Ciencias Básicas, Universidad de Medellín, Medellín, Colombia

Erna Karalija Laboratory for Plant Physiology, Department of Biology, Faculty of Science, University of Sarajevo, Sarajevo, Bosnia and Herzegovina

Harsh Kumar Department of Microbiology, School of Life Sciences, Dr. B.R. Ambedkar University, Agra, Agra, India

Adisa Parić Faculty of Science, University of Sarajevo, Sarajevo, Bosnia and Herzegovina

Manoj S. Paul Department of Botany, St. John's College, Agra, Agra, India

Selma Pilić Department of Biology, Faculty of Science, University of Sarajevo, Sarajevo, Bosnia and Herzegovina

Vinamrata Ponia Department of Botany, R.B.S. College, Agra, Agra, India

Abdul Qadeer Institute of Soil and Environmental Sciences, University of Agriculture Faisalabad, Faisalabad, Pakistan

Kakarlapudi Ramaneswari Department of Aquaculture, Adikavi Nannaya University, Rajamahendravaram, Andhra Pradesh, India

Gisela Adelina Rolón-Cárdenas Facultad de Estudios Profesionales Zona Huasteca, Universidad Autónoma de San Luis Potosí, Ciudad Valles, San Luis Potosi, Mexico

Jesús Rubio-Santiago Facultad de Ciencias Químicas, Universidad Autónoma de San Luis Potosí, San Luis Potosí, San Luis Potosí, Mexico

Tanvi Sahni Department of Soil Science, PAU, Ludhiana, Punjab, India

Mahmoud F. Seleiman Department of Plant Production, College of Food and Agriculture Sciences, King Saud University, Riyadh, Saudi Arabia
Department of Biology, Faculty of Applied Science, Umm Al-Qura University, Mecca, Saudi Arabia

Alisa Selović Laboratory for Analytical Chemistry, Department of Chemistry, Faculty of Science, University of Sarajevo, Sarajevo, Bosnia and Herzegovina

Ashish Sharma Department of Botany and Environment Science, DAV University, Jalandhar, Punjab, India

Apoorva Shekhar Department of Veterinary Physiology and Biochemistry, College of Veterinary Science, Guru Angad Dev Veterinary and Animal Sciences University, Ludhiana, Punjab, India

Laraib Shoukat Department of Biochemistry, Government College University Faisalabad, Faisalabad, Pakistan

R. Sikka Department of Soil Science, PAU, Ludhiana, Punjab, India

Annu Singh Department of Soil Science, PAU, Ludhiana, Punjab, India

Chanchal Singh Department of Veterinary Physiology and Biochemistry, College of Veterinary Science, Guru Angad Dev Veterinary and Animal Sciences University, Ludhiana, Punjab, India

Raghubir Singh Department of Veterinary Public Health, College of Veterinary Science and Animal Husbandry, Jalukie, Nagaland, India
Central Agricultural University Imphal, Imphal, Manipur, India

Ruth Elena Soria-Guerra Facultad de Ciencias Químicas, Universidad Autónoma de San Luis Potosí, San Luis Potosí, San Luis Potosí, Mexico

Dipti Srivastava Department of Biochemistry, SLSBT, CSJM University, Kanpur, Kanpur, Uttar Pradesh, India

Neerja Srivastava Department of Biochemistry, SLSBT, CSJM University, Kanpur, Kanpur, Uttar Pradesh, India

Mirel Subašić Faculty of Forestry, University of Sarajevo, Sarajevo, Bosnia and Herzegovina

Sana Ullah Department of Biochemistry, Government College University Faisalabad, Faisalabad, Pakistan

Mayank Varun Department of Botany, Hislop College, Nagpur, India

Diksha Verma Department of Soil Science, PAU, Ludhiana, Punjab, India

Neha Verma Department of Botany and Environment Science, DAV University, Jalandhar, Punjab, India

Rizwan Yasin Institute of Soil and Environmental Sciences, University of Agriculture Faisalabad, Faisalabad, Pakistan

Naqshe Zuhra Institute of Soil and Environmental Sciences, University of Agriculture Faisalabad, Faisalabad, Pakistan

Part I
Cadmium in the Environment

The Cadmium in Soil and Plants

Sana Ullah, Sadia Javed, Naheed Akhtar, Laraib Shoukat, and Shahzad Ali Shahid Chatha

Abstract The hazardous heavy cadmium (Cd) metal deposits into soil through various anthropogenic activities. It is harmful to both animals and plants, causing physiochemical, morphological, and structural changes in plants and malfunctioning of multi-organ and ultimately death in animals. Living things are often exposed to Cd directly through the food chain. Cadmium contamination in soils and groundwater causes harmful effects on the human body by consuming the affected food and water. Several factors like soil pH, electrical conductivity, organic matter, microbial activities in the soil, and root exudates influence the bioavailability of Cd. Cd toxicity in plants can cause nutrient imbalance, oxidative damage, disturbance in carbon metabolism, and reduction in photosynthetic efficiency. Cd can enter the body through oral, respiratory, and skin routes, causing harmful effects. This chapter highlights cadmium contamination in soil and plants and their potential health risks.

Keywords Cadmium · Soil · Plants · Physiochemical · Morphological · Structural changes

1 Introduction

The rapid expansion of human populations, leading to unforeseen industrialization and urbanization, exacerbates environmental degradation through the escalation of heavy metal pollution (Rehfeld et al. 2017). Day by day, the contamination of air, water, and soil due to environmental pollutants continues to escalate, magnifying the detrimental consequences it inflicts on living organisms (Akguc et al. 2010). The escalating recognition of cadmium (Cd) as a perilous and toxic heavy metal, belonging to Class 2B of the periodic table, highlights its substantial role as a

S. Ullah · S. Javed (✉) · N. Akhtar · L. Shoukat
Department of Biochemistry, Government College University Faisalabad, Faisalabad, Pakistan
e-mail: sadiajaved@gcuf.edu.pk

S. A. S. Chatha
Department of Chemistry, Government College University Faisalabad, Faisalabad, Pakistan

A. K. Jha, N. Kumar (eds.), *Cadmium Toxicity Mitigation*,
https://doi.org/10.1007/978-3-031-47390-6_1

significant industrial and environmental contaminant in recent times (Mishra et al. 2019). Cadmium (Cd) can be acquired naturally through volcanic activity and the weathering of rocks, or it can be released into the environment as a consequence of human activities, including agricultural practices and waste, livestock farming and manure, timber harvesting and wood waste, urban waste disposal, municipal sewage sludge, various organic waste materials, solid waste management, metal manufacturing for fertilizers, coal combustion byproducts such as fly ash, product waste, and deposition from the atmosphere (Akguc et al. 2008; Aslam et al. 2019; Diaz-Munoz et al. 2012; Ozyigit et al. 2016; WHO/FAO Expert Consultation 2003). The presence of these factors amplifies the significance of Cd as a contaminant.

Cadmium has garnered attention as a hazardous substance from various sustainability non-governmental organizations, including the United States Environmental Protection Agency. It is often ingested by living organisms either through inhalation or the food chain. The World Health Organization has recommended a maximum tolerable intake of 0.007 mg per kilogram of body weight and established the weekly acceptable level of Cd exposure at 50 grams (World Health Organization 2000). The soil becomes contaminated with Cd due to the output of industrial establishments, sewage waste disposal, application of manure to farmland, and aerospace deposition. Furthermore, Cd is introduced in the topsoil through the use of diesel-powered machinery, the application of organic manure and long-term treated sludge to agricultural areas, along with pesticides (Aslam et al. 2019; Ozyigit et al. 2016; Tabelin et al. 2018). Cadmium, denoted by the chemical symbol "Cd," possesses an atomic number of 48, atomic weight of 112.41 g mol^{-1}, density of 8.7 g cm^{-3}, boiling point of 766.8 °C, and melting point of 321 °C. It is a silver-colored, soft, machinable heavy metal that does not occur naturally in isolation. Among the various cadmium salts, CdS, $CdCl_2$, and $CdSO_2$ are widely recognized. From a chemical standpoint, Cd shares similarities with zinc (Zn) regarding plant uptake and metabolic functions. However, unlike zinc, cadmium exhibits toxic effects on crops, invertebrates, and human beings (Garbisu and Alkorta 2001).

Cadmium (Cd), classified as an optional trace element, is widely regarded as the fourth Deadliest element to vascular plants and the fifth most poisonous metal to vertebrates (Burr et al. 1994). Cadmium (Cd), owing to its noxious capability to human beings and its respective portability within the soil–plant system, has emerged as the foremost non-essential heavy metal of interest within fields of plant nutrition and soil science. The significance of Cd contamination in the food chain was initially highlighted by Schroeder and Balassa in 1963, who raised concerns about the presence of Cd in fertilizers and soil additives. The occurrence of severe human health issues, particularly the "Itai-itai" disease, resulting from industrial pollution of agricultural areas in Japan with Cd-rich waste streams, further fueled scientific interest in studying the impact of Cd on soils and plants, with observations dating back to the 1970s (Kobayashi et al. 1978).

Developed nations have established comprehensive monitoring and regulatory systems to record and control point source emissions, including those of cadmium (Cd). In contrast, developing nations are catching up with advancements in analytical technology and introducing emission limits for companies that release Cd into the environment. As industries adopt more efficient technologies and improve waste treatment procedures, Cd emissions from these sources are decreasing in many countries. However, accurately measuring the fluxes from natural point sources such as volcanoes and forest fires poses a greater challenge (Vander Zanden and Rasmussen 1996). Accurately quantifying these exchanges is essential for effective soil and plant management on a global scale. However, the diffuse introduction of Cd into soils through fertilizers and sewage biosolids (sludges) poses a greater challenge. Fertilizers play a vital role in sustainable agricultural production systems worldwide, and their significance is even greater in developing regions where food supplies are often limited and the soil resource is frequently depleted. In these areas, the application of fertilizers becomes essential for profitable agricultural production. Recycling human waste in the form of sewage biosolids to land is a sensible approach to utilize the valuable nutrients contained in these materials and reduce marine and aquatic pollution associated with biosolid disposal.

Crops are negatively impacted by cadmium's toxic effects because it prevents the fixation of carbon and lowers chlorophyll production, leading to diminished photosynthesis (Benjamin et al. 2018). Prolonged exposure to cadmium leads to the suppression of lateral root growth and induces Plant phenotypic and structural modifications (Cossarizza et al. 2019). Cadmium exposure in crops results in a depletion in leaves' comparative water content, stomatal conductance, and transpiration, leading to the occurrence of osmotic stress (Rizwan et al. 2016). In addition, cadmium toxicity adversely affects the absorption and transportation of mineral elements in plants, resulting in decreased yield. Excessive production of reactive oxygen species (ROS) occurs as a consequence of cadmium exposure, accelerates membrane detriment, and cell organelle death in plants (Abbas et al. 2017). Cadmium, being an extremely hazardous heavy element, exerts detrimental effects on various physiological processes, hindering development, and ultimately leading to reduced crop yields. Influences on the ability to be absorbed and lethality of Cd by the physical properties as well as the soil's chemical composition (Violante et al. 2010). As soil pH decreases, cadmium undergoes a change from an immobile form to one that is more flexible, increasing its suitability for utilize by plants (Livingston et al. 2020). Cadmium can accumulate in various organs of the human body, including the kidneys, liver, lungs, thymus, testes, heart, epididymis, prostate, and salivary glands, leading to multi-organ dysfunction along with potentially fatal outcomes. The Itai-Itai epidemic, involving 184 confirmed patients and 388 potential casualties, stands as a prominent example of the environmental risks associated with cadmium contamination. Inadequate agricultural proceedings as well as making use of dangerous agrochemicals enable cadmium to enter the human food chain. While trace elements are typically more concentrated in plant roots, many vegetables (such as lettuce and spinach) accumulate due to their swift intake and mobility throughout the crop's system (Van Hal et al. 2000).

Quantitative assessments of cadmium (Cd) volume in edibles reveal that herbaceous plants and grains play a significant role in contributing to Cd levels, even though they can also be found in low quantities in animal products. Estimated daily

Cd intake from food materials for adults in different countries ranges from 10.0 to 30.0 μg (Sahmoun et al. 2005). Studies have cited that cadmium levels in vegetables range from 0.001 to 0.124 mg/kg, and the consumption of green vegetables contributes to more than 70–90% of being exposed to Cd in people.

2 Cadmium Contamination in Soil

An extremely dangerous trace metal that is quite abundantly present in the natural world is cadmium (Cd), capable of causing significant and persistent adverse effects on human health even at very low concentrations (Khan et al. 2015a) (Gatoo et al. 2014; Kabata-Pendias 2000). Cadmium (Cd) is found in the lithosphere, sedimentary rocks, and soil at concentrations of 0.2 mg kg^{-1}, 0.3 mg kg^{-1}, and 0.53 mg kg^{-1}, respectively. However, in soil water and groundwater, the concentrations of Cd are much lower, measuring 5.0 μg L^{-1} and 1 μg L^{-1}, respectively (Cai and Huang 2016). Poisoning of soils and underground water with cadmium stems from a combination of natural processes and human activities, leading to detrimental consequences as it enters the human body through the consumption of contaminated drinking water and food sources (Cai and Huang 2016; Zhang et al. 2013). In addition, unregulated and unsafe disposal of garbage has significantly elevated cadmium (Cd) levels in soil and water bodies. A report from the late 1980s indicated that geogenic sources and anthropogenic activities mobilized approximately 24,000 and 4.5 metric tons per year, respectively, of Cd into the biosphere. This clearly demonstrates the dominant role of human-induced activities in the release of Cd into the environment (Attridge et al. 2000).

Windblown soil particles serve as the primary natural source of atmospheric cadmium (Cd) contamination, with additional contributions from ocean spritz, volcanic eruptions, blazes, and cosmic ash. The major sources of cadmium in the environment are depicted in Fig. 1. In California, it has been estimated that forest fires have doubled the normal amounts of cadmium in lakes and rivers (Burke et al. 2013). Natural emissions of cadmium (Cd) account for 1400 tones of the world's yearly emissions, whereas human sources account for about 2983 tones. Cd is naturally ubiquitous and can be found in various locations, including remote areas such as the South and North poles and ice peaks of the Himalayas (Lee et al. 2008). This suggests that human activities play a more significant role in cadmium (Cd) contamination's component in soil and the initial material's deterioration is directly associated; however, improper practices have disrupted the equilibrium between inputs (such as atmospheric precipitation and industrial or agricultural operations) and outputs (by crop absorption, erosion, and leaching) of Cd from the soil (Six and Smolders 2014). The standard global cadmium (Cd) content of uncontaminated soil is approximately 3.6%, although variations exist across continents, countries, and soil types. A soil Cd concentration exceeding 30% is considered a critical limit for Cd pollution. Interestingly, studies have shown that the level

Fig. 1 The major sources of cadmium in the environment

of Cd in soil decreases proportionally when the separation between industrial locations and urban areas widens (Belon et al. 2012; Vukmanovic-Stejic et al. 2006).

The primary source of poisoning with cadmium in soil stems from the deterioration of different mineral content and gemstones within the soil. Sedimentary rocks have been found to contain the highest levels of Cd, ranging from 0.1% to 26%. In comparison, Cd content in metamorphic rocks ranges from 1.1% to 10%, and in igneous rocks, it is typically in the range of 0.7% to 2.5% (Dutta et al. 2020; Khan and Tamer 2010).

2.1 The Behavior of Cadmium in Soil and the Rhizosphere

From a biological perspective, cadmium is rarely considered necessary for crops. However, plants can easily acquire Cd due to its association with micronutrients in the rhizosphere, which is the region of soil in close proximity to plant roots (Khan and Tamer 2010). Cadmium (Cd) has been detected in soil at concentrations ranging from 0.07 to 1.1 mg kg^{-1} soil (World Health Organization) (Joint WHO/FAO/UNU Expert Consultation 2007). The threshold level for cadmium (Cd) in agricultural soil is approximately 100 mg kg^{-1} (Khan et al. 2015b). In the soil solution, Cd primarily exists as cadmium ions forming cadmium complexes, including both biological and inorganic compounds. Cd exists in both positively and negatively charged forms and can be found in soil (Kabata-Pendias 2004). The soil mixture contains negatively charged forms of cadmium including $CdCl^{3-}$, $Cd(OH)^{3-}$, $Cd(OH)_4{}^{2-}$, and $Cd(HS)_4{}^{2-}$, in contrast positively charged forms include $CdCl^+$, $CdOH^+$, $CdHS^+$, and $CdHCO^{3+}$. Studies have shown that approximately 99% of Cd in the soil solution exists in the form of free ions (Kabata-Pendias 2004). The fractioning of cadmium in soils is influenced by various chemical processes, including the formation of cadmium ligands. The ligands primarily affect reactions, which can be of both inorganic as well as organic substances nature (Shahid et al. 2014) oxidation and reduction conditions (Shan et al. 2013), soil pH (Kunito et al. 2012), temperature, and concentration of metals (Lockshin and Corsi 2012). The fractioning of cadmium in soil plays a crucial role in regulating Cd toxicity (Rizwan et al. 2017b). The biogeochemical behavior of Cd is dependent on how many free Cd ions are available in the soil medium (Fritsche et al. 2016). The buildup of Cd in plant roots changes depending on Cd concentrations in the rhizosphere and the classes of the plant. For example, in a study, it was observed that maize (*Zea mays* L.) exhibited higher Cd accumulation in the cell wall fraction compared to broad bean (*Vicia faba* L.) seedlings (Lozano-Rodriguez et al. 1997).

2.2 The Components Impacting Cd Dynamics

The biological distribution of cadmium is affected by various factors including root exudates, cation exchange capacity (CEC), biological material, soil pH content, and microbial activity in the soil these factors collectively affect the extent to which Cd is available for uptake by plants and other organisms (Bhushan and Jung 2008; Williams and Zweig 2016) (Fig. 2).

Soil pH plays a pivotal role in regulating the partitioning of cadmium (Cd) and its bioavailability (Wang et al. 2016). The presence of cadmium (Cd) in the soil can take on different chemical forms depending on the soil pH. Acidic soil conditions have been found to have a significant impact on the solubility of Cd in the soil solution. Under such conditions, Cd can undergo a conversion from inert forms, for instance, carbonates, manganese, and iron (Fe) oxides, in order to have more versatile forms.

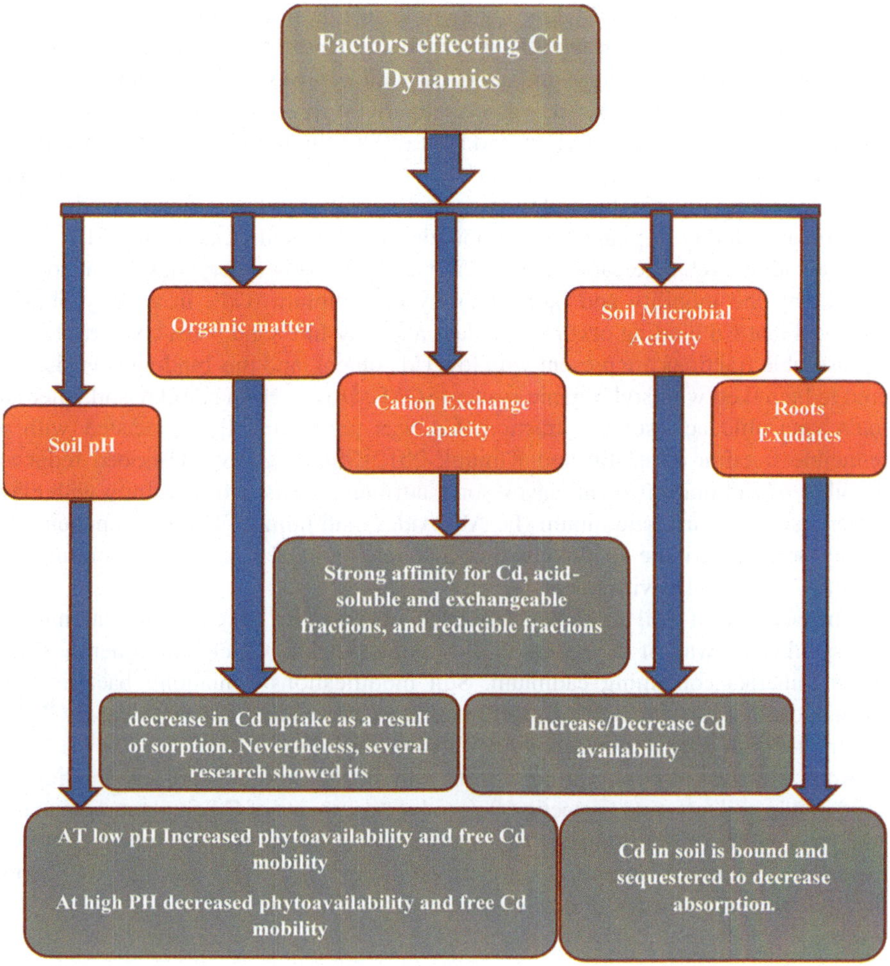

Fig. 2 Factors affecting the dynamics of cadmium

This conversion enhances the phytoavailability and mobility of free Cd in the soil (Qi et al. 2018). pH has an impact on how easily cadmium (Cd) dissolves in soil, with a starting point typically observed around pH 6.

At this pH, Cd can form complexes with organic matter and become adsorbed onto mineral surfaces, affecting its solubility in the soil (Sullivan and Artino 2013). Conversely, an increase in soil pH results in higher alkalinity, which impacts the adsorption of cadmium (Cd) onto soil particles. It has been observed that the pH of soil plays a crucial part in determining the cadmium accumulation in rice grains (Amodei et al. 2016). Elevated soil pH has a detrimental effect on the phytoavailability of cadmium (Cd) due to decreased adsorption and precipitation, leading to reduced availability of A soil mixture with free Cd (Meng et al. 2018).

Soil organic matter (SOM) plays a significant role in determining the bioavailability of cadmium (Cd) through the diverse compounds formed in the soil mixture. The source material, quantity, and chemical nature of SOM have an impact on the absorption and utilization of Cd. Additionally, SOM directly affects the binding and sequestration of Cd. According to Kirkham (2006), higher SOM content results in increased sorption potential, which can be up to 30 times higher compared to mineral soils. Application of biochar at levels exceeding 10% has been found to reduce the Cd′s bioavailability in plants by immobilizing it in the soil (Xiao et al. 2019).

The cation exchange capacity (CEC) of soil performs a very significant role in calculating the mobility and bioavailability of cadmium (Cd). In a study, it was observed that Cd binding occurred in clay- and clay-like sand soils, convertible and acid-soluble fractions are present with low Cd content, and this binding was discovered to be linked with soil organic matter (SOM). However, in silt-clay soil, after an interchangeable acid-soluble fraction, Cd was predominantly associated with a reducible fraction (Gusiatin and Klimiuk 2012). Studies have described reduced mobility of cadmium (Cd) in clayey soils, attributed to its strong affinity with clay mineral surfaces, iron-aluminum (Fe-Al) oxides, and humus. These components in clayey soils contribute to the binding and retention of Cd, thereby limiting its mobility in the soil environment.

The activity of soil microbes has been observed to increase the cadmium's accessibility by way of the secretion of biological acids, which subsequently solubilize minerals containing cadmium. Soil modifications containing bacteria that oxidize cadmium, such as Rhizobacteria that promote plant development (PGPRs) are crucial with regard to increasing Cd's bioavailability.

Root exudates also play a significant part in the confinement and binding of cadmium in soil, offering protection from adverse impacts of Cd on plant roots in the soil environment. Additionally, the uptake of cadmium (Cd) in plants by root exudates is reduced (Sarwar et al. 2010). Figure 2 illustrates the various factors that influence the kinetics of cadmium in soil.

3 Cadmium in Plants

The metal cadmium is extremely poisonous and primarily enters the surroundings by plant roots and subsequently accumulates in different plant parts, leading to reduced crop yields and deterioration in the quality of agricultural produce. It poses a significant threat to human and animal health as it enters the food chain. Ranked as the seventh most harmful toxin among the top 20, Cd finds its way into arable land through a combination of industrial processes and agricultural practices (White and Brown 2010).

3.1 Buildup of Cadmium in Plants

The building up of cadmium in plants is assisted by its acceptance, diffusion, and mobilization in different parts of the plant. Unreliable agriculture methods and the discharge of wastes from factories are the primary contributors to cadmium contamination in soil (Pan et al. 2010). Cd pollution in agricultural soil is attributed to the use of phosphoric fertilizer and sewage sludge. The Cadmium in the biodiversity of plants serves as an indicator of its soil content; however, its availability in plants is influenced by several other elements, such as soil pH and organic matter concentration, interactions additional ions, and additional ions, and specific plant species (Kirkham 2006; Tudoreanu and Phillips 2004).

A meta-data analysis conducted by Adams and colleagues on samples of 162 wheat and 215 barley grains revealed a positive correlation between grain cadmium (Cd) concentration and both soil total Cd content and soil pH (Adams 2012). Additionally, the study emphasized that increased nitrification, activity of microbes, and putting wastewater sludge to use elevated risk of cadmium poisoning. However, using bleaching to recover the soil was found to mitigate the likelihood of toxicity. Furthermore, the research conducted in Canada demonstrated biological substances exhibited nearly 30 times greater in comparison to inorganic soil, adsorption affinity for Cd. This highlights the crucial role of the ability of biological substances to attach to and collect Cd in the soil (Adams 2012).

The assumption that lowering pH promotes cadmium bioavailability for crops may not be true in soils that have high levels of biological material and a low pH. The transport of metals between plant roots and soil by means of diffusion and upwelling. As metals dissolve in the soil, their concentrations increase, leading to the formation of complex structures. Regardless of such intricate structures actively pulled through plant roots, they have been found to enhance metal transport through diffusion toward the plant roots in a significant manner. Multiple factors influence the accumulation of metals in plants, including the shifting of soil components and roots' ability to absorb them, storage in root cells, transportation through the xylem to plant components found over the surface, and with distribution of metals within such plant components (Meng et al. 2018; Song et al. 2017).

In order to comprehend the method of cadmium buildup in crops, it is important to first acknowledge the processes of cadmium uptake and translocation within plants. The ability of plants to uptake Cd is influenced by various factors such as the combined cadmium concentration in the soil humidity, temperature, biological carbon content, redox potential (Eh), and connections among various factors. Plants typically absorb cadmium by means of their roots. Once inside the roots, Cd could either be kept within the root tissues on the other hand transported via the xylem to the branches and leaves. It is noteworthy that Cd is both xylem and phloem mobile, meaning it can move upward through the xylem and be distributed to various parts of the plant via the phloem (Song et al. 2017). There are two primary mechanisms through which cadmium (Cd) is translocated into then transferred to the grains from the trees. Following mechanisms include (i) translocation mediated by the xylem,

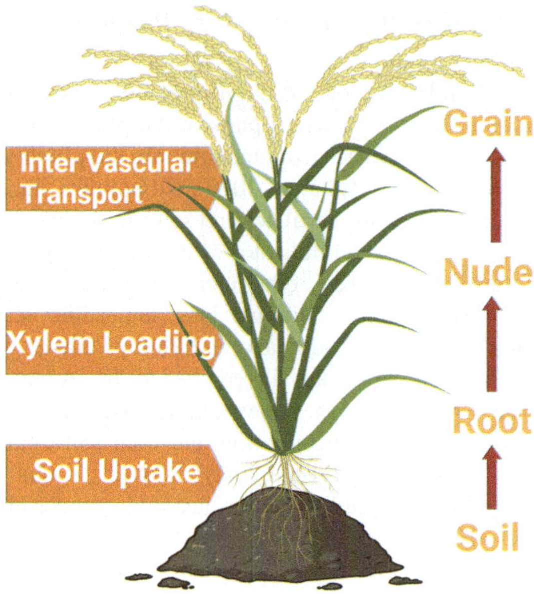

Fig. 3 The schematic representation of cadmium uptake from in rice plant

where Cd is transported through the xylem to the sink tissues such as grains and (ii) Active transportation, where Cd is actively transported to different plant components such as Culm, rachis, flag leaves, and outer panicle portions, then phloem-mediated mobilization to the cereal grains (Uraguchi et al. 2017) and Fig. 3. provides a diagram showing the absorption and subsequent translocation of Cadmium in rice. Membrane-bound transporters in the root cells perform a critical role in plants' ability to absorb cadmium uptake in plants (Wang et al. 2002).

The uptake and accumulation of cadmium (Cd) in plants are regulated by multiple genes that play a quantitative role in environment-, stage-, and tissue-specific transport, concentration, and sequestration of cadmium. These genes contribute to the intricate mechanisms involved in Cd uptake and storage in different parts of the plant (Lugon-Moulin et al. 2004). A study examining the prolonged exposure to cadmium on tomato plants (Solanumlycopersicum L.) found that cadmium poisoning has a very negative effect dependent on the dosage and strongly correlated with the nutrient status of the soil. At higher doses, Cd is severely affected changing nutritional partitioning can affect plant development and metabolic processes. These processes are regulated by several genes involved in Cd uptake, transport, and metabolic pathways.

The escalating cadmium content in agricultural soils has elevated significant consideration among scientific communities due to its detrimental effects on plant growth and crop yield. Cd accumulation in soils has become a prominent issue that requires attention and mitigation strategies (Miao et al. 2014; Rodríquez-Serrano et al. 2008; Xiong et al. 2014). Cadmium (Cd) toxicity in soils leads to various detrimental effects on plants, including the excessive creation of oxidative markers including reactive oxygen species (ROS), the oxidation of lipids, and free radicals. These oxidative stress-inducing factors in the end result in reduced crop yield (Hussain et al. 2018; Rizwan et al. 2017a).

In the present era, it is crucial to address and mitigate the toxicity of cadmium (Cd) for the well-being of both human beings and crops. To counteract the negative effects of Cd, implementing certain strategies can be beneficial. One approach is to introduce plant species that can phytoremediate heavy metals, effectively sequestering the metals in their vegetative parts. Additionally, it is important to control anthropogenic activities and enforce legal strategies at the country and government levels to prevent the contamination of crops with toxic metals. Along with adopting these measures, we can work toward minimizing the toxic effects of Cd and promoting a healthier environment for both humans and plants.

3.2 *Effects on Plant Health*

The understanding of the impact of cadmium (Cd) toxicity in plants has been an ongoing subject of study, and recent advancements in plant physiology have provided researchers with valuable insights. According to Clemens, nutrient imbalance has a considerable effect on the crop's poisoning of Cadmium, particularly in fruit-bearing plants. This imbalance affects the normal functioning of transporters involved in nutrient uptake and distribution. By examining these factors, researchers have been able to shed light on the mechanisms underlying Cd toxicity in plants (Clemens et al. 2013). An intriguing observation in the context of Cd toxicity is the substantial decline in the potassium level, zinc (Zn), and iron (Fe) in developing fruits, while calcium (Ca) and magnesium (Mg) levels increase significantly. The well-documented cadmium and K are at odds with one another is evident, as Low Potassium contents in the pericarp interfere with regular metabolic cycles, such as protein bound to the membrane, synthesized, and enzymatically active processes crucial for maintaining cellular turgidity (Clemens et al. 2013).

3.3 *Ultrastructural Changes Due to Cadmium Toxicity*

The response of plants to cadmium (Cd) exposure varies depending on the concentration of Cd. The occurrence of anatomical abnormalities in plants is highly influenced by factors such as plant species, duration of exposure, uptake amount of Cd, sequestration mechanisms, and the localization of Cd in different plant parts. These factors collectively determine the specific response of plants to Cd contamination (Benjamin et al. 2019). Cadmium (Cd) exhibits mobility within the phloem and can be distributed throughout various plant parts. Its presence often results in a decrease in biomass and yield, leading to adverse effects such as chlorosis and leaf shedding, which deviate from normal plant functions and movement (Gallego et al. 2012). Exposure to cadmium (Cd) induces significant Ceratopteris pteridoides roots stems, and leaves have altered anatomy. Abaxial stomata are closed, the size of stomata in leaves is reduced, the tracheid walls are scarified, the xylem vessels are narrowed, and the vascular bundles in roots and stems are disorganized (Avram et al. 2021). Cadmium (Cd) stress leads to a decrease in trichome length, as well as the density of abaxial and adaxial stomata in *Trigonella foenum*. Additionally, under Cd stress, *Trigonella foenum's* cortical percentage decreases. The vascular tissues of the stems and leaves of *Arundo donax* L. become heterogeneous as a result of Cd stress (Ahmad et al. 2005). Exposure of plants to cadmium (Cd) results in severe impacts, including the reduction in parenchyma size in leaves, disruption of chloroplast ultrastructure, disorganization of vascular tissue, thinning of the epidermal tissue, and the presence of phloem vessels and narrow xylem. Plants resistant to heavy metals employ various mechanisms in order to minimize the harmful effects of heavy metals, which involve modifications in their microstructures. These improvements involve a well-functioning vascular system with an increased water transfer from the vascular bundle area and nutrients. In order to save water, these plants also have thick epidermis, which is further protected by a waxy cuticle layer. Additionally, they store an immense quantity of water in their roots and shoots to prevent it from moving to their leaf. These adaptive responses may also lead to modifications in the photosynthetic apparatus of the plants.

3.4 Damages from Oxidation Caused by Cadmium Exposure

Similar to additional heavy metals, cadmium (Cd) triggers oxidative damage in plants by promoting the excessive generation of hydrogen peroxide (H_2O_2) and lipid peroxidation. This phenomenon has been well documented in Fig. 4 (Rizwan et al. 2019; Shiyu et al. 2020; Zhao et al. 2020) for further illustration. Extensive research has established that cadmium (Cd) exposure leads to the production of superoxide ($O2^-$) and hydrogen peroxide (H_2O_2) are examples of reactive oxygen species (ROS). ROS molecules generated as a result of cadmium (Cd) exposure possess the ability to scavenge and suppress the activity of antioxidant enzymes present in plants. Consequently, disruption in the antioxidant defense system occurs, compromising the plant's ability to counteract oxidative stress (Hasanuzzaman et al. 2020; Kandemirli et al. 2020; Naushad et al. 2019).

Multiple scientific investigations indicate that cadmium (Cd) does not directly contribute to the production of reactive oxygen species (ROS). Instead, Cd is known to induce temporal oxidative damage in plants (Haans et al. 2016). Cellular reactive oxygen species (ROS) consists of a combination of free radicals as well as non-radicals (Haans et al. 2016). Free radicals encompass species such as superoxide (O_2°), hydroxyl (OH°), alkoxyl (RO°), and peroxyl radicals (ROO°) (Farooq et al. 2019; Kumar et al. 2020). In addition to the previously mentioned free radicals, plants also contain other non-radicals, including hyperchlorous acid (HOCl), excited

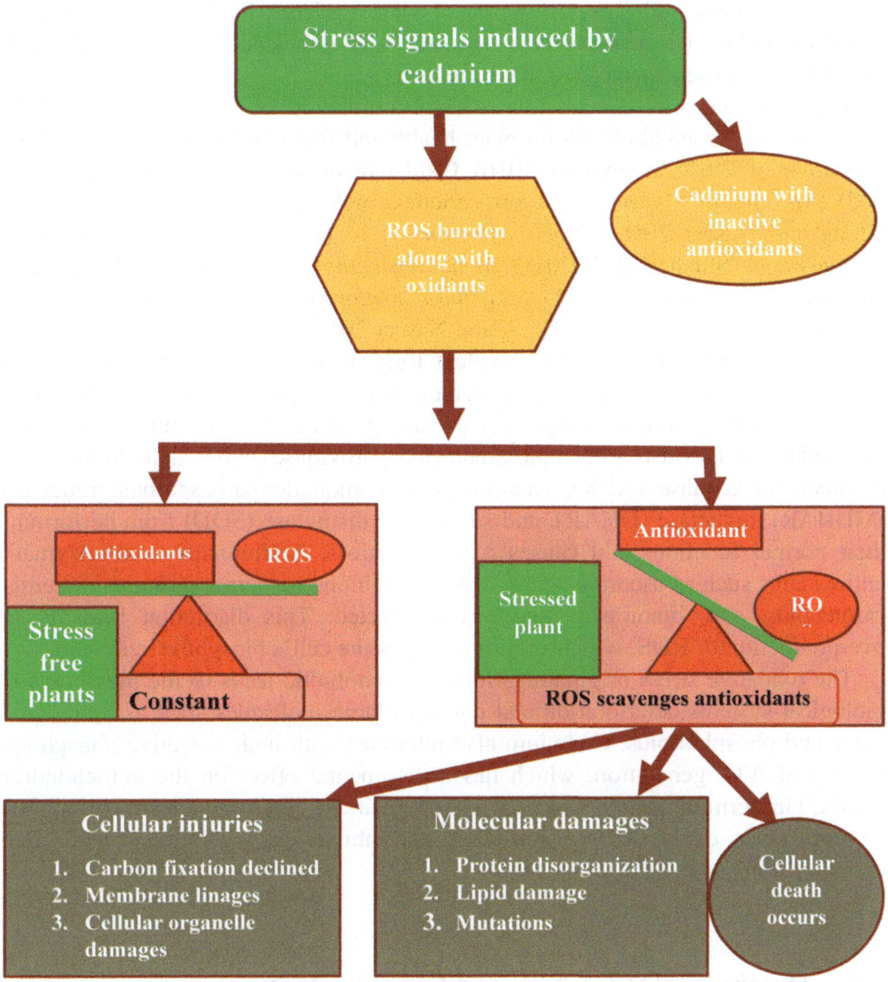

Fig. 4 The damages and oxidative stress induced by cadmium in plant

carbonyls, and hydroperoxide. These nonradical species play important roles in various cellular processes and contribute to the overall oxidative balance within plants (Kapoor et al. 2015). Under the influence of Malondialdehyde (MDA) and thiobarbituric acid reactive substances (TBARSs) have accumulated due to damage caused by oxidation in plants exposed to cadmium (Cd) stress. This accumulation subsequently leads to electrolyte leakage (Kapoor et al. 2015).

The accumulation of cadmium (Cd) in plants is facilitated by several significant mechanisms. These include (i) the structural similarity of Cd with essential nutrients such as phosphorus and zinc, which allows Cd to be absorbed by the roots of plants; (ii) cadmium's direct influence on sulfhydryl (–SH) groups, leading to the impairment of protein composition; (iii) the displacement from the binding sites of the necessary cations; and (iv) destruction of vital nutrients, lipids, proteins, cellular pigments, and nucleic acids as a consequence of an imbalance between antioxidants and reactive oxygen species (ROS) in the body.

These mechanisms contribute to the accumulation of Cd in plants and the subsequent detrimental effects on plant health and function (Choppala et al. 2014; Hossain et al. 2012; Singh et al. 2016). Oxidative stress is brought on by cadmium (Cd) exposure in a variety of crop varieties, including *Phyllostachys pubescent* (Hong and Naseer 2016) *Phoenix dactylifera* L. (Zouari et al. 2016), *Solanum lycopersicum* (Šimat et al. 2020), *Triticum aestivum* L. (Peng et al. 2010), *Salvinia auriculata* (Vestena et al. 2011), *Spartina densiflora* (Arandia Loroño et al. 2010), and *Phyllostachys pubescens* (Hong and Naseer 2016).

When cadmium (Cd) is in its bivalent form, it does not directly generate free radicals. However, reactive oxygen species such as superoxide radicals, hydrogen peroxide (H_2O_2), and hydroxyl radicals are produced significantly more when exposed to Cd. By preventing antioxidants like peroxidase (POD), dehydroascorbate ascorbate peroxidase (APX), catalase (CAT), monodehydroascorbate reductase (MDHAR), reductase (DHAR), and superoxide dismutase (SOD) from performing their normal functions, Cd causes oxidative stress. Additionally, non-enzymatic antioxidants such as ascorbic acid (ASA), glutathione reductase (GR), tocopherols, carotenoids, and vitamins C and E are affected. This disruption leads to an overproduction of ROS, which in turn damages the cell's biosynthetic system.

The oxidative stress originating from this xenobiotic leads to the membrane of biological systems deterioration and damages large molecules such as lipids, proteins, and phospholipids. Cadmium also interferes with both oxidative phosphorylation and ATP generation, which has a detrimental effect on the mitochondrial matrix. Furthermore, being subjected to Cd impairs the ability of enzyme proteins to fix themselves, damages DNA and RNA, and inhibits cell proliferation and differentiation (Shugar et al. 2021).

3.5 Metabolism of Carbon and Crop Production

Plants encounter toxic environmental conditions moreover respond by regulating carbon metabolism, ensuring a consistent supply of carbon dioxide (CO_2), maintaining the electron transport chain, and effectively assimilating CO_2 at appropriate levels (Rodrigues et al. 1993). Cadmium (Cd) toxicity disrupts carbon metabolism, resulting in a reduction in photosynthetic productivity (Gouia et al. 2003). Cadmium has a significant impact on photosynthesis as it affects various aspects of the process, including the ETC, PSI, and PSII photosystems, chlorophyll–protein complexes, and CO_2 reduction pathways are present in the stroma (Royston and Parmar 2013). Exposure to Cd-induced toxicity leads to noticeable changes in the ultrastructure of chloroplasts, characterized by swollen and disrupted thylakoid membranes (Ottman et al. 2011).

Cd-induced toxicity results in notable alterations in the chloroplast ultrastructure, including a decrease in the number of chloroplasts, a reduction in the foliage of a variety of plants including *Hordeum vulgare*, *Picris divarticata*, and *Brassica* spp., there are abnormally high amounts of plastoglobules, grana, and starch (Elhiti et al. 2012; Wang et al. 2011; Xin et al. 2010). Cd-induced toxicity led to the aggregation of grana, disruption of thylakoid membranes, and swelling of intrathylakoidal spaces, primarily caused by lipid peroxidation (LPX) in the Willow plant (Hakmaoui et al. 2007). The lipoxygenase (LOX) enzyme's activity leads to the formation of LPX (De Bruin et al. 2010). This enzyme plays a crucial role in the peroxidation of membrane fatty acids, including phosphatidylglycerol (PG), monogalactosyldiacylglycerol (MGDG), and digalactosyldiacylglycerol (DGDG), leading to lipid peroxidation (LPX). The accumulation of these activated lipids and PG, under Cd exposure causes the generation of reactive oxygen species and free radicals. The linkage between lipoxygenase activity and lipid peroxidation has been observed in various plant species, including Lupine and Barley in response to cadmium exposure (Maksymiec and Krupa 2006; Oláh et al. 2009).

Cadmium (Cd) exposure generally leads to a notable decrease in carbon metabolism, resulting in altered photosynthesis due to reduced CO_2 supply and insufficient carbon levels. These changes negatively impact the electron transport chain (ETC), thylakoid membranes, and photosynthetic enzymes. Additionally, the ultrastructure of cell components, such as chloroplasts, undergoes significant modifications, contributing to the reduced efficiency of photosynthesis in leaf cells. Inhibition and destruction of photosynthetic pigments of biosynthetic processes in both young and old leaves have been identified as major factors contributing to Cd-induced toxicity (Anjum et al. 2016; Dai et al. 2015). The biosynthesis of chlorophyll primarily relies on the availability of aminolevulinic acid (ALA). However, cadmium (Cd) inhibits ALA synthesis by interfering with the glutamate availability, and it interacts with the SH functional group of enzymes such as porphobilinogen deaminase and δ-aminolevulinic acid synthase, leading to their inhibition (Myśliwa-Kurdziel and Strzałka 2002; Skrebsky et al. 2008). In *Cucumis sativus*, it has been reported that an excessive concentration of aminolevulinic acid (ALA) can lead to the production of reactive oxygen species (ROS). These ROS can disrupt the redox potential of cells and interfere with cellular homeostatic functions (Goncalves et al. 2009) and soybean (Lluch-Cota et al. 2007). Through interactions that involve the supramolecular species and the PSII core complex, the toxicological effect of cadmium interferes with the operation of photosystems, including PSI and PSII. This interaction hinders the process of photoactivation (Lluch-Cota et al. 2007). In contrast, PSI is often regarded as being more vulnerable to the toxic effects of cadmium (Cd). This increased susceptibility could be attributed to iron shortage-induced Cd, which leads to greater destruction of PSI (Rozas et al. 2012). Damage to PSI caused by iron deficiency has been observed in *Cucumis sativus* L. plants exposed to a concentration of 10 μM of cadmium (Cd) (Hosseini-Sarvari 2008). While the effects of cadmium (Cd) toxicity on PSII have been extensively studied, less concerted effort has been made to comprehend how it would affect PSI. Nonetheless, it has been observed that Cd significantly impairs the photosynthetic yield in species such as *Thlaspi caerulescens* and *Pisum sativum* (Küpper et al. 2007; Wodala et al. 2012).

Cadmium poisoning adversely affects the Calvin cycle and exhibits inhibition of a variety of enzyme activity involved in this important metabolic pathway (Xin et al. 2010). During photosynthesis, the enzymes ribulose-1,5-bisphosphate carboxylase/oxygenase (RuBisCO) and phosphoenolpyruvate carboxylase (PEPc) play crucial roles in the fixation of carbon dioxide (CO_2) and the subsequent carbon fixation process (Xin et al. 2010). The presence of excessive Cd ions leads to a decrease in the activity of enzymes RUBP and PEP by inducing structural alterations and displacing essential cofactors, such as Mg^{2+}. These cofactors play a crucial role in the carboxylation process, but under Cd toxicity, there is a shift toward oxidation processes (Tran and Popova 2013).

Cadmium stress adversely affects photosynthetic traits, causing damage to chloroplast components and impairing the activity of essential photosynthetic enzymes (Tran and Popova 2013). The presence of cadmium significantly impairs plant growth, photosynthetic processes, and ultimately reduces the yield of grains (Rizwan et al. 2016; Tran and Popova 2013). Multiple studies have indicated that cadmium has the ability to translocate to rice grains, leading to a notable decrease in grain production and plant root nutrient absorption (He 2006; Liu et al. 2007; Rodda et al. 2011).

The harmful effects of cadmium on yield traits in rice and wheat crops are influenced by factors such as genetic makeup, cadmium volume remedy, and exposure length. Studies have shown that cadmium toxicity leads to significant reductions in yield traits including the year length, weight, and the number of spikelets per plant and grain per year in wheat harvests (Khan et al. 2007). The bridging point of cadmium poisoning can vary among different genotypes, and it is also influenced by factors such as the length of exposure and the cadmium content dose (Rizwan et al. 2016). Previous studies have documented a decrease in crop yield in various plant species, including barley (Olalde et al. 2018) pea (Ghodsi Maab et al. 2022), and tomato (Hayat et al. 2012). Cd toxicity effects on plant development and production traits vary depending on factors such as plant species, Cd concentration, and duration of exposure.

3.6 Effects of Cd on Plants

Plants have a higher tolerance for Cd compared to animals, but excessive uptake of Cd can still have detrimental results on plant health (Gill et al. 2012). Excessive Cd concentrations induce various physiological changes in plants, including alterations in nitrogen and carbohydrate metabolisms. Cd interferes with photosynthesis, which interferes with the production of green pigment and causes the closure of stomatal cells. One of the important factors contributing to the disruption of protochlorophyll reductase is the suppression of chlorophyll biosynthesis, an enzyme involved in chlorophyll synthesis, as well as aminolevulinic acid synthesis (Gill et al. 2012).

Similar to other heavy metals, Cd triggers the formation of free radicals, which contribute to the oxidative damage of lipids in the thylakoid membranes (Rai et al. 2016).

In a study examining the toxicity of Cd in tomato plants, a linear association was seen among the levels of cadmium in both nutritional supplements as well as the accumulation of cadmium in both the crop's roots and leaves. Furthermore, it was discovered that tomato plants' roots had collected Cd at a rate that was around 15 times greater than that of their shoots (Lai et al. 2020). In a separate study involving *Picea abies*, it was observed that the accumulation of Cd in the roots was directly proportional to the concentration of Cd present in the growth medium (Ozcan and Bayçu 2005). The effects of cadmium, copper, mercury, and lead on the concentrations of total protein and abscisic acid (ABA) were examined in a research on bean starting seeds (*Phaseolus vulgaris* L. cv Strike) was examined. The findings revealed that higher concentrations of heavy metals led to an increase in the production of ABA in the seedlings.

Certain studies investigating Cadmium toxicity's impacts on plants have revealed significant alterations intake, transportation, and utilization of different substances and water in crops (Wang et al. 2020). Cd toxicity in plants has been observed to reduce the absorption of nitrate, leading to limitations in the movement of nitrate between the roots and stems. This restriction in transport is attributed to the stems' metabolism of nitrate reductase is inhibited (Wang et al. 2020).

In certain studies, exposure to Cd has been found to induce stress-dependent genomic changes, leading to variations in RAPD (Random Amplified Polymorphic DNA) band profiles. For example, *Kalanchoe daigremontiana* clones were subjected to various concentrations of Cd, and it was found that new bands appeared in the RAPD profiles when treated with concentrations of 50, 200, and 400 μM (Ozyigit et al. 2016). Furthermore, changes in band intensities were observed in RAPD band profiles, with a decrease in band intensities at a concentration Band strengths increased at concentrations of 50, 200, and 400 μM and by 100 μM. Another study investigated the DNA sequences of samples of the *Arabidopsis* plant that were exposed to Cd at concentrations 4.0 and 5.0 mg L^{-1}, which revealed polymorphisms. However, whenever DNA samples were subjected to doses 0.25 and 1.0 mg L^{-1} Cd, no variations were found (Ye et al. 2016).

3.7 Cadmium Containing Foods

Cadmium and its derivatives find extensive applications within various industries including dyestuff production, ink manufacturing, glass production, fabric manufacturing, electric current generation, battery production, fungicides, insecticides, and the production of artificial polymers (Akguc et al. 2008; Diaz-Munoz et al. 2012; Ozyigit et al. 2016). The utilization of Cd in diverse industries has been associated with an elevated risk associated with this toxic metal contaminating food through air, water, and soil. Significant quantities of contamination have been detected in certain food sources as a result (Hezbullah et al. 2016). Studies have indicated that various food items, including grains, potatoes, fruits, roots and leafy vegetables, liquid–solid oils, meat, milk, and other dairy products can all become contaminated with cadmium. When paired with zinc, the usage of cadmium in galvanized zinc-coated packaging has been found to pose a risk of food poisoning, particularly in acidic foods. It is believed that organic acids present in the food can enhance the solubility of Cd within the packaging material (Vinceti et al. 2018). Furthermore, Cd that migrates into acidic foods has the potential to cause toxicity and contamination (Mohammad et al. 2018; Putra et al. 2020).

3.8 Cadmium Transportation in the Human Body

Cadmium typically passes into the human body by oral ingestion, inhalation, or skin contact. Once absorbed, it adheres to the protein albumin and the cells in the blood so it is transported into the bloodstream. It first travels via the bloodstream to the liver, where it is then delivered to the renal system by attaching to globular proteins for the process of detoxification (Zhang et al. 2019). Accumulated cadmium in the renal system can impair the filtration process in the Bowman's capsule, leading to the excretion of essential proteins and glucose in urine. Cd enters the cells through membrane carrier proteins, where it competes with essential metals like Ca, Cu, Fe, and Zn for binding to membrane receptors.

It has been observed that Cd absorption increases in individuals and animals when their diets are deficient in metals like calcium, chromium, iron, and Zinc, as well as protein inadequacy (Karahan et al. 2020; Vahter 2002). Cd enters the bloodstream through various routes, including the alveoli, intestinal lumen, and skin. Once in the bloodstream, Cd binds to proteins that contain metallothionein, albumin, and thiol groups. These proteins facilitate the transportation of cadmium is delivered to the cells via a procedure known as receptor-mediated endocytosis (Zalups and Ahmad 2003). Investigations on animals and in vitro research have consistently shown that metallothionein plays a crucial role in protecting cells rather than toxic effects of cadmium (Bobillier-Chaumont et al. 2006; Morice et al. 2012).

3.9 Cd's Consequences for Human Beings and Animals

Studies have estimated that residential areas have an average atmospheric concentration of Cadmium of around 0.001 g m^{-3}, resulting in a daily intake of approximately 0.02 mg of Cd for individuals. It has been shown that breathing cadmium oxide (CdO) over an extended period as smoke can lead to lung edema and potentially have fatal consequences. Furthermore, long-term exposure to Cd has been linked to an elevated risk of several cancer forms, including prostate cancer and particularly lung cancer (Huang et al. 2013). Intramuscular or subcutaneous injection of Cd metal or its compounds has been found to be associated with the development of sarcoma (Hunter et al. 2012).

Experiments investigating Cd′s effects on bone and calcium metabolism have revealed that Cd can disrupt calcium metabolism and lead to hypercalcemia, characterized by elevated calcium levels in the blood. Additionally, it has been reported that Cd can inhibit the the mechanism for transporting glucose and salt in renal cortex cells (Longadge and Dongre 2013). Skincare products commonly used by women, including eyeliner, blush, and lipstick, have been found to contain Cd. Prolonged exposure to Cd through these products has been linked to a higher risk of cancer of the skin (Duruibe et al. 2007; Ravindran et al. 2018). Cd has been found to induce DNA breaks and lipid peroxidation, leading to cellular damage. Researches have demonstrated that the toxic effects of Cadmium can trigger proximal tubule cell apoptosis (Fu et al. 2022; McFarland et al. 2017). Cd accumulates within cellular lysosomes, resulting in the disruption of lysosomal membrane integrity. This disruption leads to the release of phospholipases and lysosomal proteases, triggering the cascade occurrences that ultimately result in hepatocyte cytotoxicity (Huang et al. 2019). Exposure to high Cd concentrations in gas mixes can lead to acute chemical pneumonia is lethal However, the effects of long-term contact with Cd on the respiratory system appear to be more significant. Continuous and low-dose Cd exposure is primarily attributed to smoking within the community (Öztoprak and Javed 2020; Tang et al. 2020).

Workers in high-risk occupational groups who chronically uptake Cd through smoking or inhalation are susceptible to detrimental respiratory system effects. The intensity and timing of this harm are influenced by the quantity and duration of cadmium exposure, often manifesting over a prolonged period. Studies have reported the development among Cd industry employees, persistent inflammation of the nostrils, the throat, and the vocal is seen (Morabito et al. 2019; Shaban et al. 2017). Individuals who are exposed to Cd and those who smoke are commonly affected by chronic obstructive pulmonary disease (COPD). These patients often exhibit clinical and radiological evidence of emphysema, indicating the presence of lung damage (Venturutti et al. 2016).

4 Conclusion

In conclusion, the contamination of soil and plants with cadmium poses significant environmental and health risks. Human activities, such as industrialization and urbanization, have contributed to the escalation of cadmium pollution in the environment. Cadmium enters the human body through the consumption of contaminated drinking water and food sources, particularly plants. Unreliable agricultural practices and the discharge of waste from factories are major contributors to cadmium contamination in soil. The toxic nature of cadmium affects plants in various ways, leading to reduced crop yields and deterioration in the quality of agricultural produce. Cadmium accumulation in plants occurs through root uptake and subsequent mobilization to different plant parts. Cadmium toxicity in plants results in anatomical abnormalities, oxidative damage, and disruption of carbon metabolism. It interferes with photosynthesis, impairs the functioning of chloroplasts, and inhibits the production of green pigments. Cadmium-contaminated foods, including grains, vegetables, fruits, meat, and dairy products, can pose health risks to humans. Cadmium enters the human body through ingestion, inhalation, or skin contact. Once absorbed, it is transported in the bloodstream and can accumulate in organs such as the liver and kidneys causing many serious complications. In order to mitigate the risks associated with cadmium contamination, it is crucial to implement proper waste management practices, regulate the use of cadmium-containing fertilizers and industrial processes, and monitor food sources for cadmium levels. Overall, addressing cadmium pollution requires a multidisciplinary approach involving environmental regulation, sustainable agricultural practices, and public awareness.

References

Abbas AA, Assikong EB, Akeh M, Upla P, Tuluma TK (2017) Antimicrobial activity of coconut oil and its derivative (lauric acid) on some selected clinical isolates. Int J Med Sci Clin Invent 4(8): 3173–3177

Adams J (2012) The revolutionary writings of John Adams. Liberty Fund, Carmel, IN

Ahmad AS, Ansari MA, Ahmad M, Saleem S, Yousuf S, Hoda MN, Islam F (2005) Neuroprotection by crocetin in a hemi-parkinsonian rat model. Pharmacol Biochem Behav 81(4):805–813

Akguc N, Ozyigit II, Celal Y (2008) Pyracantha coccinea Roem. (Rosaceae) as a biomonitor for Cd, Pb and Zn in Mugla province (Turkey). Pak J Bot 40(4):1767–1776

Akguc N, Ozyigit I, Yasar U, Leblebici Z, Yarci C (2010) Use of Pyracantha coccinea Roem. as a possible biomonitor for the selected heavy metals. Int J Environ Sci Technol 7:427–434

Amodei D, Ananthanarayanan S, Anubhai R, Bai J, Battenberg E, Case C, et al (2016) Deep speech 2: end-to-end speech recognition in English and Mandarin. In: Paper presented at the International conference on machine learning

Anjum M, Miandad R, Waqas M, Ahmad I, Alafif Z, Aburiazaiza A et al (2016) Solid waste management in Saudi Arabia: a review. J Sci Food Agric 1:13–26

Arandia Loroño M, Alonso Olea MJ, Martínez-Domínguez IJR (2010) La metodología dialógica en las aulas universitarias

Aslam A, Thomas-Hall SR, Mughal T, Zaman Q-U, Ehsan N, Javied S, Schenk PM (2019) Heavy metal bioremediation of coal-fired flue gas using microalgae under different CO_2 concentrations. J Environ Manag 241:243–250

Attridge D, Howes M, Howes ME (2000) Semicolonial Joyce. Cambridge University Press, Cambridge

Avram S, Bologa CG, Holmes J, Bocci G, Wilson TB, Nguyen D-T et al (2021) DrugCentral 2021 supports drug discovery and repositioning. Nucleic Acids Res 49(D1):D1160–D1169

Belon E, Boisson M, Deportes I, Eglin T, Feix I, Bispo A et al (2012) An inventory of trace elements inputs to French agricultural soils. Sci Total Environ 439:87–95

Benjamin EJ, Virani SS, Callaway CW, Chamberlain AM, Chang AR, Cheng S et al (2018) Heart disease and stroke statistics—2018 update: a report from the American Heart Association. Circulation 137(12):e67–e492

Benjamin EJ, Muntner P, Alonso A, Bittencourt MS, Callaway CW, Carson AP et al (2019) Heart disease and stroke statistics—2019 update: a report from the American Heart Association. Circulation 139(10):e56–e528

Bhushan B, Jung YCJJ (2008) Wetting, adhesion and friction of superhydrophobic and hydrophilic leaves and fabricated micro/nanopatterned surfaces. J Phys Condens Matter 20(22):225010

Bobillier-Chaumont S, Maupoil V, Berthelot AJJ (2006) Metallothionein induction in the liver, kidney, heart and aorta of cadmium and isoproterenol treated rats. J Appl Toxicol 26(1):47–55

Burke EJ, Jones CD, Koven CDJJ (2013) Estimating the permafrost-carbon climate response in the CMIP5 climate models using a simplified approach. J Clim 26(14):4897–4909

Burr DC, Morrone MC, Ross JJN (1994) Selective suppression of the magnocellular visual pathway during saccadic eye movements. Nature 371(6497):511–513

Cai H, Huang JJA (2016) Leader-following attitude consensus of multiple rigid body systems by attitude feedback control. Automatica 69:87–92

Choppala G, Saifullah BN, Bibi S, Iqbal M, Rengel Z et al (2014) Cellular mechanisms in higher plants governing tolerance to cadmium toxicity. Crit Rev Plant Sci 33(5):374–391

Clemens S, Aarts MG, Thomine S, Verbruggen NJT (2013) Plant science: the key to preventing slow cadmium poisoning. Trends Plant Sci 18(2):92–99

Cossarizza A, Chang HD, Radbruch A, Acs A, Adam D, Adam-Klages S et al (2019) Guidelines for the use of flow cytometry and cell sorting in immunological studies. Eur J Immunol 49(10): 1457–1973

Dai L, Xue Y, Qu L, Choi H-J, Baek J-B (2015) Metal-free catalysts for oxygen reduction reaction. Chem Rev 115(11):4823–4892

De Bruin JL, Baas AF, Buth J, Prinssen M, Verhoeven EL, Cuypers PW et al (2010) Long-term outcome of open or endovascular repair of abdominal aortic aneurysm. N Engl J Med 362(20): 1881–1889

Diaz-Munoz MD, Osma-Garcia IC, Fresno M, Iniguez MA (2012) Involvement of PGE2 and the cAMP signalling pathway in the up-regulation of COX-2 and mPGES-1 expression in LPS-activated macrophages. Biochem J 443(2):451–461

Duruibe J, Ogwuegbu MOC, Egwurugwu J (2007) Heavy metal pollution and human biotoxic effects. Int J Phys Sci 2(5):112–118

Dutta P, Choi T-M, Somani S, Butala R (2020) Blockchain technology in supply chain operations: applications, challenges and research opportunities. Transp Res E Logist Transp Rev 142: 102067

Elhiti M, Yang C, Chan A, Durnin DC, Belmonte MF, Ayele BT et al (2012) Altered seed oil and glucosinolate levels in transgenic plants overexpressing the Brassica napus SHOOTMERISTEMLESS gene. J Exp Bot 63(12):4447–4461

Farooq MA, Niazi AK, Akhtar J, Saifullah Farooq M, Souri Z, Karimi N, Rengel Z (2019) Acquiring control: the evolution of ROS-induced oxidative stress and redox signaling pathways in plant stress responses. Plant Physiol Biochem 141:353–369

Fritsche LG, Igl W, Bailey JNC, Grassmann F, Sengupta S, Bragg-Gresham JL et al (2016) A large genome-wide association study of age-related macular degeneration highlights contributions of rare and common variants. Nat Genet 48(2):134–143

Fu R, Zhang P-S, Jiang Y-X, Sun L, Sun X-HJC (2022) Wastewater treatment by anodic oxidation in electrochemical advanced oxidation process: Advance in mechanism, direct and indirect oxidation detection methods. Chemosphere 311:136993

Gallego SM, Pena LB, Barcia RA, Azpilicueta CE, Iannone MF, Rosales EP et al (2012) Unravelling cadmium toxicity and tolerance in plants: insight into regulatory mechanisms. Environ Exp Bot 83:33–46

Garbisu C, Alkorta I (2001) Phytoextraction: a cost-effective plant-based technology for the removal of metals from the environment. Bioresour Technol 77(3):229–236

Gatoo MA, Naseem S, Arfat MY, Mahmood Dar A, Qasim K, Zubair S (2014) Physicochemical properties of nanomaterials: implication in associated toxic manifestations. Toxicity Nanomater 2014:498420

Ghodsi Maab S, Makarian H, Ghasimi Hagh Z, Gholipoor M (2022) Effects of cold plasma and salicylic acid on secondary metabolites and activity of enzymes involved in their biosynthesis in Salvia leriifolia Benth. Iran J Med Aromatic Plants Res 37(6):954–970
Gill RJ, Ramos-Rodriguez O, Raine NE (2012) Combined pesticide exposure severely affects individual-and colony-level traits in bees. Nature 491(7422):105–108
Goncalves G, Marques PA, Granadeiro CM, Nogueira HI, Singh M, Gracio J (2009) Surface modification of graphene nanosheets with gold nanoparticles: the role of oxygen moieties at graphene surface on gold nucleation and growth. Chem Mater 21(20):4796–4802
Gouia H, Suzuki A, Brulfert J, Ghorbal MH (2003) Effects of cadmium on the co-ordination of nitrogen and carbon metabolism in bean seedlings. J Plant Physiol. 160(4):367–376
Gusiatin ZM, Klimiuk E (2012) Metal (Cu, Cd, and Zn) removal and stabilization during multiple soil washing by saponin. Chemosphere 86(4):383–391
Haans RF, Pieters C, He ZL (2016) Thinking about U: theorizing and testing U-and inverted U-shaped relationships in strategy research. Strateg Manag J 37(7):1177–1195
Hakmaoui A, Ater M, Boka K, Baron M (2007) Copper and cadmium tolerance, uptake and effect on chloroplast ultrastructure. Studies on Salix purpurea and Phragmites australis. Z Naturforsch C J Biosci 62(5–6):417–426
Hasanuzzaman M, Bhuyan MB, Zulfiqar F, Raza A, Mohsin SM, Mahmud JA et al (2020) Reactive oxygen species and antioxidant defense in plants under abiotic stress: revisiting the crucial role of a universal defense regulator. Antioxidants (Basel) 9(8):681
Hayat S, Hayat Q, Alyemeni MN, Wani AS, Pichtel J, Ahmad A (2012) Role of proline under changing environments: a review. Plant Signal Behav 7(11):1456–1466
He J-H (2006) Homotopy perturbation method for solving boundary value problems. Phys Lett A 350(1–2):87–88
Hezbullah M, Sultana S, Chakraborty S, Patwary MI (2016) Heavy metal contamination of food in a developing country like Bangladesh: an emerging threat to food safety. J Toxicol Environ Health Sci 8(1):1–5
Hong K-S, Naseer NJIJ (2016) Reduction of delay in detecting initial dips from functional near-infrared spectroscopy signals using vector-based phase analysis. Int J Neural Syst 26(03): 1650012
Hossain M, Ngo H, Guo W, Setiadi T (2012) Adsorption and desorption of copper(II) ions onto garden grass. Bioresour Technol 121:386–395
Hosseini-Sarvari M (2008) TiO2 as a new and reusable catalyst for one-pot three-component syntheses of α-aminophosphonates in solvent-free conditions. ChemInform 64(23):5459–5466
Huang X, Zeng Z, Bao S, Wang M, Qi X, Fan Z, Zhang H (2013) Solution-phase epitaxial growth of noble metal nanostructures on dispersible single-layer molybdenum disulfide nanosheets. Nat Commun 4(1):1444
Huang Y, Wang Y, Wang H, Liu Z, Yu X, Yan J et al (2019) Prevalence of mental disorders in China: a cross-sectional epidemiological study. Lancet Psychiatry 6(3):211–224
Hunter S, Jones P, Mitchell A, Apweiler R, Attwood TK, Bateman A et al (2012) InterPro in 2011: new developments in the family and domain prediction database. Nucleic Acids Res 40(D1): D306–D312
Hussain N, Rigoni U, Orij RP (2018) Corporate governance and sustainability performance: analysis of triple bottom line performance. J Bus Ethics 149:411–432
Joint WHO/FAO/UNU Expert Consultation (2007) Protein and amino acid requirements in human nutrition. World Health Organ Tech Rep 935:1–265
Kabata-Pendias A (2000) Trace elements in soils and plants. CRC Press, Boca Raton, FL
Kabata-Pendias AJG (2004) Soil–plant transfer of trace elements—an environmental issue 122(2–4):143–149
Kandemirli SG, Dogan L, Sarikaya ZT, Kara S, Akinci C, Kaya D et al (2020) Brain MRI findings in patients in the intensive care unit with COVID-19 infection. Radiology 297(1):E232–E235
Kapoor DN, Bhatia A, Kaur R, Sharma R, Kaur G, Dhawan S (2015) PLGA: a unique polymer for drug delivery. Ther Deliv 6(1):41–58

Karahan F, Ozyigit II, Saracoglu IA, Yalcin IE, Ozyigit AH, Ilcim A (2020) Heavy metal levels and mineral nutrient status in different parts of various medicinal plants collected from eastern Mediterranean region of Turkey. Biol Trace Elem Res 197:316–329

Khan S, Tamer E (2010) Irregular identification, support conditions, and inverse weight estimation. Econometrica 78(6):2021–2042

Khan M, Lee H, Lee W, Kim H, Ki K, Hur T et al (2007) Structural growth, rumen development, and metabolic and immune responses of Holstein male calves fed milk through step-down and conventional methods. J Dairy Sci 90(7):3376–3387

Khan A, Khan S, Khan MA, Qamar Z, Waqas M (2015a) The uptake and bioaccumulation of heavy metals by food plants, their effects on plants nutrients, and associated health risk: a review. Environ Sci Pollut Res Int 22:13772–13799

Khan MIR, Fatma M, Per TS, Anjum NA, Khan NA (2015b) Salicylic acid-induced abiotic stress tolerance and underlying mechanisms in plants. Front Plant Sci 6:462

Kirkham MB (2006) Cadmium in plants on polluted soils: effects of soil factors, hyperaccumulation, and amendments. Geoderma 137(1–2):19–32

Kobayashi K, Kitamura K, Miura M, Sodeyama H, Murase Y, Miyashita M, Matsui H (1978) Aerobic power as related to body growth and training in Japanese boys: a longitudinal study. J Appl Physiol Respir Environ Exerc Physiol 44(5):666–672

Kumar M, Patel AK, Shah AV, Raval J, Rajpara N, Joshi M, Joshi CG (2020) First proof of the capability of wastewater surveillance for COVID-19 in India through detection of genetic material of SARS-CoV-2. Sci Total Environ 746:141326

Kunito T, Tobitani T, Moro H, Toda H (2012) Phosphorus limitation in microorganisms leads to high phosphomonoesterase activity in acid forest soils. Pedobiologia 55(5):263–270

Küpper H, Parameswaran A, Leitenmaier B, Trtílek M, Šetlík I (2007) Cadmium-induced inhibition of photosynthesis and long-term acclimation to cadmium stress in the hyperaccumulator Thlaspi caerulescens. New Phytol 175(4):655–674

Lai J, Ma S, Wang Y, Cai Z, Hu J, Wei N et al (2020) Factors associated with mental health outcomes among health care workers exposed to coronavirus disease 2019. JAMA Netw Open 3(3):e203976

Lee C, Wei X, Kysar JW, Hone J (2008) Measurement of the elastic properties and intrinsic strength of monolayer graphene. Science 321(5887):385–388

Liu J-X, Srivastava R, Che P, Howell SH (2007) An endoplasmic reticulum stress response in Arabidopsis is mediated by proteolytic processing and nuclear relocation of a membrane-associated transcription factor, bZIP28. Plant Cell 19(12):4111–4119

Liu J, Bao Z, Cui Y, Dufek EJ, Goodenough JB, Khalifah P, Zhang, J. G. (2019) Pathways for practical high-energy long-cycling lithium metal batteries. Nat Energy 4(3):180–186

Livingston G, Huntley J, Sommerlad A, Ames D, Ballard C, Banerjee S et al (2020) Dementia prevention, intervention, and care: 2020 report of the Lancet Commission. Lancet 396(10248): 413–446

Lluch-Cota SE, Aragon-Noriega EA, Arreguín-Sánchez F, Aurioles-Gamboa D, Bautista-Romero JJ, Brusca RC et al (2007) The Gulf of California: review of ecosystem status and sustainability challenges. Prog Oceanogr 73(1):1–26

Lockshin L, Corsi AM (2012) Consumer behaviour for wine 2.0: a review since 2003 and future directions. Wine Econ Policy 1(1):2–23

Longadge R, Dongre SS (2013) Class imbalance problem in data mining: review. Int J Comput Sci Netw 2(1)

Lozano-Rodriguez E, Hernandez LE, Bonay P, Carpena-Ruiz RO (1997) Distribution of cadmium in shoot and root tissues1. J Exp Bot 48(1):123–128

Lugon-Moulin N, Zhang M, Gadani F, Rossi L, Koller D, Krauss M, Wagner GJA (2004) Critical review of the science and options for reducing cadmium in tobacco (Nicotiana tabacum L.) and other plants. Adv Agron 83(1):111–118

Maksymiec W, Krupa Z (2006) The effects of short-term exposition to Cd, excess Cu ions and jasmonate on oxidative stress appearing in Arabidopsis thaliana. Environ Exp Bot 57(1–2): 187–194

McFarland J, Hussar B, De Brey C, Snyder T, Wang X, Wilkinson-Flicker S, et al (2017) The condition of education 2017. NCES 2017-144.

Meng L, Zhang Y, Wan X, Li C, Zhang X, Wang Y et al (2018) Organic and solution-processed tandem solar cells with 17.3% efficiency. Science 361(6407):1094–1098

Miao M, Xiong S, Jiang B, Jiang H, Cui SW, Zhang T (2014) Dual-enzymatic modification of maize starch for increasing slow digestion property. Food Hydrocoll 38:180–185

Mishra P, Pandey CM, Singh U, Gupta A, Sahu C, Keshri AJA (2019) Descriptive statistics and normality tests for statistical data. Ann Card Anaesth 22(1):67

Mohammad S, Bravo-Marquez F, Salameh M, Kiritchenko S (2018) Semeval-2018 task 1: affect in tweets. In: Paper presented at the proceedings of the 12th international workshop on semantic evaluation

Morabito C, Bournaud C, Maës C, Schuler M, Cigliano RA, Dellero Y et al (2019) The lipid metabolism in thraustochytrids. Prog Lipid Res 76:101007

Morice CP, Kennedy JJ, Rayner NA, Jones PD (2012) Quantifying uncertainties in global and regional temperature change using an ensemble of observational estimates: the HadCRUT4 data set. J Geophys Res Atmos 117(D8):8101

Myśliwa-Kurdziel B, Strzałka K (2002) Influence of metals on biosynthesis of photosynthetic pigments. Springer, New York, pp 201–227

Naushad M, Sharma G, Alothman ZA (2019) Photodegradation of toxic dye using Gum Arabic-crosslinked-poly (acrylamide)/Ni (OH)2/FeOOH nanocomposites hydrogel. J Clean Prod 241: 118263

Oláh S, Füle M, Komlósi G, Varga C, Báldi R, Barzó P, Tamás G (2009) Regulation of cortical microcircuits by unitary GABA-mediated volume transmission. Nature 461(7268):1278–1281

Olalde I, Brace S, Allentoft ME, Armit I, Kristiansen K, Booth T et al (2018) The beaker phenomenon and the genomic transformation of northwest Europe. Nature 555(7695):190–196

Ottman R, Lipton RB, Ettinger AB, Cramer JA, Reed ML, Morrison A, Wan GJ (2011) Comorbidities of epilepsy: results from the epilepsy comorbidities and Health (EPIC) survey. Epilepsia 52(2):308–315

Ozcan T, Bayçu G (2005) Some elemental concentrations in the acorns of Turkish Quercus L. (Fagaceae) taxa. Pak J Bot 37(2):361

Öztoprak F, Javed A (2020) Case fatality rate estimation of COVID-19 for European countries: Turkey's current scenario amidst a global pandemic; comparison of outbreaks with European countries. EJMO 21:24

Ozyigit II, Filiz E, Vatansever R, Kurtoglu KY, Koc I, Öztürk MX, Anjum NA (2016) Identification and comparative analysis of H2O2-scavenging enzymes (ascorbate peroxidase and glutathione peroxidase) in selected plants employing bioinformatics approaches. Front Plant Sci 7:301

Pan SJ, Tsang IW, Kwok JT, Yang Q (2010) Domain adaptation via transfer component analysis. IEEE Trans Neural Netw 22(2):199–210

Peng MW, Li Y, Xie E, Su Z (2010) CEO duality, organizational slack, and firm performance in China. Asia Pac J Manag 27(4):611–624

Putra P, Liriwati FY, Tahrim T, Syafrudin S, Aslan A (2020) The students learning from home experience during Covid-19 school closures policy in Indonesia. Jurnal Iqra' Kajian Ilmu Pendidikan 5(2):30

Qi CR, Liu W, Wu C, Su H, Guibas LJ (2018) Frustum pointnets for 3d object detection from RGB-d data. In: Paper presented at the Proceedings of the IEEE conference on computer vision and pattern recognition

Rai M, Bhattarai N, Dhungel N, Mandal P (2016) Isolation of antibiotic producing Actinomycetes from soil of Kathmandu valley and assessment of their antimicrobial activities. IJOMAS 2(4): 22–26

Ravindran R, Hassan SS, Williams GA, Jaiswal AK (2018) A review on bioconversion of agro-industrial wastes to industrially important enzymes. Bioengineering (Basel) 5(4):93

Rehfeld K, Müller P, Aye N, Schmicker M, Dordevic M, Kaufmann J et al (2017) Dancing or fitness sport? The effects of two training programs on hippocampal plasticity and balance abilities in healthy seniors. Front Hum Neurosci 11:305

Rizwan M, Ali S, Adrees M, Rizvi H, Zia-ur-Rehman M, Hannan F et al (2016) Cadmium stress in rice: toxic effects, tolerance mechanisms, and management: a critical review. In: Environmental science and pollution research, vol 23. Springer, pp 17859–17879

Rizwan M, Ali S, Adrees M, Ibrahim M, Tsang DC, Zia-ur-Rehman M et al (2017a) A critical review on effects, tolerance mechanisms and management of cadmium in vegetables. Chemosphere 182:90–105

Rizwan M, Ali S, Zaheer Akbar M, Shakoor MB, Mahmood A, Ishaque W, Hussain A (2017b) Foliar application of aspartic acid lowers cadmium uptake and Cd-induced oxidative stress in rice under Cd stress. Environ Sci Res Int 24:21938–21947

Rizwan M, Ali S, Ali B, Adrees M, Arshad M, Hussain A et al (2019) Zinc and iron oxide nanoparticles improved the plant growth and reduced the oxidative stress and cadmium concentration in wheat. Chemosphere 214:269–277

Rodda J, Walker Z, Carter J (2011) Depression in older adults. BMJ 343:d5219

Rodrigues M, Chaves M, Wendler R, David M, Quick W, Leegood R et al (1993) Osmotic adjustment in water stressed grapevine leaves in relation to carbon assimilation. Austr J Plant Physiol 20(3):309–321

Rodríquez-Serrano M, Martínez-de la Casa N, Romero-Puertas MC, Del Río L, Sandalio LM (2008) Toxicidad del cadmio en plantas. Ecosistemas 17(3):139–146

Royston P, Parmar MKB (2013) Restricted mean survival time: an alternative to the hazard ratio for the design and analysis of randomized trials with a time-to-event outcome. BMC Med Res Methodol 13(1):1–15

Rozas M, Bohle H, Grothusen H, Bustos P (2012) Epidemiology of amoebic gill disease (AGD) in Chilean salmon industry between 2007 and 2010. Bull Eur Assoc Fish Pathol 32(5):181–188

Sahmoun AE, Case LD, Jackson SA, Schwartz GG (2005) Cadmium and prostate cancer: a critical epidemiologic analysis. Cancer Invest 23(3):256–263

Sarwar N, Saifullah, Malhi SS, Zia MH, Naeem A, Bibi S, Farid G (2010) Role of mineral nutrition in minimizing cadmium accumulation by plants. J Sci Food Agric 90(6):925–937

Shaban A, Bansal S, Liu Z, Essa I, Boots B (2017) One-shot learning for semantic segmentation. Comput Sci Pattern Recognit

Shahid M, Dumat C, Pourrut B, Sabir M, Pinelli E (2014) Assessing the effect of metal speciation on lead toxicity to Vicia faba pigment contents. J Geochem Explor 144:290–297

Shan Q, Wang Y, Li J, Zhang Y, Chen K, Liang Z et al (2013) Targeted genome modification of crop plants using a CRISPR-Cas system. Nat Biotechnol 31(8):686–688

Shiyu Q, Hongen L, Zhaojun N, Rengel Z, Wei G, Chang L, Peng Z (2020) Toxicity of cadmium and its competition with mineral nutrients for uptake by plants: a review. Pedosphere 30(2): 168–180

Shugar DH, Jacquemart M, Shean D, Bhushan S, Upadhyay K, Sattar A et al (2021) A massive rock and ice avalanche caused the 2021 disaster at Chamoli. Indian Himalaya. Science 373(6552): 300–306

Šimat V, Elabed N, Kulawik P, Ceylan Z, Jamroz E, Yazgan H et al (2020) Recent advances in marine-based nutraceuticals and their health benefits. Mar Drugs 18(12):627

Singh JS, Koushal S, Kumar A, Vimal SR, Gupta VK (2016) Book review: microbial inoculants in sustainable agricultural productivity-Vol. II: functional application. Frontiers Media SA

Six L, Smolders E (2014) Future trends in soil cadmium concentration under current cadmium fluxes to European agricultural soils. Sci Total Environ 485:319–328

Skrebsky EC, Tabaldi LA, Pereira LB, Rauber R, Maldaner J, Cargnelutti D et al (2008) Effect of cadmium on growth, micronutrient concentration, and δ-aminolevulinic acid dehydratase and acid phosphatase activities in plants of Pfaffia glomerata. Braz J Plant Physiol 20:285–294

Song C, Wu L, Xie Y, He J, Chen X, Wang T et al (2017) Air pollution in China: Status and spatiotemporal variations. Environ Pollut 227:334–347

Sullivan GM, Artino AR (2013) Analyzing and interpreting data from Likert-type scales. J Grad Med Educ 5(4):541–542

Tabelin CB, Igarashi T, Villacorte-Tabelin M, Park I, Opiso EM, Ito M, Hiroyoshi N (2018) Arsenic, selenium, boron, lead, cadmium, copper, and zinc in naturally contaminated rocks: a review of their sources, modes of enrichment, mechanisms of release, and mitigation strategies. Sci Total Environ 645:1522–1553

Tang B, Bragazzi NL, Li Q, Tang S, Xiao Y, Wu J (2020) An updated estimation of the risk of transmission of the novel coronavirus (2019-nCov). Infect Dis Model 5:248–255

Tran TA, Popova LP (2013) Functions and toxicity of cadmium in plants: recent advances and future prospects. Turk J Bot 37(1):1–13

Tudoreanu L, Phillips CJC (2004) Modeling cadmium uptake and accumulation in plants. Adv Agron 84(4):121–157

Uraguchi S, Tanaka N, Hofmann C, Abiko K, Ohkama-Ohtsu N, Weber M et al (2017) Phytochelatin synthase has contrasting effects on cadmium and arsenic accumulation in rice grains. Plant Cell Physiol 58(10):1730–1742

Vahter M (2002) Mechanisms of arsenic biotransformation. Toxicology 181:211–217

Van Hal NL, Vorst O, Van Houwelingen AM, Kok EJ, Peijnenburg A, Aharoni A et al (2000) The application of DNA microarrays in gene expression analysis. J Biotechnol 78(3):271–280

Vander Zanden MJ, Rasmussen JB (1996) A trophic position model of pelagic food webs: impact on contaminant bioaccumulation in lake trout. Ecol Monogr 66(4):451–477

Venturutti L, Cordo Russo RI, Rivas MA, Mercogliano MF, Izzo F, Oakley R et al (2016) MiR-16 mediates trastuzumab and lapatinib response in ErbB-2-positive breast and gastric cancer via its novel targets CCNJ and FUBP1. Oncogene 35(48):6189–6202

Vestena S, Cambraia J, Ribeiro C, Oliveira JA, Oliva MA (2011) Cadmium-induced oxidative stress and antioxidative enzyme response in water hyacinth and salvinia. Braz J Plant Physiol 23:131–139

Vinceti M, Filippini T, Del Giovane C, Dennert G, Zwahlen M, Brinkman M et al (2018) Selenium for preventing cancer. Cochrane Database Syst Rev. 1(1):CD005195

Violante A, Cozzolino V, Perelomov L, Caporale A, Pigna M (2010) Mobility and bioavailability of heavy metals and metalloids in soil environments. J Soil Sci Plant Nutr 10(3):268–292

Vukmanovic-Stejic M, Zhang Y, Cook JE, Fletcher JM, McQuaid A, Masters JE et al (2006) Human CD4+ CD25 hi Foxp3+ regulatory T cells are derived by rapid turnover of memory populations in vivo. J Clin Invest 116(9):2423–2433

Wang J, Zhao F-J, Meharg AA, Raab A, Feldmann J, McGrath SP (2002) Mechanisms of arsenic hyperaccumulation in Pteris vittata. Uptake kinetics, interactions with phosphate, and arsenic speciation. Plant Physiol 130(3):1552–1561

Wang W, Wang S, Ma X, Gong J (2011) Recent advances in catalytic hydrogenation of carbon dioxide. Chem Soc Rev 40(7):3703–3727

Wang J, Yu L-C, Lai KR, Zhang X (2016) Dimensional sentiment analysis using a regional CNN-LSTM model. In: Paper presented at the proceedings of the 54th annual meeting of the association for computational linguistics (volume 2: short papers)

Wang Y, Zhang S, Li F, Zhou Y, Zhang Y, Wang Z et al (2020) Therapeutic target database 2020: enriched resource for facilitating research and early development of targeted therapeutics. Nucleic Acids Res 48(D1):D1031–D1041

White PJ, Brown P (2010) Plant nutrition for sustainable development and global health. Ann Bot 105(7):1073–1080

WHO/FAO Expert Consultation (2003) Diet, nutrition and the prevention of chronic diseases: report of a joint WHO/FAO expert consultation. 916(i–viii):1–149

Williams JD, Zweig GJ (2016) End-to-end lstm-based dialog control optimized with supervised and reinforcement learning

Wodala B, Eitel G, Gyula T, Ördög A, Horváth F (2012) Monitoring moderate Cu and Cd toxicity by chlorophyll fluorescence and P 700 absorbance in pea leaves. Photosynthetica 50:380–386

World Health Organization (2000) The world health report 2000: health systems: improving performance. World Health Organization, Geneva

Xiao R, Ali A, Wang P, Li R, Tian X, Zhang Z (2019) Comparison of the feasibility of different washing solutions for combined soil washing and phytoremediation for the detoxification of cadmium (Cd) and zinc (Zn) in contaminated soil. Chemosphere 230:510–518

Xin L, Hong-Ying H, Ke G, Ying-Xue S (2010) Effects of different nitrogen and phosphorus concentrations on the growth, nutrient uptake, and lipid accumulation of a freshwater microalga Scenedesmus sp. Bioresour Technol 101(14):5494–5500

Xiong T-T, Leveque T, Austruy A, Goix S, Schreck E, Dappe V et al (2014) Foliar uptake and metal (loid) bioaccessibility in vegetables exposed to particulate matter. Environ Geochem Health 36: 897–909

Ye Y, Wang Q, Lu J, Liu C, Yang Y (2016) High-entropy alloy: challenges and prospects. Mater Tod 19(6):349–362

Zalups RK, Ahmad S (2003) Molecular handling of cadmium in transporting epithelia. Toxicol Appl Pharmacol. 186(3):163–188

Zhang X, Shi H, Wu J, Zhang X, Sun L, Chen C, Chen ZJ (2013) Cyclic GMP-AMP containing mixed phosphodiester linkages is an endogenous high-affinity ligand for STING. Mol Cell 51(2):226–235

Zhang T, Kishore V, Wu F, Weinberger KQ, Artzi YJ (2019) BERTScore: evaluating text generation with BERT

Zhao S, Lin Q, Ran J, Musa SS, Yang G, Wang W et al (2020) Preliminary estimation of the basic reproduction number of novel coronavirus (2019-nCoV) in China, from 2019 to 2020: a data-driven analysis in the early phase of the outbreak. Int J Infect Dis 92:214–217

Zouari M, Ahmed CB, Zorrig W, Elloumi N, Rabhi M, Delmail D et al (2016) Exogenous proline mediates alleviation of cadmium stress by promoting photosynthetic activity, water status and antioxidative enzymes activities of young date palm (Phoenix dactylifera L.). Ecotoxicol Environ Saf 128:100–108

Speciation, Mobilization, and Toxicity of Cadmium in Soil–Microbe–Plant System: An Overview

Sabina Dahija, Selma Pilić, and Renata Bešta-Gajević

Abstract The extensive application of cadmium (Cd) in various industrial products results in worldwide contamination. In the first part of this chapter, we mainly focus on cadmium species, mobility, and factors influencing Cd bioavailability in soils. Furthermore, as Cd interacts with essential cellular components adversely affecting microbial biomass and diversity, we also report various sophisticated resistance mechanisms that provide microorganisms tolerance to Cd. Additionally, attention is paid to highlighting rhizodegradation as an interface between microbes and the rhizosphere that can significantly influence the increase of nutrient uptake and decline of metal toxicity. In particular, we further discuss cadmium accumulation, toxicity, and defense mechanisms in plants against Cd toxicity.

Keywords Cadmium speciation, Mobility, Toxicity, Uptake, Detoxification

1 Introduction

Cadmium (Cd) is one of the most hazardous heavy metals and an important cause of toxicological concern globally. It has a crucial role in industry and the environment due to its persistent and non-biodegradable nature, which results in toxic effects on the living world (Dutta et al. 2019; Goswami et al. 2019). It is ranked eighth out of 20 hazardous substance priorities (Li et al. 2020). Owing to increased industrial activities (mining, industrial wastewater, metallization, use of pesticides, fertilizers, and insecticides), geochemical erosion of rocks, and other environmental factors (volcanic eruptions, acid rain, and continental dust), the cadmium content (Cd) in the soil has risen sharply (Mehmood et al. 2018; Feng et al. 2021). Even in concentrations of 0.3–0.8 mg kg^{-1} in the soil, it can cause toxic effects on plants, such as reduction and inhibition of the activity of antioxidant enzymes, cell divisions, and photosynthesis, causing lipid peroxidation of the cell membrane, which ultimately

S. Dahija (✉) · S. Pilić · R. Bešta-Gajević
Department of Biology, Faculty of Science, University of Sarajevo, Sarajevo, Bosnia and Herzegovina
e-mail: sabina.dahija@pmf.unsa.ba

A. K. Jha, N. Kumar (eds.), *Cadmium Toxicity Mitigation*,
https://doi.org/10.1007/978-3-031-47390-6_2

prevents plant growth and decreases production (Chen et al. 2019). Cadmium's biologically important ionic form Cd^{2+}, interacts with a variety of biomolecules, and these bindings cause the metal to be poisonous (Moulis 2010). Increased Cd concentration in soil has adverse effects on the soil microbial community (Dutta et al. 2019). Microorganisms are essential for the healthy and normal functioning of soil ecosystems and have a key role in the rhizosphere of soils highly polluted with heavy metals (Zaidi et al. 2009; She et al. 2021). By secreting siderophores, organic acids, biosurfactants, and other compounds, they can improve the bioavailability of metals (loids) in the soil and thus impact the absorption of hazardous metals (Navarro-Torre et al. 2016; Jin et al. 2018; Kotoky and Pandey 2018; Luo et al. 2020a, b). In addition, they can change metabolic patterns, altering the toxic effect of metals on the environment (Moynahan et al. 2002; Rasool et al. 2020).

The amount of Cd in the soil and its biological availability, which are determined by the soil's physical and chemical characteristics, the local climate and the makeup of the microbial population, determine how much Cd plants can absorb (Dutta et al. 2019). Absorption of Cd in plants occurs through the plasmalemma of root cells by divalent transition metal transporters for Cd (Zhao et al. 2002). In root cells, Cd can be mobilized or immobilized by forming complexes with metal ligands like proteins, polysaccharides, organic acids, and bases in the cell wall or in vacuoles, but also translocated to metabolically active organelles such as chloroplasts and mitochondria (Deng et al. 2016; Shahid et al. 2017; Wu et al. 2016). The main defense mechanism against Cd toxicity in plants involves sequestering this metal in non-active tissues and excluding it from active tissues (Song et al. 2017). Cadmium toxicity and its detoxification are highly correlated with their chemical forms and subcellular localization (Li et al. 2020).

Cd is transported through root, stem, and leaf by apoplastic and symplastic pathways. By the apoplastic pathways, it is transported through extracellular spaces and, by the symplastic pathways, it is transported intracellularly, from cell to cell via plasmodesmata. In the symplastic pathway, many types of transporters/proteins that regulate cadmium uptake are recorded (Yuan et al. 2012; Song et al. 2017; Li et al. 2020).

Cd builds up in the roots, leaves, stems, and fruits of plants, interfering with their development (Mohanpuria et al. 2007; Feng et al. 2021). Cadmium can seriously affect physiological processes such as photosynthesis (it slows down the activity of PS I and PS II), mineral absorption, and water relations (Di Toppi and Gabbrielli 1999; Baryla et al. 2001). In plants exposed to Cd, cellular redox homeostasis is destroyed, resulting in an increase in the content of reactive oxygen species (ROS) in the plant cell (Irfan et al. 2014; Gupta et al. 2017; Çatav et al. 2020). As oxidative stress expresses its toxicity at the cellular level, the function of cell membranes is altered (Gill and Tuteja 2010). Cadmium in higher concentrations also causes damage to nucleic acids (Moura et al. 2012).

Plants have developed adaptive mechanisms to reduce damage caused by Cd stress, such as regulation of ion uptake, detoxification through chelation, intracellular sequestration, and cellular homeostasis (Hall 2002; Apel and Hirt 2004; Wang et al. 2020). In order to lessen the toxicity of the Cd, root cells produce and release small molecules that form stable complexes with the Cd ion (Fu et al. 2018; Sun et al. 2020). Some plants secrete root exudates that prevent the entry of Cd into the root cells (Bali et al. 2020). Many transporters in the cell wall of the root immobilize Cd in these places, preventing a further transport of ions to the above-ground part of the plant (Fu et al. 2018). Chelation of Cd^{2+} with different ligands (e.g., phytochelatins) and subsequent compartmentalization in the vacuole is one of the most significant Cd detoxification strategies (Xie et al. 2019).

This overview paper provides extensive information for understanding the mechanisms of accumulation, absorption, mobilization, toxicity, and detoxification of cadmium in the soil–microbe–plant system.

2 Sources of Cadmium in Soil

2.1 Natural Cadmium Sources

Cadmium (Cd) is a non-essential element, which is present in large amounts in the environment (Kubier et al. 2019). Cd concentrations in the soil can rise due to geogenic and human-caused sources, and these concentrations are crucial for preserving a healthy food supply.

Sedimentary rocks have higher Cd contents (0.01–2.6 mg/kg) than igneous rocks (0.07–0.25 mg/kg) or metamorphic rocks (0.11–1.0 mg/kg) (Mar and Okazaki 2012; Smolders and Mertens 2013). Due to its similar ionic radius to divalent cations like Ca, Fe, Zn, Pb, and Co, cadmium can replace them in a variety of minerals, including carbonate and phosphate rocks (Merkel and Sperling 1998; Smolders and Mertens 2013).

Parent materials, geochemical processes like leaching and transfer through surface water runoff, and anthropogenic inputs all frequently affect the amount of cadmium in soils (Wu et al. 2020). The average global Cd concentration in unpolluted soils is 3.6%, although certain continents, countries, and soil types may have different amounts. In soils of Europe and the United States, Cd naturally exists in quantities of 0.1–1 mg/kg. In Australia, the mean Cd value is 0.01 mg/kg, in Brazil 0.18 mg/kg, and in Japan 0.3 mg/kg (Smith et al. 2014; Taylor et al. 2016). Histosols and aridisols, which are soils rich in organic matter, have the greatest mean values of Cd content at 0.62 and 0.3 mg/kg, respectively, while Spodosols (0.2 mg/kg), Alfisols (0.11 mg/kg), and Ultisols (0.05 mg/kg) had lower Cd concentrations (Holmgren et al. 1993). However, soil cadmium content generally decreases with depth (Hiller et al. 2001).

The common opinion is that soil contamination is indicated by cadmium concentrations exceeding 3 mg/kg. However, it was shown that the amount of Cd in soil decreases proportionally as the distance between manufacturing facilities and urban regions grows (Belon et al. 2012). The Cd level in the soil is influenced by several factors such as pH, moisture content, soil texture, clay content and type, cation exchange capacity, amount and type of organic matter (OM), and hydroxides (Appel and Ma 2002; Buerge-Weirich et al. 2002; He et al. 2005).

2.2 *Anthropogenic Cadmium Sources*

The main sources of anthropogenic cadmium in soils are atmospheric deposition, industrial and municipal waste, phosphate fertilizers, sewage sludge, mining, and *smelting* of *zinc* (Zn)-bearing *ores* (Mirlean and Roisenberg 2006; Sprynskyy et al. 2011; Bigalke et al. 2017; Sidhu and Bali 2022). A common reason for elevated cadmium concentrations in soil and groundwater is the use of phosphate fertilizers that contain cadmium (Kubier et al. 2019). Studies show that the application of P fertilizers changes the chemical composition of the soil, affecting the soil biota and the microbial community (Bigalke et al. 2017; Bai et al. 2020; Dang et al. 2022). Local or diffuse Cd sources are possible. Local sources such as mines, industrial sites, or abandoned ore deposits contribute to increased Cd concentrations (Merkel and Sperling 1998; Monna et al. 2000; Cloquet et al. 2006). Agricultural practices, wastewater reusing, and atmospheric emissions can all act as diffuse sources of Cd across the environment (Knappe et al. 2008; Sprynskyy et al. 2011).

In summary, the main sources of cadmium in the world are nickel-cadmium batteries, landfills, and municipal waste (Khan et al. 2017). The main causes of cadmium toxicity in agricultural soils include improper waste management and the use of agricultural additives (Khan et al. 2017; Bali et al. 2020).

3 Cadmium Species, Mobility, and Bioavailability in Soil

In a soil solution, dissolved Cd can exist as free, hydrated cations or as species complexes with organic or inorganic ligands (Rahim et al. 2022). Cd can be found in soils in cationic and anionic forms (Kabata-Pendias and Sadurski 2004). Anionic forms of Cd are $CdCl^{3-}$, $Cd(OH)^{3-}$, $Cd(OH)_4{}^{2-}$, and $Cd(HS)_4{}^{2-}$, while cationic forms of Cd are $CdCl^+$, $CdOH^+$, $CdHS^+$, and $CdHCO^{3+}$. In the soil solution, 99% of Cd is present in free ionic form (Kabata-Pendias 1993).

Numerous processes, including sorption (adsorption/desorption), complexation (association/dissociation), precipitation/dissolution, and Cd ligand synthesis can alter Cd speciation in the soil (Degryse et al. 2006). Organic and inorganic ligands, redox conditions, soil pH, metal concentrations, and temperature are the primary factors that affect these processes (Saeki and Kunito 2012; Silber et al. 2012; Zhang et al. 2012; Shahid et al. 2014).

3.1 Factors Influencing Cd Bioavailability in Soils

The concentration and mobility of cadmium in the soil are determined by several factors that might be either naturally occurring or caused by human activity (Yang et al. 2018a, b; Kicińska et al. 2022). The bioavailability of cadmium in the soil is affected by soil pH, cation exchange capacity (CEC), organic matter, clay content, soil microbial activity, and root exudates (Fig. 1) (Bolan et al. 2003; Zhu et al. 2016: Gu et al. 2022).

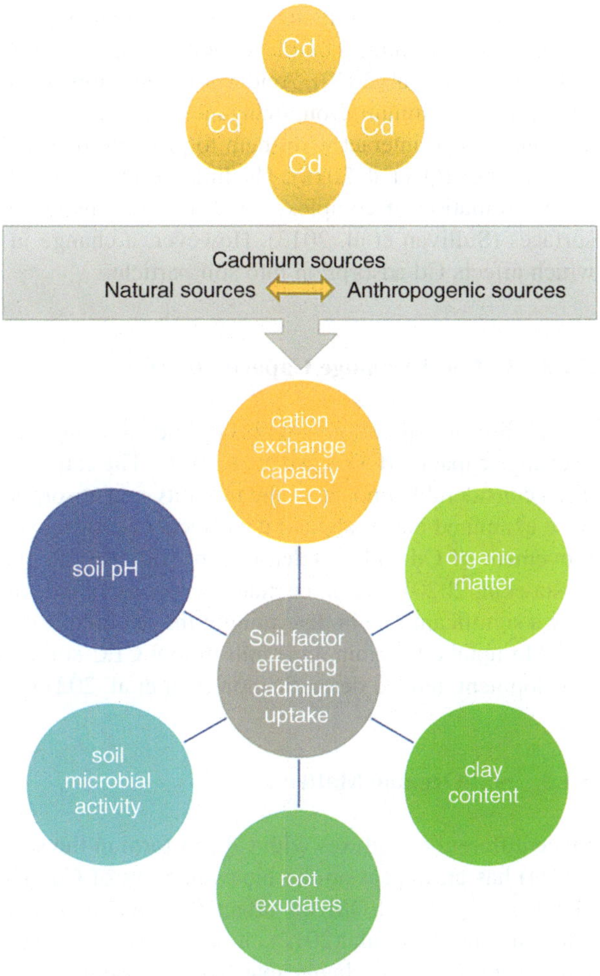

Fig. 1 Cadmium sources and factors influencing Cd bioavailability in soils

3.1.1 Soil pH

The main factor that plays a significant role in the mobility and availability of cadmium in the soil is the hydrogen potential (pH) (Oburger et al. 2020; Xu et al. 2020). The pH value represents the negative logarithm of the concentration of hydrogen ions in a certain matrix, although in the soil it indicates the activity of hydronium ions (H_3O^+) (or H_5O^{+2}).

The availability of cadmium in the soil and the uptake of cadmium into plants decreases with the increase in pH (Yang et al. 2016). Cadmium is adsorbed on the surface of soil particles when the soil pH is higher and is released into the soil solution when the soil pH is lower (Hussain et al. 2021). Plant root exudates increase the mobility of cadmium in the soil by reducing soil pH, changing soil redox potential, and forming complexes with organic ligands (Li et al. 2013; Tao et al. 2016; Mondal et al. 2020). Acidic soil conditions have been observed to affect the solubility of cadmium. Iron, manganese, and carbonate oxides are typically stationary, but Cd can interact with them to promote free mobility, phytoavailability, and Cd exchange (Qi et al. 2018). The limit point for Cd solubility in the soil is pH 6 due to the formation of complexes with organic matter and its adsorption on mineral surfaces (Sullivan et al. 2013). However, a change in pH makes it more alkaline, which affects Cd adsorption into soil particles.

3.1.2 Cation Exchange Capacity (CEC)

The ability of soil colloids to absorb and exchange cations is known as soil cation exchange capacity (CEC) (Su et al. 2021). The cation exchange capability of the soil has a considerable impact on the mobility and absorption of Cd. Numerous research have examined and evaluated the effects of cation exchange capacity (CEC) on the movement of Cd and its release from soil (Acosta et al. 2011; Zheng et al. 2013; Costa et al. 2020). Cation exchange capacity (CEC) of the root cell walls may also play a significant role in determining the net uptake of metals. The deficiency of Ca and Mg uptake by grain amaranth in low CEC soil had a negative impact on plant development and Cd detoxification (Cui et al. 2021).

3.1.3 Soil Organic Matter

Since different complexes with Cd can form in the soil solution, soil organic matter (SOM) has an impact on the bioavailability of Cd. The source, concentration, and chemical types of SOM all affect the bioavailability of Cd. In the research by Gusiatin and Klimiuk (2012), it was observed that SOM was connected to the binding of Cd to exchangeable and acid-soluble fractions in loamy and loamy sand soils with low Cd levels. Lower Cd mobility was recorded in clay soils due to the strong association of Cd with the mineral surface of the clay, Fe–Al oxides, and humus (Hong et al. 2002). Higher organic matter content soils are better able to restrict Cd uptake by plants due to Cd sorption on carboxyl, phenolic, and hydroxyl groups.

3.1.4 Soil Microbial Activity

The relationship between soil microorganisms and enzyme activity is close, where both are responsive to environmental changes. Therefore, soil enzyme activity is frequently used to assess the effect of heavy metals on the soil, microbial function, and the health of the soil ecosystem (Cepeda et al. 2016; Nannipieri et al. 2018). Previous studies have shown that soil enzyme activity decreases exponentially as heavy metal concentrations rise, either by displacing metals related to enzyme conformation and occupying the active center of the enzyme or by binding to sulfhydryl, amino, and carboxyl groups in the enzyme structure to reduce the enzyme's active site (Smolders et al. 2009; Karaca et al. 2010).

The soil microbial activity increases the availability of Cd by secreting organic acids and subsequently solubilizing minerals that contain Cd (Ahmad et al. 2015). In order to increase the bioavailability of Cd, soil amendments containing Cd-solubilizing microorganisms, such as plant growth-promoting rhizobacteria (PGPRs), are crucial. A pot experiment was used to examine how well *Micrococcus* sp. TISTR2221 promoted growth and cadmium accumulation in *Zea mays* L. The findings showed that transplanted *Z. mays* L. in both uncontaminated and cadmium-contaminated soils significantly increased root length, shoot length, and dry biomass. In comparison to uninoculated plants, *Micrococcus* sp. TISTR2221 dramatically increased cadmium accumulation in the underground and aerial parts of *Z. mays* L. (Sangthong et al. 2016). Microorganisms such as PGPR and arbuscular mycorrhizal fungi play a significant role in increasing cadmium tolerance by reducing cadmium toxicity in host plants, altering the rhizosphere microenvironment, and indirectly affecting cadmium enrichment and translocation mechanisms.

3.1.5 Clay Minerals

However, the kind of soil enzyme and other soil features, such as the amount of clay minerals in the soil, as well as the type and concentration of heavy metals, also affect how much soil enzyme activity occurs in response to heavy metals (Aponte et al. 2020). The interaction of enzymes with heavy metals and substrates is influenced by the molecular structure and catalytic characteristics of the enzyme complexes formed by clay minerals (Rakhsh and Golchin 2018). In addition, heavy metal–enzyme–clay mineral complexes can be created when heavy metals interact with the surfaces of clay minerals and compete with enzymes (Huang et al. 2009; Zimmerman and Ahn 2011). As a result, the responses of free enzymes, enzymes immobilized by clay minerals, and soil enzymes to heavy metal pollution may vary.

3.1.6 Root Exudates

Root exudates play a significant role in the sequestration and binding of cadmium in the soil (Liao and Xie 2004). Owing to complexation, ionic strength, and competition for soil exchange sites or root surface exchange sites, the presence of other ions significantly affects the availability of Cd in soil (Sarwar et al. 2010). For exchange sites in soils and uptake by plants, cations (such as Ca, Mg, Zn, and Mn) compete with Cd. Root exudates of plants are secondary products that include organic acids, proteins, peptides, amino acids, sugars, and polysaccharides (Zulfiqar et al. 2019).

Some plants, including wheat and buckwheat, release organic acids from their roots (such as oxalic, malic, and citric acids) that bind Cd^{2+} and keep it from penetrating the roots (Dong et al. 2007). According to the authors Zhi et al. (2020), chemical reactions in the root exudates demonstrated the defensive mechanisms of the plants against Cd stress, including the up-regulation of amino acids to sequester/exclude Cd, regulation of citric acid on chelation/complexation, and precipitation of cadmium ions.

4 Microbial Community in Metal Contaminated Soil

Microorganisms, as the most prevalent and various organisms on Earth, significantly influence environmental functioning and homeostasis. Soil contains enormous numbers of bacteria and archaea, primarily in the form of microbial aggregates, with an estimated total cell number of 3×10^{29} cells, ranging from 1×10^{27} for tropical rainforests to 6.3×10^{28} for desert scrub. However, data on the number of bacterial species and their ecological functions are still lacking and are incomplete (Flemming and Wuertz 2019). Hence, the soil microbiome is one of the most diverse environments, which is largely an unexplored and unused natural resource (Brown et al. 2022). By actively participating in the circulation of matter in the soil, microbes are directly influenced by various stimulating and inhibitory substances, including heavy metals. It is well-established that metals have significant consequences on various bacterial metabolic processes, either directly or indirectly (Nanda et al. 2019). Thus, metal ions such as iron, copper, and manganese are constituent parts of enzyme complexes, active centers of enzymes, cytochromes, pigments, and other compounds, and can act very positively on the growth and abundance (biomass) of different groups of microorganisms (Kaur et al. 2019). Therefore, some metals, such as calcium, cobalt, chromium, copper, iron, potassium, magnesium, manganese, sodium, nickel, and zinc are essential nutrients (Fuhrmann 2021). However, essential metals are required nutrients but, at high levels, they can be harmful to microorganisms and can cause serious risks such as destruction of cell membranes, inhibiting enzyme activity, damaging DNA, and interfering with cell function (Patel et al. 2021). On the other hand, silver, aluminum, cadmium, gold, lead, and mercury have no biological role and nutritive values; indeed, they are toxic to microorganisms (Sandeepa et al. 2023). Furthermore, contamination with heavy metals reduces the diversity of bacteria in the soil, indicating a very toxic effect of metal contamination, especially for specific taxonomic species (Gans et al. 2005). Thus, increased amounts of heavy metals in the soil can have a toxic effect on soil microbial communities. They have an adverse effect on soil microorganisms by significantly altering their growth and development, as well as their overall diversity (Khandaker et al. 2021).

5 Resistance Mechanisms of Cadmium Detoxification in Microorganisms

The toxic effects of heavy metals can be various. Heavy metals have an impact on microbial populations by unfavorably affecting their morphology, growth, and metabolic activities, which ultimately results in a decrease in biomass and diversity (Abdu et al. 2017).

Mostly through adsorption, heavy metals bind to the surface of the cell, i.e., the cell membrane, thereby reducing permeability and preventing the exchange of matter and energy with the environment, which eventually results in the death of the microorganisms themselves. The cell wall of Gram-negative bacteria includes phosphorus, carboxyl, hydroxyl, and amino groups that show an affinity for heavy metals, especially cadmium. Toxic metals exert noxiousness on microbes by disturbing the osmotic balance and oxidative phosphorylation, along with the alterations in the polypeptides and nucleic acids which result in a decline in the number of microorganisms in the soil (Sodhi et al. 2022). As microorganisms have always encountered various metals in the environment, it is therefore not surprising that this selective pressure has contributed to the appearance of toxic metal resistance in almost all bacterial types. In order to withstand heavy metals contaminated environments, bacteria have certain resistance mechanisms and metabolism to transform heavy metals into less dangerous forms (Pal et al. 2022). This also resulted in the emergence of different types of bacteria resistant to heavy metals. This specific resistance is very important for microbial ecology. Otherwise, many microbial strains may decrease or disappear under heavy metal stress. As a response to metal toxicity influences, various resistance sophisticated mechanisms of adaption and resistance occur in microorganisms. Numerous chromosomal, transposon, and plasmid-mediated resistance mechanisms enable bacteria to adapt to metals (Bruins et al. 2000; Pal et al. 2022).

Cadmium is a non-essential metal that is toxic even in small concentrations. It enters the bacterial cell via the transport system for divalent ions. When found in cells, cadmium can bind to sulfhydryl groups on essential proteins, thereby disrupting major cellular functions and causing DNA single-strand breaks. Many studies have confirmed decreased culturable counts, decreased biomass, composition and function, growth, morphology change, and biochemical activities, due to the adverse effects of cadmium on the soil microbial community (Diaconu et al. 2020; Peng et al. 2020).

Despite these toxic effects, eukaryotic organisms detoxify cadmium, as well as some other heavy metals, mainly by binding to polythiols, while bacteria have developed several and highly efficient detoxification mechanisms to tolerate heavy metals (Rudakiya and Patel 2021).

These mechanisms can be so effective that high levels of relevant toxic metals have no visible consequence on the growth of cells of resistant strains. Genes encoding determinants of cadmium tolerance can be located on chromosomes, plasmids, or transposons (Vera-Bernal and Martínez-Espinosa 2021). In most cases, the controlling genes for heavy metals are commonly located on plasmids (Biswas et al. 2021). These plasmids ensure that the bacteria have an advantage over competitors in the environment where heavy metals are found. Furthermore, soil isolated Cd-resistant bacteria also typically show various resistances to antibiotics, chemicals, and organic solvents (Hui et al. 2022).

Several microbes produce carbonate, H_2S, phosphate, and oxalate in order to form insoluble cadmium salts in reaction with Cd (II). In urease-producing bacteria, urease breaks down urea into carbonate to precipitate Cd (II). Furthermore, some sulfate-reducing bacteria produce H_2S to form an intracellular or extracellular precipitate of CdS (Xia et al. 2021). Nevertheless, mechanisms by which bacteria provide tolerance to heavy metals are the following: (a) removal of metals by permeability change and biosorption, (b) metal ions extracellular sequestration, (c) active efflux of metals by transport, (d) intracellular sequestration, and (e) enzymatic detoxification/reduction of metals (Tayang and Songachan 2021). Any bacteria can possess one or several tolerance mechanisms. Through mechanisms that cells combine, microorganisms are enabled to live in unfavorable living conditions and in metal-polluted habitats. Primary four mechanisms have been proven to apply in the case of cadmium. Tolerance to cadmium can be achieved by the mentioned mechanisms except for enzymatic detoxification (Nkosi 2020). Covalent modification of the divalent Cd form is not biologically positive because this form is more unstable and toxic (Bruins et al. 2000).

5.1 Removal of Cd by Membrane Permeability Change

Examples of this resistance mechanism are any changes in the cell wall, membrane, or envelope of the microorganism in an effort to protect metal sensitive, essential cellular components from the toxic effects of heavy metals (Cleophas et al. 2022). An example of this mechanism is in the case of *Staphylococcus aureus* bacteria, when changing the conformation of the membrane, it changes the permeability for cadmium and other metals. In some species such as *Staphylococcus aureus*, penicillinase plasmids can mediate resistance by changing the cell membrane permeability to Cd(II), as well as to other metals. In this case, there appear to be conformational changes in the membrane that prevent metal ions from entering (Bruins et al. 2000).

5.2 Extracellular Cd Sequestration

On the other hand, bacteria can prevent metal ions from interacting with cellular components binding and adsorbing them non-specifically onto the cellular surface and the cell walls of microorganisms. Without the utilization of energy through a metabolism-independent process, heavy metals are adsorbed via the microbial cell wall at binding sites (Sreedevi et al. 2022). Biosorption or uptake between the positively charged heavy metal and the negatively charged cell wall is the most commonly used method in biological systems. Therefore, the bacterial cell wall played a crucial role in Cd^{2+} binding. The cell wall and surface of Gram-positive bacteria have a negative charge density because of the peptidoglycan network made up of alternating strands of glucosamine and muramic acid residues, which consequently results in the binding of positively charged cadmium ions to the cell surface (Rivera-Utrilla et al. 2003). Thus, Zhu et al. (2022) noticed morphological changes in bacteria such as elongation after undergoing Cd^{2+} stress. This occurrence can be explained as a self-defense mechanism that enables cells to reduce the adverse effects of environmental stress factors. Extracellular polymeric substances are complex compounds such as lipopolysaccharides, proteins, carbohydrates, siderophores, glutathione, and biosurfactants. These chemical compounds have a large number of metal-binding functional groups. As the cell wall consists of several functional groups, such as thiol, thioether, sulfonate, phosphoryl, sulfhydryl, hydroxyl, ester, amine, and carboxyl groups, various cations and anions of heavy metals bind to it through electrostatic interactions and hydrogen bonding. Extracellular polymeric substances (EPS) have a major influence, depending on their chemical characteristics, on the metal adsorption and acid–base properties of microbes. This protective layer prevents the uptake of metals, keeping metal ions away from delicate cellular components, therefore exhibiting an excellent Cd (II) adsorption ability (Monachese et al. 2012; Xia et al. 2021).

The ability to absorb metal ions and prevent their interaction with delicate and crucial cellular components is demonstrated by bacteria that produce an extracellular polysaccharide coating. Actually, this exopolysaccharide coating of bacteria can provide sites for the attachment of metal cations. Gram-positive *B. subtilis, Lactobacillus rhamnosus* GG, and some *Bifidobacterium longum* strains are known to produce exopolysaccharides, which have more negatively charged groups such as carboxyl, hydroxyl, and phosphate, enhance the number of ligands that can bind cationic metals like cadmium and lead (Kumar et al. 2018). *Klebsiella aerogenes* strain has demonstrated the ability to excrete sulfur to eliminate Cd (II) ions from the environment to limit metal influx by external precipitation (Bruins et al. 2000). This process known as biosorption can find application in the bioremediation of polluted environments. Both living and non-living microbial biomasses have been used as adsorbents for decontaminating Cd (II) from wastewater.

5.3 Cd Uptake Efflux System

In order to achieve their physiological or toxic role, heavy metals must first enter the bacterial cell. Active transport, or efflux transport system, represents the predominant and the most widespread mechanism of resistance/adaptation to heavy metals. Microorganisms use active transport mechanisms to reduce Cd (II) uptake and increase Cd (II) efflux out of cells (Xia et al. 2021). This heavy metal ions removal mechanism is mostly represented in various microorganisms, especially those isolated from metal-polluted environments (Bazzi et al. 2020).

It is well known that high intracellular accumulation of Cd exerts obvious cytotoxicity on bacteria. Excess Cd is usually excreted outside the cells via several efflux mechanisms (Hui et al. 2022). Generally, heavy metal ions and bacterial species both affect the expression of metal ion transporters in the efflux. Special resistance genes that encoded membrane transporters that control the uptake and exclusion of heavy metal ions are located on plasmids or chromosomes (Bruins et al. 2000). Moreover, the most famous mechanism of cadmium tolerance is the removal of cadmium from cells by efflux pumps.

Cadmium utilizes, through low-specificity, transportation systems for manganese and zinc in Gram-positive and Gram-negative bacteria (Bruins et al. 2000). In Gram-negative bacteria, Cd^{2+} enters into bacterial cells as a toxic alternative substrate for the Zn^{2+} divalent cation uptake, based on the similar chemical properties among Cd (II) and Zn (II), respectively. Both these systems are chromosomally coded, nutritionally required cation transport systems. A negative consequence of this is the cotransport of other cations that may be toxic to the microorganism. The accumulation of cadmium in the cytoplasm induces synthesize efflux systems to expel excess metal from the cell, maintain homeostasis, and prevent the toxic effect (Xia et al. 2021).

Various efflux systems, namely resistance nodulation cell division (RND), P-type ATPases, and cation diffusion facilitator (CDF), have been discovered to be active in bacteria, removing Cd^{+2} and other heavy metals divalent from the bacterial cells. These efflux systems can be dependent on and independent of ATP molecules, and they can also be specific for the anion or cation they transport. Metal-exporting has been attributed to efflux pump systems, including ATP-binding cassette transporter (ABC transporter) and CBA efflux transporter, CDF (cation diffusion facilitator), and P-type ATPases that are involved in eliminating heavy metal divalent cations, including Cd^{+2} from the bacterial cells. Moreover, P-type efflux ATPase can play a role in the Cd export of Gram-positive bacteria. Cd (II) efflux membrane protein includes the ABC superfamily's Cad system and the RND superfamily's Czc systems. CzcD can transmit the extracellular signal of Cd (II) to CzcR by slow uptake of Cd (II) (Xia et al. 2021). ABC transporters can mediate membrane translocation of heavy metal ions, which help microorganisms overcome heavy metal stress (Khan et al. 2022).

5.4 Intracellular Cd Sequestration

Another mechanism which heavy metal ions pass the cell membrane into the cytoplasm is through bioaccumulation. Intracellular traps within the cytoplasm can sequester heavy metals that pass through the cell wall and enter microorganisms, thus preventing them from reaching toxic levels. This mechanism known as intracellular sequestration is the accumulation of metals within the cytoplasm to prevent interactions and exposure to essential cellular components (Pal et al. 2022). Frequently sequestered metals are Cd (II), Cu (II), and Zn (II). A large number of microorganisms can transform heavy metals with the help of sulfides, cytosolic polyphosphates, and cysteine-rich proteins. The metallothionein-like proteins, low molecular mass polypeptides, and thiol-rich, cysteine-abundant proteins show a high binding ability to heavy metal ions, which is an effective bacterial protective strategy against metal toxicity (Mathivanan et al. 2021).

Most often, bacteria produce cysteine-rich proteins like metallothioneins, which enable the "trapping" of metals within the cytoplasm. In circumstances where it is exposed to heavy metal ions, metallothionein can be overexpressed to overcome this stress. Cysteine residues in metallothionein can remove an excess of toxic heavy metal ions (Dave et al. 2020). For example, cyanobacteria *Synechococcus* sp. can excrete metallothionein to bind Cd (II) to reduce toxicity to cells (Bontidean et al. 2000). Significant quantities of Cd (II) induce smtA and smtB genes which encode metallothionein that binds to Cd (II) in *Synechococcus* sp. (Bruins et al. 2000). On the other hand, the harmful effect of metals is eliminated by the production of cysteine-rich proteins in *Pseudomonas putida* (Odokuma and Abah 2003). In addition, phytochelatin, produced by the enzyme phytochelatin synthase in cyanobacteria, provides another approach for immobilizing Cd (II) (Xia et al. 2021). Therefore, delicate cellular components can be protected from exposure to heavy metal ions, and chelation substrates exhibit great potential in Cd (II) immobilization (Yin et al. 2019).

6 Rhizodegradation of Cd

Plants and specific microbes interact in relationships known as symbiotic and this species-specific phenomenon is known as rhizodegradation (Kumar et al. 2021). Increased nutrient uptake and decreased metal toxicity may both be significantly influenced by the rhizosphere, the interaction between microorganisms and plant roots (Zhao et al. 2023).

The amelioration of heavy metal toxicity in plants by microorganisms may be through a reduction in the metal uptake by plants or through a reduction in the amounts of detrimental stress ethylene induced by heavy metals with no effect on their uptake (Zhou et al. 2022). On the other hand, solubilizing or mobilizing heavy metals by metal-resistant microorganisms with properties that promote plant growth could increase the availability of heavy metals for plant uptake and be useful for phytoremediation (Yu et al. 2023).

The plant growth-promoting rhizobacteria (PGPR) can be utilized successfully in Cd bioremediation with bacterial assistance. Kim et al. (2010) established that *Echinochloa crus-galli* citric acid and oxalic acid compounds of root exudates significantly improve the Cd translocation and bioaccumulation. According to Kumar et al. (2021), increased populations of Cd-resistant bacterial species of *Pseudomonas*, *Cupriavidus*, *Bacillus,* and *Acinetobacter are* associated with *Boehmeria nivea*. On the other hand, Jiang et al. (2022) demonstrated that the plant symbiotic fungus *Metarhizium robertsii* promoted beneficial modulation of plant health and growth under Cd stress and reduced Cd accumulation in plant tissues. They also reported that *M. robertsii* colonization increased the total antioxidant capacity and production of Cd-binding proteins and chlorophyll in *Arabidopsis thaliana* and *Oryza sativa*. Therefore, it can be noticed that the symbiotic association between plants and microorganisms is a very attractive candidate for metal decontamination.

7 Cadmium Accumulation in Plants

Cd is primarily found insoluble in soil and is not bioavailable to plants. By releasing root exudates that alter the pH of the rhizosphere, plants can make Cd more soluble (Dalvi and Bhalerao 2013). Most Cd is absorbed via the apoplastic and symplastic pathways. The apoplastic pathway is a process of passive diffusion and involves the non-living spaces between the cells and the cell membranes. The symplastic pathway is a transport mechanism through the plasma membrane that relies on gradients of electrochemical potential and concentration and requires energy expenditure (Fig. 2). When Cd reaches the root, it forms complexes with chelating agents that lose their toxicity when they become immobilized in the cell wall, cytoplasm, or vacuoles (Ali et al. 2013). Cadmium toxicity and long-distance transport to the shoot are reduced when it is stored in root vacuoles (Thakur et al. 2016). Shoots are where Cd is sequestered, while plant vacuoles or cell walls are where it is detoxified (Tong et al. 2004).

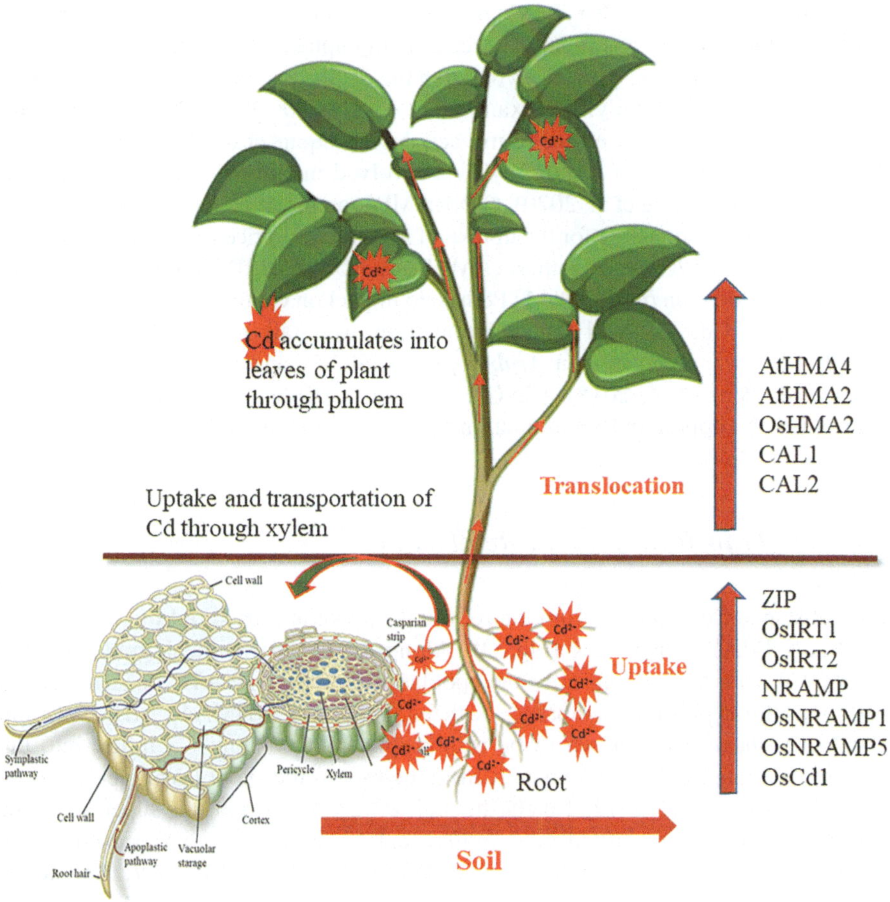

Fig. 2 Cadmium uptake and translocation in plant

7.1 Cd Uptake Transporters in Plant

Transporters involved in Cd transport are zinc-regulated transporters, iron-regulated transporter proteins (ZIP), natural resistance-associated macrophage proteins (NRAMP), and metal tolerance proteins (MTP) (Fig. 2).

Transporters from the ZIP family are involved in the accumulation, absorption, and transport of many cations from roots to shoots (Guerinot 2000). For instance, AtIRT1 has a higher level of expression in the root epidermis and a lower level in the root cortex. Iron deficiency increases the expression of OsIRT1 and OsIRT2 transporters and promotes the absorption and transport of cadmium (Ishimaru et al. 2006; Nakanishi et al. 2006). Double mutations of two zinc transporter genes in rice, OsZIP5 and OsZIP9, which are expressed mainly in the cortex and are located on the cell membrane, reduce Zn and Cd uptake in plants (Tan et al. 2020). The findings imply that these genes synergistically regulate the uptake of Zn and Cd.

Another group of transporters involved in the uptake and transport of Cd in plants is from the NRAMP family. For example, in *A. thaliana*, AtNRAMP3/4 is responsible for Fe, Mn, and Cd efflux from vacuoles (Lanquar et al. 2010; Pottier et al. 2015). The Fe transporter OsNRAMP1 is involved in Cd and Mn absorption and transport in rice (Chang et al. 2020). OsNRAMP5, on the other hand, is a transporter that is primarily responsible for the uptake of Mn and Cd in rice roots (Ishikawa et al. 2012; Sasaki et al. 2012). Together, OsMTP8 and OsNRAMP5 regulate the absorption and translocation of Mn. OsMTP8 has no impact on Cd transport, though (Ueno et al. 2015). Cd, Co, and Mn accumulation were all increased when wheat TpNRAMP5 was expressed in *Arabidopsis*, while Zn and Fe accumulation were unaffected (Peng et al. 2018). RING E3 ligase OsHIR1 and MicroRNA166 also control Cd absorption and accumulation in rice (Ding et al. 2018).

7.2 Cd Translocation and Reallocation

Long-distance Cd transport from the root to the shoot is carried out by heavy metal ATPase (HMA) transporters (Fig. 2). Along an electrochemical gradient, they release Cd from the cytoplasm to the apoplast and hydrolyze ATP. In *Arabidopsis* and rice, three plasma membrane-localized proteins—AtHMA4, AtHMA2, and OsHMA2—have been shown to be responsible for mediating Cd influx into the stele to facilitate Zn and Cd transport from the root to the shoot (Wong et al. 2009; Satoh-Nagasawa et al. 2012; Takahashi et al. 2012). As a result, the super tolerance and hyperaccumulation of Cd in *A. halleri* and *N. caerulescens* are caused by the enhanced expression and triplication of HMA4 (Courbot et al. 2007; Hanikenne et al. 2008).

Cd accumulation in shoots is regulated by CAL1 and CAL2 proteins that can chelate cytoplasmic Cd to form complexes and secrete them into the xylem sap via long-distance transport (Luo et al. 2018, 2020a, b). OsCd1 transporter, which is expressed mainly in the root cortex and is located on the cell membrane, mediates the differential accumulation of cadmium as demonstrated in rice (Yan et al. 2019).

7.3 Vacuolar Sequestration of Cd

Plants store many ions in vacuoles. Free Cd and phytochelatin (PC)–Cd complexes can be transported to the vacuole by several transporters such as ATP-binding cassette transporters (ABCCs), NRAMPs, H+ /cation exchangers (CAXs), and P1B-type heavy metal ATPases (HMAs) (Park et al. 2012; Brunetti et al. 2015). Three ABCC vacuolar membrane transporters, ABCC1, ABCC2, and ABCC3, are particularly crucial for the sequestration of the PC–Cd complex. Furthermore, two

NRAMP tonoplast transporters, NRAMP3 and NRAMP4 play a key role in the efflux of free cadmium from the vacuole to the cytosol (Lanquar et al. 2005). The tonoplast-localized transporters, AtCAX2 and AtCAX4, not only have the specificity for Ca but also the ability to transport other metals, such as Cd, by employing the proton gradient to regulate Ca and Cd accumulation in the vacuole (Korenkov et al. 2009). The expression of AhCAX1 has been found to promote Cd tolerance in the Cd hyperaccumulator *Arabidopsis helleri* (Baliardini et al. 2015). Within the HMA transporter family, AtHMA3 is responsible for cadmium storage in the vacuole (Morel et al. 2009; Chao et al. 2012). Cd hyperaccumulation in *Noccaea caerulescens* and *Sedum alfredii* is correlated with higher NcHMA3 and SaHMA3 expression (Zhang et al. 2016). According to a recent study by Liu et al. (2017), SpHMA3 is responsible for Cd vacuolar sequestration and detoxification in early leaf cells of *Sedum plumbizincicola* plants.

8 Toxic Effects of Cadmium on Plants

A significant reduction in plant growth and yield due to increased soil cadmium concentrations has become a primary concern of scientists (Romero-Puertas et al. 2004; Goix et al. 2014; Zhang et al. 2014). Because of the excessive generation of oxidative markers, which results in damaging biomolecules such as carbohydrates, lipids, proteins, and DNA, the buildup of cadmium in soils has harmful consequences on plants (Foyer and Noctor 2005; Qayyum et al. 2017; Hussain et al. 2018). Strategies that can reduce the toxicity of cadmium in order to preserve human and plant health are the development of heavy metal hyperaccumulators that can extract metals in their organs and avoid the introduction of Cd and other heavy metals into the above-ground parts. When exposed to different cadmium concentrations, plants exhibit varying reactions. The response of plants to Cd stress depends on the plant species, duration of exposure, amount of uptake, capturing, removal, and storage in different plant compartments (Shah et al. 2019).

After the addition of different concentrations of cadmium, anatomical changes were observed in the root, stem, and leaves of *Ceratopteris pteridoides*. Abaxial stomata closure, reduced stomatal size in leaves, scarification of tracheid walls, small xylem vessels, and disorganization of vascular bundles in roots and stems are some of these changes (Bora and Sarma 2021). In the species *Trigonella foenum* exposed to Cd stress, a decrease in trichome length, abaxial and adaxial stomatal density, and cortex fraction was recorded (Ahmad et al. 2005).

In plants, oxidative damage from cadmium is caused by high quantities of H_2O_2 and lipid peroxidation (Rizwan et al. 2019; Shiyu et al. 2020). Several authors indicated that Cd causes an increase in ROS (H_2O_2, O-2) generation (Kapoor et al. 2019; Hasanuzzaman et al. 2020; Unsal et al. 2020). Free radicals and non-radicals make up the majority of cellular ROS (Hasanuzzaman et al. 2020). Free radicals include O_2•−, •OH, RO•, peroxyl radical (ROO•) and non-radicals, H_2O_2, $1O_2$ and ozone (O_3) (Maurya 2020). Excited carbonyls, hypochlorous acid (HOCl), and hydroperoxides (ROOH) are additional non-radicals found in plants (Kapoor et al. 2015). When plants are exposed to Cd, there is an increase in the formation of ROS, which results in oxidative stress and produces an accumulation of reactive chemicals such as thiobarbituric acid (TBARS) and malondialdehyde (MDA), which causes electrolyte leakage (Younis et al. 2016).

Some of the key processes that contribute to Cd accumulation include structural similarities with the phosphorus and zinc nutrients absorbed by roots, Cd interaction with the sulfhydryl (-SH) group compromises protein structure, removal of necessary cations from binding sites, and damage important nutrients, nucleic acids, lipids, proteins, and cell pigments (Hossain et al. 2012; Choppala et al. 2014; Singh et al. 2016). The oxidative stress caused by cadmium is counteracted by enzymatic antioxidants (superoxide dismutase, catalase, ascorbate peroxidase, glutathione reductase, monodehydroascorbate reductase, dehydroascorbate reductase, etc.), and non-enzymatic antioxidants (ascorbic acid, glutathione, phenolic acids, alkaloids, flavonoids, carotenoids, α-tocopherol, etc.).

Cd stress interferes with metabolic processes in the cell, damages the respiratory chain of mitochondria, leads to DNA and RNA damage, and breakdown of proteins and lipids (Guo et al. 2013; Xie et al. 2018; Akhter et al. 2021).

9 Mechanisms of Cadmium Detoxification in Plants

Exposure of plants to stress due to Cd accumulation triggers a number of strategies to reduce Cd toxicity, including sequestration in vacuoles, cytoplasmic chelation, and binding Cd to the cell wall.

9.1 Glutathione and Phytochelatin

Mechanisms of cadmium detoxification and redox balance maintenance include the non-protein peptides Glutathione (GSH) and phytochelatin (PC) (Zare et al. 2023). Phytochelatins (PCs) are small metal-binding cysteine-rich peptides found in plants. The common structure of PCs is (γGluCys)nGly where n is limited to 2–11, but is usually between 2 and 5. It is synthesized under the action of enzymes phytochelatin synthase (PCS). PC are metal-binding agents due to their repeating thiol units and can participate in the detoxification of Cd and other heavy metals (Seregin and Kozhevnikova 2023). Several studies (Vatamaniuk et al. 2000; Zientara et al. 2009; Song et al. 2017; Jalmi 2022) report that the main mechanism of PC-mediated detoxification is Cd chelation by PC to form a complex, which is then transported to the vacuole by ABC transporters.

On the other hand, glutathione (GSH) is described as the first line of defense in many plant cell types. One of the main roles of GSH in the cell is that it can act as a Cd chelator because Cd has a high affinity for the thiol-containing molecules such as glutathione, which leads to the formation of a Cd–GSH complex and reduces the content of toxic Cd^{2+} in plant cells. Furthermore, GSH serves as a precursor for phytochelatin (PC) biosynthesis and acts as an important antioxidant that scavenges excessive reactive oxygen species in Cd-stressed plants (Ding et al. 2017).

9.2 Metallothionein

Metallothioneins (MTs) are small peptides (size 6–7 kDa) with a primary structure characterized by the absence of aromatic amino acids and histidine, and a high content of SH groups derived from cysteine (30%) (Nordberg and Nordberg 2009). MTs are involved in *cellular heavy metal ion homeostasis* and are widely distributed in the living world (Cobbett and Goldsbrough 2002). According to Peng et al. (2017), Cd hyperaccumulation and hypertolerance in *S. plumbizincicola* are caused by the MT-like protein SpMTL, which functions as a cytoplasmic Cd chelation protein. When heterologously expressed in yeast cells and *A. thaliana*, the MT-like gene *DcCDT1* from *Digitaria ciliaris* and its rice counterpart, *OsCDT1*, improve Cd tolerance by lowering cytoplasmic Cd concentration (Kuramata et al. 2009). Rice and wheat HsfA4a increased MT gene expression to increase Cd tolerance (Shim et al. 2009).

9.3 Plant Defensins

Plant defensins have an important role in cadmium allocation and detoxification and are considered natural immune agents with extensive biological activities (Parisi et al. 2019). They constitute a large family of cationic proteins of ~45–54 amino acid residues with eight cysteine residues (C1–C8), forming a conserved pattern of disulfide bonds: C1–C8, C2–C5, C3–C6, and C4–C7 (Bukhteeva et al. 2022). Research in the past few years has shown that human defensin 5 (HD5) has Zn and Cd chelating activity (Zn/Cd binding activity) (Zhang et al. 2013). Human defensin 5 (HD5) was demonstrated through in vitro research to have Zn and Cd chelating activity. Furthermore, it has been noted that Zn supply lowered PDF1.2 expression in the Zn/Cd hyperaccumulator *Noccaea caerulescens*, but Cd stress greatly increased PDF1.2 expression in *Arabidopsis* (Cabot et al. 2013; Gallego et al. 2016). *Arabidopsis* defensin AtPDF2.5, which is located in the cell wall of vascular xylem bundles, plays a significant role in cytoplasmic chelation of cadmium and secretion of the AtPDF2.5–Cd complex into the apoplast (Luo et al. 2019a). AtPDF2.6 is found in the cytoplasm of xylem vascular bundles and is significantly activated by Cd stress. AtPDF2.6 can improve the tolerance of *Arabidopsis* to Cd by chelating the cytoplasmic form of the heavy metal (Luo et al. 2019a, b).

9.4 *Cell Wall Detoxification*

Cellulose, hemicellulose, and pectin are the primary ingredients of the cell wall. The sorption of cations in the cell walls is made possible by electronegative sites on the chains of cellulose, hemicellulose, and pectin (Haynes 1980; Grignon and Sentenac 1991). An essential defense mechanism for plants under Cd stress is the cell wall, which binds Cd and prevents its buildup in the cytoplasm (Vázquez et al. 2006; Nocito et al. 2011). According to subcellular localization, most of the Cd is found in the apoplast, particularly in the cell wall, and only a small amount is found in the protoplast (Vatehová-Vivodová et al. 2018). The cell wall promotes plant Cd tolerance by blocking Cd from entering root cells (Sharma et al. 2016; Peng et al. 2017; Gutsch et al. 2018).

The structural element of the cell wall that binds to heavy metals is pectin (Meychik et al. 2014; Gutsch et al. 2018). Homogalacturonic acids (HGAs) make up the majority of the complex polysaccharide known as pectin. The response to Cd stress is significantly influenced by HGAs, which are linear sugar chains created by the polymerization of D-galacturonic acid (Harholt et al. 2010).

References

Abdu N, Abdullahi AA, Abdulkadir A (2017) Heavy metals and soil microbes. Environ Chem Lett 15:65–84

Acosta JA, Jansen B, Kalbitz K, Faz A, Martínez-Martínez S (2011) Salinity increases mobility of heavy metals in soils. Chemosphere *85*(8):1318–1324

Ahmad SH, Reshi Z, Ahmad J, Iqbal M (2005) Morpho-anatomical responses of Trigonella foenum graecum Linn. to induced cadmium and lead stress. J Plant Biol *48*:64–84

Ahmad I, Akhtar MJ, Zahir ZA, Mitter B (2015) Organic amendments: effects on cereals growth and cadmium remediation. Int J Environ Sci Technol 12:2919–2928

Akhter Z, Bi Z, Ali K, Sun C, Fiaz S, Haider FU, Bai J (2021) In response to abiotic stress, DNA methylation confers epigenetic changes in plants. Plants 10(6):1096

Ali H, Khan E, Sajad MA (2013) Phytoremediation of heavy metals—concepts and applications. Chemosphere 91(7):869–881

Apel K, Hirt H (2004) Reactive oxygen species: metabolism, oxidative stress, and signal transduction. Annu Rev Plant Biol 55:373–399

Aponte H, Meli P, Butler B, Paolini J, Matus F, Merino C, Cornejo P, Kuzyakov Y (2020) Meta-analysis of heavy metal effects on soil enzyme activities. Sci Total Environ 737:139744

Appel C, Ma L (2002) Concentration, pH, and surface charge effects on cadmium and lead sorption in three tropical soils. J Environ Qual 31(2):581–589

Bai YC, Chang YY, Hussain M, Lu B, Zhang JP, Song XB, Lei XS, Pei D (2020) Soil chemical and microbiological properties are changed by long-term chemical fertilizers that limit ecosystem functioning. Microorganisms 8(5):694

Bali AS, Sidhu GPS, Kumar V (2020) Root exudates ameliorate cadmium tolerance in plants: a review. Environ Chem Lett 18:1243–1275

Baliardini C, Meyer CL, Salis P, Saumitou-Laprade P, Verbruggen N (2015) CATION EXCHANGER1 cosegregates with Cadmium tolerance in the metal hyperaccumulator Arabidopsis halleri and plays a role in limiting oxidative stress in Arabidopsis spp. Plant Physiol 169(1):549–559

Baryla A, Carrier P, Franck F, Coulomb C, Sahut C, Havaux M (2001) Leaf chlorosis in oilseed rape plants (Brassica napus) grown on cadmium-polluted soil: causes and consequences for photosynthesis and growth. Planta 212:696–709

Bazzi W, Abou Fayad AG, Nasser A, Haraoui LP, Dewachi O, Abou-Sitta G, Nguyen VK, Abara A, Karah N, Landecker H, Knapp C (2020) Heavy metal toxicity in armed conflicts potentiates AMR in A. baumannii by selecting for antibiotic and heavy metal co-resistance mechanisms. Front Microbiol 11:68

Belon E, Boisson M, Deportes IZ, Eglin TK, Feix I, Bispo AO, Galsomies L, Leblond S, Guellier CR (2012) An inventory of trace elements inputs to French agricultural soils. Sci Total Environ 439:87–95

Bigalke M, Ulrich A, Rehmus A, Keller A (2017) Accumulation of cadmium and uranium in arable soils in Switzerland. Environ Pollut 221:85–93

Biswas R, Halder U, Kabiraj A, Mondal A, Bandopadhyay R (2021) Overview on the role of heavy metals tolerance on developing antibiotic resistance in both Gram-negative and Gram-positive bacteria. Arch Microbiol 203:2761–2770

Bolan NS, Adriano DC, Naidu R (2003) Role of phosphorus in (Im)mobilization and bioavailability of heavy metals in the soil-plant system. Rev Environ Contam Toxicol 177(177):1–44

Bontidean I, Lloyd JR, Hobman JL, Wilson JR, Csöregi E, Mattiasson B, Brown NL (2000) Bacterial metal-resistance proteins and their use in biosensors for the detection of bioavailable heavy metals. J Inorg Biochem 79(1–4):225–229

Bora MS, Sarma KP (2021) Anatomical and ultrastructural alterations in Ceratopteris pteridoides under cadmium stress: a mechanism of cadmium tolerance. Ecotoxicol Environ Saf 218:112285

Brown MD, Shinn LM, Reeser G, Browning M, Schwingel A, Khan NA, Holscher HD (2022) Fecal and soil microbiota composition of gardening and non-gardening families. Sci Rep 12(1): 1595

Bruins MR, Kapil S, Oehme FW (2000) Microbial resistance to metals in the environment. Ecotoxicol Environ Saf 45(3):198–207

Brunetti P, Zanella L, De Paolis A, Di Litta D, Cecchetti V, Falasca G, Barbieri M, Altamura MM, Costantino P, Cardarelli M (2015) Cadmium-inducible expression of the ABC-type transporter AtABCC3 increases phytochelatin-mediated cadmium tolerance in Arabidopsis. J Exp Bot 66(13):3815–3829

Buerge-Weirich D, Hari R, Xue H, Behra P, Sigg L (2002) Adsorption of Cu, Cd, and Ni on goethite in the presence of natural groundwater ligands. Environ Sci Technol 36(3):328–336

Bukhteeva I, Hrunyk NI, Yusypovych YM, Shalovylo YI, Kovaleva V, Nesmelova IV (2022) Structure, dynamics, and function of PsDef2 defensin from Pinus sylvestris. Structure 30(5): 753–762

Cabot C, Gallego B, Martos S, Barceló J, Poschenrieder C (2013) Signal cross talk in Arabidopsis exposed to cadmium, silicon, and Botrytis cinerea. Planta 237:337–349

Çatav ŞS, Genç TO, Oktay MK, Küçükakyüz K (2020) Cadmium toxicity in wheat: impacts on element contents, antioxidant enzyme activities, oxidative stress, and genotoxicity. Bull Environ Contam Toxicol 104:71–77

Cepeda CT, Sotres FG, Bello D (2016) Use of enzyme activities to monitor pollution of agricultural land. EQA Int J Environ Qual 22:15–24

Chang JD, Huang S, Yamaji N, Zhang W, Ma JF, Zhao FJ (2020) OsNRAMP1 transporter contributes to cadmium and manganese uptake in rice. Plant Cell Environ 43(10):2476–2491

Chao DY, Silva A, Baxter I, Huang YS, Nordborg M, Danku J, Lahner B, Yakubova E, Salt DE (2012) Genome-wide association studies identify heavy metal ATPase3 as the primary determinant of natural variation in leaf cadmium in Arabidopsis thaliana. PLoS Genet 8(9):e1002923

Chen D, Chen D, Xue R, Long J, Lin X, Lin Y, Jia L, Zeng R, Song Y (2019) Effects of boron, silicon and their interactions on cadmium accumulation and toxicity in rice plants. J Hazard Mater 367:447–455

Choppala G, Saifullah, Bolan N, Bibi S, Iqbal M, Rengel Z, Kunhikrishnan A, Ashwath N, Ok YS (2014) Cellular mechanisms in higher plants governing tolerance to cadmium toxicity. Crit Rev Plant Sci 33(5):374–391

Cleophas FN, Zahari NZ, Murugayah P, Rahim SA, Mohd Yatim AN (2022) Phytoremediation: a novel approach of bast fiber plants (Hemp, Kenaf, Jute and Flax) for heavy metals decontamination in soil. Toxics 11(1):5

Cloquet C, Carignan J, Libourel G, Sterckeman T, Perdrix E (2006) Tracing source pollution in soils using cadmium and lead isotopes. Environ Sci Technol 40(8):2525–2530

Cobbett C, Goldsbrough P (2002) Phytochelatins and metallothioneins: roles in heavy metal detoxification and homeostasis. Annu Rev Plant Biol 53(1):159–182

Costa M, Henriques B, Pinto J, Fabre E, Viana T, Ferreira N, Amaral J, Vale C, Pinheiro-Torres J, Pereira E (2020) Influence of salinity and rare earth elements on simultaneous removal of Cd, Cr, Cu, Hg, Ni and Pb from contaminated waters by living macroalgae. Environ Pollut 266:115374

Courbot M, Willems G, Motte P, Arvidsson S, Roosens N, Saumitou-Laprade P, Verbruggen N (2007) A major quantitative trait locus for cadmium tolerance in Arabidopsis halleri colocalizes with HMA4, a gene encoding a heavy metal ATPase. Plant Physiol 144(2):1052–1065

Cui X, Mao P, Sun S, Huang R, Fan Y, Li Y, Li Y, Zhuang P, Li Z (2021) Phytoremediation of cadmium contaminated soils by *Amaranthus Hypochondriacus* L.: the effects of soil properties highlighting cation exchange capacity. Chemosphere 283:131067

Dalvi AA, Bhalerao SA (2013) Response of plants towards heavy metal toxicity: an overview of avoidance, tolerance and uptake mechanism. Ann Plant Sci 2(9):362–368

Dang P, Li C, Lu C, Zhang M, Huang T, Wan C, Wang H, Chen Y, Qin X, Liao Y, Siddique KH (2022) Effect of fertilizer management on the soil bacterial community in agroecosystems across the globe. Agric Ecosyst Environ 326:107795

Dave D, Sarma S, Parmar P, Shukla A, Goswami D, Shukla A, Saraf M (2020) Microbes as a boon for the bane of heavy metals. Environ Sustain 3(3):233–255

Degryse F, Smolders E, Merckx R (2006) Labile Cd complexes increase Cd availability to plants. Environ Sci Technol 40(3):830–836

Smolders E, Mertens J (2013) Cadmium. In: Alloway JB (ed) Heavy metals in soils – trace metals and metalloids in soils and their bioavailability, 3rd edn. Springer, Dordrecht, pp 283–299

Deng S, Wang P, Zhang G, Dou Y (2016) Polyacrylonitrile-based fiber modified with thiosemicarbazide by microwave irradiation and its adsorption behavior for Cd(II) and Pb(II). J Hazard Mater 307:64–72

Di Toppi LS, Gabbrielli R (1999) Response to cadmium in higher plants. Environ Exp Bot 41(2):105–130

Diaconu M, Pavel LV, Hlihor RM, Rosca M, Fertu DI, Lenz M et al (2020) Characterization of heavy metal toxicity in some plants and microorganisms—a preliminary approach for environmental bioremediation. New Biotechnol 56:130–139

Ding S, Ma C, Shi W, Liu W, Lu Y, Liu Q, Luo ZB (2017) Exogenous glutathione enhances cadmium accumulation and alleviates its toxicity in *Populus canescens*. Tree Physiol 37(12):1697–1712

Ding Y, Gong S, Wang Y, Wang F, Bao H, Sun J, Cai C, Yi K, Chen Z, Zhu C (2018) MicroRNA166 modulates cadmium tolerance and accumulation in rice. Plant Physiol 177(4):1691–1703

Dong J, Mao WH, Zhang GP, Wu FB, Cai Y (2007) Root excretion and plant tolerance to cadmium toxicity - a review. Plant Soil Environ 53(5):193

Dutta M, Kushwaha A, Kalita S, Devi G, Bhuyan M (2019) Assessment of bioaccumulation and detoxification of cadmium in soil-plant-insect food chain. Bioresour Technol Rep 7:100242

Feng Z, Ji S, Ping J, Cui D (2021) Recent advances in metabolomics for studying heavy metal stress in plants. TrAC Trends Anal Chem 143:116402

Flemming HC, Wuertz S (2019) Bacteria and archaea on Earth and their abundance in biofilms. Nat Rev Microbiol 17(4):247–260

Foyer CH, Noctor G (2005) Oxidant and antioxidant signalling in plants: a re-evaluation of the concept of oxidative stress in a physiological context. Plant Cell Environ 28(8):1056–1071

Fu H, Yu H, Li T, Zhang X (2018) Influence of cadmium stress on root exudates of high cadmium accumulating rice line (Oryza sativa L.). Ecotoxicol Environ Saf 150:168–175

Fuhrmann JJ (2021) Microbial metabolism. In: Principles and applications of soil microbiology. Elsevier, pp 57–87

Gallego B, Martos S, Cabot C, BarcelÓ J, Poschenrieder C (2016) Zinc hyperaccumulation substitutes for defense failures beyond salicylate and jasmonate signaling pathways of *Alternaria brassicicola* attack in *Noccaea caerulescens*. Physiol Plant 159:401–415. https://doi.org/10.1111/ppl.12518

Gans J, Wolinsky M, Dunbar J (2005) Computational improvements reveal great bacterial diversity and high metal toxicity in soil. Science 309(5739):1387–1390

Gill SS, Tuteja N (2010) Reactive oxygen species and antioxidant machinery in abiotic stress tolerance in crop plants. Plant Physiol Biochem 48(12):909–930

Goix S, Lévêque T, Xiong TT, Schreck E, Baeza-Squiban A, Geret F, Uzu G, Austruy A, Dumat C (2014) Environmental and health impacts of fine and ultrafine metallic particles: assessment of threat scores. Environ Res 133:185–194

Goswami L, Kumar RV, Pakshirajan K, Pugazhenthi G (2019) A novel integrated biodegradation—microfiltration system for sustainable wastewater treatment and energy recovery. J Hazard Mater 365:707–715

Grignon C, Sentenac H (1991) pH and ionic conditions in the apoplast. Annu Rev Plant Biol 42(1):103–128

Gu F, Zhang J, Shen Z, Li Y, Ji R, Li W, Zhang L, Han J, Xue J, Cheng H (2022) A review for recent advances on soil washing remediation technologies. Bull Environ Contam Toxicol 109(4):651–658

Guerinot ML (2000) The ZIP family of metal transporters. Biochim Biophys Acta 1465(1–2):190–198

Guo C, Sun L, Chen X, Zhang D (2013) Oxidative stress, mitochondrial damage and neurodegenerative diseases. Neural Regen Res 8(21):2003

Gupta DK, Pena LB, Romero-Puertas MC, Hernández A, Inouhe M, Sandalio LM (2017) NADPH oxidases differentially regulate ROS metabolism and nutrient uptake under cadmium toxicity. Plant Cell Environ 40(4):509–526

Gusiatin ZM, Klimiuk E (2012) Metal (Cu, Cd and Zn) removal and stabilization during multiple soil washing by saponin. Chemosphere 86(4):383–391

Gutsch A, Zouaghi S, Renaut J, Cuypers A, Hausman JF, Sergeant K (2018) Changes in the proteome of Medicago sativa leaves in response to long-term cadmium exposure using a cell-wall targeted approach. Int J Mol Sci 19(9):2498

Hall JÁ (2002) Cellular mechanisms for heavy metal detoxification and tolerance. J Exp Bot 53(366):1–11

Hanikenne M, Talke IN, Haydon MJ, Lanz C, Nolte A, Motte P, Kroymann J, Weigel D, Krämer U (2008) Evolution of metal hyperaccumulation required cis-regulatory changes and triplication of HMA4. Nature 453(7193):391–395

Harholt J, Suttangkakul A, Vibe Scheller H (2010) Biosynthesis of pectin. Plant Physiol 153(2):384–395

Hasanuzzaman M, Bhuyan MB, Zulfiqar F, Raza A, Mohsin SM, Mahmud JA, Fujita M, Fotopoulos V (2020) Reactive oxygen species and antioxidant defense in plants under abiotic stress: revisiting the crucial role of a universal defense regulator. Antioxidants 9(8):681

Haynes RJ (1980) Ion exchange properties of roots and ionic interactions within the root apoplasm: their role in ion accumulation by plants. Bot Rev 46:75–99

He ZL, Xu HP, Zhu YM, Yang XE, Chen GC (2005) Adsorption-desorption characteristics of cadmium in variable charge soils. J Environ Sci Health 40(4):805–822

Hiller DA, Winzig G, Dornauf C (2001) Bodenchemische Untersuchungen von Versickerungsanlagen als Grundlage für eine nachhaltige Niederschlagswaserbewirtschaftung im Sinne des Boden-und Grundwasserschutzes. University of Duisburg-Essen, p 94
Holmgren GGS, Meyer MW, Chaney RL, Daniels RB (1993) Cadmium, lead, zinc, copper, and nickel in agricultural soils of the United States of America. J Environ Qual 22(2):335–348
Hong KJ, Tokunaga S, Kajiuchi T (2002) Evaluation of remediation process with plant-derived biosurfactant for recovery of heavy metals from contaminated soils. Chemosphere 49(4): 379–387
Hossain MA, Piyatida P, da Silva JAT, Fujita M (2012) Molecular mechanism of heavy metal toxicity and tolerance in plants: central role of glutathione in detoxification of reactive oxygen species and methylglyoxal and in heavy metal chelation. J Bot 2012:872875
Huang Q, Zhu J, Qiao X, Cai P, Rong X, Liang W, Chen W (2009) Conformation, activity and proteolytic stability of acid phosphatase on clay minerals and soil colloids from an Alfisol. Colloids Surf B 74(1):279–283
Hui CY, Guo Y, Liu L, Yi J (2022) Recent advances in bacterial biosensing and bioremediation of cadmium pollution: a mini-review. World J Microbiol Biotechnol 38(1):9
Hussain A, Ali S, Rizwan M, ur Rehman MZ, Javed MR, Imran M, Chatha SAS, Nazir R (2018) Zinc oxide nanoparticles alter the wheat physiological response and reduce the cadmium uptake by plants. Environ Pollut 242:1518–1526
Hussain B, Ashraf MN, Abbas A, Li J, Farooq M (2021) Cadmium stress in paddy fields: effects of soil conditions and remediation strategies. Sci Total Environ 754:142188
Irfan M, Ahmad A, Hayat S (2014) Effect of cadmium on the growth and antioxidant enzymes in two varieties of Brassica juncea. Saudi J Biol Sci 21(2):125–131
Ishikawa S, Ishimaru Y, Igura M, Kuramata M, Abe T, Senoura T, Hase Y, Arao T, Nishizawa NK, Nakanishi H (2012) Ion-beam irradiation, gene identification, and marker-assisted breeding in the development of low-cadmium rice. Proc Natl Acad Sci 109(47):19166–19171
Ishimaru Y, Suzuki M, Tsukamoto T, Suzuki K, Nakazono M, Kobayashi T, Wada Y, Watanabe S, Matsuhashi S, Takahashi M, Nakanishi H (2006) Rice plants take up iron as an Fe3+-phytosiderophore and as Fe2+. Plant J 45(3):335–346
Jalmi SK (2022) The role of ABC transporters in metal transport in plants. Springer Nature Singapore, In Plant metal and metalloid transporters Singapore, pp 55–71
Jiang X, Dai J, Zhang X, Wu H, Tong J, Shi J, Fang W (2022) Enhanced Cd efflux capacity and physiological stress resistance: the beneficial modulations of Metarhizium robertsii on plants under cadmium stress. J Hazard Mater 437:129429
Jin Y, Luan Y, Ning Y, Wang L (2018) Effects and mechanisms of microbial remediation of heavy metals in soil: a critical review. Appl Sci 8(8):1336
Kabata-Pendias A (1993) Behavioural properties of trace metals in soils. Appl Geochem 8:3–9
Kabata-Pendias A, Sadurski W (2004) Trace elements and compounds in soil. In: Elements and their compounds in the environment: occurrence, analysis and biological relevance. Springer, pp 79–99
Kapoor D, Sharma R, Handa N, Kaur H, Rattan A, Yadav P, Gautam V, Kaur R, Bhardwaj R (2015) Redox homeostasis in plants under abiotic stress: role of electron carriers, energy metabolism mediators and proteinaceous thiols. Front Environ Sci 3:13
Kapoor D, Singh S, Kumar V, Romero R, Prasad R, Singh J (2019) Antioxidant enzymes regulation in plants in reference to reactive oxygen species (ROS) and reactive nitrogen species (RNS). Plant Gene 19:100182
Karaca A, Cetin SC, Turgay OC, Kizilkaya R (2010) Effects of heavy metals on soil enzyme activities. In: Soil heavy metals. Springer, pp 237–262
Kaur K, Sharma A, Capalash N, Sharma P (2019) Multicopper oxidases: biocatalysts in microbial pathogenesis and stress management. Microbiol Res 222:1–13
Khan MA, Khan S, Khan A, Alam M (2017) Soil contamination with cadmium, consequences and remediation using organic amendments. Sci Total Environ 601:1591–1605

Khan Z, Elahi A, Bukhari DA, Rehman A (2022) Cadmium sources, toxicity, resistance and removal by microorganisms-a potential strategy for cadmium eradication. J Saudi Chem Soc 26:101569

Khandaker MM, Abdullahi UA, Elyni N, Alias N (2021) Biological processes of heavy metals-contaminated environmental remediation: a review. J Environ Treat Tech 9(3):601–608

Kicińska A, Pomykała R, Izquierdo-Diaz M (2022) Changes in soil pH and mobility of heavy metals in contaminated soils. Eur J Soil Sci 73(1):e13203

Kim S, Lim H, Lee I (2010) Enhanced heavy metal phytoextraction by Echinochloa crus-galli using root exudates. J Biosci Bioeng 109(1):47–50

Knappe F, Möhler S, Ostermayer A, Lazar S, Kaufmann C (2008) Vergleichende Auswertung von Stoffeinträgen in Böden über verschiedene Eintragspfade. Umweltforschungsplan des Bundesministeriums für Umwelt, Naturschutz und Reaktorsicherheit Forschungsbericht 203(74):275

Korenkov V, King B, Hirschi K, Wagner GJ (2009) Root-selective expression of AtCAX4 and AtCAX2 results in reduced lamina cadmium in field-grown Nicotiana tabacum L. Plant Biotechnol J 7(3):219–226

Kotoky R, Pandey P (2018) Plant-microbe symbiosis as an instrument for the mobilization and removal of heavy metals from contaminated soils - a realistic approach. Curr Biotechnol 7(2): 71–79

Kubier A, Wilkin RT, Pichler T (2019) Cadmium in soils and groundwater: a review. Appl Geochem 108:104388

Kumar N, Kumari V, Ram C, Thakur K, Tomar SK (2018) Bio-prospectus of cadmium bioadsorption by lactic acid bacteria to mitigate health and environmental impacts. Appl Microbiol Biotechnol 102:1599–1615

Kumar A, Subrahmanyam G, Mondal R, Cabral-Pinto MMS, Shabnam AA, Jigyasu DK, Malyan SK, Fagodiya RK, Khan SA, Yu ZG (2021) Bio-remediation approaches for alleviation of cadmium contamination in natural resources. Chemosphere 268:128855

Kuramata M, Masuya S, Takahashi Y, Kitagawa E, Inoue C, Ishikawa S, Youssefian S, Kusano T (2009) Novel cysteine-rich peptides from Digitaria ciliaris and Oryza sativa enhance tolerance to cadmium by limiting its cellular accumulation. Plant Cell Physiol 50(1):106–117

Lanquar V, Lelièvre F, Bolte S, Hamès C, Alcon C, Neumann D, Vansuyt G, Curie C, Schröder A, Krämer U, Barbier-Brygoo H (2005) Mobilization of vacuolar iron by AtNRAMP3 and AtNRAMP4 is essential for seed germination on low iron. EMBO J 24(23):4041–4051

Lanquar V, Ramos MS, Lelièvre F, Barbier-Brygoo H, Krieger-Liszkay A, Krämer U, Thomine S (2010) Export of vacuolar manganese by AtNRAMP3 and AtNRAMP4 is required for optimal photosynthesis and growth under manganese deficiency. Plant Physiol 152(4):1986–1999

Li XG, Zhang TL, Wang XX, Hua K, Zhao L, Han ZM (2013) The composition of root exudates from two different resistant peanut cultivars and their effects on the growth of soil-borne pathogen. Int J Biol Sci 9(2):164

Li X, Cui X, Zhang X, Liu W, Cui Z (2020) Combined toxicity and detoxification of lead, cadmium and arsenic in Solanum nigrum L. J Hazard Mater 389:121874

Liao M, Xie XM (2004) Cadmium release in contaminated soils due to organic acids. Pedosphere 14(2):223–228

Liu H, Zhao H, Wu L, Liu A, Zhao FJ, Xu W (2017) Heavy metal ATPase 3 (HMA3) confers cadmium hypertolerance on the cadmium/zinc hyperaccumulator Sedum plumbizincicola. New Phytol 215(2):687–698

Luo JS, Huang J, Zeng DL, Peng JS, Zhang GB, Ma HL, Guan Y, Yi HY, Fu YL, Han B, Lin HX (2018) A defensin-like protein drives cadmium efflux and allocation in rice. Nat Commun 9(1): 645

Luo JS, Yang Y, Gu T, Wu Z, Zhang Z (2019a) The Arabidopsis defensin gene AtPDF2 5 mediates cadmium tolerance and accumulation. Plant Cell Environ 42(9):2681–2695

Luo JS, Gu T, Yang Y, Zhang Z (2019b) A non-secreted plant defensin AtPDF2. 6 conferred cadmium tolerance via its chelation in Arabidopsis. Plant Mol Biol 100:561–569

Luo JS, Xiao Y, Yao J, Wu Z, Yang Y, Ismail AM, Zhang Z (2020a) Overexpression of a defensin-like gene CAL2 enhances cadmium accumulation in plants. Front Plant Sci 11:217

Luo Y, Zhou M, Zhao Q, Wang F, Gao J, Sheng H, An L (2020b) Complete genome sequence of Sphingomonas sp. Cra20, a drought resistant and plant growth promoting rhizobacteria. Genomics 112(5):3648–3657

Mar SS, Okazaki M (2012) Investigation of Cd contents in several phosphate rocks used for the production of fertilizer. Microchem J 104:17–21

Mathivanan K, Chandirika JU, Vinothkanna A, Yin H, Liu X, Meng D (2021) Bacterial adaptive strategies to cope with metal toxicity in the contaminated environment – a review. Ecotoxicol Environ Saf 226:112863

Maurya AK (2020) Oxidative stress in crop plants. Agronomic crops: volume 3: stress responses and tolerance, pp 349–380

Mehmood S, Saeed DA, Rizwan M, Khan MN, Aziz O, Bashir S, Ibrahim M, Ditta A, Akmal M, Mumtaz MA, Ahmed W (2018) Impact of different amendments on biochemical responses of sesame (Sesamum indicum L.) plants grown in lead-cadmium contaminated soil. Plant Physiol Biochem 132:345–355

Merkel, B.J. and Sperling, B., 1998. Hydrogeochemische Stoffsysteme Teil II, Schriftenreihe des Deutschen Verbandes für Wasserwirtschaft und Kulturbau e. V. Wirtschafts-und Verl.-Ges.

Meychik N, Nikolaeva Y, Kushunina M, Yermakov I (2014) Are the carboxyl groups of pectin polymers the only metal-binding sites in plant cell walls? Plant Soil 381:25–34

Mirlean N, Roisenberg A (2006) The effect of emissions of fertilizer production on the environment contamination by cadmium and arsenic in southern Brazil. Environ Pollut 143(2):335–340

Mohanpuria P, Rana NK, Yadav SK (2007) Cadmium induced oxidative stress influence on glutathione metabolic genes of Camellia sinensis (L.) O. Kuntze. Environ Toxicol *22*(4): 368–374

Monachese M, Burton JP, Reid G (2012) Bioremediation and tolerance of humans to heavy metals through microbial processes: a potential role for probiotics? Appl Environ Microbiol 78(18): 6397–6404

Mondal SC, Sarma B, Farooq M, Nath DJ, Gogoi N (2020) Cadmium bioavailability in acidic soils under bean cultivation: role of soil additives. Int J Environ Sci Technol 17:153–160

Monna F, Hamer K, Lévêque J, Sauer M (2000) Pb isotopes as a reliable marker of early mining and smelting in the Northern Harz province (Lower Saxony, Germany). J Geochem Explor 68(3): 201–210

Morel M, Crouzet J, Gravot A, Auroy P, Leonhardt N, Vavasseur A, Richaud P (2009) AtHMA3, a P1B-ATPase allowing Cd/Zn/co/Pb vacuolar storage in Arabidopsis. Plant Physiol 149(2): 894–904

Moulis JM (2010) Cellular mechanisms of cadmium toxicity related to the homeostasis of essential metals. Biometals 23:877–896

Moura DJ, Péres VF, Jacques RA, Saffi J (2012) Heavy metal toxicity: oxidative stress parameters and DNA repair. In: Metal toxicity in plants: perception, signaling and remediation. Springer, pp 187–205

Moynahan OS, Zabinski CA, Gannon JE (2002) Microbial community structure and carbon-utilization diversity in a mine tailings revegetation study. Restor Ecol 10(1):77–87

Nakanishi H, Ogawa I, Ishimaru Y, Mori S, Nishizawa NK (2006) Iron deficiency enhances cadmium uptake and translocation mediated by the Fe2+ transporters OsIRT1 and OsIRT2 in rice. Soil Sci Plant Nutr 52(4):464–469

Nanda M, Kumar V, Sharma DK (2019) Multimetal tolerance mechanisms in bacteria: the resistance strategies acquired by bacteria that can be exploited to 'clean-up' heavy metal contaminants from water. Aquat Toxicol 212:1–10

Nannipieri P, Trasar-Cepeda C, Dick RP (2018) Soil enzyme activity: a brief history and biochemistry as a basis for appropriate interpretations and meta-analysis. Biol Fertil Soils 54:11–19

Navarro-Torre S, Mateos-Naranjo E, Caviedes MA, Pajuelo E, Rodríguez-Llorente ID (2016) Isolation of plant-growth-promoting and metal-resistant cultivable bacteria from Arthrocnemum

macrostachyum in the Odiel marshes with potential use in phytoremediation. Mar Pollut Bull 110(1):133–142
Nkosi N (2020) Characterisation of extracellular polymeric substances (EPS) from bacteria isolated from acid mine decant. University of Johannesburg
Nocito FF, Lancilli C, Dendena B, Lucchini G, Sacchi GA (2011) Cadmium retention in rice roots is influenced by cadmium availability, chelation and translocation. Plant Cell Environ 34 (6):994–1008
Nordberg M, Nordberg GF (2009) Metallothioneins: historical development and overview. Met Ions Life Sci 5(2009):1–29
Oburger E, Cid CV, Schwertberger D, Roschitz C, Wenzel WW (2020) Response of tungsten (W) solubility and chemical fractionation to changes in soil pH and soil aging. Sci Total Environ 731:139224
Odokuma LO, Abah AE (2003) Heavy metal biosorption by three bacteria isolated from a tropical river. Glob J Environ Sci 2(2):98–101
Pal A, Bhattacharjee S, Saha J, Sarkar M, Mandal P (2022) Bacterial survival strategies and responses under heavy metal stress: a comprehensive overview. Crit Rev Microbiol 48(3): 327–355
Parisi K, Shafee TM, Quimbar P, van der Weerden NL, Bleackley MR, Anderson MA (2019, April) The evolution, function and mechanisms of action for plant defensins. In: Seminars in cell & developmental biology, vol 88. Academic Press, pp 107–118
Park J, Song WY, Ko D, Eom Y, Hansen TH, Schiller M, Lee TG, Martinoia E, Lee Y (2012) The phytochelatin transporters AtABCC1 and AtABCC2 mediate tolerance to cadmium and mercury. Plant J 69(2):278–288
Patel HK, Kalaria RK, Vasava DK (2021) Prospects for exploiting microbes and plants for bioremediation of heavy metals. In: Removal of refractory pollutants from wastewater treatment plants. CRC Press, pp 33–60
Peng JS, Wang YJ, Ding G, Ma HL, Zhang YJ, Gong JM (2017) A pivotal role of cell wall in cadmium accumulation in the Crassulaceae hyperaccumulator Sedum plumbizincicola. Mol Plant 10(5):771–774
Peng F, Wang C, Zhu J, Zeng J, Kang H, Fan X, Sha L, Zhang H, Zhou Y, Wang Y (2018) Expression of TpNRAMP5, a metal transporter from Polish wheat (Triticum polonicum L.), enhances the accumulation of Cd, Co and Mn in transgenic Arabidopsis plants. Planta 247: 1395–1406
Peng D, Qiao S, Luo Y, Ma H, Zhang L, Hou S et al (2020) Performance of microbial induced carbonate precipitation for immobilizing Cd in water and soil. J Hazard Mater 400:123116
Pottier M, Oomen R, Picco C, Giraudat J, Scholz-Starke J, Richaud P, Carpaneto A, Thomine S (2015) Identification of mutations allowing natural resistance associated macrophage proteins (NRAMP) to discriminate against cadmium. Plant J 83(4):625–637
Qayyum MF, Liaquat F, Rehman RA, Gul M, ul Hye MZ, Rizwan M, Rehaman MZU (2017) Effects of co-composting of farm manure and biochar on plant growth and carbon mineralization in an alkaline soil. Environ Sci Pollut Res 24:26060–26068
Qi F, Lamb D, Naidu R, Bolan NS, Yan Y, Ok YS, Rahman MM, Choppala G (2018) Cadmium solubility and bioavailability in soils amended with acidic and neutral biochar. Sci Total Environ 610:1457–1466
Rahim HU, Akbar WA, Alatalo JM (2022) A comprehensive literature review on cadmium (Cd) status in the soil environment and its immobilization by biochar-based materials. Agronomy 12(4):877
Rakhsh F, Golchin A (2018) Carbohydrate concentrations and enzyme activities as influenced by exchangeable cations, mineralogy and clay content. Appl Clay Sci 163:214–226
Rasool A, Nasim W, Xiao T, Ali W, Shafeeque M, Sultana SR, Fahad S, Munis MFH, Chaudhary HJ (2020) Microbial diversity response in thallium polluted riverbank soils of the Lanmuchang. Ecotoxicol Environ Saf 187:109854

Rivera-Utrilla J, Bautista-Toledo I, Ferro-Garcıa MA, Moreno-Castilla C (2003) Bioadsorption of Pb(II), Cd(II), and Cr(VI) on activated carbon from aqueous solutions. Carbon 41(2):323–330
Rizwan M, Ali S, Rehman MZU, Maqbool A (2019) A critical review on the effects of zinc at toxic levels of cadmium in plants. Environ Sci Pollut Res 26:6279–6289
Romero-Puertas MC, Rodríguez-Serrano M, Corpas FJ, Gomez MD, Del Rio LA, Sandalio LM (2004) Cadmium-induced subcellular accumulation of O_2^- and H_2O_2 in pea leaves. Plant Cell Environ 27(9):1122–1134
Rudakiya DM, Patel Y (2021) Bioremediation of metals, metalloids, and nonmetals. In: Microbial rejuvenation of polluted environment, vol 2. Springer, pp 33–49
Saeki K, Kunito T (2012) Influence of chloride ions on cadmium adsorptions by oxides, hydroxides, oxyhydroxides, and phyllosilicates. Appl Clay Sci 62:58–62
Sandeepa GM, Lakshmanna B, Madakka M (2023) Recent advancements in the bioremediation of heavy metals from the polluted environment by novel microorganisms. In: Microbiology for cleaner production and environmental sustainability. CRC Press, pp 383–394
Sangthong C, Setkit K, Prapagdee B (2016) Improvement of cadmium phytoremediation after soil inoculation with a cadmium-resistant Micrococcus sp. Environ Sci Pollut Res 23:756–764
Sarwar N, Saifullah, Malhi SS, Zia MH, Naeem A, Bibi S, Farid G (2010) Role of mineral nutrition in minimizing cadmium accumulation by plants. J Sci Food Agric 90(6):925–937
Sasaki A, Yamaji N, Yokosho K, Ma JF (2012) Nramp5 is a major transporter responsible for manganese and cadmium uptake in rice. Plant Cell 24(5):2155–2167
Satoh-Nagasawa N, Mori M, Nakazawa N, Kawamoto T, Nagato Y, Sakurai K, Takahashi H, Watanabe A, Akagi H (2012) Mutations in rice (Oryza sativa) heavy metal ATPase 2 (OsHMA2) restrict the translocation of zinc and cadmium. Plant Cell Physiol 53(1):213–224
Seregin IV, Kozhevnikova AD (2023) Phytochelatins: sulfur-containing metal (loid)-chelating ligands in plants. Int J Mol Sci 24(3):2430
Shah K, Nahakpam S, Chaturvedi V, Singh P (2019) Cadmium-induced anatomical abnormalities in plants. In: Cadmium toxicity and tolerance in plants. Academic Press, pp 111–139
Shahid M, Austruy A, Echevarria G, Arshad M, Sanaullah M, Aslam M et al (2014) EDTA-enhanced phytoremediation of heavy metals: a review. Soil Sediment Contam 23:389–416. https://doi.org/10.1080/15320383.2014.831029
Shahid M, Dumat C, Khalid S, Schreck E, Xiong T, Niazi NK (2017) Foliar heavy metal uptake, toxicity and detoxification in plants: a comparison of foliar and root metal uptake. J Hazard Mater 325:36–58
Sharma SS, Dietz KJ, Mimura T (2016) Vacuolar compartmentalization as indispensable component of heavy metal detoxification in plants. Plant Cell Environ 39(5):1112–1126
She J, Wang J, Wei X, Zhang Q, Xie Z, Beiyuan J, Xiao E, Yang X, Liu J, Zhou Y, Xiao T (2021) Survival strategies and dominant phylotypes of maize-rhizosphere microorganisms under metal (loid)s contamination. Sci Total Environ 774:145143
Shim D, Hwang JU, Lee J, Lee S, Choi Y, An G, Martinoia E, Lee Y (2009) Orthologs of the class A4 heat shock transcription factor HsfA4a confer cadmium tolerance in wheat and rice. Plant Cell 21(12):4031–4043
Shiyu QIN, Hongen LIU, Zhaojun NIE, Rengel Z, Wei GAO, Chang LI, Peng ZHAO (2020) Toxicity of cadmium and its competition with mineral nutrients for uptake by plants: a review. Pedosphere 30(2):168–180
Sidhu GPS, Bali AS (2022) Cd in the environment: uptake, toxicity and management. In: Appraisal of metal(loids) in the ecosystem. Elsevier, pp 283–300
Silber A, Bar-Yosef B, Suryano S, Levkovitch I (2012) Zinc adsorption by perlite: Effects of pH, ionic strength, temperature, and pre-use as growth substrate. Geoderma 170:159–167
Singh S, Parihar P, Singh R, Singh VP, Prasad SM (2016) Heavy metal tolerance in plants: role of transcriptomics, proteomics, metabolomics, and ionomics. Front Plant Sci 6:1143
Smith DBC, Woodruff WF, Solano LG, Ellefsen F, Karl J (2014) Geochemical and mineralogical maps for soils of the conterminous United States, Open-File Report 2014–1082. U.S. Geological Survey, Reston, p 386

Smolders E, Oorts K, Van Sprang P, Schoeters I, Janssen CR, McGrath SP, McLaughlin MJ (2009) Toxicity of trace metals in soil as affected by soil type and aging after contamination: using calibrated bioavailability models to set ecological soil standards. Environ Toxicol Chem 28(8): 1633–1642

Sodhi KK, Mishra LC, Singh CK, Kumar M (2022) Perspective on the heavy metal pollution and recent remediation strategies. Curr Res Microb Sci 3:100166

Song Y, Jin L, Wang X (2017) Cadmium absorption and transportation pathways in plants. Int J Phytoremediation 19(2):133–141

Sprynskyy M, Kowalkowski T, Tutu H, Cozmuta LM, Cukrowska EM, Buszewski B (2011) The adsorption properties of agricultural and forest soils towards heavy metal ions (Ni, Cu, Zn, and Cd). Soil Sediment Contam 20(1):12–29

Sreedevi PR, Suresh K, Jiang G (2022) Bacterial bioremediation of heavy metals in wastewater: a review of processes and applications. J Water Process Eng 48:102884

Su J, Weng X, Luo Z, Huang H, Wang W (2021) Impact of biochar on soil properties, pore water properties, and available cadmium. Bull Environ Contam Toxicol 107(3):544–552

Sullivan TS, McBride MB, Thies JE (2013) Soil bacterial and archaeal community composition reflects high spatial heterogeneity of pH, bioavailable Zn, and Cu in a metalliferous peat soil. Soil Biol Biochem 66:102–109

Sun L, Cao X, Tan C, Deng Y, Cai R, Peng X, Bai J (2020) Analysis of the effect of cadmium stress on root exudates of Sedum plumbizincicola based on metabolomics. Ecotoxicol Environ Saf 205:111152

Takahashi R, Ishimaru Y, Shimo H, Ogo Y, Senoura T, Nishizawa NK, Nakanishi H (2012) The OsHMA2 transporter is involved in root-to-shoot translocation of Zn and Cd in rice. Plant Cell Environ 35(11):1948–1957

Tan L, Qu M, Zhu Y, Peng C, Wang J, Gao D, Chen C (2020) ZINC TRANSPORTER5 and ZINC TRANSPORTER9 function synergistically in zinc/cadmium uptake. Plant Physiol 183(3): 1235–1249

Tao X, Wang J, Liu C, Wang H, Yao H, Zheng G, Seh ZW, Cai Q, Li W, Zhou G, Zu C (2016) Balancing surface adsorption and diffusion of lithium-polysulfides on nonconductive oxides for lithium–sulfur battery design. Nat Commun 7(1):11203

Tayang A, Songachan LS (2021) Microbial bioremediation of heavy metals. Curr Sci 120(6): 00113891

Taylor M, Kim N, Smidt G, Busby C, McNally S, Robinson B, Kratz S, Schnug E (2016) Trace element contaminants and radioactivity from phosphate fertiliser. In: Phosphorus agriculture 100% zero, pp 231–266

Thakur S, Singh L, Wahid ZA, Siddiqui MF, Atnaw SM, Din MFM (2016) Plant-driven removal of heavy metals from soil: uptake, translocation, tolerance mechanism, challenges, and future perspectives. Environ Monit Assess 188:1–11

Tong YP, Kneer R, Zhu YG (2004) Vacuolar compartmentalization: a second-generation approach to engineering plants for phytoremediation. Trends Plant Sci 9(1):7–9

Ueno D, Sasaki A, Yamaji N, Miyaji T, Fujii Y, Takemoto Y, Moriyama S, Che J, Moriyama Y, Iwasaki K, Ma JF (2015) A polarly localized transporter for efficient manganese uptake in rice. Nat Plants 1(12):1–8

Unsal V, Dalkiran T, Cicek M (2020) The role of natural antioxidants against reactive oxygen species produced by cadmium toxicity: a review. Adv Pharm Bull 10(2):184–202

Vatamaniuk OK, Mari S, Lu YP, Rea PA (2000) Mechanism of heavy metal ion activation of phytochelatin (PC) synthase: blocked thiols are sufficient for PC synthase-catalyzed transpeptidation of glutathione and related thiol peptides. J Biol Chem 275(40):31451–31459

Vatehová-Vivodová Z, Kollárová K, Malovíková A, Lišková D (2018) Maize shoot cell walls under cadmium stress. Environ Sci Pollut Res 25:22318–22322

Vázquez S, Goldsbrough P, Carpena RO (2006) Assessing the relative contributions of phytochelatins and the cell wall to cadmium resistance in white lupin. Physiol Plant 128 (3):487–495

Vera-Bernal M, Martínez-Espinosa RM (2021) Insights on cadmium removal by bioremediation: the case of Haloarchaea. Microbiol Res 12(2):354–375

Wang J, Song L, Gong X, Xu J, Li M (2020) Functions of jasmonic acid in plant regulation and response to abiotic stress. Int J Mol Sci 21(4):1446

Wong CKE, Jarvis RS, Sherson SM, Cobbett CS (2009) Functional analysis of the heavy metal binding domains of the Zn/Cd-transporting ATPase, HMA2, in Arabidopsis thaliana. New Phytol 181(1):79–88

Wu S, Shen C, Yang Z, Lin B, Yuan J (2016) Tolerance of Ricinus communis L. to Cd and screening of high Cd accumulation varieties for remediation of Cd contaminated soils. Int J Phytoremediation *18*(11):1148–1154

Wu W, Qu S, Nel W, Ji J (2020) The impact of natural weathering and mining on heavy metal accumulation in the karst areas of the Pearl River Basin, China. Sci Total Environ 734:139480

Xia X, Wu S, Zhou Z, Wang G (2021) Microbial Cd(II) and Cr(VI) resistance mechanisms and application in bioremediation. J Hazard Mater 401:123685

Xie L, Hao P, Cheng Y, Ahmed IM, Cao F (2018) Effect of combined application of lead, cadmium, chromium and copper on grain, leaf and stem heavy metal contents at different growth stages in rice. Ecotoxicol Environ Saf 162:71–76

Xie M, Chen W, Lai X, Dai H, Sun H, Zhou X, Chen T (2019) Metabolic responses and their correlations with phytochelatins in Amaranthus hypochondriacus under cadmium stress. Environ Pollut 252:1791–1800

Xu T, Nan F, Jiang X, Tang Y, Zeng Y, Zhang W, Shi B (2020) Effect of soil pH on the transport, fractionation, and oxidation of chromium(III). Ecotoxicol Environ Saf 195:110459

Yan H, Xu W, Xie J, Gao Y, Wu L, Sun L, Feng L, Chen X, Zhang T, Dai C, Li T (2019) Variation of a major facilitator superfamily gene contributes to differential cadmium accumulation between rice subspecies. Nat Commun 10(1):2562

Yang YJ, Xiong J, Chen RJ, Fu GF, Chen TT, Tao LX (2016) Excessive nitrate enhances cadmium (Cd) uptake by up-regulating the expression of OsIRT1 in rice (Oryza sativa). Environ Exp Bot 122:141–149

Yang P, Zhou XF, Wang LL, Li QS, Zhou T, Chen YK, Zhao ZY, He BY (2018a) Effect of phosphate-solubilizing bacteria on the mobility of insoluble cadmium and metabolic analysis. Int J Environ Res Public Health 15(7):1330

Yang Q, Li Z, Lu X, Duan Q, Huang L, Bi J (2018b) A review of soil heavy metal pollution from industrial and agricultural regions in China: Pollution and risk assessment. Sci Total Environ 642:690–700

Yin K, Wang Q, Lv M, Chen L (2019) Microorganism remediation strategies towards heavy metals. Chem Eng J 360:1553–1563

Younis U, Malik SA, Rizwan M, Qayyum MF, Ok YS, Shah MHR, Rehman RA, Ahmad N (2016) Biochar enhances the cadmium tolerance in spinach (Spinacia oleracea) through modification of Cd uptake and physiological and biochemical attributes. Environ Sci Pollut Res 23:21385–21394

Yu J, Xie R, Yu J, He H, Deng S, Ding S, Sun X, Hllah H (2023) Enhanced dandelion phytoremediation of Cd-contaminated soil assisted by tea saponin and plant growth-promoting rhizobacteria. J Soils Sediments 23(4):1745–1759

Yuan L, Yang S, Liu B, Zhang M, Wu K (2012) Molecular characterization of a rice metal tolerance protein, OsMTP1. Plant Cell Rep 31:67–79

Zaidi A, Khan M, Ahemad M, Oves M (2009) Plant growth promotion by phosphate solubilizing bacteria. Acta Microbiol Immunol Hung 56(3):263–284

Zare L, Ronaghi A, Ghasemi R, Zarei M, Sepehri M (2023) Alterations in glutathione, phytochelatin and micronutrients of corn plants exposed to cadmium stress at different time periods. Commun Soil Sci Plant Anal 54(9):1185–1197

Zhang M, Liu X, Yuan L, Wu K, Duan J, Wang X, Yang L (2012) Transcriptional profiling in cadmium-treated rice seedling roots using suppressive subtractive hybridization. Plant Physiol Biochem 50:79–86

Zhang Y, Cougnon FB, Wanniarachchi YA, Hayden JA, Nolan EM (2013) Reduction of human defensin 5 affords a high-affinity zinc-chelating peptide. ACS Chem Biol 8(9):1907–1911

Zhang C, Yu ZG, Zeng GM, Jiang M, Yang ZZ, Cui F, Zhu MY, Shen LQ, Hu L (2014) Effects of sediment geochemical properties on heavy metal bioavailability. Environ Int 73:270–281

Zhang J, Zhang M, Shohag MJI, Tian S, Song H, Feng Y, Yang X (2016) Enhanced expression of SaHMA3 plays critical roles in Cd hyperaccumulation and hypertolerance in Cd hyperaccumulator Sedum alfredii Hance. Planta 243:577–589

Zhao FJ, Hamon RE, Lombi E, McLaughlin MJ, McGrath SP (2002) Characteristics of cadmium uptake in two contrasting ecotypes of the hyperaccumulator Thlaspi caerulescens. J Exp Bot 53(368):535–543

Zhao W, Chen Z, Yang X, Sheng L, Mao H, Zhu S (2023) Metagenomics reveal arbuscular mycorrhizal fungi altering functional gene expression of rhizosphere microbial community to enhance Iris tectorum's resistance to Cr stress. Sci Total Environ 895:164970

Zheng S, Chen C, Li Y, Li S, Liang J (2013) Characterizing the release of cadmium from 13 purple soils by batch leaching tests. Chemosphere 91(11):1502–1507

Zhi Y, Zhou Q, Leng X, Zhao C (2020) Mechanism of remediation of cadmium-contaminated soil with low-energy plant Snapdragon. Front Chem 8:222

Zhou X, Zhang X, Ma C, Wu F, Jin X, Dini-Andreote F, Wei Z (2022) Biochar amendment reduces cadmium uptake by stimulating cadmium-resistant PGPR in tomato rhizosphere. Chemosphere 307:136138

Zhu H, Chen C, Xu C, Zhu Q, Huang D (2016) Effects of soil acidification and liming on the phytoavailability of cadmium in paddy soils of central subtropical China. Environ Pollut 219: 99–106

Zhu G, Tan W, Xie L, Ma C, Chen X, Zhang S, Wei Y (2022) Mechanisms underlying the inhibitory effects of Cd2+ on prodigiosin synthesis in Serratia marcescens KMR-3. J Inorg Biochem 236:111978

Zientara K, Wawrzyńska A, Łukomska J, López-Moya JR, Liszewska F, Assunção AG, Aarts MG, Sirko A (2009) Activity of the AtMRP3 promoter in transgenic Arabidopsis thaliana and Nicotiana tabacum plants is increased by cadmium, nickel, arsenic, cobalt and lead but not by zinc and iron. J Biotechnol 139(3):258–263

Zimmerman, A.R. and Ahn, M.Y., 2011. Organo-mineral–enzyme interaction and soil enzyme activity. In Soil enzymology, pp.271-292. Springer

Zulfiqar U, Farooq M, Hussain S, Maqsood M, Hussain M, Ishfaq M, Ahmad M, Anjum MZ (2019) Lead toxicity in plants: Impacts and remediation. J Environ Manag 250:109557

Part II
Cadmium Toxicity and Health

Human Health Effects of Chronic Cadmium Exposure

Naqshe Zuhra, Tayyaba Akhtar, Rizwan Yasin, Iqra Ghafoor, Muhammad Asad, Abdul Qadeer, and Sadia Javed

Abstract Cadmium (Cd) is a hazardous unnecessary transition metal that is harmful to people and animals. Cadmium is naturally prevalent in the environment and is frequently obtained from agriculture and industrial processes. Humans are typically exposed to Cd through filthy food and water, inhalation, and cigarette smoking. Cadmium is stored in plants and animals and has a lengthy half-life of 25–30 years. Observational l evidence suggests that occupational and environmental Cd exposure may be connected to malignancies of the breast, lung, prostate, nasopharynx, pancreas, and kidney, as well as an increased risk of osteoporosis. Because of their capacity to generate metallothioneins (MT), which are Cd-inducible proteins that protect cells, the liver and kidneys are particularly sensitive to the harmful effects of Cd. Cadmium-induced oxidative stress is likely to contribute to a variety of liver and kidney illnesses, and mitochondrial damage is a possible mechanism, as these organelles play an important role in the generation of reactive oxygen species (ROS) and are important intracellular targets for Cd. The determination of dietary Cd consumption is a critical step in estimating Cd body burden and associated health consequences. Chronic Cd exposure is known to have a substantial influence on the kidneys, which are the most vulnerable to Cd toxicity. Urinary Cd (UC), the proportional relationship between Cd accumulation in the kidneys and Cd excretion via urine, is a reliable indicator of Cd exposure. This chapter outlines the numerous pathways of Cd exposure, their impact on human health, and the use of several biomarkers to measure Cd exposure.

Keywords Cadmium · Human health · Toxicity · Oxidative stress · Diseases · Biomarkers

---

N. Zuhra · T. Akhtar · R. Yasin · I. Ghafoor · M. Asad · A. Qadeer
Institute of Soil and Environmental Sciences, University of Agriculture Faisalabad, Faisalabad, Pakistan

S. Javed (✉)
Department of Biochemistry, Government College University Faisalabad, Faisalabad, Pakistan
e-mail: sadiajaved@gcuf.edu.pk

A. K. Jha, N. Kumar (eds.), *Cadmium Toxicity Mitigation*,
https://doi.org/10.1007/978-3-031-47390-6_3

1 Introduction

Cadmium (Cd) is a transition metal that is non-essential and known for its toxic characteristics. It unveils serious health hazards to humans as well as animals. The presence of this entity can be observed in our natural surroundings, indicating a widespread distribution, and can be derived from both natural and anthropogenic sources. Cd exposure in humans can occur through multiple pathways, such as the consumption of contaminated food and water, inhaling Cd-containing particles, and cigarette smoking, with food being the primary source of exposure (Kim et al. 2023; Niede and Benbi 2022). Cd exhibits a prolonged biological half-life of roughly 25–30 years, encouraging its gradual accumulation within the human body (Lordan and Zabetakis 2022).

Epidemiological research shows an association between Cd exposure to work-related and environmental variables that have several adverse health outcomes, including increased cancer risks and bone disorders such as osteoporosis (Satarug et al. 2023). Exposure to Cd has been linked to developing multiple types of cancer, such as breast, lung, prostate gland, nasopharynx, pancreas, and kidney cancers (Satarug et al. 2023). The mechanisms responsible for Cd-induced carcinogenesis are complex and involve multiple pathways, including oxidative stress, DNA damage, epigenetic alterations, and disruption of cellular signaling (Hernández-Cruz et al. 2022).

The liver and kidneys are particularly susceptible to the deadly effects of Cd. Metallothioneins (MTs), a protein provoked by Cd that protects cells by forcefully binding Cd ions, are synthesized by the liver (Yang et al. 2015). However, prolonged exposure to Cd can overwhelm the liver's detoxification capacity, leading to Cd accumulation and subsequent hepatotoxicity. Cd-induced liver damage is thought to involve oxidative stress, mitochondrial dysfunction, and inflammation (Alshehri et al. 2022). Similarly, the kidneys play a vital role in Cd handling and are the prime victim of Cd toxicity. Cd accumulates in the proximal tubular cells of the kidneys, where it disrupts normal cellular functions and causes renal tubular dysfunction (Wang et al. 2022). Cd-induced nephrotoxicity involves oxidative stress, apoptosis, inflammation, and impaired reabsorption and excretion processes (ElMahdy et al. 2022).

Accessing one's Cd intake through food is crucial in order to estimate the amount of Cd in the body and its possible impact on health. Chronic Cd exposure, even at relatively low levels, can have significant impacts on kidney function. Cd level in urine (urinary Cd) is a reliable biomarker of Cd exposure due to its direct reflection of Cd accumulation in the kidneys (Gallagher and Meliker 2010; Laouali et al. 2023). The relationship between urinary Cd (UC) levels and Cd-induced renal damage has been extensively studied, and various threshold values for UC have been proposed as indicators of adverse health effects (Laouali et al. 2023).

This book chapter aims to deliver a comprehensive review of the *Homo sapiens* health effects of chronic Cd exposure. Here examine the current understanding of Cd toxicity, including its carcinogenic potential, impacts on bone health, and specific organ toxicity, focusing on the liver and kidneys. Furthermore, explore the mechanisms underlying Cd-induced toxicity, including oxidative stress, mitochondrial dysfunction, and inflammation. Additionally, it discussed the assessment of Cd exposure through biomarkers such as UC and highlighted the importance of monitoring Cd levels in vulnerable populations. By elucidating the health risks associated with chronic Cd exposure, this chapter objects to contributing to the expansion of preventive strategies and public health interventions to mitigate the adverse effects of Cd on human health.

2 Sources of Cadmium Exposure

The discovery of Cadmium (Cd) dates back to 1817 when it was first observed as a by-product of the zinc refining industry, resulting from extracting zinc ores. In comparison to zinc, cadmium is around 700 times more prevalent in the Earth's crust with an amount of approximately 0.09–0.18 g/ton in the lithosphere (El Rasafi et al. 2022; Heinrichs et al. 1980). The occurrence of cadmium is closely associated with zinc ores and is mainly obtained through the refining process of mining zinc sulfide (Rao and Kashifuddin 2016). Thus, worldwide Cd production is influenced by zinc purification, and the Cd to Zn ratio varies between 0.07% and 0.83%.

In addition, around 10–15% of the cadmium supply comes from steel waste, making it the second major cause of cadmium. Various elements, including cadmium, are released into the environment due to the decomposition of surface materials and weathering of parent rock minerals, leading to an increase in cadmium concentrations. In agricultural settings, the application of phosphorus-containing fertilizers leads to an increase in cadmium content in the soil. Additionally, cadmium levels in agricultural land are amplified by activities such as landfill disposal, sewage sludge application, and other forms of pollution (Rao and Kashifuddin 2016). Cadmium input, on the other hand, is significantly contributed by mine tailings, compost inputs, and the irrigation of mine effluents (Bolan et al. 2013).

3 Pathway of Cadmium Exposure

Cadmium, an environmental pollutant, is primarily introduced through industrial and technological advancements (Satarug 2019). Its absorption occurs through water, food, and air contamination. Some types of aquatic foods including crustaceans, cephalopods, bivalve mollusks, oysters, crabs, offal products (specifically liver and kidney), cocoa beans, certain wild mushrooms, and oil seeds have been identified as having higher levels of Cd (Satarug 2018). Plant-derived foods, depending on soil contamination levels, generally exhibit higher Cd concentrations compared to milk, eggs, meat, and dairy products. Plant-based foods such as potatoes, wheat, rice, green leafy vegetables, carrots, and corn have the potential to exhibit increased levels of Cd. For those who follow a vegetarian diet or enjoy consuming shellfish could potentially have higher Cd intake than omnivores (Sirot et al. 2008). Rice consumption is a major route of Cd exposure in humans (Shi et al. 2020). Consequently, flooding rice fields during harvest has been recommended in Cd-contaminated soil areas, particularly in Japan, as a water-controlling approach to reduce Cd accumulation in rice, although this practice may increase arsenic (As) accumulation (Arao 2019; Horiguchi 2019). Cd possesses unique hydrochemical characteristics, allowing for its solubility to be maintained at a pH level close to neutral (<6.5), unlike other heavy metals that are typically fixed. Consequently, drinking water and rivers in the vicinity of factories or mines may become contaminated with Cd (Järup and Åkesson 2009). Tobacco use is another major source of Cd exposure for the general population in terms of environmental factors (ATSDR 2012). Cd exposure from cigarette smoking varies depending on the category of cigarette, with 1–2 μg of Cd in every cigarette. Individuals who smoke 20 cigarettes on a daily basis are estimated to assimilate approximately 1 μg Cd, with a rate of absorption ranging from 40% to 50% within the pulmonary system. The inhalation of cadmium (Cd) has been linked to pulmonary disease, respiratory diseases, and respiratory cancer. Mona et al. reported that smokers have significantly higher levels of Cd in their serum and urine compared to non-smokers. Additionally, smokers have a higher likelihood of experiencing bone pain, with a prevalence of 95% compared to 37.7% in non-exposed individuals. They concluded that prolonged Cd exposure may lead to osteotoxic effects that ultimately lead to bone tissue loss (Taha et al. 2018). Furthermore, Cd, widely used in electroplating and certain industrial paints, poses a threat when used in mist form. Soil contamination and subsequent uptake by vegetables and crops intended for human consumption can be attributed to a variety of factors, including the atmospheric release of airborne Cd, mining activities, and agricultural utilization of Cd phosphatic fertilizer and sewage sludge. Cd Sources and food chain exposure are illustrated in Fig. 1.

4 Importance of Studying Human Health Effects of Cadmium

The importance of studying the health effects of cadmium on humans is significant due to the potential risks and adverse effects associated with cadmium exposure. Understanding these effects is crucial for implementing appropriate preventive measures and regulations to protect public health. Some key reasons:

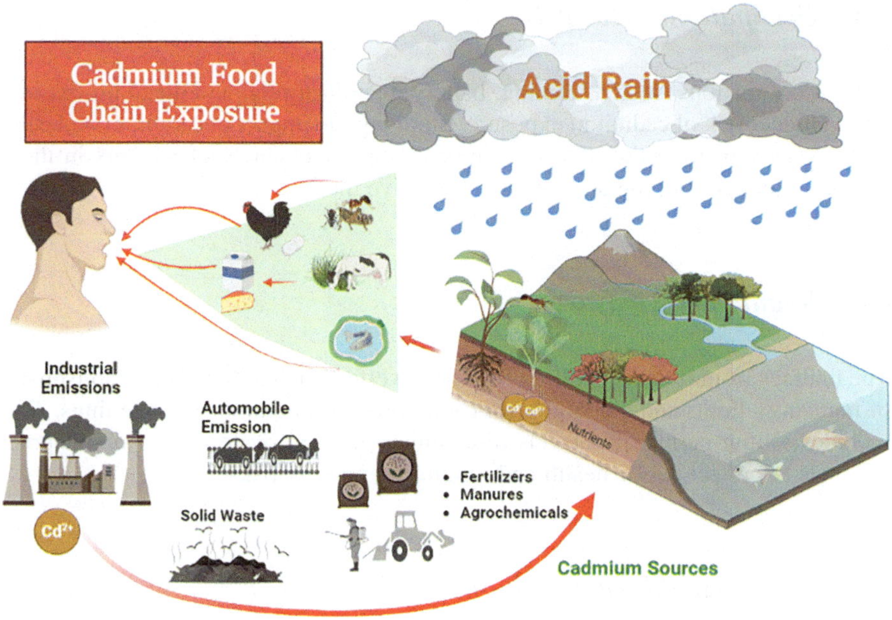

Fig. 1 Cadmium sources and food chain exposure

4.1 Toxicity and Health Risks

Over time, cadmium, a lethal heavy metal, can build up in the human body, leading to various health risks. It is associated with organ-specific effects, including kidney damage, pulmonary diseases (such as respiratory cancer and chronic obstructive pulmonary disease), liver damage, bone disorders (like osteoporosis), and cardiovascular diseases. Cadmium exposure has also been linked to carcinogenicity, reproductive effects, immune system dysregulation, and neurological disorders (Saleh et al. 2020).

4.2 Environmental Exposure

Exposure to cadmium can happen in different ways, including consuming food and water that has been contaminated, inhalation of air pollutants, and dermal contact. Understanding the health effects of cadmium is essential to assess and manage exposure risks in different populations, especially those living in industrial areas or regions with high cadmium pollution levels.

4.3 Occupational Hazards

Certain occupations, such as mining, battery manufacturing, and metal processing, pose a higher risk of cadmium exposure. Studying the impact of cadmium on health is important for protecting the occupational health and safety of workers in these industries and developing appropriate preventive measures.

4.4 Regulatory Measures

The findings from studies on cadmium's health effects help inform the establishment of regulatory limits and guidelines for cadmium exposure in different settings. This includes setting permissible levels of cadmium in food, water, air, and consumer products to protect public health and minimize exposure risks.

4.5 Public Health Interventions

Studying the health effects of cadmium provides the necessary evidence to develop public health interventions and awareness programs. These interventions can focus on educating the public about potential sources of cadmium exposure, promoting healthy dietary practices, and encouraging preventive measures to reduce exposure risks.

4.6 Risk Assessment and Management

Understanding the health effects of cadmium is crucial for conducting accurate risk assessments and implementing effective risk management strategies. This includes identifying vulnerable populations, evaluating exposure pathways, and determining safe exposure limits to minimize health risks.

Studying the human health effects of cadmium is essential for identifying the risks associated with cadmium exposure, implementing preventive measures, establishing regulatory guidelines, and protecting public health. It helps in promoting informed decision-making, raising awareness, and developing strategies to minimize cadmium-related health risks in various settings and populations.

5 Chemical Form and Properties of Cadmium

Cadmium, designated as Cd, has the periodic arrangement's atomic number (Z) 48 and atomic mass (A) 112.41. This silvery-white metal is fragile and has comparable physical and chemical characteristics to zinc and mercury. Cd's atomic mass is determined by the fusion of eight stable isotopes. It belongs to the category of transition metals and has two electrons in the s orbital and one in the d orbital. Like zinc, cadmium, possesses the properties of flexibility, malleability, and ductility, and exhibits a preference for the +2 oxidation state in most of its chemical compounds. Cd exhibits corrosion resistance properties and is commonly employed as a protective plating material. It possesses insolubility in water and is non-flammable in nature. It burns in oxygen to produce cadmium oxide. Cadmium can be dissolved by hydrochloric, sulfuric, and nitric acids, which produce cadmium chloride, cadmium sulfate, and cadmium nitrate, respectively (Genchi et al. 2020).

In nature, Cd is found at modest concentrations, primarily in lead, zinc and copper sulfide ores Cadmium ores are rare, but it is possible to locate Cd in most zinc rocks because of its isomorphic nature with zinc. Because of its ubiquitous presence, however, cadmium is detectable in food, drink, and breath (Sinicropi et al. 2010). Anthropogenic actions, which include the burning of fossil fuels, alongside the release of leachate from landfills, agricultural areas, and mining waste, particularly from zinc and lead mining operations, are significant contributors to the presence of cadmium in the environment (Thompson and Bannigan 2008). Cd is used as a dye, a stabilizer for PVC and metal alloys, and as an anti-corrosion coating for steel (Hayat et al. 2019). Cadmium compounds are frequently employed for coloring lampworking and borosilicate glass. Cd is additionally produced during the production of Nickle-Cadmium batteries (Genchi et al. 2020). Due to its toxicity, Cd is used less frequently today, and Nickle-Cadmium batteries have been supplanted by nickel-metal hydride and lithium-ion batteries. The novel utilization of this technology can be observed in the implementation of semiconducting infrared optical windows and cadmium telluride solar panels (Karunakaran and Dhanalakshmi 2009). It is employed in various applications such as cadmium plating solutions, catalysts, ceramic glazes, and storage battery electrodes. The main application of cadmium oxide involves its utilization as an additive in electroplating solutions and pigments. Cadmium telluride and Cadmium oxide thin films have found applications in various optoelectronic devices, such as phototransistors, photodiodes, photovoltaic solar cells, transparent electrodes, and anti-reflection coatings (Lokhande et al. 2004).

The presence of Cd-dependent carbonic anhydrase was observed in the unicellular microalga *Thalassiosira weissflogii*, a marine diatom. It is worth noting that cadmium does not possess any recognized biological role in higher organisms. The diatoms under consideration reside in an environment characterized by a scarcity of zinc, resulting in the substitution of zinc for cadmium as a catalytic metallic ion within the enzyme's active site (Alterio et al. 2015).

6 Mechanisms of Cadmium Toxicity

Cadmium is a prominent hazardous pollutant that raises occupational and environmental concerns due to its wide ranging toxicological impacts. These effects encompass a remarkably protracted living half-life of approximately 20–30 years in human beings, a partial rate of removal from the body, and a tendency to accumulate mainly in tissues, particularly the kidneys and liver (Rani et al. 2014). Cadmium is found in significant quantities within the environment due to anthropogenic activities, including the utilization of fossil fuels, the incineration of metal ore, and the disposal of waste materials. The potential transfer of cadmium compounds, which have been absorbed by plants from the discharge of sewage sludge to agricultural soil is believed to have a noteworthy impact on the food chain and the subsequent accumulation of this element in various human organs. Cadmium exposure is also significantly attributed to the inhalation of cigarette smoke. The examination of cadmium levels in blood revealed that smokers exhibited cadmium levels that were 4–5 times greater than those observed in non-smokers (Munisamy et al. 2013). The sequential progression of heavy metal pollution typically adheres to the following cyclic pattern: originating from industrial sources, it subsequently permeates the atmosphere, soil, water bodies, and ultimately enters the human food chain (Keshav Krishna and Rama Mohan 2016). Cd harms gene expression, apoptosis, autophagy, oxidative stress, and bioelements by altering gene expression and preventing DNA repair. Cd in water, air, and soil may increase after industrial activity, exposing humans to Cd. Consuming contaminated food also increases Cd exposure (Cao et al. 2018). Cd is widely recognized as the most harmful trace element for both land and aquatic organisms, including other living creatures and humans (Nasirzadeh et al. 2022). Cd influences the balance of bioelements (Se, Fe, Mn, Zn, Cu, Ca, and Mg), whereas higher ingestion of particular bioelements affects Cd destiny in the body and reduces Cd harmful effects (Andrulewicz-Botulińska et al. 2018). The high mobility of cadmium facilitates its rapid absorption by plant roots and subsequent translocation to aerial tissues, thereby posing significant challenges for humans and animals (Nasirzadeh et al. 2022). It is observed that the substance tends to accumulate in various organs within the human body, with particular emphasis on the kidneys (Wang et al. 2021). Because Cd is analogous to calcium (Ca) in terms of ionic radius, charge, and chemical activity, it may readily transmission to the human body and be kept at a high level in many organs (Gutsch et al. 2018). The contact of Cd with bioelements at places where bioelements have a physiological role is an essential mechanism of Cd toxicity. Cadmium exhibits competition for binding sites of essential bioelements, such as zinc in copper-zinc superoxide dismutase (CuZnSOD) and zinc finger proteins, manganese in manganese superoxide dismutase (MnSOD), iron in catalase (CAT), and various other binding sites. Cadmium exhibits a diverse array of deleterious effects, encompassing nephrotoxicity, carcinogenicity, teratogenicity, as well as endocrine and reproductive toxicity. Cadmium exerts effects on various biological functions including cell proliferation, differentiation, and apoptosis (Rani et al. 2014). Cadmium poisoning can be caused by ingesting contaminated food (such as crustaceans, organ meats, green vegetables, and rice from particular regions of Japan and China) or water (like that which comes from ancient Zn/Cd sealed water pipes or industrial contamination). Medications and nutritional supplements are susceptible to contamination, which might have serious health consequences (Abernethy et al. 2010). Although rice, cereals, and marine food have been reported to be contaminated by Cd (Chunhabundit 2016).

After absorption, Cd is transported all across the body, usually bound to a sulfhydryl-containing protein like metallothionein. About 30% of the deposition is in the liver, 30% in the kidneys, and the rest are found elsewhere in the body (Bernhoft 2012). Cd accumulation in the proximal tubule epithelial cells impairs reabsorption ability, resulting in polyuria, glucosuria, and low molecular weight proteinuria (Yang et al. 2015). Cd enters the GI tract via the actions of several distinct transporters. Excessive Cd exposure causes increased intestine accumulation, particularly in the duodenum and in the two metals' molecular similarities. Cuypers et al. (2010) found experimental evidence for a change in cellular redox balance caused by cadmium. Cadmium causes tissue damage by causing oxidative stress (Matovi et al. 2011). Changes in DNA expression epigenetically, inhibition or overexpression of transport pathways that Cadmium substitutes Ca in minerals due to its equal charge, ionic radius, and chemical behavior (Kubier et al. 2019). Cd bioavailability for plants is improved by soil acidity, while root exudates boost its solubility. Cadmium is usually found in soil as Cd^{2+}, but it also appears in the form of Cd chelates (Abedi and Mojiri 2020). At low Cd activities, the roots' energy for Cd absorption exhibits biphasic features of saturable components in the absorption solution and a linear portion at advanced Cd activities (Nazar et al. 2012). Cd absorption in roots can take place as inorganic complexes (i.e., Cd^{2+} SO_4, $CdCl^+$, and CdCl) or organic forms (i.e., phytometallophore complexes) (Kubier et al. 2019). The apoplastic and symplastic adsorption routes are involved in the absorption of metal ions such as Cd (Ismael et al. 2019). Cadmium has higher flexibility and assimilation; therefore, it enters plants through the roots and is subsequently translocated into vascular bundles (i.e., xylem and phloem) by ascent of sap and/or transporters through shoots (Dong et al. 2019). Cadmium ions have the ability to traverse the cellular membranes of electrogenic cells in the human body by utilizing L- or N-type voltage-dependent calcium channels. In contrast, non-electrogenic cells may employ store-operated calcium channels (Zhou et al. 2015). Cadmium (Cd) initially binds to blood cells and albumin, followed by subsequent binding to metallothionein within kidney and liver tissues. The diffusion of manganese vapor occurs across the lung membrane, facilitating its transportation from the bloodstream to the lungs and subsequently to the Central Nervous System (Florea and Büsselberg 2006). Cd toxicity has detrimental effects on the liver and bones, and it can lead to impaired calcium absorption (Saleh et al. 2020).

6.1 Oxidative Stress

Oxidative stress has been linked to the etiology of numerous disease states and toxicities in animals, at least in part. Following xenobiotic exposure, the cells' inherent capacity to eliminate ROS and free radicals is exceeded, resulting in oxidative stress and lipid, protein, and DNA damage. Lead and cadmium are the most frequent heavy metal contaminants in the environment, and are widely distributed (Patra et al. 2011). There exists conclusive evidence of the occurrence of free radical generation in animals subsequent to acute Cd overload. Additionally, there is indirect evidence suggesting the involvement of ROS in persistent Cd toxicity and the development of carcinogenic effects (Matovi et al. 2011).

A growing body of research suggests that Cd toxicity is linked to ROS generation at the mitochondrial level (Wang et al. 2019). Reactive oxygen species formation has been observed in mitochondrial impairment in human osteoblasts. Additionally, exposure to cadmium (Cd) treatment led to significant inhibition of enzymes involved in the respiratory chain of the mitochondria and citrate synthase in these cells (Monteiro et al. 2018). The iron–sulfur clusters of the ETC complexes have been acknowledged as possible Cd target sites, leading to electron transfer inhibition (Kurochkin et al. 2011). All biological macromolecules are vulnerable to oxidative stress. The assessment of oxidized forms of lipids, proteins, and DNA has proven to be a valuable approach for evaluating oxidative damage, which is commonly associated with both acute and chronic illnesses (Dalle-Donne et al. 2006). Lipid peroxidation has emerged as a prominent consequence of Cd-induced oxidative stress and has been observed to be positively correlated with the levels of Cd exposure. This pathway is among the identified mechanisms through which Cd exerts its damaging effects (Valko et al. 2006).

Liver damage is a result of Cd's primary targeting of cellular and intracellular membranes. Lipid peroxidation is a crucial underlying process involved in this damage. The process of cellular macromolecule oxidation, specifically lipids, results in the impairment of hepatocytes, leading to subsequent changes in the structure and function of the liver. The adverse impacts of these detrimental effects have far-reaching implications across various physiological systems within the body (Mężyńska et al. 2018).

Many investigations have found that Cd can change a variety of cellular activities like cellular structural components. The impact of Cd has often been associated with alterations in redox status, commonly assessed through cellular GSH levels, the antioxidant activity of enzymes, or lipid peroxidation (Patra et al. 2011). Cd exhibits the capacity to orchestrate apoptosis via ROS-dependent pathways, while also engendering necrotic toxicity, forming a tight bond with the intricate process of lipid peroxidation across diverse organs including the liver, kidney, lungs, and breast (Wang et al. 2009).

Recent investigations have concentrated on studying the effects of Cd-induced oxidative stress and cellular injury specifically in human breast cancer cells, which are known for their lipid-rich composition (Wang et al. 2019). Exposure of human

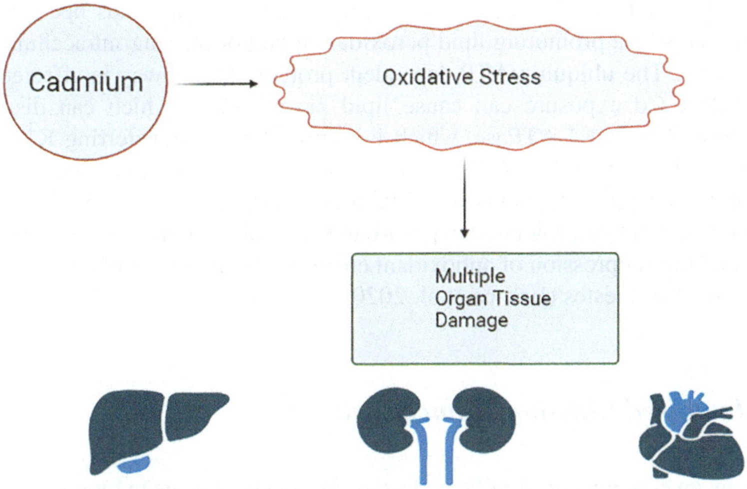

Fig. 2 Cadmium induces oxidative stress that affects human organs

breast cancer cells to Cd resulted in elevated concentrations of lipid hydro-peroxides, which have been associated with cytotoxic effects through various mechanisms. These mechanisms encompass alterations in the organization, composition, configuration, and mobility of lipid membranes and the making of reactive oxygen species capable of inducing crosslinking in DNA and proteins (Gaschler and Stockwell 2017). Cadmium induces oxidative stress that affects human organs is depicted in Fig. 2.

The breakdown of cadherin-mediated cell–cell adhesion is one of the first harmful consequences of Cd. This may result in mild oxidative stress. The effect of Cd on cadherin-mediated cell–cell junctions may be due to contact with the macromolecule's extracellular Ca-binding domains (Messner et al. 2016). Cadmium toxicity has long been linked to zinc homeostasis and oxidative stress in mammalian cells. The reasons for this link can be found in zinc's ability to protect against the negative effects of cadmium exposure, as recently reaffirmed (Rogalska et al. 2009). As a result, Cd initiates cellular toxicity by activating pathways related to oxidative stress. Upon cellular entry, Cd induces protein unfolding and dysfunction, leading to the development of endoplasmic reticulum stress and eventual cellular death (Mao et al. 2011). The cytotoxicity induced by cadmium is attributed to the occurrence of oxidative stress, leading to detrimental effects on mitochondria, inhibition of cellular respiration, and peroxidation of membrane lipids (Nazima et al. 2016). Cd also provokes oxidative stress by interfering with the thiol groups of antioxidant enzymes such as superoxide dismutase, catalase, glutathione reductase, and glutathione peroxidase (You et al. 2018). Oxidative DNA damage is caused by oxidative stress. The creation of 8 hydroxy-2′-deoxyguanosine (8-OHdG), a major marker for oxidative stress and carcinogenesis, is the critical connection between Cd-induced ROS generation and DNA damage. Furthermore, Cd inhibits the repair of DNA damage

(Bishak et al. 2015). Cadmium is a powerful cell toxin that has been linked to oxidative stress via promoting lipid peroxidation and/or altering intracellular glutathione levels. The ubiquitin/ATP-dependent proteolytic pathway is affected (Patra et al. 2011). Cd exposure can cause lipid peroxidation, which can disrupt the functioning of Na^+, K^+-ATPase, which is responsible for transferring K^+ and Na^+ ions into cells in the CNS. Cd exposure reduces membrane potential, which can affect neurotransmitter transmission (Malin and Wright 2018). A significant study has demonstrated that prolonged exposure to Cd resulted in heightened lipid peroxidation and the suppression of antioxidant enzymes, leading to oxidative harm in the liver, kidney, and testes (Branca et al. 2020).

6.2 *Impaired Calcium Homeostasis*

Cadmium, an ecologically significant hazardous metal, appears in biological systems as the Cd ion, which physically mimics Ca^{+2} ion. Additionally, Cd has a wide range of negative effects on cells due to its affinity for some protein thiol groups, It is widely recognized that Cd plays a role in various pathways that are dependent on Ca. Calmodulin and Ca/calmodulin-dependent protein kinase II (CaMK-II) are known to influence cytoskeletal dynamics and apoptotic cell death (Choong et al. 2014). Calcium and cadmium are both divalent metals that exhibit similar chemical properties. Both substances have the ability to enter cells through multiple channels or by permeating the cell membrane with the assistance of proteins. In contrast, cadmium disrupts the maintenance of calcium equilibrium by inhibiting the function of calcium channels and/or their associated proteins. Cadmium can also change membrane phospholipid concentrations, causing a calcium homeostasis issue. Cadmium-induced changes in calcium homeostasis cause cell death, autophagy, or cancer (Zhou et al. 2015). Cadmium disturbs the calcium homeostasis by decreasing ATPase activity and increasing calcium (Ca) content, and upregulated the expression levels of calcium homeostasis-related factors Ca^{+2}/CAM-dependent kinase II (CaMKII), calcineurin (CaN), inositol-1,4,5-triphosphate receptor (IP3R), and calreticulin (CRT) (Lin et al. 2021). Cadmium interacts with cell surface receptors and ion channels, as well as the intracellular estrogen receptor, where it binds competitively to Ca($^{+2}$)-shared residues. It raises cytosolic [Ca($^{+2}$)] by many pathways, but it also lowers the transcript levels of certain Ca($^{+2}$)-transporter genes (Choong et al. 2014).

Cadmium absorption and accumulation in the body is mostly performed by a divalent metal transporter, although the molecular basis for this mechanism is unknown (Begg et al. 2015). Cd exposure reduces membrane potential, which can affect neurotransmitter signaling. Neurotransmitter release is dependent on calcium (Ca^{+2}) channels, which can become blocked owing to Cd toxicity (Cd^{+2}), interfering with the flow of Ca^{+2} into and out of the cell and disrupting homeostasis (Forcella et al. 2020). This can subsequently lead to a decrease in the intracellular influx of calcium ions (Ca^{+2}) or an increase in the release of intracellular calcium ions (Ca^{+2})

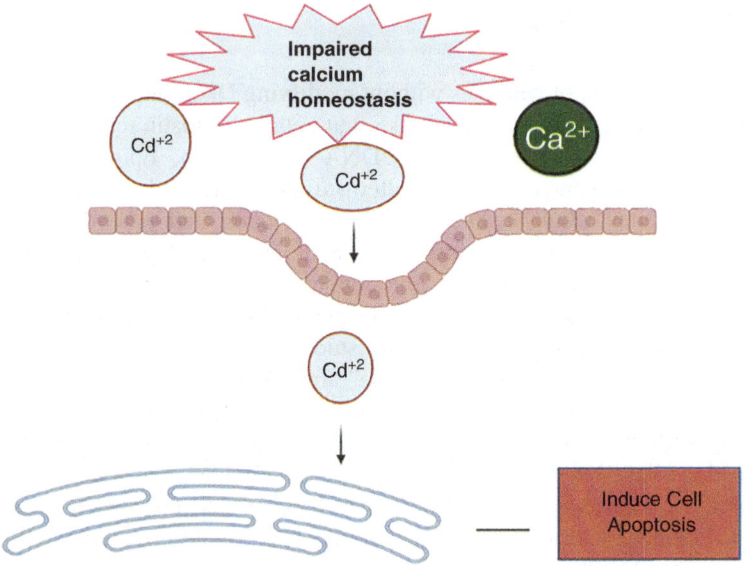

Fig. 3 Cadmium induces cell apoptosis through elevating the intracellular Ca^{+2}

from intracellular stores. An enduring increase in intracellular calcium ions (Ca^{+2}) levels may potentially contribute to cellular demise. During the initial phase of development, exposure to Cd^{+2} induces changes in the functioning of particular neurotransmitter systems, namely the monoaminergic, cholinergic, glutaminergic, and GABAergic systems (Đukić-Ćosić et al. 2020). Cadmium induces cell apoptosis through elevating the intracellular Ca^{+2} is illustrated in Fig. 3.

The kidney is a vital organ for maintaining Ca^{+2} homeostasis in the body. Acute kidney injury (AKI), autosomal dominant polycystic kidney disease (ADPKD), renal ischemia/reperfusion (I/R) damage, chronic kidney disease (CKD), podocytopathy, and diabetic nephropathy are all caused by a change in Ca^{+2} homeostasis. Ca^{+2} signaling is important in several cell actions that occur throughout the course of kidney disease, including necrosis, apoptosis, eryptosis, and autophagy (Ning et al. 2021). The endoplasmic reticulum (ER) is the primary intracellular calcium (Ca^{+2}) storage and is required for Ca^{+2}-mediated cell signaling. Intracellular Ca^{+2} is a ubiquitous and adaptable signaling molecule that regulates several essential cellular functions such as differentiation, proliferation, secretion, contraction, metabolism, and cell death. As a result, maintaining calcium homeostasis is essential for cell function and survival (Park et al. 2021).

7 Epigenetic Changes

The reversible change of genomes without modifying DNA sequences is referred to as epigenetics; epigenetic modifications often entail chromatin remodeling, histone modification, non-coding RNAs, and DNA methylation. Epigenetic control is required for normal gene expression patterns to be maintained; disruption of epigenetic changes may result in altered cellular function and potentially malignant transformation. As a result, abnormal epigenetic changes play a significant role in metal-induced cancer formation, progression, and angiogenesis (Zhao et al. 2022b). When compared to genomic changes, epigenetic changes are frequently reversible and more rapidly controlled. Numerous studies have revealed that cancer cells contain significant "epigenomic landscape" abnormalities, which may result in the worldwide deregulation of gene expression patterns (Della Chiara et al. 2021).

During cadmium exposure, epigenetic alterations in RNA, protein, and DNA modifications have been found, implying scientific evidence as a possible route of Cd toxicity for developmental origins of well-being and diseases (Lawless et al. 2023). Cd epigenetic effects have been shown in vitro and in vivo that epigenetics happened in conjunction with Cd induction of malignant cell transformation and pathological tissue proliferation, indicating that epigenetic effects may play a role in Cd harmful, notably carcinogenic consequences Cadmium's Epigenetic Impact (Wang et al. 2012) (Fig. 4). Cd might be acting indirectly via epigenetic mechanisms by altering the molecular signals upstream of DNA repair and apoptosis (Nzengue et al. 2012). Cd has been demonstrated to serve as a metalloestrogenic carcinogen and to be linked to breast carcinogenesis, and abnormal epigenetic

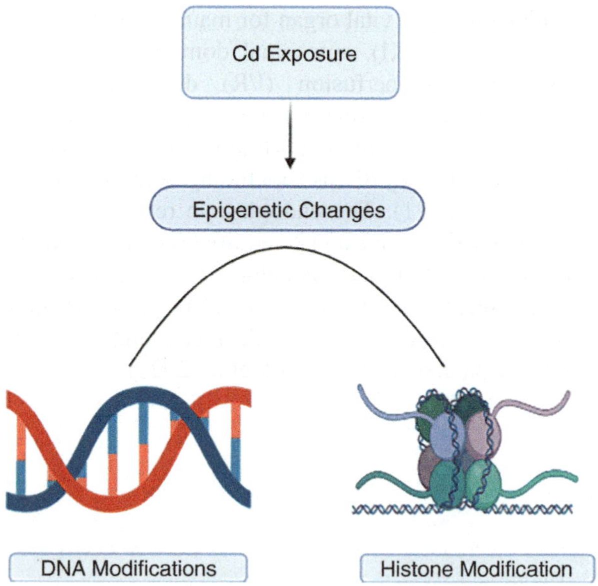

Fig. 4 Cd exposure causes epigenetic changes in humans

changes are implicated in Cd-induced breast malignancies (Filippini et al. 2020). Cadmium exposure during pregnancy has been shown to affect thyroid function and reproductive system development in children by interfering with the generation of progesterone, testosterone, and leptin (Banzato et al. 2012). Prenatal cadmium exposure induces adverse effects on the developing fetus, including increased embryonic mortality, fragmentation, and developmental blockades in mice. These outcomes are accompanied by significant epigenetic changes, highlighting the detrimental impact of cadmium on early embryonic development (Zhu et al. 2021).

The mechanism of cadmium reproductive harm in females, particularly on egg maturation and early embryo development, is unknown. We previously found that maternal cadmium exposure hindered oocyte meiotic maturation and subsequent embryonic development by altering cytoskeletal structure and histone modification in the oocyte (Zhu et al. 2018). Furthermore, paternal cadmium exposure disrupts their offspring's neurobehavioral system, suggesting that cadmium has transgenerational consequences (Zhao et al. 2015). Epigenetic changes emphasize these negative illness effects mechanistically. Environmental and maternal lifestyle factors have been demonstrated to play a main role in the development of metabolic syndrome, obesity, cardiovascular disease, and hypertension in children (Ryznar et al. 2021).

8 Overview of Cd Concentrations in Food Items

Food items play a significant role in contributing to cadmium exposure in different populations. Several studies have identified the percentage contributions of cadmium concentration from various food sources. In the United States, cereals and bread account for 34% of cadmium intake, followed by vegetables at 26%, potatoes at 11%, fruit at 5% nuts, and legumes at 7% (Kim et al. 2018). In European states grains and grain products account for 26.6%, vegetables and vegetable-based products for 17.4%, fish and shellfish for 8.9%, starchy roots and tubers for 12.3%, and meat and edible offal for 8.9% of dietary cadmium exposure, according to EFSA (2012). Rice is the most important crop in Asia, accounting for 58% of total income in Bangladesh (Kim and Wolt 2011; Watanabe et al. 2000; Song et al. 2017; Al-Rmalli et al. 2012). These findings highlight the importance of considering specific food sources when assessing and managing cadmium exposure in different regions (Table 1).

Table 1 highlights the significant role of various food categories in contributing to cadmium exposure in various regions. Cereals, grains, vegetables, and seafood emerge as consistent sources of cadmium intake across different countries. Rice, in particular, stands out as a prominent contributor to cadmium exposure in Bangladesh.

Table 1 Major food items contributing to cadmium exposure

Country/region	Food items	Cd concentrations (mg/kg)	References
United States	Cereal and bread	1.57	Kim et al. (2018)
	Leafy vegetables	0.94	
	Potatoes	0.52	
European countries	Shelfish/ mollusks	0.01–4.8	EFSA (2012), WHO (2011)
	Coffee	0.0001–1.8	
	Vegetables	0.006–1.0	
	Meat	0.006–0.5	
	Spices	0.006–0.2	
China	Aquatic food	0.017	Zhao et al. (2022a)
	Legumes	0.017	
	Cereals	0.015	
	Vegetables	0.013	
	Potatoes	0.011	
Bangladesh	Rice	0.001–0.18	Shahriar et al. (2020)

9 Variation in Dietary Cadmium Intake Across Countries and Associated Health Risks

Cadmium (Cd) is a transition metal that is toxic and can be harmful to both humans and animals, posing a potential health risk. The daily consumption of Cadmium (Cd) through the diet varies across different countries due to variations in pollution levels. The highest ingestion of this metal, ranging from 188 to 224 μg per day, has been observed in Cd-polluted regions of Thailand (Satarug et al. 2013). Conversely, the lowermost intake, amounting to 10 μg per day (1 μg/kg body weight/week), is reported in Sweden (Sand and Becker 2012), with the first percentile at 0.45 μg/kg body weight/week and the 99th percentile at 2.1 μg/kg body weight/week. In European countries, Poland stands out with one of the highest Cd dietary intakes, ranging from 11 μg to as high as 120 μg/day (Koch et al. 2016; Marzec and Łukasiewicz 2010; Pysz et al. 2016). The average Cd intake in the Polish population does not exceed the Provisional Tolerable Weekly Intake (PTWI) and Provisional Tolerable Monthly Intake (PTMI), but it exceeds or approaches the Tolerable Weekly Intake (TWI) limit. Polish consumers consume more Cd per day through cereal items and milk (Poland: 5 g, rest of Europe: 4 g), but less through vegetables (Poland: 5 g, rest of Europe: 6 g) (European Food Safety Authority 2012; Wojciechowska-Mazurek et al. 2008).

Studies by Marzec and Łukasiewicz (2010) indicate that Cd intake through the food in residents of social welfare homes and pupils in educational centers in the Lubelskie Voivodeship region of Poland reaches approximately 40% of the PTWI but significantly exceeds the TWI. Pysz et al. (2016) also found that the TWI for Cd was surpassed in 4–20 year old residents of orphanages in Krakow, with Cd intake ranging from 3.09 ± 0.21 to 20.36 ± 2.21 μg Cd/kg body weight/week. A study conducted among adult residents of Lubelskie Voivodeship revealed that Cd intake in men's diet was close to the TWI value for this metal (Koch et al. 2016).

Studies have indicated that exposure to environmental cadmium could increase the risk of osteoporosis. Cadmium has a highly toxic effect on the liver and kidneys, possibly because these organs can produce metallothioneins (MT), proteins that respond to the presence of cadmium by binding to it tightly and preventing its harmful effects on cells. The toxicity of cadmium is thought to induce oxidative stress, which may be a contributing factor to various liver and kidney diseases (Ma et al. 2022). Cadmium has the potential to enter the human body through the skin, respiratory tract, and gastrointestinal tract, leading to chronic kidney damage. The proximal tubule is the primary location where cadmium accumulates and causes harm within the nephrons (Yan and Allen 2021).

(i) **Organ-Specific Effects**

(a) **Kidney**

The kidney is a vital organ that helps the body function normally by filtering the blood and getting rid of waste products like urea, uric acid, and creatinine that would otherwise build up if left in the body (Rennke and Denker 2020). In addition, kidneys play a crucial role in regulating various bodily functions, such as red blood cell production, hemodynamics, and vitamin D maturation, through the secretion of hormones (Li et al. 2020). Cadmium primarily accumulates in the proximal tubule of the nephrons in the kidney (Fig. 5). Electron leakage and reactive oxygen species generation can occur from mitochondrial electron transport chain dysfunction brought on by Cd buildup in the proximal tubule (Sun et al. 2022). Furthermore, cadmium may also hinder NADPH oxidase, an additional ROS source. When these ROS combine their damaging effects, they can oxidize DNA, lipids, and proteins, causing epithelial cell death and a loss in kidney function (Paithankar et al. 2021).

A Case Study on Turtle

As human exposure to cadmium typically occurs over a prolonged period and at low rates, studying the renal toxicity of cadmium in humans would require extensive monitoring and long-term follow-up studies. Consequently, researchers have frequently employed animal models, such as rats and turtles to simulate the pathophysiological processes of cadmium-induced kidney toxicity (Chen et al. 2018).

The aim of the study was to examine the effects of cadmium exposure on kidney function in freshwater turtles specifically *Mauremys reevesii*. The experimental procedure involved administering different doses of cadmium through injections to the turtles (0, 7.5, 15, and 30 mg per kg of body weight) through the abdominal cavity once. Following the cadmium injections, the researchers collected samples from the turtles to assess several factors including kidney index, total protein (TP) content in kidney tissue, as well as levels of creatinine (CRE) and blood urea nitrogen (BUN) in the plasma. These analyses were done by a microplate reader.

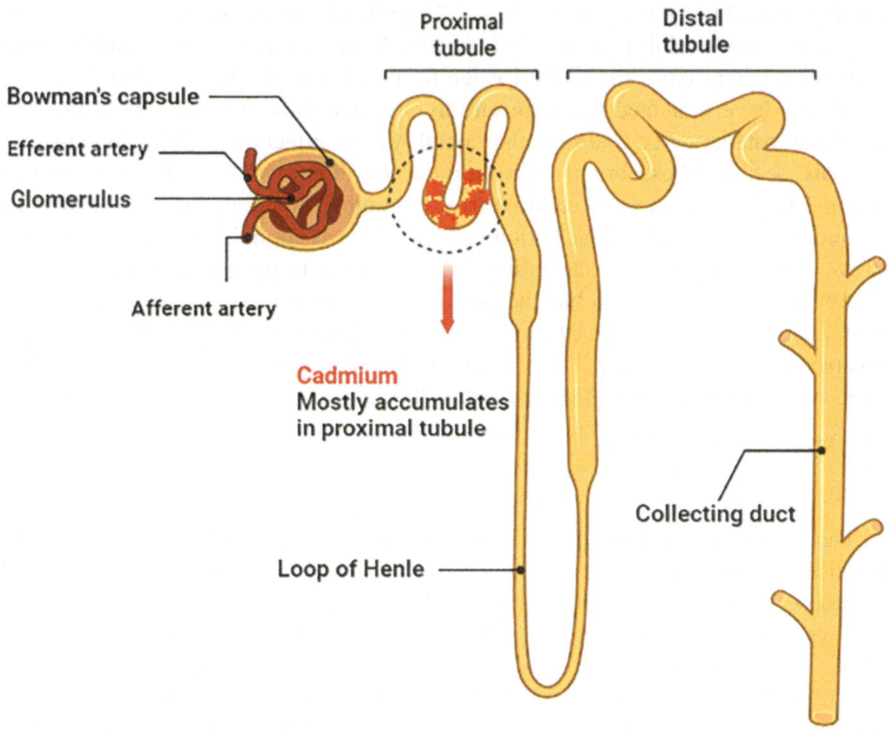

Fig. 5 Cadmium accumulation in Nephron

The results of the study showed that the TP conc. in kidney tissue decreased with the increase of the Cd exposure time and with an increasing exposure dose of Cd. As the duration of exposure time rose, so did the CRE concentration in the plasma of all treatment groups. In addition, there was a consistent increase in BUN levels across all groups that had been exposed to toxin. The treated turtles' renal index also increased (Dong et al. 2023).

(b) **Lungs**

Worldwide, 9.6 million people fall victim to cancer every year, with lung cancer being the biggest contributor to cancer-related deaths. In particular, lung cancer accounts for 18.4% of annual cancer-related mortality, or about 1.7 million deaths annually (Bray et al. 2018).

The significant carcinogenic effects of heavy metals, specifically cadmium, lead, arsenic, and mercury, are particularly notable in relation to the development of lung cancer (Pietrzak et al. 2021). Cd primarily enters in lungs through cigarette smoking since tobacco readily absorbs cadmium from soil (Iradukunda et al. 2021). The mean concentration of cadmium in the urine and blood of individuals who smoke is approximately twice as high as that of individuals who do not smoke since the

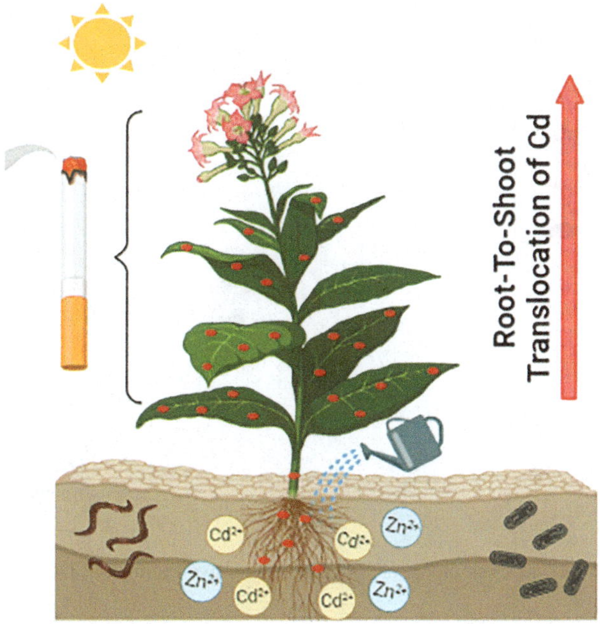

Fig. 6 Cadmium accumulation in Tobacco plant

biological half-life of cadmium in humans is believed to be between 10 and 30 years (Repić et al. 2020) (Fig. 6).

Chronic obstructive pulmonary disease (COPD) is a significant contributor to mortality and morbidity both in the United States and globally. This condition is primarily attributed to the toxic effects of Cd. It impacts a considerable proportion of individuals aged 45 years and above, estimated to be around 4–6%. The incidence of this health problem is on the rise, and it currently ranks as the third prominent contributor to mortality (Knoell and Wyatt 2021). COPD is characterized by a partially irreversible obstruction of the airways, which is attributed to the malfunctioning of immune system in response to the inhalation of more than 7000 harmful substances, including Cd, found in cigarette smoke (Baptist and Busse 2018) (Fig. 7).

Cadmium leads to Pulmonary fibrosis a chronic respiratory condition characterized by the gradual thickening, stiffening, and scarring of lung tissue, resulting in respiratory distress and diminished pulmonary capacity (Confalonieri et al. 2022). The fibrosis of pulmonary tissue can be attributed to a range of factors, including exposure to environmental pollutants (Cd), infections, autoimmune disorders, or idiopathic origins (Salton et al. 2019). The formation of scars within the lungs hinders their physiological function, thereby preventing the efficient transfer of oxygen into the bloodstream. This can result in a range of symptoms such as breathing difficulties, persistent cough, exhaustion, and chest discomfort. Pulmonary fibrosis is a chronic condition that can result in respiratory failure and increased vulnerability to lung infections as it advances it can be devasting. Although a definitive cure for pulmonary fibrosis is yet to be discovered, various interventions

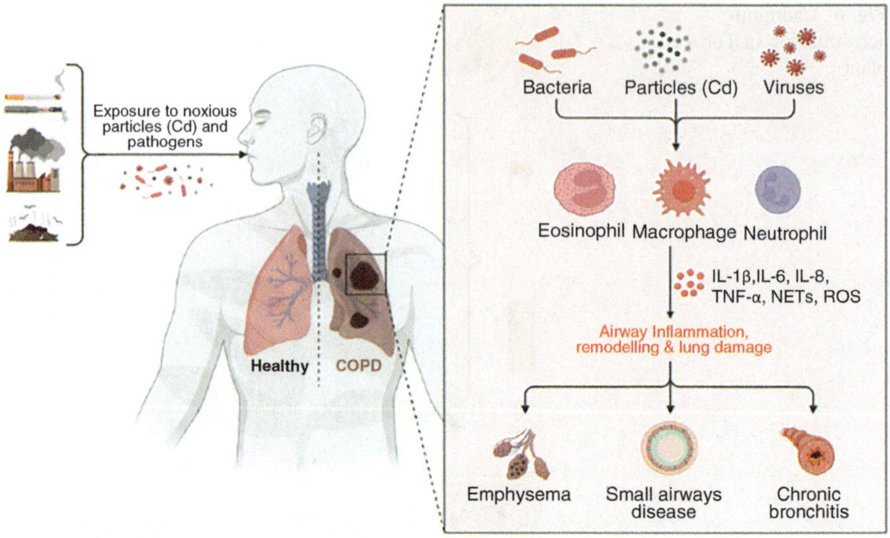

Fig. 7 Development of Chronic Obstructive Pulmonary Disease (COPD)

including oxygen therapy, pulmonary rehabilitation, and pharmaceutical treatment have been shown to effectively mitigate symptoms and decelerate the advancement of the condition (Li et al. 2022).

(c) **Liver**

Cadmium possesses a distinctive toxicological characteristic whereby it exhibits a prolonged half-life thereby enabling it to persist within the human body for an extended duration. Upon absorption, Cd has a tendency to accumulate in various parts of the body, especially the liver (Reshma and Mohanan 2019). The liver plays an important role in the filtration of blood from the gastrointestinal tract and the detoxification of xenobiotics and pharmaceuticals that may potentially pose a threat to the body. The process involves the metabolic breakdown and subsequent elimination of toxic chemicals to ensure the secure distribution of blood to other internal organs and tissues (Messelmani et al. 2022). Cd enters the body in ionic form hence, transportation of Cd^{2+} ions across the hepatocyte cell membrane is predominantly facilitated by two mechanisms: (1) association with iron (Fe^{2+}) and zinc (Zn^{2+}) transporters; and (2) transversing Ca^{2+} channels that are voltage gated (Niture et al. 2021). The other way, Protein-bound Cd usually forms a complex with the liver synthesized metallothionein (MT) protein, leading to the formation of Cd-MT. Receptor-mediated endocytosis allows the Cd–MT complex to enter cells, where it is then digested by lysosomes. The result of this reaction is that Cd ions are released from the Cd–MT complex (Udi et al. 2022).

Several mechanisms are used by the liver to remove cadmium from the body. But Cd exposure causes a variety of cellular and physiological reactions that may severely damage the liver, leading to steatosis, steatohepatitis, and ultimately

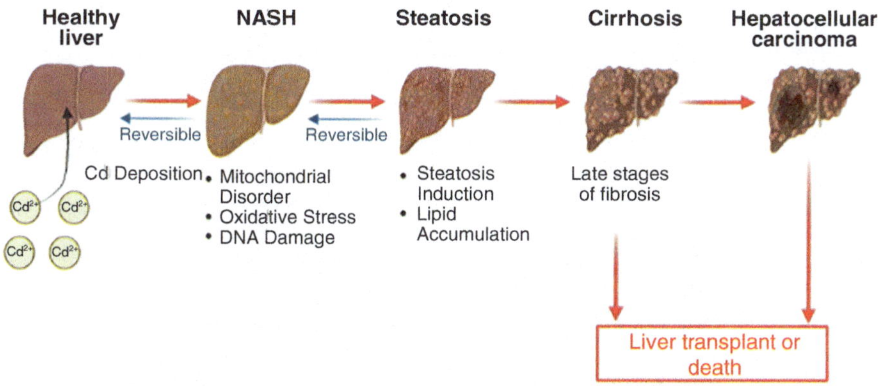

Fig. 8 Role of Cd in live Inflammation

hepatocellular carcinoma (Fig. 8). According to the length of exposure to the metal, different cellular reactions are involved in the advancement of cadmium-induced liver damage. Oxidative stress, apoptosis, adipogenesis, and decreased autophagy are a few of these reactions (Souza-Arroyo et al. 2022).

Alkaline phosphatase (ALP), alanine aminotransferase (ALT), and aspartate aminotransferase (AST) are all biomarkers of liver damage that rise in response to acute Cd poisoning in blood. Nonalcoholic hepatitis and fatty liver are also more common in those who have been exposed to Cd (Al-Janabi et al. 2020). The outcomes of the histopathological examination indicate that exposure to Cd has the potential to initiate an acute liver injury, which may result in tissue necrosis, apoptosis, hyperplasia, and an enlargement of the liver sinus and hilum (Balali-Mood et al. 2021). Severe necrosis can be followed by the infiltration of neutrophils and their chemotaxis toward the affected area, which may lead to severe liver damage (Nguyen et al. 2023) (Fig. 8).

Results of numerous studies indicate a significant relationship between raised urinary Cd concentrations and amplified serum levels of hepatic enzymes, such as ALT, AST, and GGT. Long-term Cd exposure has been associated with nonalcoholic fatty liver disease NAFLD, nonalcoholic steatohepatitis NASH, and neuroinflammation in humans. Chronic Cd exposure causes several health issues and harms liver and brain function (Park et al. 2021). Exposure to Cd has been found to impact liver function through molecular mechanisms that involve changes in the methylation patterns of the Klotho gene. Lower concentrations of Cd exposure have been observed to cause a reduction in Klotho methylation, whereas higher doses of Cd have been found to induce an increase in Klotho methylation (Yu et al. 2020). There is a positive correlation between the level of Klotho methylation and β2-microglobulin levels, while a negative correlation is observed with albumin levels. This suggests that the effects of Cd on the liver are mediated through this particular mechanism (Niture et al. 2021).

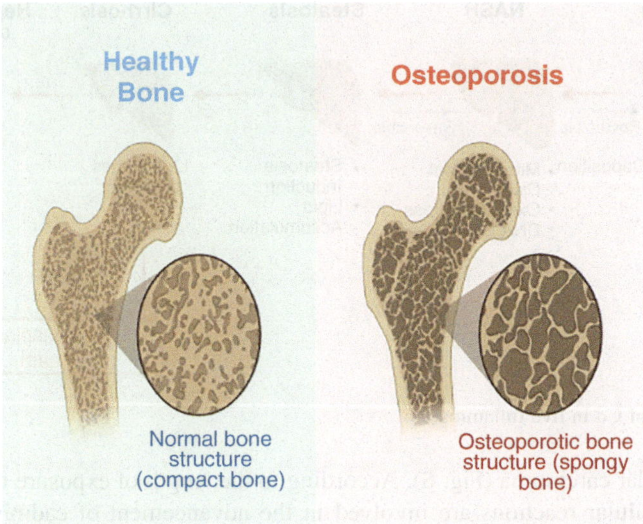

Fig. 9 Cd causes Osteoporosis in bone

(d) **Bone**

Cadmium can effect bones both ways directly and indirectly. Direct effects of cadmium on bones can occur through kidney dysfunction, as the kidneys play a critical role in regulating calcium balance in the body. Cd can accumulate in the kidneys and cause damage to the tubules which can lead to impaired calcium reabsorption and increased excretion of calcium in urine. This can result in a decrease in calcium levels in the blood which can in turn lead to bone loss (He et al. 2020). Cd affects osteoblasts and osteoclasts, the cells accountable for bone formation and resorption, and can lead to indirect consequences on bones. The disturbance of bone formation and resorption balance by cadmium can result in reduced bone density and heightened fracture susceptibility (He et al. 2019). Cd has been identified to possess nephrotoxic properties and has the potential to disrupt vitamin D metabolism. The association between cadmium and various health outcomes, such as osteoporosis and osteomalacia has been attributed to the disruption of the vitamin D metabolic pathway (Agbadua et al. 2020) (Fig. 9). Recent studies reveal that itai-itai disease is worst case scenario of chronic Cd intoxication (Imura et al. 2019).

Some epidemiological studies enlighten that exposure to toxic metals such as Cd can activate molecular mechanisms of inflammation and oxidative stress at the bone level, and could potentially contribute to the onset of musculoskeletal disorders, including osteoporosis, rheumatoid arthritis (RA), and osteoarthritis (OA). Bone calcium absorption, osteoclast activity, resorption, and renal function are all impacted by cadmium exposure making it more likely that osteoporosis may develop. Cadmium (Cd) has been observed to hinder the activity of the antioxidant enzymes, namely superoxide dismutase (SOD) and catalase (CAT), in individuals diagnosed with rheumatoid arthritis (RA). It also induces citrullination, which modifies immune response proteins, which contributes to an inflammatory condition (Reyes-Hinojosa et al. 2019).

(e) **Cardiovascular System**

The exact effect of Cd on cardiovascular system is yet under discussion. However, some studies enlighten that even at very low doses it has extremely adverse effects on cardiovascular system (Fatima et al. 2019). Cd causes oxidative stress in cardiovascular system and is associated with many diseases, when it targets vessels it encourages endothelial disinfection, which results in atherosclerosis (thickening and hardening of arteries) and hypertension (Almenara et al. 2020).

Cadmium is transported across the cardiac cell membrane by various cellular carriers and ion channels, like calcium channel (Hocaoğlu-Özyiğit and Genç 2020). Immune system cells may also take up and transported by Cadmium-Laden monocyte. Excessive production of monocytes and macrophages as a result of cadmium exposure can potentially contribute to the initiation and progression of atherosclerosis, primarily through their transdifferentiation into foam cells and induction of necrotic foam cell death. This underscores the critical role of cadmium in promoting endothelial dysfunction, a key factor in the development of atherosclerosis (Lin et al. 2021). Cd accumulates in smooth muscle cells after being transported across endothelial cells. It has a variety of impacts on smooth muscle cells, including cytotoxicity, disruption of calcium ion flow and ion homeostasis, promotion of smooth muscle cell proliferation, and interactions with these processes. Low cadmium concentrations are particularly essential for these responses, which promote lipid deposition in vessel walls and change lipid profiles to a more atherogenic state (Aleksandrov et al. 2021). This further leads to inflammation of vessels. The presence of Cd has been found to stimulate the synthesis of crucial pro-inflammatory cytokines, such as interleukin (IL)-6, IL-8, IL-1β, and tumor necrosis factor-alpha (TNF-α). These cytokines play significant roles in promoting inflammation, and their increased expression as a result of cadmium exposure may have implications for the inflammatory processes involved in atherosclerosis. The upregulation of these cytokines suggests that cadmium accumulation has the potential to influence and contribute to the inflammatory response associated with atherosclerosis (Bonaventura et al. 2018; Hossein-Khannazer et al. 2020) which ultimately leads to high blood pressure and cardiac arrest.

(ii) **Non-Organ-Specific Effects**

(a) **Carcinogenicity**

Cadmium has been designated a human carcinogen by the International Agency for Research on Cancer and the US National Toxicology Program (Goering et al. 1995; Program 2000)and is clearly a potent, multi-tissue animal carcinogen (Waalkes 2000). Occupational exposure to cadmium is associated with lung cancers in humans, while other sites, potentially including the prostate, are not definitively established (Goering et al. 1995; Program 2000; Waalkes 2000) (Fig. 10). Rodent

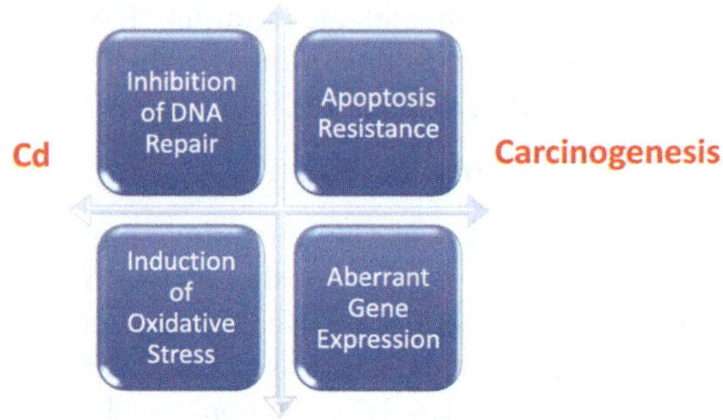

Fig. 10 Carcinogenesis effects of Cadmium

studies have shown that chronic inhalation of cadmium causes pulmonary adenocarcinomas, in clear support of human data (Oldiges et al. 1984; Waalkes 2000; Waalkes et al. 1999). Cadmium can also cause prostatic proliferative lesions, including adenocarcinomas, after systemic or direct exposure (Hoffmann et al. 1985; Waalkes and Rehm 1994; Waalkes et al. 1988). Other target tissues of cadmium carcinogenesis in animals include injection sites, adrenals, testes, and the hemopoietic system (Waalkes 2000; Waalkes and Rehm 1992). The exposure of cells to Cadmium (Cd) leads to their cellular transformation. The administration of Cadmium (Cd) in animals has been observed to lead to the development of tumors in various organs and tissues. Furthermore, a correlation has been observed between exposure to Cd and the prevalence of lung cancer in humans, indicating a causal relationship. Multiple studies have shown that Cd can lead to the development of cancer through various mechanisms. Among these mechanisms, the most significant ones include the disruption of gene expression, hindrance of DNA damage repair, generation of oxidative stress, and suppression of apoptosis. The extant body of evidence suggests that oxidative stress may be a pivotal factor in the process of Cd carcinogenesis, owing to its participation in the induction of anomalous gene expression, hindrance of DNA damage repair, and apoptosis (Joseph 2009). Cd exposure in humans can cause a variety of negative effects, including renal and hepatic dysfunction, testicular damage, pulmonary edema, osteomalacia, and damage to the adrenals and hemopoietic system (Tinkov et al. 2018). There was also a link found between Cd exposure indicators (blood and urine) and coronary heart disease, stroke, peripheral artery disease, and atherogenic alterations in lipid profile (Malhotra et al. 2016). Cadmium is a confirmed human carcinogen (group I of the International Agency for Research on Cancer classification) in addition to its cytotoxic effects that may result in apoptotic or necrotic events (Roe 1995; Tinkov et al. 2018). Occupational or environmental cadmium exposure has been related to breast, prostate, lung, pancreas, urinary bladder, and nasopharynx cancers.

(b) **Reproductive Effects**

Over the past five decades, international literature has observed a notable trend toward a global decline in human fertility (Adsera 2004); the advanced decline of couple fertility cannot be justified solely by variations in lifestyle (increase in

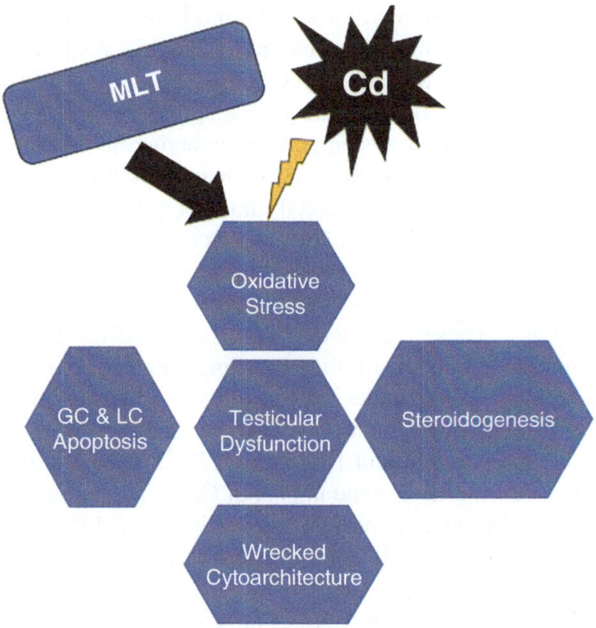

Fig. 11 Cadmium effect on the male reproductive system

women's occupational rate, increase in women's age at pregnancy), prevention of unwanted parenthood (use of contraceptives, intentional abortions), or economic issues (decrease in income, increase in expense). A considerable general drop in male fertility has been documented during the early 1990s, in a meta-analysis of research published between 1938 and 1991 (Carlsen et al. 1992).

Nonetheless, a recent re-analysis of the same trials found varied findings, owing mostly to significant regional differences in sperm quality (Fisch and Braun 2013). These contradictory and ambiguous findings revealed a probable influence of local chronic environmental contamination on male fertility (Nordkap et al. 2012) As a result, the contribution of contemporaneous, global, and local variables should be considered to explain the discrepancies between research. Environmental contact with toxic pollutants in the background environment, as well as occupational contact with particular toxic pollutants as a result of job-related actions, have been shown to reduce male fertility in humans (Jenardhanan et al. 2016; Knez 2013). Cadmium effect on the male reproductive system is illustrated in Fig. 11.

(c) **Immune System Effects**

The primary character of the immune system is to protect the body from harmful microbes and respond to tissue damage caused by various factors such as mechanical injuries and chemical insults. Immune cells can be found in the blood, lymph, primary lymphoid organs (thymus and bone marrow), secondary lymphoid organs (spleen and lymph nodes), and nearly all tissues. The initial answer is mediated by innate immunity, which swiftly reacts to microbial invasion and tissue injury through cellular and biochemical mechanisms. Subsequently, the adaptive immune system comes into play, mounting specific defense mechanisms against different insults in various parts of the body, including tissues, circulation, and intracellular environments. Both innate and adaptive immune responses work together as an integrated defense system, involving numerous cells and molecules that cooperate to combat threats (Boehm 2008; Franchimont 2004; Ruddle and Akirav 2009).

Cadmium (Cd) can directly interact with immune cells, leading to alterations in their status and functionality, which can be assessed in vitro. These direct impacts can vary from cell death to signaling system interference, which results in alterations in cell surface marker expression, cell activation, cytokine production, and differentiation. These changes can eventually result in immunosuppression/immunodepression or immunostimulation/immunoenhancement. Though in vitro immunotoxicity studies cannot fully reproduce the intricacy of an immune response, they are frequently employed to test chemical compounds for their immunotoxic potential (Corsini and Roggen 2017). In vitro studies using human immune cells provide insights into the impact of Cd on the human immune system. However, the most informative approach to investigating Cd immunotoxicity is the use of animal models that replicate natural routes of exposure and allow evaluation of effects on tissue and organ physiology.

Cd immunotoxicity studies have used various dosages (high, moderate, or low), modes of exposure (systemic, pulmonary, oral), and durations (acute, subchronic, chronic, long-term). Immune endpoints have been assessed, including immune cell activation, proliferation, effector function, and survival. Even when analyzing the same type of cell or activity, these studies have produced diverse outcomes in terms of the effects seen (stimulation, inhibition, or no impact).

(d) **Neurological Effects**

Cadmium (Cd) is a heavy metal that is widely distributed in the environment due to both natural and anthropogenic sources. It is a known toxicant that can have detrimental effects on various organ systems in humans and animals. In particular, cadmium exposure has been associated with a range of neurological effects.

Neurodevelopmental Effects Early-life exposure to cadmium has been linked to neurodevelopmental effects. Animal studies have shown that prenatal and early postnatal exposure to cadmium can lead to impaired learning and memory, altered motor function, and changes in behavior in offspring (Wang et al. 2019). Furthermore, epidemiological studies have reported associations between maternal cadmium exposure during pregnancy and adverse neurodevelopmental outcomes in children, including reduced cognitive function and increased risk of attention deficit hyperactivity disorder (ADHD) (Skogheim et al. 2021).

Neurotoxicity and Neurodegenerative Disorders Cadmium exposure has been implicated in the development or progression of neurodegenerative disorders, such as Parkinson's disease (PD) and Alzheimer's disease (AD). Animal studies have demonstrated that chronic exposure to cadmium can induce neuronal damage, oxidative stress, and neuroinflammation, which are hallmarks of neurodegeneration

(Pyatha et al. 2022). Additionally, epidemiological studies have reported associations between cadmium exposure and increased risk of PD and AD in human populations (Wang et al. 2018).

Disruption of Neurotransmitter Systems Cadmium exposure can interfere with neurotransmitter systems in the brain. Animal studies have shown that cadmium can disrupt the balance of neurotransmitters such as dopamine, serotonin, and glutamate, which are crucial for normal brain function (Kodavanti and Loganathan 2017). These alterations in neurotransmitter levels and signaling pathways may contribute to the neurological effects observed following cadmium exposure, including changes in behavior and mood disorders.

Blood–Brain Barrier (BBB) Dysfunction Cadmium has been shown to disrupt the integrity of the blood–brain barrier (BBB), which is responsible for maintaining a tightly controlled environment in the brain. Animal studies have demonstrated that cadmium exposure can increase BBB permeability, allowing the entry of harmful substances into the brain (Gonçalves et al. 2017). BBB dysfunction can lead to neuroinflammation, neuronal damage, and cognitive impairment.

Oxidative Stress and Mitochondrial Dysfunction Cadmium exposure induces oxidative stress in the brain, leading to an imbalance between the production of reactive oxygen species (ROS) and the antioxidant defense system. Increased ROS production can result in oxidative damage to lipids, proteins, and DNA in neurons, contributing to neurotoxicity (Karazivan et al. 2015). Furthermore, cadmium exposure has been shown to disrupt mitochondrial function, impairing energy production and leading to neuronal dysfunction (Gonçalves et al. 2017).

In conclusion, cadmium exposure has been associated with a range of neurological effects, including neurodevelopmental disorders, neurodegenerative diseases, disruption of neurotransmitter systems, BBB dysfunction, oxidative stress, and mitochondrial dysfunction. These effects have been observed in both animal studies and human populations. It is crucial to minimize cadmium exposure in order to protect neurological health. Further research is needed to elucidate the underlying mechanisms and develop effective strategies for prevention and intervention.

10 Biomarkers of Cd Exposure (Blood, Urine, Nail, Hair)

Cadmium (Cd) exposure can be assessed by measuring its concentrations in various biological samples such as urine, blood, hair, or nails, which serve as biomarkers. Blood Cd levels are considered valuable indicators of recent Cd exposure and are commonly measured in whole blood. Hair and nail Cd concentrations have also been used to evaluate Cd exposure, but standardized protocols for analyzing these matrices are lacking, and there is a high risk of significant contamination from external sources.

Chronic, low-dose Cd exposure primarily leads to renal injury. Population-based studies have demonstrated that the burden of Cd in the kidneys is directly proportional to the amount of Cd excreted in urine. Therefore, urinary Cd (UCd) is a suitable biomarker for assessing long-term Cd exposure (Menke et al. 2009; Vacchi-Suzzi et al. 2016). One of the foremost rewards of using urine as a matrix is its non-invasive collection method, allowing for the analysis of large volumes of samples. This makes urine particularly useful in human surveys, particularly in surveys involving a substantial number of participants, where monitoring low levels of chemicals is necessary.

However, it is vital to remember that UCd concentrations vary between spot pee samples. UCd concentrations (reported in g/L) are often accustomed by urinary creatinine concentrations (expressed in g/L) to accommodate for varying dilutions among spot samples when scheduled the elimination of urine or 24-h samples are not obtained. The adjusted values are given in micrograms per gram of creatinine (g/g creatinine). By employing this correction method, the variability in UCd measurements can be mitigated and provide more reliable data for assessing Cd exposure levels in individuals.

(a) **Urinary Cd Levels for Different Diseases**

Urinary cadmium (UCd) levels have been extensively studied worldwide to assess the overall burden of cadmium in the human body. In the United States, data from the National Health and Nutrition Examination Survey III showed that the average urine Cd concentrations were 0.28 μgg^{-1} creatinine for men and 0.40 μg/g creatinine for women (Menke et al. 2009). Similar UCd levels were observed in European countries like Sweden (mean: 0.26 μgg^{-1} creatinine), France (mean: 0.29 μgg^{-1} creatinine), and Spain (median: 0.39 μgg^{-1} creatinine), all of which were below the EFSA threshold of 1 μgg^{-1} creatinine (Grau-Perez et al. 2017; Fréry et al. 2010; Olsson et al. 2002). The Food and Agriculture Organization/World Health Organization (FAO/WHO) has established a threshold of 5.24 μgg^{-1} creatinine for UCd concentration to safeguard against kidney damage (FAO/WHO 2010). However, the European Food Safety Authority (EFSA) has set a lower threshold of 1 μgg^{-1} creatinine for UCd based on a meta-analysis of 54 epidemiological studies that examined the relationship between UCd and b2-microglobulin, an indicator of tubular impairment (European Food Safety Authority 2012).

Comparatively higher UCd levels have been reported in Asian countries compared to the United States and Europe. For example, a study in eastern China found UCd concentrations ranging from 0.36 to 0.45 μgg^{-1} creatinine among the general population (Sun et al. 2016). In Korea, a literature survey covering the period from 1975 to 2015 showed a time-dependent decrease in UCd, with levels stabilizing around or below 1 μgg^{-1} creatinine after 2010 (Moon et al. 2016). In the Boso Peninsula of Japan, where no known environmental cadmium pollution existed, the average UCd concentration was 1.3 μgg^{-1} creatinine among men and women (Moon et al. 2016). It is important to note that these UCd levels are well below the FAO/WHO threshold of 5.24 μgg^{-1} creatinine. However, several studies have challenged this limit by suggesting that even low-level UCd exposure can have multiple adverse health effects These findings suggest that there may be no safety margin between current environmental cadmium exposure levels and the threshold for adverse health effects (Suwazono et al. 2011; Gallagher et al. 2008; Gallagher and Meliker 2010; He et al. 2013).

11 Summary

Cadmium (Cd) is a toxic metal that can have harmful effects on human health. The daily intake of Cd through the diet varies across different countries, with Thailand having the highest intake and Sweden having the lowest. In European countries, Poland stands out with high Cd dietary intakes. Cd exposure has been linked to various health risks, including kidney damage, liver dysfunction, bone disorders, and cardiovascular diseases.

The kidney is particularly susceptible to Cd toxicity as it accumulates in the proximal tubules, leading to oxidative stress and damage to the nephrons. Studies on animals, such as turtles, have shown that Cd exposure can cause kidney dysfunction, indicated by decreased total protein in kidney tissue and increased levels of creatinine and blood urea nitrogen in the plasma.

Cd exposure through cigarette smoking is a major contributor to lung cancer, as Cd readily absorbs from tobacco. Cd toxicity can also lead to chronic obstructive pulmonary disease (COPD) characterized by airway obstruction and inflammation. Pulmonary fibrosis, a condition characterized by the scarring of lung tissue, can also result from Cd exposure.

The liver is another target organ for Cd accumulation. Cd can disrupt liver function and lead to hepatocellular damage, fatty liver, and liver inflammation. Increased levels of hepatic enzymes, such as alkaline phosphatase, alanine aminotransferase, and aspartate aminotransferase, are observed in response to Cd exposure. Cd exposure has also been associated with nonalcoholic fatty liver disease (NAFLD) and changes in gene methylation patterns in the liver.

Cd exposure can affect bone health by disrupting calcium balance and impairing bone formation and resorption processes. It has been linked to osteoporosis, osteomalacia, and itai-itai disease, a severe condition resulting from chronic Cd intoxication. Cd exposure can lead to bone loss, reduced bone density, and increased fracture susceptibility.

The cardiovascular system can also be affected by Cd exposure. Cd can induce oxidative stress, inflammation, and endothelial dysfunction, leading to atherosclerosis and hypertension. Cd exposure has been associated with an increased risk of cardiovascular diseases such as coronary heart disease and stroke.

Biomarkers such as blood, urine, hair, and nails can be used to assess Cd exposure. Blood Cd levels reflect recent exposure, while urinary Cd levels indicate long-term exposure. Urinary Cd concentrations are commonly adjusted by urinary creatinine levels to account for variable dilutions among spot samples. The general population in different countries has varying levels of Cd exposure, with higher levels reported in Asian countries.

Cd exposure has been associated with various adverse health effects, including renal injury, neurodevelopmental disorders, neurodegenerative diseases, immune system dysfunction, and disruption of neurotransmitter systems. Further research is needed to better understand the mechanisms of Cd toxicity and develop effective strategies for prevention and intervention.

References

Abedi T, Mojiri A (2020) Cadmium uptake by wheat (triticum aestivum l.): an overview. Plants 9:500

Abernethy DR, Destefano AJ, Cecil TL, Zaidi K, Williams RL, UMIA P (2010) Metal impurities in food and drugs. Pharm Res 27:750–755

Adsera A (2004) Changing fertility rates in developed countries. The impact of labor market institutions. J Popul Econ 17:17–43

Agbadua OG, Obi F, Amadi CS, Taiwo S, Ehikhamhen D, Malio-Oliseh G (2020) Heavy metal interactions on cadmium-induced renal osteodystrophy and anaemia in rats. NISEB J 19(3)

Aleksandrov AP, Mirkov I, Tucovic D, Kulas J, Zeljkovic M, Popovic D et al (2021) Immunomodulation by heavy metals as a contributing factor to inflammatory diseases and autoimmune reactions: cadmium as an example. Immunol Lett 240:106–122

Al-Janabi AIH, Khadairi MM, Al-Amari MJ, Hirallah AAK (2020) Curative role of vitamin (C) in reduction of cadmium toxicity on the levels of some liver functions, lipid peroxidation and antioxidants enzymes in in vivo condition. Plant Arch 20(2):936–940

Almenara CC, Oliveira TF, Padilha AS (2020) The role of antioxidants in the prevention of cadmium-induced endothelial dysfunction. Curr Pharm Des 26(30):3667–3675

Al-Rmalli SW, Jenkins R, Haris PI (2012) Dietary intake of cadmium from Bangladeshi foods. J Food Sci 77:T26–T33

Alshehri AS, El-Kott AF, El-Gerbed MS, El-Kenawy AE, Albadrani GM, Khalifa HS (2022) Kaempferol prevents cadmium chloride-induced liver damage by upregulating Nrf2 and suppressing NF-κB and keap1. Environ Sci Pollut Res 29:1–13

Alterio J, Masson J, Diaz J, Chachlaki K, Salman H, Areias J, Al Awabdh S, Emerit MB, Darmon M (2015) Yif1b is involved in the anterograde traffic pathway and the golgi architecture. Traffic 16:978–993

Andrulewicz-Botulińska E, Wiśniewska R, Brzóska MM, Rogalska J, Galicka A (2018) Beneficial impact of zinc supplementation on the collagen in the bone tissue of cadmium-exposed rats. J Appl Toxicol 38:996–1007

Arao T (2019) Mitigation strategies for cadmium and arsenic in rice. In: Cadmium toxicity: new aspects in human disease, rice contamination, and cytotoxicity. Springer, pp 125–138

ATSDR (2012) Toxicological profile for cadmium. ATSDR, Atlanta, GA

Balali-Mood M, Naseri K, Tahergorabi Z, Khazdair MR, Sadeghi M (2021) Toxic mechanisms of five heavy metals: mercury, lead, chromium, cadmium, and arsenic. Front Pharmacol 227

Banzato TP, Godinho AF, Da Silva Zacarin ECM, Perobelli JE, Fernandez CDB, Favareto APA, Kempinas WDG (2012) Sperm quality in adult male rats exposed to cadmium in utero and lactation. J Toxicol Environ Health Part A 75:1047–1058

Baptist AP, Busse PJ (2018) Asthma over the age of 65: all's well that ends well. J Allergy Clin Immunol Pract 6(3):764–773

Begg SL, Eijkelkamp BA, Luo Z, Counago RM, Morey JR, Maher MJ, Ong C-LY, Mcewan AG, Kobe B, O'mara ML (2015) Dysregulation of transition metal ion homeostasis is the molecular basis for cadmium toxicity in streptococcus pneumoniae. Nat Commun 6:6418

Bernhoft RA (2012) Mercury toxicity and treatment: a review of the literature. J Environ Public Health

Bishak YK, Payahoo L, Osatdrahimi A, Nourazarian A (2015) Mechanisms of cadmium carcinogenicity in the gastrointestinal tract. APJCP 16:9–21

Boehm T (2008) Thymus development and function. Curr Opin Immunol 20(2):178–184

Bolan NS, Makino T, Kunhikrishnan A, Kim P-J, Ishikawa S, Murakami M et al (2013) Cadmium contamination and its risk management in rice ecosystems. Adv Agron 119:183–273

Bonaventura P, Lamboux A, Albarede F, Miossec P (2018) Differential effects of TNF-α and IL-1β on the control of metal metabolism and cadmium-induced cell death in chronic inflammation. PLoS One 13(5):e0196285

Branca JJ, Fiorillo C, Carrino D, Paternostro F, Taddei N, Gulisano M, Pacini A, Becatti M (2020) Cadmium-induced oxidative stress: focus on the central nervous system. J Antioxid Act 9:492

Bray F, Ferlay J, Soerjomataram I, Siegel RL, Torre LA, Jemal A (2018) Global cancer statistics 2018: GLOBOCAN estimates of incidence and mortality worldwide for 36 cancers in 185 countries. CA Cancer J Clin 68(6):394–424

Cao Z, Cui S, Lu X, Chen X, Yang X, Cui J, Zhang G (2018) Effects of occupational cadmium exposure on workers' cardiovascular system. Zhonghua lao dong wei sheng zhi ye bing za zhi=Zhonghua laodong weisheng zhiyebing zazhi=Chin J Ind Hyg Occup Dis 36:474–477

Carlsen E, Giwercman A, Keiding N, Skakkebæk NE (1992) Evidence for decreasing quality of semen during past 50 years. Br Med J 305(6854):609–613

Chen S, Liu G, Long M, Zou H, Cui H (2018) Alpha lipoic acid attenuates cadmium-induced nephrotoxicity via the mitochondrial apoptotic pathways in rat. J Inorg Biochem 184:19–26

Choong G, Liu Y, Templeton DM (2014) Interplay of calcium and cadmium in mediating cadmium toxicity. Chem Biol Interact 211:54–65

Chunhabundit R (2016) Cadmium exposure and potential health risk from foods in contaminated area, Thailand. Toxicol Res 32:65–72

Confalonieri P, Volpe MC, Jacob J, Maiocchi S, Salton F, Ruaro B et al (2022) Regeneration or repair? The role of alveolar epithelial cells in the pathogenesis of idiopathic pulmonary fibrosis (IPF). Cells 11(13):2095

Corsini E, Roggen EL (2017) Overview of in vitro assessment of immunotoxicity. Curr Opin Toxicol 5:13–18

Cuypers A, Plusquin M, Remans T, Jozefczak M, Keunen E, Gielen H, Opdenakker K, Nair AR, Munters E, Artois TJ (2010) Cadmium stress: an oxidative challenge. Biometals 23:927–940

Dalle-Donne I, Aldini G, Carini M, Colombo R, Rossi R, Milzani A (2006) Protein carbonylation, cellular dysfunction, and disease progression. J Cell Mol Med 10:389–406

Della Chiara G, Gervasoni F, Fakiola M, Godano C, D'oria C, Azzolin L, Bonnal RJP, Moreni G, Drufuca L, Rossetti G (2021) Epigenomic landscape of human colorectal cancer unveils an aberrant core of pan-cancer enhancers orchestrated by yap/taz. Nat Commun 12:2340

Dong Q, Fang J, Huang F, Cai K (2019) Silicon amendment reduces soil cd availability and cd uptake of two pennisetum species. IJERPH 16:1624

Dong A, Dong H, He H, Dong A, Yan J, Huo J (2023) Effects of cadmium on kidney function of the freshwater turtles Mauremys reevesii. Biol Trace Elem Res 201(6):3000–3005. https://doi.org/10.1007/s12011-022-03397-y

Đukić-Ćosić D, Baralić K, Javorac D, Djordjevic AB, Bulat Z (2020) An overview of molecular mechanisms in cadmium toxicity. Curr Opin Toxicol 19:56–62

El Rasafi T, Oukarroum A, Haddioui A, Song H, Kwon EE, Bolan N et al (2022) Cadmium stress in plants: a critical review of the effects, mechanisms, and tolerance strategies. Crit Rev Environ Sci Technol 52(5):675–726

ElMahdy MK, Zaki MO, Al-Karmalawy AA, Abdo W, Alnasser SM, Antar SA (2022) Glimepiride ameliorates renal toxicity induced by cadmium in mice: modulation of Jun N terminal kinase

(JNK)/nuclear factor kappa B (NF-κB) and phosphatidylinositol 3-kinases (PI3K)/protein kinase (AKT) pathways. Life Sci 311:121184
European Food Safety Authority (2012) Cadmium dietary exposure in the European population. EFSA J 10(1):2551
FAO/WHO (2010) Expert meeting on the application of nanotechnologies in the food and agriculture sectors: potential food safety implications: Meeting report: World Health Organization
Fatima G, Raza AM, Hadi N, Nigam N, Mahdi AA (2019) Cadmium in human diseases: it's more than just a mere metal. Indian J Clin Biochem 34:371–378
Filippini T, Torres D, Lopes C, Carvalho C, Moreira P, Naska A, Kasdagli M-I, Malavolti M, Orsini N, Vinceti M (2020) Cadmium exposure and risk of breast cancer: a dose-response meta-analysis of cohort studies. Environ Int 142:105879
Fisch H, Braun SR (2013) Trends in global semen parameter values. Asian J Androl 15(2):169
Florea A-M, Büsselberg D (2006) Occurrence, use and potential toxic effects of metals and metal compounds. Biometals 19:419–427
Forcella M, Lau P, Oldani M, Melchioretto P, Bogni A, Gribaldo L, Fusi P, Urani C (2020) Neuronal specific and non-specific responses to cadmium possibly involved in neurodegeneration: a toxicogenomics study in a human neuronal cell model. Neurotoxicology 76:162–173
Franchimont D (2004) Overview of the actions of glucocorticoids on the immune response: a good model to characterize new pathways of immunosuppression for new treatment strategies. Ann N Y Acad Sci 1024(1):124–137
Fréry N, Saoudi A, Garnier R, Zeghnoun A, Falq G, Guldner L (2010) Exposure of the French population to environmental pollutants. Environmental components of the French National Survey on Nutrition and Health–Initial results. French Institute for Public Health Surveillance, Saint-Maurice, France
Gallagher CM, Meliker JR (2010) Blood and urine cadmium, blood pressure, and hypertension: a systematic review and meta-analysis. Environ Health Perspect 118(12):1676–1684
Gallagher CM, Kovach JS, Meliker JR (2008) Urinary cadmium and osteoporosis in US women ≥ 50 years of age: NHANES 1988–1994 and 1999–2004. Environ Health Perspect 116(10): 1338–1343
Gaschler MM, Stockwell BR (2017) Lipid peroxidation in cell death. BBRC 482:419–425
Genchi G, Sinicropi MS, Lauria G, Carocci A, Catalano A (2020) The effects of cadmium toxicity. Int J Environ Res Public Health 17(11):3782
Goering P, Waalkes M, Klaassen C (1995) Toxicology of cadmium. In: Toxicology of metals: biochemical aspects. Springer, pp 189–214
Gonçalves AL, Pires JC, Simões M (2017) A review on the use of microalgal consortia for wastewater treatment. Algal Res 24:403–415
Grau-Perez M, Pichler G, Galan-Chilet I, Briongos-Figuero LS, Rentero-Garrido P, Lopez-Izquierdo R et al (2017) Urine cadmium levels and albuminuria in a general population from Spain: a gene-environment interaction analysis. Environ Int 106:27–36
Gutsch A, Zouaghi S, Renaut J, Cuypers A, Hausman J-F, Sergeant K (2018) Changes in the proteome of medicago sativa leaves in response to long-term cadmium exposure using a cell-wall targeted approach. Int J Mol Sci 19:2498
Hayat SMG, Bianconi V, Pirro M, Sahebkar A (2019) Stealth functionalization of biomaterials and nanoparticles by cd47 mimicry. Int J Pharm 569:118628
He P, Lu Y, Liang Y, Chen B, Wu M, Li S et al (2013) Exposure assessment of dietary cadmium: findings from Shanghainese over 40 years, China. BMC Public Health 13:1–11
He T, Shen H, Zhu J, Zhu Y, He Y, Li Z, Lu H (2019) Geniposide attenuates cadmium-induced oxidative stress injury via Nrf2 signaling in osteoblasts. Mol Med Rep 20(2):1499–1508
He S, Zhuo L, Cao Y, Liu G, Zhao H, Song R, Liu Z (2020) Effect of cadmium on osteoclast differentiation during bone injury in female mice. Environ Toxicol 35(4):487–494
Heinrichs H, Schulz-Dobrick B, Wedepohl K (1980) Terrestrial geochemistry of Cd, Bi, Tl, Pb, Zn and Rb. Geochim Cosmochim Acta 44(10):1519–1533

Hernández-Cruz EY, Arancibia-Hernández YL, Loyola-Mondragón DY, Pedraza-Chaverri J (2022) Oxidative stress and its role in Cd-induced epigenetic modifications: use of antioxidants as a possible preventive strategy. Oxygen 2(2):177–210

Hocaoğlu-Özyiğit A, Genç BN (2020) Cadmium in plants, humans and the environment. Front Life Sci Relat Technol 1(1):12–21

Hoffmann L, Putzke H-P, Kampehl H-J, Russbült R, Gase P, Simonn C et al (1985) Carcinogenic effects of cadmium on the prostate of the rat. J Cancer Res Clin Oncol 109:193–199

Horiguchi H (2019) Cadmium exposure and its effects on the health status of rice farmers in Akita prefecture. In: Cadmium toxicity: new aspects in human disease, rice contamination, and cytotoxicity. Springer, pp 75–83

Hossein-Khannazer N, Azizi G, Eslami S, Alhassan Mohammed H, Fayyaz F, Hosseinzadeh R et al (2020) The effects of cadmium exposure in the induction of inflammation. Immunopharmacol Immunotoxicol 42(1):1–8

Imura J, Tsuneyama K, Ueda Y (2019) Novel pathological study of cadmium nephropathy of Itai-Itai disease. In: Cadmium toxicity: new aspects in human disease, rice contamination, and cytotoxicity. Springer, pp 39–50

Iradukunda A, Zhang D, Proshad R, Mperejekumana P (2021) A review on cadmium contamination in soil and bioaccumulation by tobacco, its source, toxicity and health risk. Asian J Plant Sci Res 11(5):154–163

Ismael MA, Elyamine AM, Moussa MG, Cai M, Zhao X, Hu C (2019) Cadmium in plants: uptake, toxicity, and its interactions with selenium fertilizers. Metallomics 11:255–277

Järup L, Åkesson A (2009) Current status of cadmium as an environmental health problem. Toxicol Appl Pharmacol 238(3):201–208

Jenardhanan P, Panneerselvam M, Mathur PP (2016) Effect of environmental contaminants on spermatogenesis.. In: Paper presented at the Seminars in cell & developmental biology

Joseph P (2009) Mechanisms of cadmium carcinogenesis. Toxicol Appl Pharmacol 238(3): 272–279

Karazivan P, Dumez V, Flora L, Pomey M-P, Del Grande C, Ghadiri DP et al (2015) The patient-as-partner approach in health care: a conceptual framework for a necessary transition. Acad Med 90(4):437–441

Karunakaran C, Dhanalakshmi R (2009) Selectivity in photocatalysis by particulate semiconductors. Cent Eur J Chem 7:134

Keshav Krishna A, Rama Mohan K (2016) Distribution, correlation, ecological and health risk assessment of heavy metal contamination in surface soils around an industrial area, Hyderabad, India. Environ Earth Sci 75:1–17

Kim M, Wolt J (2011) Probabilistic risk assessment of dietary cadmium in the South Korean population. Food Addit Contam Part A 28:62–70

Kim K, Melough MM, Vance TM, Noh H, Koo SI, Chun OK (2018) Dietary cadmium intake and sources in the US. Nutrients 11(1):2

Kim J, Song H, Lee J, Kim YJ, Chung HS, Yu JM et al (2023) Smoking and passive smoking increases mortality through mediation effect of cadmium exposure in the United States. Sci Rep 13(1):3878

Knez J (2013) Endocrine-disrupting chemicals and male reproductive health. Reprod Biomed Online 26(5):440–448

Knoell DL, Wyatt TA (2021) The adverse impact of cadmium on immune function and lung host defense. In: Paper presented at the Seminars in Cell & Developmental Biology

Koch W, Karim MR, Marzec Z, Miyataka H, Himeno S, Asakawa Y (2016) Dietary intake of metals by the young adult population of eastern Poland: results from a market basket study. J Trace Elem Med Biol 35:36–42

Kodavanti PRS, Loganathan BG (2017) Organohalogen pollutants and human health. In: Quah SR, Cockerham WC (eds) The international encyclopedia of public health, vol 5. Springer, pp 359–366

Kubier A, Wilkin RT, Pichler T (2019) Cadmium in soils and groundwater: a review. J Appl Geochem 108:104388
Kurochkin IO, Etzkorn M, Buchwalter D, Leamy L, Sokolova IM (2011) Top-down control analysis of the cadmium effects on molluscan mitochondria and the mechanisms of cadmium-induced mitochondrial dysfunction. Am J Physiol Regul 300:R21–R31
Laouali N, Benmarhnia T, Lanphear BP, Oulhote Y (2023) Associations with blood lead and urinary cadmium concentrations in relation to mortality in the US population: A causal survival analysis with G-computation. Toxics 11(2):133
Lawless L, Xie L, Zhang K (2023) The inter-and multi-generational epigenetic alterations induced by maternal cadmium exposure. Front Cell Dev Biol 11:1148906
Li X, Zheng S, Wu G (2020) Amino acid metabolism in the kidneys: nutritional and physiological significance. In: Amino acids in nutrition and health: amino acids in systems function and health. Springer, pp 71–95
Li R, Jia Y, Kong X, Nie Y, Deng Y, Liu Y (2022) Novel drug delivery systems and disease models for pulmonary fibrosis. J Control Release 348:95–114
Lin H-C, Hao W-M, Chu P-H (2021) Cadmium and cardiovascular disease: an overview of pathophysiology, epidemiology, therapy, and predictive value. Rev Port Cardiol (Engl Ed) 40(8):611–617
Lokhande B, Patil P, Uplane M (2004) Studies on cadmium oxide sprayed thin films deposited through nonaqueous medium. Mater Chem Phys 84:238–242
Lordan R, Zabetakis I (2022) Cadmium: A focus on the Brown crab (Cancer pagurus) industry and potential human health risks. Toxics 10(10):591
Ma Y, Su Q, Yue C, Zou H, Zhu J, Zhao H et al (2022) The effect of oxidative stress-induced autophagy by cadmium exposure in kidney, liver, and bone damage, and neurotoxicity. Int J Mol Sci 23(21):13491
Malhotra J, Malvezzi M, Negri E, La Vecchia C, Boffetta P (2016) Risk factors for lung cancer worldwide. Eur Respir J 48(3):889–902
Malin AJ, Wright RO (2018) The developmental neurotoxicity of cadmium, Handbook of developmental neurotoxicology. Elsevier, pp 407–412
Mao W, Zhang N, Zhou F, Li W, Liu H, Feng J, Zhou L, Wei C, Pan Y, He Z (2011) Cadmium directly induced mitochondrial dysfunction of human embryonic kidney cells. HET 30:920–929
Marzec Z, Łukasiewicz M (2010) Cadmium, lead and nickel in food rations of the canteens of welfare institutions. Bromat Chem Toksykol 3:281–286
Matović V, Buha A, Bulat Z, Đukić-Ćosić D (2011) Cadmium toxicity revisited: focus on oxidative stress induction and interactions with zinc and magnesium. Arhiv za higijenu rada i toksikologiju 62:65–75
Menke A, Muntner P, Silbergeld EK, Platz EA, Guallar E (2009) Cadmium levels in urine and mortality among US adults. Environ Health Perspect 117(2):190–196
Messelmani T, Morisseau L, Sakai Y, Legallais C, Le Goff A, Leclerc E, Jellali R (2022) Liver organ-on-chip models for toxicity studies and risk assessment. Lab Chip 22:2423
Messner B, Türkcan A, Ploner C, Laufer G, Bernhard D (2016) Cadmium overkill: autophagy, apoptosis and necrosis signalling in endothelial cells exposed to cadmium. Cell Mol Life Sci 73:1699–1713
Mężyńska M, Brzóska MM, Rogalska J, Piłat-Marcinkiewicz B (2018) Extract from aronia melanocarpa l. Berries prevents cadmium-induced oxidative stress in the liver: a study in a rat model of low-level and moderate lifetime human exposure to this toxic metal. Nutrients 11:21
Monteiro C, Ferreira De Oliveira JMP, Pinho F, Bastos V, Oliveira H, Peixoto F, Santos C (2018) Biochemical and transcriptional analyses of cadmium-induced mitochondrial dysfunction and oxidative stress in human osteoblasts. J Toxicol Environ Health Part A 81:705–717
Moon C-S, Yang H-R, Nakatsuka H, Ikeda M (2016) Time trend of cadmium intake in Korea. Environ Health Prev Med 21:118–128
Munisamy R, Ismail SNS, Praveena SM (2013) Cadmium exposure via food crops: a case study of intensive farming area. Am J Appl Sci 10:1252–1262

Nasirzadeh L, Kvarnheden A, Sorkhilaleloo B, Hervan EM, Fatehi F (2022) Foliar-applied selenium nanoparticles can alleviate soil-cadmium stress through physio-chemical and stomatal changes to optimize yield, antioxidant capacity, and fatty acid profile of wheat (triticum aestivum l.). J Soil Sci Plant Nutr 22:2469–2480

Nazar R, Iqbal N, Masood A, Khan MIR, Syeed S, Khan NA (2012) Cadmium toxicity in plants and role of mineral nutrients in its alleviation

Nazima B, Manoharan V, Miltonprabu S (2016) Oxidative stress induced by cadmium in the plasma, erythrocytes and lymphocytes of rats: attenuation by grape seed proanthocyanidins. HET 35:428–447

Nguyen NT, Umbaugh DS, Smith S, Adelusi OB, Sanchez-Guerrero G, Ramachandran A, Jaeschke H (2023) Dose-dependent pleiotropic role of neutrophils during acetaminophen-induced liver injury in male and female mice. Arch Toxicol 97:1–16

Niede R, Benbi DK (2022) Integrated review of the nexus between toxic elements in the environment and human health. AIMS Public Health 9(4):758

Ning B, Yu T, Zhang S, Huang Z, Tian D, Lin Z, Niu A, Golden N, Hensley K, Threeton B (2021) A smartphone-read ultrasensitive and quantitative saliva test for covid-19. Sci Adv 7:3703

Niture S, Lin M, Qi Q, Moore JT, Levine KE, Fernando RA, Kumar D (2021) Role of autophagy in cadmium-induced hepatotoxicity and liver diseases. J Toxicol 2021:9564297

Nordkap L, Joensen UN, Jensen MB, Jørgensen N (2012) Regional differences and temporal trends in male reproductive health disorders: semen quality may be a sensitive marker of environmental exposures. Mol Cell Endocrinol 355(2):221–230

Nzengue Y, Steiman R, Rachidi W, Favier A, Guiraud P (2012) Oxidative stress induced by cadmium in the c6 cell line: role of copper and zinc. Biol Trace Elem Res 146:410–419

Oldiges H, Hochrainer D, Takenaka S, Oberdörster G, König H (1984) Lung carcinomas in rats after low level cadmium inhalation. Toxicol Environ Chem 9(1):41–51

Olsson I-M, Bensryd I, Lundh T, Ottosson H, Skerfving S, Oskarsson A (2002) Cadmium in blood and urine--impact of sex, age, dietary intake, iron status, and former smoking--association of renal effects. Environ Health Perspect 110(12):1185–1190

Paithankar JG, Saini S, Dwivedi S, Sharma A, Chowdhuri DK (2021) Heavy metal associated health hazards: an interplay of oxidative stress and signal transduction. Chemosphere 262: 128350

Patra R, Rautray AK, Swarup D (2011) Oxidative stress in lead and cadmium toxicity and its amelioration. Vet Med Int

Park E, Kim J, Kim B, Park EY (2021) Association between environmental exposure to cadmium and risk of suspected non-alcoholic fatty liver disease. Chemosphere 266:128947

Pietrzak S, Wójcik J, Baszuk P, Marciniak W, Wojtyś M, Dębniak T et al (2021) Influence of the levels of arsenic, cadmium, mercury and lead on overall survival in lung cancer. Biomol Ther 11(8):1160

Program NT (2000) Ninth report on carcinogens. National Toxicology Program, Research Triangle Park, NC

Pyatha S, Kim H, Lee D, Kim K (2022) Association between heavy metal exposure and Parkinson's disease: a review of the mechanisms related to oxidative stress. Antioxidants 11(12):2467

Pysz K, Leszczyńska T, Bieżanowska-Kopeć R, Kopeć A (2016) Chemical assessment of lead, cadmium, nitrate, and nitrite intakes with daily diets of children and adolescents from orphanages in Krakow, Poland. Environ Sci Pollut Res 23:25200–25209

Rani A, Kumar A, Lal A, Pant M (2014) Cellular mechanisms of cadmium-induced toxicity: a review. Int J Environ Health Res 24:378–399

Rao RAK, Kashifuddin M (2016) Adsorption studies of Cd(II) on ball clay: comparison with other natural clays. Arab J Chem 9:S1233–S1241

Rennke H, Denker B (2020) Renal pathology: the essentials. Wolters Kluwer, New York

Repić A, Bulat P, Antonijević B, Antunović M, Džudović J, Buha A, Bulat Z (2020) The influence of smoking habits on cadmium and lead blood levels in the Serbian adult people. Environ Sci Pollut Res 27:751–760

Reshma V, Mohanan P (2019) Quantum dots: applications and safety consequences. J Lumin 205: 287–298

Reyes-Hinojosa D, Lozada-Pérez C, Cuevas YZ, López-Reyes A, Martínez-Nava G, Fernández-Torres J et al (2019) Toxicity of cadmium in musculoskeletal diseases. Environ Toxicol Pharmacol 72:103219

Roe FJ (1995) IARC monographs on the evaluation of carcinogenic risk to humans: beryllium, cadmium, mercury and exposures in the glass manufacturing industry, volume 58: International Agency for Research on Cancer, Lyon, France, 1993 (ISBN 92 832 1258 4.), 444 pp

Rogalska J, Brzóska MM, Roszczenko A, Moniuszko-Jakoniuk J (2009) Enhanced zinc consumption prevents cadmium-induced alterations in lipid metabolism in male rats. Chem Biol Interact 177:142–152

Ruddle NH, Akirav EM (2009) Secondary lymphoid organs: responding to genetic and environmental cues in ontogeny and the immune response. J Immunol 183(4):2205–2212

Ryznar RJ, Phibbs L, Van Winkle LJ (2021) Epigenetic modifications at the center of the barker hypothesis and their transgenerational implications. IJERPH 18:12728

Saleh SR, Kandeel MM, Ghareeb D, Ghoneim TM, Talha NI, Alaoui-Sossé B, Aleya L, Abdel-Daim MM (2020) Wheat biological responses to stress caused by cadmium, nickel and lead. Sci Total Environ 706:136013

Salton F, Volpe MC, Confalonieri M (2019) Epithelial–mesenchymal transition in the pathogenesis of idiopathic pulmonary fibrosis. Medicina 55(4):83

Sand S, Becker W (2012) Assessment of dietary cadmium exposure in Sweden and population health concern including scenario analysis. Food Chem Toxicol 50(3–4):536–544

Satarug S (2018) Dietary cadmium intake and its effects on kidneys. Toxics 6(1):15

Satarug S (2019) Cadmium sources and toxicity, vol 7. MDPI, p 25

Satarug S, Swaddiwudhipong W, Ruangyuttikarn W, Nishijo M, Ruiz P (2013) Modeling cadmium exposures in low-and high-exposure areas in Thailand. Environ Health Perspect 121(5): 531–536

Satarug S, Vesey DA, Gobe GC, Phelps KR (2023) Estimation of health risks associated with dietary cadmium exposure. Arch Toxicol 97(2):329–358

Shahriar S, Rahman MM, Naidu R (2020) Geographical variation of cadmium in commercial rice brands in Bangladesh, Human health risk assessment. Sci Total Environ 716,:137049

Shi Z, Carey M, Meharg C, Williams PN, Signes-Pastor AJ, Triwardhani EA et al (2020) Rice grain cadmium concentrations in the global supply-chain. In: Exposure and health, vol 12. Springer, pp 869–876

Sinicropi MS, Amantea D, Caruso A, Saturnino C (2010) Chemical and biological properties of toxic metals and use of chelating agents for the pharmacological treatment of metal poisoning. Arch Toxicol 84:501–520

Sirot V, Samieri C, Volatier J-L, Leblanc J-C (2008) Cadmium dietary intake and biomarker data in French high seafood consumers. J Expo Sci Environ Epidemiol 18(4):400–409

Skogheim TS, Weyde KVF, Engel SM, Aase H, Surén P, Øie MG et al (2021) Metal and essential element concentrations during pregnancy and associations with autism spectrum disorder and attention-deficit/hyperactivity disorder in children. Environ Int 152:106468

Song C, Wu L, Xie Y, He J, Chen X, Wang T, Lin Y, Jin T, Wang A, Liu Y (2017) Air pollution in china: status and spatiotemporal variations. Environ Pollut 227:334–347

Souza-Arroyo V, Fabián JJ, Bucio-Ortiz L, Miranda-Labra RU, Gomez-Quiroz LE, Gutiérrez-Ruiz MC (2022) The mechanism of the cadmium-induced toxicity and cellular response in the liver. Toxicology 2022:153339

Sun H, Wang D, Zhou Z, Ding Z, Chen X, Xu Y et al (2016) Association of cadmium in urine and blood with age in a general population with low environmental exposure. Chemosphere 156: 392–397

Sun Q, Li Y, Shi L, Hussain R, Mehmood K, Tang Z, Zhang H (2022) Heavy metals induced mitochondrial dysfunction in animals: molecular mechanism of toxicity. Toxicology 2022: 153136

Suwazono Y, Nogawa K, Uetani M, Kido T, Nakagawa H (2011) Reassessment of the threshold of urinary cadmium by using hybrid approach in a cadmium non-polluted area in Japan. Int J Hyg Environ Health 214(2):175–178

Taha MM, Mahdy-Abdallah H, Shahy EM, Ibrahim KS, Elserougy S (2018) Impact of occupational cadmium exposure on bone in sewage workers. Int J Occup Environ Health 24(3–4):101–108

Thompson J, Bannigan J (2008) Cadmium: toxic effects on the reproductive system and the embryo. Reprod Toxicol 25:304–315

Tinkov AA, Filippini T, Ajsuvakova OP, Skalnaya MG, Aaseth J, Bjørklund G et al (2018) Cadmium and atherosclerosis: A review of toxicological mechanisms and a meta-analysis of epidemiologic studies. Environ Res 162:240–260

Udi OA, Efekemo O, Orororo OC (2022) Changes in liver Histomorphology, hematological parameters and lipid profile of cadmium-exposed rats treated with combined leaf extract of Vernonia amygdalina and Occimum gratissimum. Asian J Med Health 20(11):195–203

Vacchi-Suzzi C, Kruse D, Harrington J, Levine K, Meliker JR (2016) Is urinary cadmium a biomarker of long-term exposure in humans? A review. Curr Environ Health Rep 3:450–458

Valko M, Rhodes C, Moncol J, Izakovic M, Mazur M (2006) Free radicals, metals and antioxidants in oxidative stress-induced cancer. Chem Biol Interact 160:1–40

Waalkes MP (2000) Cadmium carcinogenesis in review. J Inorg Biochem 79(1–4):241–244

Waalkes MP, Rehm S (1992) Carcinogenicity of oral cadmium in the male Wistar (WFNCr) rat: effect of chronic dietary zinc deficiency. Fundam Appl Toxicol 19(4):512–520

Waalkes MP, Rehm S (1994) Cadmium and prostate cancer. J Toxicol Environ Health 43(3): 251–269

Waalkes MP, Rehm S, Riggs CW, Bare RM, Devor DE, Poirier LA et al (1988) Cadmium carcinogenesis in male Wistar [Crl:(WI) BR] rats: dose-response analysis of tumor induction in the prostate and testes and at the injection site. Cancer Res 48(16):4656–4663

Waalkes MP, Anver M, Diwan BA (1999) Carcinogenic effects of cadmium in the noble (NBL/Cr) rat: induction of pituitary, testicular, and injection site tumors and intraepithelial proliferative lesions of the dorsolateral prostate. Toxicol Sci 52(2):154–161

Wang L, He L, Mishra A, Li C (2009) Active contours driven by local gaussian distribution fitting energy. IEEE Signal Process Lett 89:2435–2447

Wang B, Li Y, Shao C, Tan Y, Cai L (2012) Cadmium and its epigenetic effects. Curr Med Chem 19:2611–2620

Wang X, Mukherjee B, Park SK (2018) Associations of cumulative exposure to heavy metal mixtures with obesity and its comorbidities among US adults in NHANES 2003–2014. Environ Int 121:683–694

Wang H, Abel GM, Storm DR, Xia Z (2019) Cadmium exposure impairs adult hippocampal neurogenesis. Toxicol Sci 171(2):501–514

Wang C, Cheng T, Liu H, Zhou F, Zhang J, Zhang M, Liu X, Shi W, Cao T (2021) Nano-selenium controlled cadmium accumulation and improved photosynthesis in indica rice cultivated in lead and cadmium combined paddy soils. J Environ Sci 103:336–346

Wang W, Silva LM, Wang HH, Kavanaugh MA, Pottorf TS, Allard BA et al (2022) Ttc21b deficiency attenuates autosomal dominant polycystic kidney disease in a kidney tubular-and maturation-dependent manner. Kidney Int 102(3):577–591

Watanabe K, Koguchi N, Gotoh Y (2000) Fabrication of GaAs quantum dots by modified droplet epitaxy. Jpn J Appl Phys 39:L79

WHO (2011) Who report on the global tobacco epidemic: warning about the dangers of tobacco: World Health Organization

Wojciechowska-Mazurek M, Starska K, Brulińska-Ostrowska E, Plewa M, Biernat U, Karłowski K (2008) Monitoring of contamination of foodstuffs with elements noxious to human health. Part I. Wheat cereal products, vegetable products, confectionery and products for infants and children (2004 year). Rocz Panstw Zakl Hig 59(3):251–266

Yan L-J, Allen DC (2021) Cadmium-induced kidney injury: oxidative damage as a unifying mechanism. Biomol Ther 11(11):1575

Yang H, Wu Z, Li X, Li H, Wang L (2015) Metal binding feature of copper-induced metallothionein from fresh water crab sinopotamon henanense reveals its Cu-thionein character. Available at SSRN 4500261

You X, Liu L, Li X, Du H, Nie D, Zhang X, Tong H, Wu M, Gao Y, Liao Z (2018) Immune response of interferon-γ-inducible lysosomal thiol reductase (gilt) from chinese sturgeon (acipenser sinensis) to microbial invasion and its antioxdative activity in lipopolysaccharides-treated mammalian dentritic cells. Fish Shellfish Immunol 72:356–366

Yu D, Zhang L, Yu G, Nong C, Lei M, Tang J et al (2020) Association of liver and kidney functions with klotho gene methylation in a population environment exposed to cadmium in China. Int J Environ Health Res 30(1):38–48

Zhao X, Cheng Z, Zhu Y, Li S, Zhang L, Luo Y (2015) Effects of paternal cadmium exposure on the sperm quality of male rats and the neurobehavioral system of their offspring. Exp Ther Med 10:2356–2360

Zhao L, Islam R, Wang Y, Zhang X, Liu L-Z (2022a) Epigenetic regulation in chromium-, nickel- and cadmium-induced carcinogenesis. J Cancer 14:5768

Zhao Y, Dong BR, Hao Q (2022b) Probiotics for preventing acute upper respiratory tract infections. Cochrane database of systematic reviews

Zhou X, Hao W, Shi H, Hou Y, Xu Q (2015) Calcium homeostasis disruption-a bridge connecting cadmium-induced apoptosis, autophagy and tumorigenesis. Oncol Res Treat 38:311–315

Zhu J-Q, Liu Y, Zhang J-H, Liu Y-F, Cao J-Q, Huang Z-T, Yuan Y, Bian J-C, Liu Z-P (2018) Cadmium exposure of female mice impairs the meiotic maturation of oocytes and subsequent embryonic development. Toxicol Sci 164:289–299

Zhu J, Huang Z, Yang F, Zhu M, Cao J, Chen J, Lin Y, Guo S, Li J, Liu Z (2021) Cadmium disturbs epigenetic modification and induces DNA damage in mouse preimplantation embryos. Ecotoxicol Environ Saf 219:112306

Cadmium-Induced Neurotoxicity

Sridhar Dumpala, Kakarlapudi Ramaneswari, and Vivek Chintada

Abstract Cadmium-induced neurotoxicity has become a growing concern due to its potential adverse effects on the central nervous system. This chapter provides a comprehensive exploration of the mechanisms and implications of cadmium exposure on neuronal function and behavior. The chapter begins with an introduction to the topic, highlighting the significance of cadmium-induced neurotoxicity and the need for further understanding. It then delves into the various sources of cadmium, including contaminated soil, water, and industrial emissions, as well as the different routes of exposure, such as inhalation, ingestion, and dermal contact. The accumulation of cadmium in the central nervous system is discussed in detail. The cellular and molecular mechanisms underlying cadmium neurotoxicity are explored, including the entry of cadmium into neuronal cells, its accumulation in different brain regions, and the induction of oxidative stress. The disruption of calcium homeostasis, mitochondrial dysfunction, activation of apoptotic pathways, and the role of inflammation are examined as key contributors to the neurotoxic effects of cadmium. The neurobehavioral consequences of prolonged cadmium exposure are presented, with a focus on cognitive impairment, learning and memory deficits, and behavioral abnormalities observed in both animals and humans. The impact of cadmium on neurodevelopment during critical periods is discussed, including alterations in brain structure and function. The vulnerability to cadmium neurotoxicity in specific populations, such as children, pregnant women, and the elderly, is investigated. The lifelong implications of cadmium exposure during critical brain development stages are examined, along with the interaction between cadmium exposure and other risk factors that increase susceptibility to neurotoxic effects. Emerging strategies for preventing or mitigating cadmium neurotoxicity are presented, including the potential use of chelating agents to enhance cadmium excretion. The importance of environmental and occupational regulations in minimizing exposure risks is

S. Dumpala · K. Ramaneswari
Department of Aquaculture, Adikavi Nannaya University, Rajamahendravaram, Andhra Pradesh, India

V. Chintada (✉)
Department of Zoology, Sri Venkateswara University, Tirupati, Andhra Pradesh, India

A. K. Jha, N. Kumar (eds.), *Cadmium Toxicity Mitigation*,
https://doi.org/10.1007/978-3-031-47390-6_4

discussed, as well as the role of nutritional interventions, such as antioxidants and essential nutrients, in attenuating cadmium-induced neurotoxicity.

Keywords Cadmium sources · Cadmium toxicity · Bioavailability · Target organs · Neurotoxicity

1 Introduction

Cadmium-induced neurotoxicity is a subject of increasing significance and concern due to the potential adverse effects it can have on human health. Cadmium, a heavy metal, is a widespread environmental pollutant that is released into the air, water, and soil through various industrial processes, such as mining, smelting, and battery production (ATSDR 2012; Shayan et al. 2023). The accumulation of cadmium in the environment, coupled with its long biological half-life, raises concerns about the impact of cadmium exposure on human health, particularly the nervous system. Numerous studies have highlighted the neurotoxic effects of cadmium exposure, leading to a growing concern among scientists, health professionals, and policymakers. Cadmium is known to readily accumulate in the central nervous system and has been associated with the impairment of cognitive function, altered behavior, and neurodevelopmental disorders (ATSDR 2012; Julvez et al. 2016; Srivastava et al. 2023). The persistent levels of cadmium found in the environment and its ability to cross the blood–brain barrier pose significant risks to neurological health.

This chapter aims to provide a comprehensive understanding of cadmium-induced neurotoxicity, its significance, and the potential mechanisms underlying its adverse effects on the nervous system. We will explore the various factors associated with cadmium exposure, such as occupational exposure, contaminated food and water sources, and environmental pollution. Additionally, we will examine the pathways through which cadmium enters the body, including inhalation, ingestion, and dermal absorption. Furthermore, this chapter will delve into the potential mechanisms of cadmium-induced neurotoxicity. Cadmium is known to exert its toxic effects through oxidative stress, disruption of calcium homeostasis, mitochondrial dysfunction, and induction of inflammation (ATSDR 2012; Julvez et al. 2016; Rajabian et al. 2023). Understanding these mechanisms is crucial for identifying potential targets for intervention and developing effective preventive strategies.

The neurological effects of cadmium exposure, both acute and chronic, will also be discussed comprehensively in this chapter. Acute cadmium exposure can lead to symptoms such as headaches, dizziness, and irritability, while chronic exposure has been associated with cognitive impairment, neurobehavioral changes, and neurodevelopmental disorders (ATSDR 2012; Julvez et al. 2016). The underlying mechanisms for these effects will be explored in detail, providing insights into the potential risk factors and consequences of cadmium-induced neurotoxicity.

In conclusion, cadmium-induced neurotoxicity is a vital area of research due to the potential neurological impairments associated with cadmium exposure. This chapter aims to provide a comprehensive overview of the subject, including the significance, growing concern, potential mechanisms, and neurological effects of cadmium-induced neurotoxicity.

2 Sources and Routes of Cadmium Exposure

Cadmium is a highly toxic heavy metal that can be found in various environmental sources and can enter the human body through different exposure routes. This discussion will explore the sources of cadmium, including contaminated soil, water, and industrial emissions, as well as the different routes of exposure, such as inhalation, ingestion, and dermal contact. Furthermore, we will delve into how these exposure pathways can contribute to the accumulation of cadmium in the body, particularly in the central nervous system.

2.1 Sources of Cadmium

Contaminated Soil: One of the significant sources of cadmium exposure is contaminated soil. Cadmium naturally occurs in the Earth's crust but has also been introduced into the environment through anthropogenic activities, such as mining, smelting operations, and the use of fertilizers containing cadmium. Contaminated soil can subsequently contaminate crops and plants, thus entering the food chain and becoming a potential source of exposure for humans (Dooyema et al. 2012; Al Marshoudi et al. 2023).

Water: Cadmium can leach into water sources from contaminated soil or through industrial discharges. Industrial activities like mining, battery manufacturing, and metal plating can release cadmium into water bodies, making them susceptible to contamination. Consuming water contaminated with cadmium or using it for irrigation purposes can introduce cadmium into the human body (Agency for Toxic Substances and Disease Registry 2012).

Industrial Emissions: Industrial emissions, particularly from combustion processes such as fossil fuel combustion and waste incineration, can release cadmium particles into the air. These particles can travel long distances in the atmosphere and eventually settle onto plants, soil, and water bodies. Therefore, communities living in proximity to industrial areas or those in urban environments may experience higher exposure to cadmium through inhalation routes (Bernhoft and Hauge 2013).

2.2 *Routes of Cadmium Exposure*

Inhalation: Inhalation of cadmium particles or fumes is a significant route of exposure, particularly in occupational settings. Occupations involving metal processing, battery manufacturing, welding, and the production of pigments, plastics, and ceramics carry a higher risk of airborne cadmium exposure. Workers inhaling cadmium particles can absorb it directly into their bloodstream, allowing it to circulate throughout the body (ATSDR 2012).

Ingestion: Ingestion of cadmium-contaminated food and water is another primary route of exposure. As mentioned earlier, cadmium can contaminate crops, especially those grown on cadmium-contaminated soil or irrigated with cadmium-contaminated water. Certain seafoods, including shellfish and liver, are known to accumulate high levels of cadmium due to bioaccumulation in marine ecosystems. Additionally, the consumption of food and drinks stored or cooked in cadmium-coated containers can lead to exposure (Järup and Åkesson 2009).

Dermal Contact: Dermal contact with cadmium can occur through direct contact with cadmium-containing substances or contaminated soil. Workers handling cadmium-containing materials without proper protective measures may face dermal absorption of the heavy metal. Moreover, individuals coming into contact with cadmium-contaminated soil during activities like gardening or playing in contaminated areas can also absorb cadmium through their skin (Radauceanu et al. 2015).

3 Contribution to Cadmium Accumulation in the Central Nervous System

Cadmium accumulates in various organs, with the kidneys being the primary target due to their high affinity for cadmium. However, cadmium is also known to accumulate in the central nervous system (CNS), including the brain. The precise mechanisms through which cadmium accumulates in the CNS are complex and not yet fully understood. However, studies have shown that after entering the bloodstream, cadmium can cross the blood–brain barrier, a protective barrier separating the circulating blood from the brain, and accumulate in the brain tissue (Klaassen 2019).

Once in the CNS, cadmium can interfere with various cellular processes, disrupt neurotransmitter signaling, and promote oxidative stress, leading to neurotoxicity. Chronic exposure to cadmium has been linked to neurobehavioral disorders, cognitive impairments, and an increased risk of neurodegenerative diseases such as Parkinson's and Alzheimer's diseases (Tinkov et al. 2018a, b; Al Marshoudi et al. 2023).

In conclusion, cadmium can be sourced from contaminated soil, water, and industrial emissions. Inhalation, ingestion, and dermal contact are the primary routes through which cadmium enters the human body. These exposure pathways contribute to the accumulation of cadmium in various organs, including the central nervous system. Therefore, it is crucial to mitigate the sources of cadmium and minimize exposure to prevent adverse health effects.

4 Cellular and Molecular Mechanisms of Cadmium Neurotoxicity: A Comprehensive Analysis

Cadmium neurotoxicity is a significant concern due to its detrimental effects on the central nervous system. Understanding the cellular and molecular mechanisms associated with cadmium-induced neurotoxicity is crucial for developing effective preventive and therapeutic strategies. This note aims to provide an in-depth analysis of the cellular and molecular mechanisms underlying cadmium neurotoxicity, including cadmium entry and accumulation, oxidative stress, disrupted calcium homeostasis, mitochondrial dysfunction, apoptotic pathways, and inflammation.

4.1 Cadmium Entry and Accumulation

Cadmium gains entry into neuronal cells through various uptake mechanisms, including divalent metal transporters (DMTs) and calcium channels. Research suggests that certain proteins, such as ZIP and ZnT transporters, may also contribute to cadmium uptake in neurons (Liu et al. 2020a, b). Once inside the cells, cadmium can accumulate in different brain regions due to its affinity for binding to cytoplasmic components, including metallothioneins and heat shock proteins (Goyer and Cherian 1983; da Silva et al. 2023). Studies have shown that cadmium accumulates specifically in regions associated with cognitive functions, such as the hippocampus and prefrontal cortex (Kalia and Kumar 2020).

4.2 Oxidative Stress

Cadmium-induced oxidative stress is a critical mechanism underlying neurotoxicity. Upon entering neuronal cells, cadmium stimulates the production of reactive oxygen species (ROS) while inhibiting antioxidant defense mechanisms (Valko et al. 2006). The resulting imbalance between ROS production and elimination leads to oxidative damage to lipids, proteins, and DNA (Wang et al. 2016). Notably, cadmium-induced ROS generation occurs via various pathways, including the activation of nicotinamide adenine dinucleotide phosphate (NADPH) oxidase, dysregulation of mitochondrial electron transport chain, and depletion of glutathione (Li et al. 2017).

4.3 Disrupted Calcium Homeostasis

Cadmium disrupts calcium homeostasis in neuronal cells, resulting in impaired intracellular signaling and cellular functions. It can interfere with calcium channels, modulating calcium influx into the cytoplasm and affecting neurotransmitter release, synaptic plasticity, and neuronal activity (Aschner et al. 2001). Moreover, cadmium impairs the activity of calcium pumps and transporters, leading to increased cytoplasmic calcium levels and disturbances in intracellular calcium dynamics (Fränkenhuysen et al. 2012; Olaolu et al. 2023).

4.4 Mitochondrial Dysfunction

Cadmium-induced mitochondrial dysfunction significantly contributes to neurotoxicity. Cadmium disrupts mitochondrial structure and impairs energy metabolism, leading to decreased ATP production and increased generation of ROS (Fu et al. 2021). Moreover, cadmium-induced mitochondrial dysfunction affects mitochondrial membrane potential, calcium handling, and electron transport chain complexes, further enhancing oxidative stress and cellular damage (Toullec et al. 2017).

4.5 Apoptotic Pathways and Neuronal Cell Death

The activation of apoptotic pathways is a notable consequence of cadmium neurotoxicity. Cadmium triggers intrinsic apoptotic pathways, involving the release of cytochrome c from mitochondria, the activation of caspase cascades, DNA damage, and nuclear fragmentation (Zhang et al. 2007; Rajabian et al. 2023). Additionally, cadmium can activate extrinsic apoptotic pathways through death receptors such as Fas, leading to apoptosis via the activation of caspase-8 (Chen et al. 2019). These apoptotic processes ultimately contribute to the loss of neuronal cells and neurodegeneration.

4.6 Role of Inflammation

Inflammation plays a crucial role in cadmium-induced neurotoxicity. Cadmium exposure activates microglial cells, the resident immune cells in the brain, promoting the production of pro-inflammatory cytokines (Liu et al. 2020a, b). These cytokines, such as interleukin-1 beta (IL-1β), tumor necrosis factor-alpha (TNF-α), and interleukin-6 (IL-6), further exacerbate oxidative stress, contribute to neuronal damage, and perpetuate the neuroinflammatory response (Kim et al. 2018).

Understanding the cellular and molecular mechanisms of cadmium neurotoxicity provides valuable insights into the pathogenesis of neurological disorders associated with cadmium exposure. The processes involve cadmium entry and accumulation in neuronal cells, oxidative stress, disrupted calcium homeostasis, mitochondrial dysfunction, activation of apoptotic pathways, and inflammation. Further research is essential to uncover additional molecular events and develop targeted interventions to protect against cadmium-induced neurotoxicity.

5 Neurobehavioral Effects of Cadmium Exposure

Cadmium is a toxic heavy metal present in various environmental and occupational settings, posing a significant risk to human health. Prolonged exposure to cadmium has been linked to a wide range of health issues, including neurobehavioral consequences. In this note, we will present and analyze the evidence surrounding the neurobehavioral effects of cadmium exposure, focusing on cognitive impairment, learning and memory deficits, and behavioral abnormalities. We will also discuss the impact of cadmium on neurodevelopmental processes, including alterations in brain structure and function during early life stages.

5.1 Cognitive Impairment

Numerous studies have documented the negative impact of cadmium exposure on cognitive function. A study conducted by Tellez-Plaza et al. (2011) found that higher levels of urinary cadmium were associated with decreased cognitive function in adults. Similarly, a study by Chen et al. (2014) revealed that cadmium exposure significantly impaired cognitive performance in rats.

5.2 Learning and Memory Deficits

Exposure to cadmium has also been linked to learning and memory deficits. A study by Eto et al. (2017) investigated the effects of long-term cadmium exposure on spatial learning and memory in mice. The results demonstrated that cadmium-exposed mice displayed significant impairments in spatial memory compared to control mice. Additionally, a study by Wang et al. (2019) reported that chronic cadmium exposure in rats resulted in impaired learning and memory performance, suggesting that cadmium exposure may disrupt these crucial cognitive processes.

5.3 Behavioral Abnormalities

Studies have indicated a correlation between cadmium exposure and behavioral abnormalities. A study conducted by Faroon et al. (2012) examined the association between cadmium exposure and hyperactivity in children. The findings revealed that higher urinary cadmium levels were associated with increased hyperactivity. Furthermore, a study by Kobal et al. (2015) demonstrated that cadmium-exposed mice exhibited increased anxiety-like behavior and reduced exploratory behavior.

5.4 Impact on Neurodevelopmental Processes

Cadmium exposure during early life stages has been shown to disrupt neurodevelopmental processes. A study by Goyer (2017) highlighted the teratogenic effects of cadmium exposure on the developing brain, leading to various anatomical and functional alterations. The research revealed that cadmium exposure during critical periods of brain development resulted in structural changes in the hippocampus, cortex, and other brain regions.

Furthermore, studies in animals have shown that cadmium exposure during early life stages leads to deficits in neurobehavioral development. Hernández et al. (2019) conducted a study on rats, demonstrating that cadmium exposure during infancy resulted in impaired motor coordination and altered exploratory behavior. These findings suggest that exposure to cadmium during crucial developmental periods can have long-lasting effects on neurobehavioral outcomes.

The evidence regarding the neurobehavioral effects of cadmium exposure supports the notion that prolonged exposure to this toxic heavy metal can lead to cognitive impairment, learning and memory deficits, and behavioral abnormalities. Additionally, cadmium exposure during early life stages can cause significant disruptions in neurodevelopmental processes and result in long-lasting consequences on brain structure and function. These findings underscore the importance of minimizing cadmium exposure to safeguard neurological health.

6 Cadmium-Induced Neurotoxicity in Specific Populations

Cadmium is a toxic heavy metal that is widely distributed in the environment due to both natural and anthropogenic sources. Exposure to cadmium has been linked to various adverse health effects, including neurotoxicity. However, certain populations, such as children, pregnant women, and the elderly, may be more vulnerable to the neurotoxic effects of cadmium due to a combination of genetic, developmental, and physiological factors (Yousuf et al. 2023). This note aims to investigate the vulnerability of specific populations to cadmium neurotoxicity, discuss the developmental and lifelong implications of cadmium exposure during critical periods of brain development, and examine the interaction between cadmium exposure and other risk factors, leading to increased susceptibility to neurotoxic effects.

Vulnerability of Specific Populations

A. **Children:**

Children are particularly susceptible to the neurotoxic effects of cadmium due to their higher intake and absorption rates, immature metabolic systems, and longer duration of exposure. Their hand-to-mouth behavior, combined with the presence of cadmium in various consumer products, contaminated soils, and some food sources, increases the risk of exposure (Patel et al. 2020). Furthermore, the developing brain is more sensitive to toxic insults, and exposure to cadmium during critical periods of brain development may have profound and long-lasting effects on neurodevelopment (Sara et al. 2016).

B. **Pregnant women:**

Pregnant women represent another vulnerable population due to the transfer of cadmium across the placenta. The placenta does not efficiently regulate cadmium levels, allowing the metal to penetrate the fetal brain. Cadmium accumulation in the developing brain can disrupt normal neurodevelopment and lead to cognitive impairments in offspring (Taweel et al. 2017). Additionally, cadmium exposure during pregnancy has been associated with adverse birth outcomes, such as low birth weight and preterm delivery (Antonisamy et al. 2019).

C. **The elderly:**

The elderly may also be at increased risk of cadmium-induced neurotoxicity due to age-related changes in metabolism and reduced detoxification mechanisms. Additionally, the accumulation of cadmium over a lifetime can result in a higher body burden and increased susceptibility to neurotoxic effects (Apostoli et al. 2019). Elderly individuals may also have compromised blood–brain barrier function, which can facilitate the entry of cadmium into the brain (Senanayake and Lu 2010).

D. **Developmental and lifelong implications:**

Exposure to cadmium during critical periods of brain development can have long-term consequences on cognitive function and behavior. Animal studies have shown that cadmium exposure during gestation and lactation can impair learning and memory, alter neurotransmitter levels, and disrupt the structure and function of the hippocampus, a brain region crucial for memory formation (Sara et al. 2016). In humans, prenatal and early life cadmium exposure has been associated with decreased cognitive performance and increased risk of neurodevelopmental disorders, such as attention-deficit/hyperactivity disorder (ADHD) and autism spectrum disorder (Mukherjee et al. 2017; Navas-Acien et al. 2007; Yousuf et al. 2023).

E. **Interaction with other risk factors:**

The interaction between cadmium exposure and other risk factors can further increase susceptibility to the neurotoxic effects. Genetic factors play a significant role in determining individual susceptibility to cadmium toxicity. Polymorphisms in genes involved in cadmium metabolism and detoxification pathways can affect an individual's ability to eliminate cadmium and mitigate its adverse effects (Schaumlöffel 2000). Additionally, dietary factors, such as inadequate intake of essential minerals like calcium and zinc, can exacerbate the neurotoxic effects of cadmium (Bernard 2015a, b). The combination of genetic predisposition, poor nutrition, and exposure to environmental cadmium can lead to a higher risk of neurotoxicity.

Cadmium-induced neurotoxicity poses a significant concern, particularly in vulnerable populations such as children, pregnant women, and the elderly. Exposure to cadmium during critical periods of brain development can have lifelong implications on cognitive function and behavior. Furthermore, the interaction between cadmium exposure and other risk factors can further increase susceptibility to its neurotoxic effects. Understanding the vulnerability and adverse consequences of cadmium exposure in specific populations is crucial for implementing effective preventive strategies and minimizing the burden of cadmium-induced neurotoxicity.

7 Emerging Strategies for Preventing or Mitigating Cadmium Neurotoxicity

Cadmium is a highly toxic heavy metal and a common environmental pollutant that poses significant risks to human health. Prolonged exposure to cadmium can result in a diverse range of adverse effects, including neurotoxicity. In recent years, extensive research efforts have been directed toward developing strategies to prevent or mitigate cadmium-induced neurotoxicity. This note will discuss some of the emerging strategies that show promise in this regard.

7.1 Recent Advances in Preventing or Reducing Neurotoxic Effects

Several recent advances have demonstrated potential for preventing or decreasing the neurotoxic effects of cadmium exposure. One such approach involves the use of natural or synthetic chelating agents to enhance cadmium excretion from the body (Flora et al. 2008). Chelating agents such as ethylenediaminetetraacetic acid (EDTA) and dimercaptosuccinic acid (DMSA) have shown efficacy in reducing cadmium accumulation in the brain and preventing associated neurological damage.

7.2 *Importance of Environmental and Occupational Regulations*

Minimizing cadmium exposure risks is crucial in preventing neurotoxicity. The establishment of stringent environmental regulations and occupational safety guidelines plays a vital role in reducing cadmium exposure. Proper waste management and disposal practices, as well as the implementation of workplace control measures, are essential for preventing occupational cadmium exposure (Järup 2003). Strong regulatory frameworks, coupled with effective monitoring and enforcement, are necessary to ensure compliance and reduce the prevalence of cadmium-induced neurotoxicity.

7.3 *Role of Chelating Agents in Enhancing Cadmium Excretion*

Chelating agents have been extensively studied for their ability to enhance the excretion of cadmium from the body. By forming stable complexes with cadmium ions, these agents facilitate the elimination of cadmium via urine or feces. Studies have shown that chelation therapy with agents like DMSA can effectively reduce the body burden of cadmium, thus mitigating neurotoxicity (Wang et al. 2014). However, caution should be exercised in the administration of chelating agents, as their indiscriminate use may lead to potential side effects.

7.4 *Nutritional Interventions in Attenuating Cadmium-Induced Neurotoxicity*

Nutritional interventions hold significant promise in mitigating cadmium-induced neurotoxicity. Studies have suggested that supplementation with antioxidants, such as vitamin C, vitamin E, and selenium, may protect against the neurotoxic effects of cadmium (Chen et al. 2020). These antioxidants scavenge free radicals generated by cadmium exposure, thereby reducing oxidative stress and neuronal damage. Furthermore, adequate intake of essential nutrients, such as calcium, iron, and zinc, may help prevent cadmium absorption and reduce its neurotoxic effects (Gupta and Gupta 1998). Targeted nutritional interventions, combined with a well-balanced diet, can contribute to the protection of neuronal health in the presence of cadmium exposure.

The prevention and mitigation of cadmium neurotoxicity require a multifaceted approach that includes the implementation of environmental and occupational regulations, the use of chelating agents, and the adoption of nutritional interventions. Advances in these strategies offer promising opportunities to reduce the burden of cadmium-induced neurotoxicity on both individuals and society as a whole.

Continued research, collaboration between scientific and regulatory bodies, and public awareness are essential for successful implementation and improvement of these emerging strategies.

8 Cadmium-Induced Neurotoxicity: Key Findings, Implications, and Urgent Need for Further Research

Cadmium (Cd) is a heavy metal pollutant that is widely distributed in the environment and poses a significant threat to human health. Its neurotoxic effects have been a subject of increasing concern, as research reveals its ability to induce damage to the nervous system. This note aims to summarize the key findings and implications of cadmium-induced neurotoxicity, emphasizing the urgent need for further research to better understand the underlying mechanisms and long-term consequences of cadmium exposure on the nervous system.

8.1 Key Findings

1. Neurobehavioral and cognitive impairments: Numerous studies have demonstrated that exposure to cadmium is associated with neurobehavioral and cognitive impairments, including deficits in memory, learning, attention, and motor skills (Tinkov et al. 2018a, b; Saiyed et al. 2019). These findings highlight the detrimental impact of cadmium on the functioning of the nervous system.
2. Oxidative stress: Cadmium exposure leads to the generation of reactive oxygen species (ROS) in the brain, resulting in oxidative stress. Increased oxidative stress can damage essential cellular components, disrupt signaling pathways, and contribute to neurodegenerative processes (Jin et al. 2019; Salama et al. 2019).
3. Neuroinflammation: Cadmium exposure can induce neuroinflammatory responses, characterized by the activation of microglia and the release of pro-inflammatory cytokines (Levine et al. 2019; Tang et al. 2020). Prolonged activation of the immune system in the brain can lead to neuronal dysfunction and contribute to the progression of neurodegenerative diseases.
4. Disruption of neurotransmitter systems: Cadmium has been shown to disrupt the balance of neurotransmitters in the brain, including glutamate, dopamine, and serotonin (Suwalsky et al. 2018; Hernández-Franco et al. 2020; Feruke-Bello, 2023). Dysregulation of these systems can impact various neurophysiological processes, leading to neuronal dysfunction and cognitive impairments.

8.2 Implications

1. Human health risks: The findings on cadmium-induced neurotoxicity highlight the need to assess the potential health risks associated with chronic exposure to low levels of cadmium, as individuals are commonly exposed to this heavy metal through food, water, and environmental contamination (Bernard 2015a, b; Shayan et al. 2023). It is particularly important to protect vulnerable populations, such as children, pregnant women, and occupational workers, who may be more susceptible to the detrimental effects of cadmium on the nervous system.
2. Environmental policy and regulation: Understanding the adverse effects of cadmium on the nervous system calls for the implementation of stricter environmental policies and regulations to minimize cadmium pollution. Stronger regulations on industrial emissions, waste management, and the use of cadmium in consumer products can help prevent further exposure and protect public health (ATSDR 2012; Elsaid and Faheem 2023).

8.3 Urgent Need for Further Research

1. Mechanisms of cadmium-induced neurotoxicity: Despite the significant progress made in understanding cadmium-induced neurotoxicity, there is still much to be discovered regarding the precise mechanisms by which cadmium exerts its detrimental effects on the nervous system. Further research is needed to elucidate the complex pathways involved in oxidative stress, neuroinflammation, and neurotransmitter dysregulation, among others.
2. Long-term consequences: Long-term follow-up studies are necessary to evaluate the lasting consequences of cadmium exposure on the nervous system. Assessing the potential link between cadmium exposure and the development of neurodegenerative diseases, such as Alzheimer's or Parkinson's disease, can provide valuable insights into the long-term effects of this heavy metal.
3. Protective strategies: Research is needed to identify and develop effective protective strategies against cadmium-induced neurotoxicity. This could include the investigation of potential antioxidants or chelating agents that can counteract the oxidative stress caused by cadmium exposure and mitigate its neurotoxic effects.

Cadmium-induced neurotoxicity poses a significant public health concern due to its ability to induce neurobehavioral and cognitive impairments, oxidative stress, neuroinflammation, and disruption of neurotransmitter systems. It is imperative to prioritize further research efforts to better understand the underlying mechanisms and long-term consequences of cadmium exposure on the nervous system. This knowledge will inform the development of public health measures, policies, and interventions aimed at minimizing cadmium exposure and protecting vulnerable populations.

References

Agency for Toxic Substances and Disease Registry (2012) Toxicological profile for cadmium. https://www.atsdr.cdc.gov/toxprofiles/tp.asp?id=48&tid=15

Agency for Toxic Substances and Disease Registry (ATSDR) (2012) Toxicological profile for cadmium. U.S. Department of Health and Human Services, Public Health Service

Al Marshoudi M, Al Reasi HA, Al-Habsi A, Barry MJ (2023) Additive effects of microplastics on accumulation and toxicity of cadmium in male zebrafish. Chemosphere 334:138969

Antonisamy B et al (2019) Association of maternal cadmium exposure during pregnancy with birth outcomes and risk of preeclampsia: a population-based birth cohort study. Environ Res 170: 370–376. https://doi.org/10.1016/j.envres.2018.12.040

Apostoli P et al (2019) Neurotoxic effects in humans. In: Cadmium: from toxicity to essentiality. Springer, Cham. https://doi.org/10.1007/978-3-030-18834-3_14

Aschner M et al (2001) Involvement of DMT1 in uptake of Cd2+ at the apical membrane of rat brain capillary endothelial cells. J Neurochem 78(4):854–867

Bernard A (2015a) Cadmium and its adverse effects on human health. Indian J Med Res 141(3): 364–368. https://doi.org/10.4103/0971-5916.159309

Bernard A (2015b) Cadmium & its adverse effects on human health. Indian J Med Res 142(2): 235–246

Bernhoft RA, Hauge SJ (2013) Airborne cadmium and lead particles in urban environments: inter- and intra-object variability of Cd/Pb ratio. Biol Trace Elem Res 152(2):231–238

Chen J, Xu L, Yang X, Liang Y, Wang X, Shao J et al (2014) Cadmium exposure affects the learning and memory of rats by impairing the synaptic plasticity of hippocampus. Int J Occup Med Environ Health 27(5):829–839

Chen Y et al (2019) Cadmium induces death of dopaminergic cells through oxidative stress-mediated apoptosis pathway in a neural cell line. Neurotoxicology 70:1–7

Chen L, Xu B, Liu L, Luo Y, Zhou H (2020) Mitigating effects of antioxidants on neurotoxicity induced by cadmium in adult zebrafish. Ecotoxicol Environ Saf 189:109955

da Silva AOF, Bezerra V, Meletti PC, Simonato JD, dos Reis Martinez CB (2023) Cadmium effects on the freshwater teleost Prochilodus lineatus: accumulation and biochemical, genotoxic, and behavioural biomarkers. Environ Toxicol Pharmacol 99:104121

Dooyema CA, Neri A, Lo YC, Durant J, Dargan PI, Swarthout T et al (2012) Outbreak of fatal childhood lead poisoning related to artisanal gold mining in northwestern Nigeria, 2010. Environ Health Perspect 120(4):601–607

Elsaid A, Faheem NM (2023) Ameliorative effects of Omega-3 on the formaldehyde-induced damage in rat olfactory bulb. Egypt J Histol 46(1):78–90

Eto K, Takamine S, Nagaoka K (2017) Repeated exposure to cadmium enhances memory of spatial learning in mice. Toxicol Rep 4:395–400

Faroon O, Ashizawa A, Wright S, Tucker P, Jenkins K, Ingerman L (2012) Toxicological profile for cadmium. Agency for Toxic Substances and Disease Registry

Feruke-Bello YM (2023) Contamination of fermented foods with heavy metals. In: Indigenous fermented foods for the tropics. Academic Press, pp 549–559

Flora SJ, Mittal M, Mehta A (2008) Heavy metal induced oxidative stress & its possible reversal by chelation therapy. Indian J Med Res 128(4):501–523

Fränkenhuysen M et al (2012) Effects of cadmium on intracellular Ca2+ in neuroblastoma neuro-2a cells. Environ Toxicol Pharmacol 34(1):262–269

Fu X et al (2021) Cadmium-induced mitochondrial dysfunction and mitochondrial dynamics disruption in neurodegenerative diseases. Environ Pollut 278:116856

Goyer RA (2017) Cadmium and lead toxicity and their role in metal-induced neurotoxicity. Oxidative Med Cell Longev 2017:1–11

Goyer RA, Cherian MG (1983) Metallothionein in human brain: evidence for a role in heavy metal-induced neurotoxicity. Neurosci Lett 36(3):293–298

Gupta UD, Gupta SC (1998) Sources and toxicity of lead: a review. Environ Monit Assess 51(1–2): 1–26

Hernández LT, Tacchi R, Gorojod RM, Romano A (2019) Long-lasting cognitive deficits in rats after early-life cadmium exposure: relationship to changes in cholinergic and glutamatergic receptors expression in the cerebral cortex and hippocampus. Neurotox Res 35(4):942–955

Hernández-Franco P et al (2020) Cadmium neurotoxicity: the role of glial cells. Int J Mol Sci 21(11):3819

Järup L (2003) Hazards of heavy metal contamination. Br Med Bull 68(1):167–182

Järup L, Åkesson A (2009) Current status of cadmium as an environmental health problem. Toxicol Appl Pharmacol 238(3):201–208

Jin T et al (2019) The role of neuroinflammation in the pathogenesis of Parkinson's disease. Acta Neurobiol Exp 79(4):328–339

Julvez J, Smith GD, Carsin AE, Ballester F, Lertxundi N, Ibarluzea J et al (2016) Environmental risk factors for attention-deficit hyperactivity disorder in children born between 2000 and 2007: a case-control study. Environ Int 92:635–641

Kalia K, Kumar P (2020) Targeting oxidative stress and inflammation in cadmium toxicity: a review of pharmacological interventions. Biomed Pharmacother 130:110649

Kim YH et al (2018) Neuroinflammation-induced interactions between protease-activated receptor 1 and proprotein convertases in HIV-associated neurocognitive disorder. Mol Neurodegener 13(1):1–13

Klaassen CD (2019) Casarett & Doull's toxicology: the basic science of poisons. McGraw-Hill Education, New York

Kobal AB, Rabah N, Avanzo JL, Diez R (2015) Neurotoxic effects of cadmium in Sprague-Dawley rats (Rattus norvegicus): protective effect of quercetin. J Med Toxicol 11(1):100–108

Levine H et al (2019) Cadmium, lead, and mercury disrupt neural stem cell activity and neurodevelopment in human-derived 3D brain organoids. ACS Chem Neurosci 10(12): 4514–4526

Li Y et al (2017) Cadmium-induced oxidative stress and apoptosis in the testis and the protective role of melatonin. Environ Toxicol Pharmacol 49:141–149

Liu M et al (2020a) Cadmium-induced neuroinflammation and Alzheimer's-like symptoms in rats. Neuroscience 429:31–45

Liu Z et al (2020b) Mechanisms underlying cadmium uptake and accumulation in neurons. Environ Sci Technol 54(13):7923–7934

Mukherjee S et al (2017) Childhood cadmium exposure and neurodevelopmental outcomes in school-aged children: a systematic review and meta-analysis. Neurotoxicology 59:101–109. https://doi.org/10.1016/j.neuro.2016.12.012

Navas-Acien A et al (2007) Lead, cadmium, smoking, and increased risk of peripheral arterial disease. Circulation 115(21):3233–3239. https://doi.org/10.1161/CIRCULATIONAHA.106.647627

Olaolu T, Jibulu F, Damilare R (2023) Hepatotoxic effect of cadmium and available therapeutic options. Thai J Pharm Sci 46(6):631–637

Patel AB et al (2020) Cadmium exposure and neurodevelopmental outcomes in children: a systematic review and meta-analysis. JAMA Netw Open 3(7):e207812. https://doi.org/10.1001/jamanetworkopen.2020.7812

Radauceanu A, Popescu C, Bacalbasa N (2015) Dermal implications at long-term exposure with medium and low cadmium concentration. Metal Int 20(10):78–82

Rajabian F, Rajabian A, Tayarani-Najaran Z (2023) Protective effect of betanin on cadmium-induced toxicity in MRC-5 cells

Saiyed HN et al (2019) Effects of chronic low-level exposure to cadmium on cognitive and neurobehavioral functioning in humans: a systematic review. Environ Res 177:108598

Salama M et al (2019) Role of oxidative stress in the neurotoxicity of heavy metals. In: Neurotoxicity of metals. Springer, Cham, pp 3–42

Sara F et al (2016) Neurotoxicity of cadmium on memory and learning ability via activating Nrf2-ARE signaling pathway in the hippocampus of rats. Biomed Res Int 2016:1980534. https://doi.org/10.1155/2016/1980534

Schaumlöffel D (2000) Heavy metal toxicity of the brain: possible involvement of divalent cations and of divalent metal transporter 1. Environ Health Perspect 108(Suppl 3):429–432. https://doi.org/10.1289/ehp.00108s3429

Senanayake V, Lu L (2010) Toxicity of cadmium and its health risks from long-term exposure to food crops through food chain. Ecotoxicol Environ Saf 73(1):1–8. https://doi.org/10.1016/j.ecoenv.2009.05.001

Shayan M, Mehri S, Razavi BM, Hosseinzadeh H (2023) Minocycline protects PC12 cells against cadmium-induced neurotoxicity by modulating apoptosis. Biol Trace Elem Res 201(4):1946–1954

Srivastava A, Kumari A, Jagdale P, Ayanur A, Pant AB, Khanna VK (2023) Potential of quercetin to protect cadmium induced cognitive deficits in rats by modulating NMDA-R mediated downstream signaling and PI3K/AKT—Nrf2/ARE signaling pathways in hippocampus. NeuroMolecular Med 25:1–15

Suwalsky M et al (2018) Effect of cadmium on the stability of cell membrane models involving complex lipids. Cell Mol Biol Lett 23:5

Tang YC et al (2020) Effects of environmental pollutants on mitochondrial function and innate immunity in neural cells: implications for neurological disorders. Environ Int 134:105162

Taweel A et al (2017) Maternal cadmium exposure—a possible risk factor for neurodevelopmental disorders in children. Front Genet 8:38. https://doi.org/10.3389/fgene.2017.00038

Tellez-Plaza M, Navas-Acien A, Crainiceanu CM, Guallar E, Tellez-Plaza M (2011) Cadmium exposure and cognitive function in US adults. Environ Health Perspect 120(4):557–562

Tinkov AA, Gritsenko VA, Skalnaya MG, Cherkasov SV, Skalny AV (2018a) Serum and tissue markers of hepatotoxicity as markers of cadmium exposure in occupationally exposed population. J Trace Elem Med Biol 48:207–214

Tinkov AA et al (2018b) Cadmium-induced neurotoxicity: focus on blood-brain barrier disruption and inflammatory response. Chem Res Toxicol 31(9):894–905

Toullec A et al (2017) Cadmium impairs mitochondrial ATP production through different pathways. Toxicol In Vitro 40:293–303

Valko M et al (2006) Free radicals and antioxidants in normal physiological functions and human disease. Int J Biochem Cell Biol 39(1):44–84

Wang J, Wang X, Li X et al (2014) Efficacy of DMSA treatment for attenuating low-level cadmium-induced nephrotoxicity: an eight-year follow-up study. J Trace Elem Med Biol 28(2):177–182

Wang X et al (2016) Oxidative stress and mitochondrial dysfunction in Alzheimer's disease. Biochim Biophys Acta 1862(5):878–887

Wang J, Zhang T, Du X, Xu Y, Xu T, Tian Y et al (2019) Protective effect of taurine against cadmium-induced learning and memory deficits associated with the inhibition of NMDAR-mediated calcium influx and CaMKII activation in the hippocampus. J Toxicol Sci 44(1):23–35

Yousuf R, Verma PK, Sharma P, Sood S, Aït-Kaddour A, Bhat ZF (2023) Ameliorative potential of quercetin and catechin against sodium arsenite and mancozeb-induced oxidative renal damage in Wistar rats. J Trace Elem Miner 2023:100079

Zhang Y et al (2007) Cadmium-induced apoptosis in primary rat cerebral cortical neurons culture is mediated by a calcium signaling pathway. Brain Res 1162:9–20

Part III
Sustainable Nexus Solution of Cadmium Toxicity

Biological Interventions in Bioremediation of Cadmium Poisoning

Neha Verma and Ashish Sharma

Abstract In recent years, owing to the excessive urbanization and industrial development heavy metals have emerged as pollutants of concern for the environment. There are many sources that are responsible for the addition of heavy metals to the environment but anthropogenic activities are the biggest contributors. Out of all the heavy metals released into the environment, one of the most notorious elements is cadmium (Cd), owing to its variety of adverse effects on plants, animals as well as humans. Hence, there arises a requirement to remove the said element from the environment in order to protect the same. Out of the many strategies that can be used for the removal of Cd from environment, bioremediation techniques can prove to be most beneficial because apart from being an answer to the pollutants, these processes can also benefit the environment. Therefore, in the present manuscript, we try to present evidences in favor that bioremediation as the best technique for Cd removal. We shall explore different agencies that can be utilized in the process of Cd removal and maintenance.

Keywords Cadmium · Cadmium poisoning · Fungal bioremediation · Bacterial Bioremediation · Biosorption

1 Introduction

In the past few decades, excessive urbanization and developmental activities to support the growing populations have been accompanied by introduction of several different pollutants in the environment, including a plethora of heavy metals (Ozyigit and Genc 2020). Of all the problems being faced today environmental damage due to contamination by different agents has been the most significant constraint for modern human society as we know it (Afzal et al. 2019). Given their ability to accumulate in the environment, heavy metals pose the greatest threat to different life

---

N. Verma · A. Sharma (✉)
Department of Botany and Environment Science, DAV University, Jalandhar, Punjab, India
e-mail: ashish10210@davuniversity.org

A. K. Jha, N. Kumar (eds.), *Cadmium Toxicity Mitigation*,
https://doi.org/10.1007/978-3-031-47390-6_5

forms along with the environment (Karahan et al. 2020). Plants being the primary source of food, raw materials, and energy for humans, the masses mostly depend on the plants for a variety of requirements. These plants are greatly affected by the soil in which they grow and the contaminants present in the soil can find their way into plant parts and thus can affect the human and animal populations dependent on them (Meena et al. 2020). Of the most damaging nature of all these contaminants present in the soil are heavy metals because there exist mechanisms in plants for the uptake of these chemical species. Different metal species have been classified as heavy metals mostly belonging to transition series of the periodic table and including elements such as iron, zinc, manganese, lead, selenium, and cadmium. Most of these metals, although potentially toxic show negligible effects on plants or animals because their natural concentration in soil remains sufficiently low. However, many industrial and anthropogenic practices result in significant contributions to the addition of heavy metals to the soil and environment (Yuan et al. 2019).

The metal ions required by the plants can be classified as essential or non-essential based on their concentration required by the plants, where essential metals are indispensable from plant function but non-essential that include metals like lead, cadmium, silver can be extremely toxic to the plants when present in the soil. These elements, apart from affecting plants, can contaminate food chains resulting in ecological imbalances and giving rise to lethal health issues for both plants as well as animals (Haider et al. 2021). Out of all the heavy metals affecting plant growth and function, one that has caught particular attention is cadmium. Given the fact that it is easily taken up by the plants and gets accumulated in higher trophic levels, Cd is highly harmful as far as toxic materials are concerned. In fact, in the USA, Faroon et al. (2022) has classified Cd at seventh rank in the list of substance priority list for most noxious substances due to its adverse health effects. Both natural as well as anthropogenic activities have resulted in the increase of cadmium concentration in the soils and environment. Amongst natural causes, withering of cadmium containing rocks remains the major cause for its introduction into the soil. However, due to its inherent low concentration in the lithosphere, anthropogenic causes remain the major source for the introduction of cadmium in the soil. Among the various activities, reliance on phosphatic fertilizers, production of nickel-cadmium batteries, mining activities, agricultural activities such as composting, and use of pesticides and industrial emissions are the major contributors to the increase of cadmium concentration in the soils (Manzoor et al. 2019).

2 Cadmium and Plants

Cadmium, a non-essential element, exerts detrimental effects on the growth and development of plants. It is released into the environment from various sources including power stations, heating systems, metal-working industries, and urban traffic. The element finds widespread application in electroplating, pigments, plastic stabilizers, and nickel-cadmium batteries. Its significant solubility in water and high toxicity mark cadmium as a notably influential pollutant (Pinto et al. 2004). The element's genotoxicity and ecotoxicity extend to animals as well, with reports of adverse effects in both terrestrial and aquatic ecosystems. Cadmium's introduction into marine environments originates from atmospheric deposition, domestic wastewater, and industrial discharges. Baker et al. (1990) highlighted that cadmium rarely exists in isolation in natural environments, often occurring alongside other metals like lead and zinc in mineralization settings. Wagner (1993) estimated that non-polluted soil solutions carry cadmium concentrations ranging from 0.04 to 0.32 mM, with concentrations varying from 0.32 to about 1 mM considered moderately polluted (Sanita di Toppi and Gabbrielli 1999).

Concerning its potential harm to soil organisms and microbial processes, Duxbury (1985) classified cadmium as a moderately toxic element. Despite numerous studies documenting the toxic effects of cadmium on biological systems (Das et al. 1997; Sanita di Toppi and Gabbrielli 1999), the precise mechanisms of cadmium toxicity remain incompletely understood. Cadmium's impact on plants includes alterations in mineral uptake due to effects on soil mineral availability and soil microbial populations (Moreno et al. 1999). Although cadmium influences stomatal behavior, transpiration, and photosynthesis in nutrient solutions, it is more readily taken up by plants from these solutions compared to soil. Easily visible symptoms of cadmium toxicity in plants comprise chlorosis, leaf rolling, and stunting. These symptoms can be confused with other metal ions particularly iron, phosphorus, and manganese deficiencies, however, chemical analysis can be used to confirm presence of Cd in the plant system for confirmation of these occurrences. The inhibition of root Fe(III) reductase by cadmium triggers Fe(II) deficiency, which shows a significant effect on the photosynthetic ability of the plants. Besides iron, it has been shown that cadmium exhibits antagonistic roles in the uptake, transport, and assimilation of other essential elements particularly calcium, magnesium, phosphorus, and potassium. Moreover, not only nutrients, cadmium has been shown to affect water absorption in plants as well (Das et al. 1997). It has also been shown to interfere with the nitrogen uptake and metabolism in the plants ultimately leading to nitrogen deficiency in plants exposed to cadmium. It was also shown that following cadmium treatments, nitrogen fixation and primary ammonia assimilation reduced in root nodules of soybean (Balestrasse et al. 2003).

Cadmium toxicity can also affect the plants at membrane level. It can disrupt plasma membrane permeability, leading to reduced water content. Cadmium interacts with the water balance of plants leading to additional problems. Numerous scientific investigations have proposed that oxidative stress might play a pivotal role in the toxicity of cadmium. This stress can arise through two main mechanisms: either by inducing the production of oxygen-free radicals or by diminishing the levels of both enzymatic and non-enzymatic antioxidants. Several studies have contributed to this understanding:

- Oxygen-Free Radical Production: Cadmium exposure is linked to the generation of oxygen-free radicals, highly reactive molecules that can cause damage to cellular components. Somashekaraiah et al. (1992) and Stohs and Bagchi

(1995) observed that cadmium triggers the production of these damaging radicals, leading to cellular harm.

- Antioxidant Depletion: Cadmium's adverse effects on antioxidant systems have been noted by various researchers. Sandilo et al. (2001) highlighted the element's ability to reduce the levels of both enzymatic and non-enzymatic antioxidants, compromising the cell's defense mechanisms against oxidative stress.
- Role in Senescence: The process of senescence, or aging, can be accelerated by cadmium-induced oxidative stress. Studies by Balestrasse et al. (2001) and Fornazier et al. (2002) emphasized that cadmium exposure could lead to premature senescence in plants due to the overwhelming oxidative stress caused by the metal.
- Nodule Senescence in Soybean: Specifically, in soybean plants, the accelerated senescence observed in nodules (the structures that house nitrogen-fixing bacteria) has been attributed to the oxidative stress induced by cadmium. Balestrasse et al. (2004) found that the metal's impact on nodules could be traced back to its ability to generate oxidative stress, contributing to the observed senescence.

3 Cadmium Homeostasis

Cadmium homeostasis in plants can be broken down into three distinct phases that include uptake from the soil, immobilization in the cytosol, and transport inside the plants (as shown in Fig. 1). The collaborative action of all these three components affects the sensitivity of the plants toward the presence of cadmium.

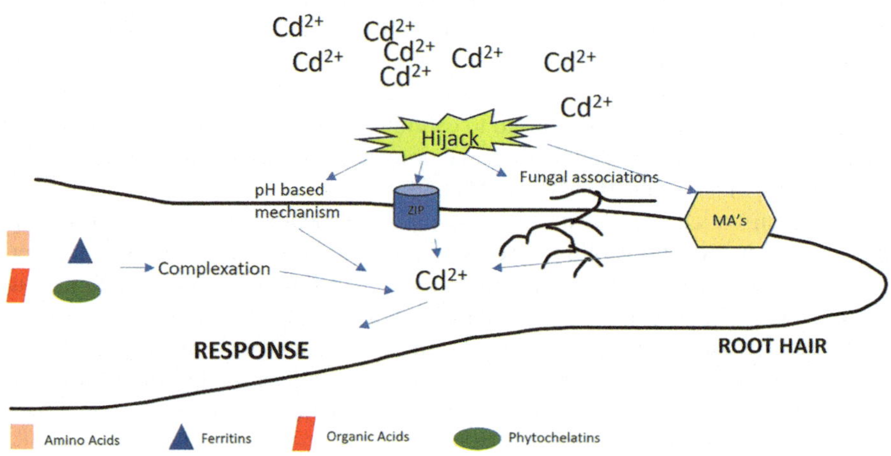

Fig. 1 Demonstration of cadmium homeostasis in plant roots. Instead of directly entering the roots and showing its response cadmium hijacks the mechanisms available for uptake of other metal ions particularly zinc and iron. After its entry into root cells, it can form complexes with different compounds like amino acids, organic acids, ferritins and phytochelatins and then result in the response observed in the plants

4 Cadmium Uptake

Soil particles exhibit a tendency to bind with metal ions due to the inherent negative charges present on them which result in the non-availability and reduced mobility of many positively charged metal ions in the soil (as shown in Fig. 2).

Besides affecting their bioavailability, interaction of metal ions and soil particles also reduces their solubility in the soil thereby further affecting their availability to the plants. This condition can be overcome by decreasing the pH of soil in which root exudates play an important role. This process of acidification and its role in metal uptake has been explained by Sharma et al. (2013) in relation to zinc (another positively charged heavy metal). The extent to which higher plants have the capacity to absorb Cadmium (Cd) is contingent upon its concentration in the soil and its bioavailability, a factor influenced by a multitude of conditions such as the presence of organic matter, pH levels, redox potential, temperature, and concentrations of other chemical elements. Apart from Iron (Fe), which is solubilized by the action of phytosiderophores secreted in root exudates and demonstrated by many workers (Vaid et al. 2014), the understanding of the active mobilization of trace elements by plant roots remains limited. Notably, the uptake of Cd ions appears to be in competition for the same transmembrane carrier with essential nutrients like potassium, calcium, magnesium, iron, manganese, copper, and zinc (Rivetta et al. 1997).

The cellular membrane assumes a pivotal role in maintaining metal homeostasis, acting as a gatekeeper to prevent or minimize the entry of these substances into the cell. However, instances of exclusion or mechanisms leading to reduced uptake in higher plants are relatively scarce. The elevated mobility of Cadmium, coupled with its ability to elicit adverse effects in plants even at low concentrations, designates it as one of the most hazardous metals. The continuum of the apoplast—a network of interconnected spaces within the root epidermis and cortex—readily permits the passage of solutes. However, the cell walls of the endodermal cell layer function as a barricade, impeding apoplastic diffusion into the vascular system. As a general rule, solutes must first permeate the root symplasm before gaining access to the xylem, the plant's water-conducting tissue (Tester and Leigh 2001). Once the ions enter into the root cells, their further movement from root to the xylem for transportation to the

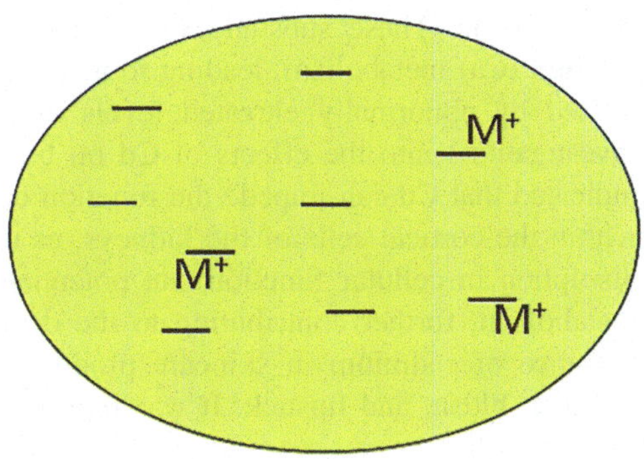

Fig. 2 Negatively charged soil particles bind to positively charged metal ions

shoot, is affected by three distinct processes that include sequestration within root cells, symplastic transport into the stele (the central column of vascular tissue), and eventual release into the xylem (Clemens et al. 2002).

Despite variations in the mobility of different metal ions within plants, there is a prevailing trend of higher metal accumulation in roots compared to above-ground tissues, as outlined by Ramos et al. (2002). Under most environmental conditions, Cadmium (Cd) predominantly enters roots first, making them susceptible to Cd-induced damage at an early stage (Sanita di Toppi and Gabbrielli 1999). Upon incorporation into the roots, Cd can reach the xylem either by apoplastic or symplastic route which can be facilitated by the interaction of Cd with various ligands on offer inside the cytosol of root cells. Generally, roots are the primary site of Cd ion accumulation and only a small amount of these can find their way into the shoots. If a comparison can be made amongst the various plant parts highest amount of Cd accumulates in roots and least amount in seeds with stems, leaves, and fruits falling in between in the same sequence as mentioned (Blum 1997).

5 Cadmium, Animals, and Humans

Many anthropogenic activities and contamination of agricultural lands with cadmium have led to an unprecedented rise in Cd pollution particularly in the residential areas, with the result that residential areas have to contend with Cd levels of 0.001 g m^{-3}. Creation of such drastic conditions has inevitably led to the ingestion of 0.02 mg of Cd daily by the people living in these areas. This has affected the populations in such a manner that many cases of lung edema have been linked to inhalation of cadmium oxide ultimately resulting in the death of the sufferer (Ozyigit and Genc 2020). Furthermore, prolonged exposure to Cd has been associated with severe health implications. Notably, research by Person et al. (2013) has underscored the connection between extended Cd exposure and the heightened risk of certain malignancies, particularly lung cancer and even prostate cancer.

Research has demonstrated that both cadmium metal and its various compounds can trigger the development of sarcoma when introduced into the body through intramuscular or subcutaneous injections, as highlighted in the study by Yongming et al. (2011). These substances have been associated with negative impacts on calcium (Ca) metabolism, leading to a condition known as hypercalcemia, characterized by abnormally elevated levels of calcium in the bloodstream. In-depth investigations into the effects of Cd on bone health and calcium regulation have indicated that Cd can impede the function of the sodium–glucose transport system within the cortical cells of the kidneys, as reported by Dongre et al. (2013). This disruption in cellular function can potentially have cascading effects on calcium metabolism, further contributing to the disturbances observed. Of concern is the presence of cadmium in skincare products commonly used by women, such as eyeliner, blush, and lipstick. It was reported that prolonged use of these products containing Cd can significantly increase the risk of skin cancer (Pratinidhi et al. 2018).

Cadmium rapidly undergoes a transformation into cadmium oxide (CdO) upon exposure to the air. Notably, inorganic salts like Cd $(NO_3)_2$, $CdSO_4$, and $CdCl_2$ possess water solubility. When the concentration of Cd in the air surpasses the threshold of 1 mg m^{-3}, it leads to acute respiratory effects. It is important to note that due to the body's limited ability to excrete Cd efficiently, the element tends to accumulate over time, thereby giving rise to its progressively detrimental health impacts. Within industrial contexts, instances of Cd poisoning frequently occur in connection with alloy compounds utilized during welding procedures, electrochemical coatings, dyes containing Cd, and batteries incorporating Cd components. This observation is highlighted in the research conducted by Aslan in 2020. The cumulative nature of Cd toxicity underscores the necessity for stringent precautions and regulations to mitigate its adverse effects on both human health and the environment.

Among workers who encounter Cd exposure, the initial manifestation of Cd-induced kidney toxicity, a prevalent concern, is the onset of tubular dysfunction, characterized specifically by low molecular weight proteinuria. This condition involves the presence of small proteins in the urine due to impaired tubular function. It is noteworthy that the extent of proteinuria can fluctuate across time intervals. Compelling evidence suggests that the tubular proteinuria provoked by Cd exposure often persists for a significant duration, spanning at least a couple of years, in many affected workers (Fay et al. 2018; Chen et al. 2020). Importantly, this protracted proteinuria underscores the challenging and potentially irreversible nature of Cd-induced tubular damage, emphasizing the critical need for vigilance and preventive measures in occupational settings to safeguard workers' renal health.

Cd is associated with significant biological impacts, including DNA damage and the initiation of lipid peroxidation. Scientific investigations have illuminated that Cd-induced toxicity prompts the fragmentation of DNA strands and the oxidation of lipids within the cellular environment. A notable consequence of Cd toxicity is the induction of apoptosis, a programmed cell death process, within proximal tubule cells (Wang et al. 2017). Furthermore, Cd's detrimental effects on the kidneys extend to the induction of interstitial fibrosis. This outcome emerges as a result of the interplay between tubular cell necrosis, the death of these kidney cells, and the ensuing inflammatory responses. Fujiki et al. (2019) emphasize the role of Cd in driving these pathological processes, leading to the development of interstitial fibrosis—a condition characterized by the excessive accumulation of fibrous tissue in the kidney's interstitial spaces. It is important to recognize that Cd does not solely impact the kidneys. Given its tendency to accumulate most prominently in the liver after the kidneys, the assessment of its hepatotoxic effects becomes a matter of great significance. This aspect of Cd toxicity has been explored in the study by Cao et al. (2017). Understanding the implications of Cd exposure on liver health is pivotal in comprehensively addressing the potential health risks associated with this toxic metal.

Evidence from studies, such as those conducted by Lavryshyn and Gutyj (2019) and Yaseen (2019), has indicated that instances of environmental Cd exposure do not typically lead to heightened liver enzyme levels or reduced serum albumin levels. This suggests that such exposure might not directly correlate with significant liver dysfunction. However, the onset of Cd toxicity in the liver can be marked by early signs of impaired mitochondrial function (Okoye et al. 2019). This perturbation in mitochondrial activity is attributed to Cd-induced alterations in the balance of glutathione levels and a decrease in glutathione peroxidase activity within liver cells. These changes in cellular components, rather than increased permeability of the mitochondrial inner membrane, contribute to the toxic impact of Cd on mitochondria (Abarikwu et al. 2017).

Cadmium's deleterious effects extend to the cellular lysosomes, where its accumulation leads to the disruption of the lysosomal membrane's integrity. This disruption triggers the release of lysosomal enzymes, including proteases and phospholipases, initiating a chain of events that ultimately results in hepatocyte cytotoxicity (Wang et al. 2019). While acute exposure to gas mixtures containing elevated Cd concentrations can induce fatal chemical pneumonia, more significant concerns arise from chronic Cd exposure's impact on the respiratory system. Notably, community members who smoke are at particular risk due to ongoing and low-level Cd exposure (Tao et al. 2020).

Continual exposure to Cd through activities like smoking or inhalation of particles in occupational settings can lead to detrimental impacts on the respiratory system. The extent and timing of these damages are influenced by the concentration and duration of Cd exposure, often manifesting gradually over the course of several years. Research indicates that chronic inflammation can develop in the nasal passages, pharynx, and larynx of workers in the cadmium industry, as highlighted in studies by Hassanin et al. (2017) and Prokopowicz et al. (2019). This chronic inflammation can contribute to various respiratory issues and discomfort among affected individuals. Furthermore, chronic obstructive pulmonary disease (COPD) is prevalent among workers who are exposed to cadmium, particularly when combined with smoking. COPD is a progressive lung condition characterized by restricted airflow, making it difficult to breathe. Notably, both clinical examinations and radiological assessments have shown evidence of emphysema in these patients (Sundblad et al. 2016). Emphysema involves the destruction of the air sacs in the lungs, leading to reduced lung function and increased breathlessness. The combined effects of cadmium exposure and smoking can synergistically exacerbate respiratory issues, leading to more severe and rapid deterioration of lung function over time. These findings emphasize the importance of minimizing exposure to cadmium, particularly among occupational risk groups, and promoting smoking cessation to mitigate the risk of developing respiratory disorders. Occupational safety measures, regular health screenings, and proper respiratory protection are crucial in preventing these adverse health outcomes in individuals who work in cadmium-related industries.

The impact of Cd exposure on cellular health and its association with various health conditions, such as bone disorders and organ dysfunction, has been studied extensively. One significant aspect of Cd's toxicity is its ability to induce apoptosis, a programmed cell death process. Research suggests that Cd-mediated apoptosis is particularly sensitive to certain lung cells, which are critical for respiratory function. In a study conducted by Lee et al. (2016), it was observed that Cd exposure led to an increase in the expression of the *p53* and *Bax* genes. These genes play essential roles in regulating apoptosis. The overexpression of *p53* can trigger apoptosis as part of the cell's defense mechanism against genetic damage, while *Bax* is a pro-apoptotic protein that promotes the release of apoptotic factors from mitochondria. The observed apoptotic effect in Clara cells and type II cells suggests that Cd exposure can have detrimental effects on lung health and function. Beyond its effects on lung cells, Cd exposure has also been associated with bone metabolism disorders and bone diseases. Cd's impact on bone health is linked to its disruption of calcium phosphate and vitamin D metabolism. Calcium and phosphate are essential components of bone structure, and vitamin D is crucial for their absorption and utilization. Cd's interference with these processes can lead to the weakening of bones, contributing to osteomalacia and osteoporosis. Workers exposed to Cd have reported bone metabolism disorders, including osteomalacia, which underscores the widespread effects of Cd on bone health. Long-term exposure to Cd has been found to have significant consequences on bone mass and fracture incidence (Nishijo et al. 2017; Reyes-Hinojosa et al. 2019).

Numerous experimental studies have consistently demonstrated the detrimental effects of Cd exposure during pregnancy. These studies have revealed that the application of Cd salts in the later stages of pregnancy can lead to damage in the placenta, while teratogenic properties of Cd can result in severe developmental anomalies such as exencephaly (a condition where the brain is located outside the skull) and hydrocephalus (abnormal accumulation of cerebrospinal fluid in the brain) when exposure occurs early in pregnancy. In animal embryo cultures, Cd has been observed to cause the reopening of the neural tube, a significant developmental event. Neural tube closure is a crucial step in embryonic development that ultimately leads to the formation of the brain and spinal cord. The fact that Cd can disrupt this process underscores its potent teratogenic effects (Santoyo-Sánchez et al. 2018; Geng and Wang 2019; Kmecick et al. 2019). These detrimental effects are not limited to animals; the human placenta also exhibits sensitivity to the toxic impact of Cd. The toxic activity of Cd on the human placenta has been reported in various studies (Geng and Wang 2019). The metal's ability to damage the fetus goes beyond direct effects on placental tissues. It can interfere with the metabolism of essential elements like copper (Cu), iron (Fe), selenium (Se), and zinc (Zn), which are vital for proper fetal development.

Smoking further exacerbates the harmful effects of Cd during pregnancy. Mothers who smoke are at an increased risk of Cd toxicity, and this is reflected in adverse pregnancy outcomes. Notably, the incidence of low birth weight has been observed to rise in infants born to mothers who smoke. Additionally, smoking is associated with an increase in placental mass, which can be indicative of stress or dysfunction in the placenta caused by Cd exposure. Surprisingly, while Cd is known to be harmful, it has not been detected in significant amounts in breast milk from smoking mothers, as highlighted by studies (Ali et al. 2018). Hence, Cd exposure during pregnancy has been extensively investigated through experimental studies.

The application of Cd salts at different stages of pregnancy has revealed its damaging effects on placental health and its potential to cause severe developmental abnormalities in the fetus. Cd's teratogenic properties are exemplified by its ability to disrupt neural tube closure. Furthermore, Cd's interference with essential element metabolism and its synergy with smoking emphasizes the complexity of its impact on pregnancy outcomes. Efforts to mitigate Cd exposure and its associated risks are crucial to safeguarding maternal and fetal health.

6 Biological Interventions for Cadmium Removal from Environment

Bioremediation stands as an ecologically conscientious and cutting-edge technique that harnesses the inherent capabilities of natural biological processes to entirely eliminate harmful contaminants from various environments. It involves employing microorganisms, fungi, green plants, or their enzymes to restore environments altered by toxic substances back to their original, uncontaminated state. Bioremediation technologies can be broadly categorized into two main types: in situ and ex situ. In Situ Bioremediation entails treating contaminated materials right at the site where the contamination occurred. This can involve introducing suitable microorganisms, nutrients, or other agents to the contaminated area to facilitate the breakdown or neutralization of pollutants. Ex Situ Bioremediation involves removing the contaminated material from its original location and treating it at a different site. This can be particularly useful when the contamination is severe or when the site's conditions are not conducive to effective bioremediation.

Various techniques fall under the umbrella of bioremediation, each tailored to specific situations and types of contaminants (Kensa 2011):

Bioventing: This technique introduces air and nutrients to enhance the activity of naturally occurring microorganisms in the soil, thereby accelerating the degradation of contaminants.

Landfarming: Contaminated soil is spread over a designated area and periodically turned over to promote the growth of indigenous microorganisms that break down pollutants.

Bioreactors: These are controlled environments where conditions are optimized for microbial degradation of contaminants. Bioreactors can be used for various types of waste, such as wastewater treatment.

Composting: Organic waste contaminated with pollutants is mixed with materials like sawdust and turned to promote decomposition by microorganisms.

Bioaugmentation: Selected microorganisms are added to a contaminated site to enhance the existing microbial community's ability to break down pollutants.

Rhizofiltration: Plants are used to absorb and accumulate contaminants from water or soil, offering a natural means of remediation.

Biostimulation: This involves supplying nutrients or other growth-promoting factors to the environment to stimulate the growth and activity of indigenous microorganisms capable of breaking down contaminants.

However, not all types of contaminants can be effectively treated through bioremediation using microorganisms. For example, heavy metals like cadmium and lead are challenging to be absorbed or captured by organisms. Moreover, the incorporation of certain metals, such as mercury, into the food chain could exacerbate environmental problems.

7 Fungal Interventions

Fungi have displayed remarkable adaptability, enabling them to flourish across a wide range of habitats. The intricate composition of soil, serving as a primary location for fungal colonization, has been a significant site for their proliferation. Additionally, fungi have established stable colonies in freshwater and marine environments. This adaptability is integral to the crucial role they play in ecosystem dynamics. Fungi exhibit the capacity to thrive in diverse soil types and climatic conditions, including extreme environments. They have been observed to endure and propagate through the dispersal of spores in the air. This dissemination method not only aids in their survival but also contributes to the dispersion of fungi across various ecosystems. Consequently, fungi play a pivotal role in maintaining ecological equilibrium and influencing the structure and function of ecosystems. Furthermore, fungi have demonstrated their ability to survive within effluent treatment plants (ETPs), where they are involved in treating a wide array of wastewater types. Their presence in these treatment facilities underscores their resilience and potential to contribute to the breakdown and removal of pollutants from different waste streams. One of the distinctive qualities that positions fungi as valuable candidates for bioremediation is their capacity to secrete a diverse array of enzymes. These enzymes play a pivotal role in breaking down complex organic compounds and pollutants (as shown in Fig. 3). The combination of their enzymatic capabilities and their propensity to colonize diverse habitats makes fungi well-suited for various bioremediation applications (Deshmukh et al. 2016).

Fungi have emerged as key players in bioremediation, displaying their effectiveness in addressing a wide array of pollutants, including persistent organic pollutants (POPs), textile dyes, petroleum hydrocarbons, pulp and paper industry effluents, leather tanning effluents, polycyclic aromatic hydrocarbons (PAHs), pesticides, and pharmaceuticals and personal care products (PPCPs). Filamentous fungi, such as *Aspergillus, Curvularia, Acrimonium*, and *Pithium*, have been investigated for their ability to tolerate heavy metals (Akhtar et al. 2013). Basidiomycota members like *Trametes versicolor* and white-rot fungi *Pleurotus ostreatus* have been noted for their capability to degrade model PAHs in solid-state fermentation, using agro-industrial residues like orange peels as substrates (Rosales et al. 2013).

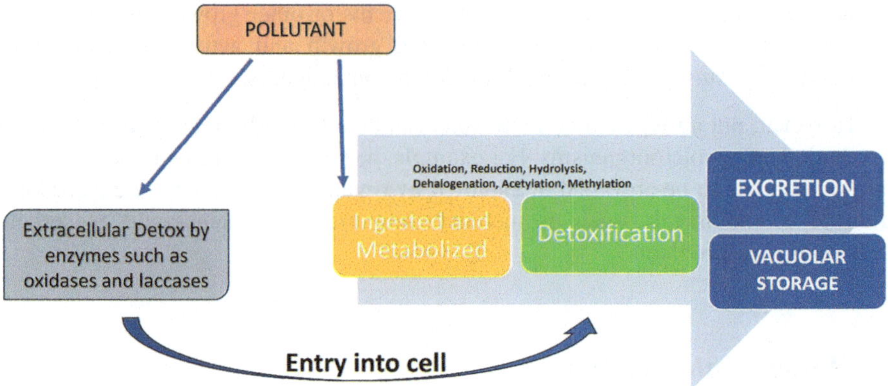

Fig. 3 Fungal strategies for the removal of pollutants from the environment

Fungi have been found to be effective in decolorizing and treating colored effluents from industries like sugar, textiles, bleached kraft pulp mills, and leather tanneries. Various fungal groups, including *Aspergillus* and *Penicillium* species, as well as alkalophilic white-rot fungi, have shown proficiency in this regard, showcasing their versatility in targeting different pollutants (Jebapriya and Gnanadoss 2013; Huang et al. 2014). Fungi have also been harnessed for specific applications like the decaffeination of coffee pulp under controlled conditions with added nutrients. This process renders the decaffeinated pulp suitable for animal feed preparation or bioethanol production, with fungi such as *Aspergillus restrictus, Chrysosporium keratinophilum, Fusarium solani, Gliocladium roseum, Penicillium*, and *Stemphylium* showing promise (Nayak et al. 2013). In the realm of hydrocarbon degradation, fungi have demonstrated their potential. *A. niger* and *P. chrysosporium*, for instance, have exhibited significant removal of petroleum hydrocarbons from soil contaminated with petrol and diesel within short incubation periods. This is evidenced by enhanced total organic carbon (TOC) removal. Even the removal of challenging contaminants like chlorpyrifos and its metabolite 3,5,6-trichloro-2-pyridinol (TCP) from contaminated soils has been achieved using fungi like *A. niger* JAS1. Importantly, this process was accomplished without additional nutrients and underscores the fungi's ability to address substances with antimicrobial properties and catabolite repression traits (Silambarasan and Abraham 2013). In essence, fungi have demonstrated their versatile capabilities in bioremediation across diverse pollutants and industries. Their capacity to break down and remove contaminants, even in challenging conditions, highlights their potential as efficient and environmentally friendly tools for tackling various forms of pollution.

The widespread presence of heavy metals in the environment, stemming from their diverse applications, raises concerns due to their systemic toxicity to human health even at low concentrations. Due to their high degree of toxicity, which can lead to severe health effects like multiple organ failure and carcinogenic effects, certain heavy metals—namely arsenic, cadmium, chromium, lead, and mercury—

are classified as priority metals requiring removal from the environment to mitigate their impact on both public health and ecosystems. Tchounwou et al. (2012) have conducted analyses on the role and production of various heavy metals, their occurrence in the environment, and their relation to potential human exposure. Their focus extends to the molecular mechanisms underlying the toxic effects of these metals. Various microorganisms exhibit the ability to tolerate heavy metal presence and employ diverse mechanisms for their removal from the environment. Notably, filamentous fungi exhibit high tolerance and significant remediation capabilities for heavy metals like cadmium (Cd), copper (Cu), and nickel (Ni), with some strains demonstrating resilience even at concentrations as high as 1500 mg/L (Akhtar et al. 2013). This characteristic of fungi is particularly promising for bioremediation efforts aimed at cleaning up heavy metal-contaminated soil and wastewater (Table 1).

Genus *Aspergillus* showcases remarkable versatility in degrading an array of toxic compounds, spanning heavy metals, textile dyes, aromatic compounds, and pesticides. Species like *A. flavus* and *A. niger* have been identified for their capacity to reduce hexavalent chromium (Cr^{6+}) to trivalent chromium (Cr^{3+}), which is less toxic (Bennett et al. 2013). *A. foetidus*, isolated from a wastewater treatment plant, exhibited tolerance to high lead (Pb) concentrations, up to 200 mg/L, and removed it through a process called biosorption. Similarly, *A. flavus* displayed the ability to remove about 98% of mercury from aqueous substrates containing 10 mg/L of mercury (Kurniati et al. 2014). Researchers, such as Mumtaz et al. (2013), have demonstrated the potential of fungi including *Aspergillus, Cryptococcus, Penicillium*, and *Curvularia* in bioremediating uranium-contaminated soils, attributing their efficacy to their ability to bind uranium. Symbiotic associations of arbuscular mycorrhizal (AM) fungi with plant roots contribute to the immobilization of heavy metals, enhancing plant growth in metal-contaminated soils by improving Cd tolerance (Yang et al. 2015; Garg and Bhandari 2014). Enhancements in fungi's remediation potential can be achieved through specific pre-treatments. For instance, Das et al. (2013) exposed *Aspergillus* sp. to gamma rays in media supplemented with cadmium, resulting in increased growth and improved Cd removal compared to unirradiated controls.

The applicability of arbuscular mycorrhizal fungi (AMF) has been evaluated for alleviation of cadmium stress in an experiment conducted on a medicinal plant *Cassia italica*. Authors observed in the experiment that cadmium exposure resulted in reduction of chlorophyll and protein which was reversed by addition of AMF in the growth medium. Inoculation with AMF resulted in an increase of 19.3% and 38.3% in chlorophyll and protein content, respectively, in Cd stressed plants. AMF inoculation also reduced lipid peroxidation in Cd exposed plants along with enhancing the proline and phenol contents in the same. Therefore, the presented research gives the evidentiary proofs that AMF inoculation can mitigate negative impact of stress by reducing lipid peroxidation and enhancing antioxidant activity in the presence of Cd (Hashem et al. 2016).

Table 1 Fungal species used in remediation of Cd poisoning in environment

Fungal species	Nature of use	Effect exerted	References
Glomus mosseae Glomus etunicatum	Used for alleviation of Cd stress in *Cassia italica* plants	Reduction of lipid peroxidation, enhancement of phenol and proline content, enhancement of antioxidant system	Hashem et al. (2016)
Coprinopsis atramentaria	Used as mycoremediant to remove Cd in medium	Can remove 76% of cadmium at a concentration of 1 mg L^{-1} of cadmium	Cobas et al. (2013)
Aspergillus foetidus	Can effectively tolerate 25 mM and 63 mM Cd in the medium and remove it from the media	Production of oxalic acid for bioleaching of Cd and increase cellular proline and antioxidants to combat increased free radicals	Chakraborty et al. (2014)
Saccharomyces cerevisiae	Transformed cells with metallothionine was used for removal of Cd from growth media	Removal of approximately 80% Cd from the growth media	De Oliveira et al. (2020)
Aspergillus niger	Used for reduction of Cd toxicity in wheat plants	Increase in total sugars, total proteins and ascorbic acid; and reduction in total flavonoids, total phenolics and H_2O_2 in Cd stressed plants	Bashir et al. (2022)
Aspergilus versicolor	Isolated from Cd-contaminated site and their Cd removal potential studied in the liquid media	Able to bioaccumulate over 7 mg of cadmium per 1 g of mycelium	Fazli et al. (2015)
Aspergilus fumigatus	Cd removal potential tested in the culture media	Accumulated almost 5.02 mg Cd per g mycelium	Tulakdar et al. (2020)
Penicillium chrysogenum FMS2	Cd removal potential tested in the culture media	Able to tolerate upto 1000 mg L^{-1} Cd and bioaccumulated Cd in cells clearly visible under electron microscope	Din et al. (2022)
Trichoderma asperellum, T. harzianum and *T. tomentosum*	Studied for Cd removal potential at various pH and concentrations in media	Removed Cd at all the pH ranges tested with maximum removal being 10.75 mg/g of fungal dry weight	Mohsenzadeh and Shahrokhi (2014)
Cladosporium sphaerospermum and *Fusarium oxysporum*	Biosorption of Cd in waste waters collected form factories was tested	The fungal species were able to remove upto 65% and 56% Cd in different samples, respectively	Ali et al. (2018)
Paecilomyces lilacinus	Isolated and tested for Cd uptake in media containing 100 mg L^{-1} Cd	Was able to biosorb Cd upto a concentration of 24.23 mg/g during growth phase	Zeng et al. (2010)
Penicellium sp. XK10	Isolated and absorption characteristics for Cd studied in different environmental conditions	Maximum absorption of Cd observed at pH 4 of upto 35% in 7 days time	He et al. (2023)

(continued)

Table 1 (continued)

Fungal species	Nature of use	Effect exerted	References
Aspergillus terreus LCM8	Melanin production ability of the species tested for Cd removal	Species was able to tolerate 2000 mg /L Cd and melanin produced was able to reduce 40 mg Cd in medium initially containing 100 mg/L Cd in 7 days	Hayat et al. (2023)
Penicillium oxalicum ZP6	Cd bioimmobilization by the fuingi was tested in phosphate mining wastelands	81.6% Cd removed or bioimmobilized by formation of Cd-phosphate or Cd-oxalate	Zheng et al. (2022)
Glomus mosseae	Cd removal was tested in association with Rosemary, Amaranth and ornamental cabbage	Inoculated AMF significantly increased P uptake along with Cd removal with highest efficiency in Rosemary	Nasiri et al. (2022)
Trametes pubescens	Cocultured with rice seedlings in pot experiment	Upregulated MAPK and phosphatidyl inositol signalling pathways and rewired energy metabolism and N uptake under Cd stress	Liu et al. (2022)

8 Bacterial Interventions

Bioremediation is a process that relies on the use of biological mechanisms to transform, degrade, and detoxify pollutants, converting them into harmless forms, ultimately leading to the mineralization of these pollutants into compounds like carbon dioxide, nitrogen, and water. This process can involve the use of living or dead biomass from various sources, such as plants, fungi, bacteria, algae, and cyanobacteria. Among these, microorganisms, particularly bacteria, have gained significant attention for their ability to remediate heavy metal pollution (Sreedevi et al. 2022). Microorganisms are particularly well-suited for bioremediation due to their ease of handling, culturing, and implementation (Table 2). They offer unique advantages such as omnipresence, abundance, diversity, small size, and the capacity to grow and reproduce under both controlled and uncontrolled environmental conditions (Kapahi and Sachdeva 2019).

In the context of heavy metal bioremediation, bacteria have shown remarkable capabilities. They can employ various strategies, including adsorption (or biosorption), bioaccumulation, bioleaching, and bioprecipitation, to mitigate heavy metal pollution (as shown in Fig. 4). Adsorption occurs primarily on the bacterial cell surface and acts as the first line of defence against metal toxicity. The bacterial cell wall contains negatively charged functional groups like carboxyl, amino, phosphoryl, and sulfo groups, which serve as ion exchange sites and metal

Table 2 Various bacteria used in bioremediation of Cadmium

Bacteria	Nature of use	Effect exerted	References
Pseudomonas aeruginosa	Cd removal checked in media as well as inoculated with Pea plants	Adsorbed and immobilized Cd in media and enhanced root and shoot biomass of Pea plants and reduced tissue Cd concentration	Huang et al. (2016)
Pseudomonas sp.	Isolated from industrial waste and Cd tolerance and removal ability tested	Able to tolerate Cd upto 2100 μg/ml and showed high efficacy for removal of Cd from medium	Al-Dhabi et al. (2019)
Bacillus subtilis	Used for reduction of Cd toxicity in wheat plants	Increase in total sugars, total proteins and ascorbic acid; and reduction in total flavonoids, total phenolics and H_2O_2 in Cd stressed plants	Bashir et al. (2022)
Desulfovibrio sp.	Sodium alginate immobilized bacteria tested for heavy metal removal in continuous batch process	Removed Cd upto 92% from the medium	Kiran et al. (2018)
Paenarthrobactor nitroguajacolicus, Lysinibacillus fusiformis, Bacillus licheniformis, and *Methyllobacium brachiatum*	Inoculated to Cd-contaminated rape plants	Changed soil physical and chemical properties; reduced Cd uptake and transport in plants and reduced Cd availability in soil	Ma et al. (2023)
Escherichia coli	Transgenic engineered *E. coli* cells were used overexpressing metallothionine protein on surface	Almost 5 time more Cd adsorption than wild-type *E. coli*	Lu et al. (2023)
Bacillus vietnamensis 151–6	Rice plants inoculated with the strain	Inoculation increased panicle number to 114% and reduced grain Cd content by 52%	Yu et al. (2023)
Sporosarcina pasteurii	Added to contaminated soil coupled with rice straw	Enhanced Cd immobilization and improved available nitrogen, phosphorus, and potassium in the soil by a significant amount	Cai et al. (2023)
Myxotrphic acidophiles	Used as soil treatment in rice cultivation system	Cd removal efficiency was enhanced by 40%, improved microbial diversity and nutrient content of soils, better rice growth	Yuan et al. (2022)

(continued)

Table 2 (continued)

Bacteria	Nature of use	Effect exerted	References
Curtobacterium oceanosedimentum	Inoculated to rhizosphere of chili plants as a plant growth-promoting rhizosphere additive	Improved phytostabilization potential of chili plants and improved growth and productivity of chili plants	Patel et al. (2022)
Paenibacillus, Burkholderia, Ensifer, and *Cupriavidus*	Isolated from Cd, Cr, and Ni-contaminated soil and used in maize cultivation system	Removed upto 33% Cd and upto 60% Cr from medium and were also able to improve maize growth significantly	Minari et al. (2020)
Bacillus sp. TZ5	Inoculated along with biochar in ryegrass experimental system	Cd mobility in soil was reduced by 11% and Cd content in ryegrass was reduced by almost 45%	Ma et al. (2020)

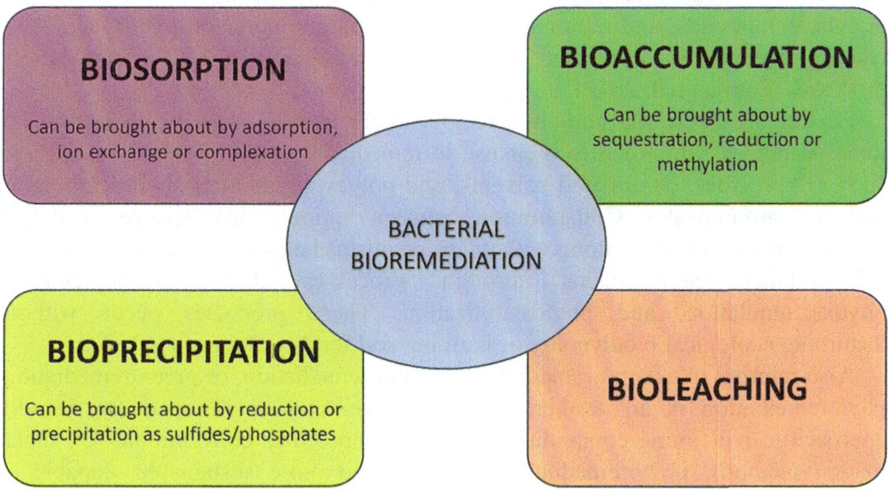

Fig. 4 Mechanisms of bacteria mediated bioremediation processes

binding sites. This process involves redox reactions, ion exchange, complexation, electrostatic attraction, and precipitation. Adsorption is an energy-independent process, where heavy metals are passively bound to the cell surface until equilibrium is reached.

Bioaccumulation, on the other hand, is an active process where living bacterial biomass takes up heavy metal ions. Bacteria can bind these ions using exopolymers or polysaccharides on their cell surface, and these metal species can further be internalized into the cell for various metabolic functions. The bacteria can also produce metabolic ligands that chelate the metals, forming stable complexes. Bioleaching involves the release or excretion of ligands, such as organic acids, cyanide, thiosulfate, and phenol derivatives, by bacteria. These ligands interact with metals in their vicinity, forming soluble complexes that can be easily mobilized. Bioprecipitation is a metal-sequestering process performed by metabolically active cells. Anaerobic bacteria, in particular, are known to convert metals present or dissolved in water into solid precipitates. This is achieved through processes such as sulfate, carbonate, phosphate, or hydroxide precipitation (Srivastava et al. 2015; Kumari et al. 2016; Ilyas et al. 2018).

9 Phytoremediation

The term "phytoremediation" originates from the combination of the Greek word "phyto," meaning plant, and the Latin word "remedium," signifying correction or removal of something harmful (Ali et al. 2013). Phytoremediation is a promising green technology that involves using plants to restore soil, sediments, groundwater, and surface water contaminated with environmental pollutants like toxic metals, organic compounds, and radionuclides. This technique relies on plants' ability to take up contaminants from their surroundings and detoxify them through various processes (Sarma et al. 2019).

Phytoremediation was initially recognized for plants' capacity to accumulate toxic pesticides, which led to their use in remediating a wide range of pollutants such as pesticides, chlorinated solvents, and polycyclic aromatic hydrocarbons in soil and groundwater. Collaboration between various plant species including grasses, shrubs, and trees along with microorganisms helps facilitate the purification of polluted environments through processes like phytodegradation, phytoaccumulation, and phytostabilization. These processes occur without disturbing ecological biodiversity or harming soil fertility (Rajput et al. 2021).

Also referred to as agro-remediation, botano-remediation, or green remediation, phytoremediation is an aesthetic, cost-effective, and environmentally friendly approach to reclaiming contaminated areas. Compared to conventional soil reclamation methods, phytoremediation is considered more sustainable, capable of preventing the spread of pollutants, and reducing their movement in the environment. The advantages of phytoremediation include:

- **Economic Feasibility**: Phytoremediation is driven by solar energy and thus easy to manage, with low installation and maintenance costs.
- **Environmental and Eco-Friendly**: It reduces exposure of pollutants to the environment and ecosystems.
- **Scalability**: Applicable on a large-scale field and easy to implement and dispose of.
- **Preventing Erosion and Metal Leaching**: By stabilizing heavy metals, it prevents erosion and leaching, curbing the risk of contaminant spreading.
- **Soil Fertility Improvement**: It releases organic matter into the soil, potentially enhancing soil fertility (Arthur et al. 2005).

Phytoremediation has emerged as a cost-effective, non-invasive, and socially acceptable approach to addressing environmental contaminant removal. This method avoids the need for excavation of contaminated sites, reduces the risk of contaminant dispersion, and can be employed for remediating sites contaminated with various pollutants. The efficiency and mechanisms of phytoremediation depend on factors such as the type of contaminant, its bioavailability, soil characteristics, and the specific plant species used (Cristaldi et al. 2017).

10 Conclusion

Presence of cadmium in the environment is one of the most critical problem that has to be dealt with in a manner such that environment can be remediated in a friendly manner. Removal of such hazards using the biological systems is considered as bioremediation. Bioremediation by fungi, bacteria, and plants can treat heavy metal problem by mechanisms that benefit the environment. Biosorption, bioimmobilization, bioleaching, bioaccumulation, etc. can help to remove and/or stabilize Cd in the soil or the organism involved. In conclusion it is safe to say that bioremediation is most versatile and eco-friendly mechanism to deal with heavy metal pollution case in point cadmium.

References

Abarikwu SO, Olufemi PD, Lawrence CJ, Wekere FC, Ochulor AC, Barikuma AM (2017) Rutin, an antioxidant flavonoid, in-duces glutathione and glutathione peroxidase activities to protect against ethanol effects in cadmium-induced oxidative stress in the testis of adult rats. Andrologia 49(7):e12696

Afzal J, Hu C, Imtiaz M, Elyamine AM, Rana MS, Imran M, Farag MA (2019) Cadmium tolerance in rice cultivars associated with antioxidant enzymes activities and Fe/Zn concentrations. Int J Environ Sci Technol 16:4241–4252. https://doi.org/10.1007/s13762-018-2018-y

Akhtar S, Mahmood-ul-Hassan M, Ahmad R, Suthor V, Yasin M (2013) Metal tolerance potential of filamentous fungi isolated from soils irrigated with untreated municipal effluent. Soil Environ 32:55–62

Al-Dhabi NA, Esmail GA, Ghilan AKM, Arasu MV (2019) Optimizing the management of cadmium bioremediation capacity of metal-resistant *Pseudomonas* sp. strain Al-Dhabi-126 isolated from the industrial city of Saudi Arabian environment. Int J Environ Res Public Health 16:4788. https://doi.org/10.3390/ijerph16234788

Ali H, Khan E, Sajad MA (2013) Phytoremediation of heavy metals—concepts and applications. Chemosphere 91(7):869–881

Ali F, Kazi TG, Afridi HI, Baig JA (2018) Exposure of cadmium via smoking and drinking water on zinc levels of biological samples of malnutrition pregnant women: a prospective cohort study. Environ Toxicol Pharmacol 63:48–54

Arthur EL, Rice PJ, Rice PJ, Anderson TA, Baladi SM, Henderson KLD, Coats JR (2005) Phytoremediation—an overview. Crit Rev Plant Sci 24(2):109–122. https://doi.org/10.1080/07352680590952496

Baker AJM, Ewart K, Hendry GAF, Thorpe PC, Walker PL (1990) The evolutionary basis of Cd tolerance in higher plants. In: 4th international conference on environmental contamination. Barcelona, Spain, pp 23–29

Balestrasse KB, Gardey L, Gallego SM, Tomaro ML (2001) Response to antioxidant defence system in soybean nodules and roots subjected to cadmium stress. Aust J Plant Physiol 28:497–504

Balestrasse KB, Benavides MP, Gallego SM, Tomaro ML (2003) Effect of cadmium stress on nitrogen metabolism in nodules and roots of soybean plants. Funct Plant Biol 30:57–64

Balestrasse KB, Gallego SM, Tomaro ML (2004) Cadmium induced senescence in nodules of soybean (*Glycine max* L.) plants. Plant Soil 262:373–381

Bashir S, Javed S, Al-Anazi KM, Farah MA, Ali S (2022) Bioremediation of cadmium toxicity in wheat (*Triticum aestivum* L.) plants primed with L-Proline, *Bacillus subtilis* and *Aspergillus niger*. Int J Environ Res Public Health 19:12683. https://doi.org/10.3390/ijerph191912683

Bennett RM, Cordero PRF, Bautista GS, Dedeles GR (2013) Reduction of hexavalent chromium using fungi and bacteria isolated from contaminated soil and water samples. Chem Ecol 29:320–328. https://doi.org/10.1080/02757540.2013.770478

Blum WH (1997) Cadmium uptake by higher plants. In: Proceedings of extended abstracts from the fourth international conference on the biogeochemistry of trace elements. University of California, Berkely, USA, pp 109–110

Cai Q, Xu M, Ma J, Zhang X, Yang G, Long L, Chen C, Wu J, Song C, Xiao Y (2023) Improvement of cadmium immobilization in contaminated paddy soil by using ureolytic bacteria and rice straw. Sci Tot Environ 874:162594. https://doi.org/10.1016/j.scitotenv.2023.162594

Cao Z, Fang Y, Lu Y, Tan D, Du C, Li Y, Pei L (2017) Melatonin alleviates cadmium-induced liver injury by inhibiting the TXNIP-NLRP3 inflammasome. J Pineal Res 62(3):e12389

Chakraborty S, Mukherjee A, Khuda-Buksh AR, Das TK (2014) Cadmium-induced oxidative stress tolerance in cadmium resistant *Aspergillus foetidus*: its possible role in cadmium bioremediation. Ecotoxicol Environ Saf 146:46–53

Chen X, Cui W, Duan N, Zhu G, Jin T, Wang Z (2020) The as-sociation between alcohol consumption and renal tubular dysfunction induced by cadmium exposure. Biol Trace Elem Res 194(1):58–65

Clemens S, Palmgren MG, Kramer U (2002) A long way ahead: understanding and engineering plant metal accumulation. Trends Plant Sci 7:309–315

Cobas M, Ferreira L, Tavares T, Sanroman MA, Pazos M (2013) Development of permeable reactive biobarrier for the removal of PAHs by Trichoderma longibrachiatum. Chemosphere 91:711–716. https://doi.org/10.1016/j.chemosphere.2013.01.028

Cristaldi A, Conti GO, Jho EH, Zuccarello P, Grasso A, Copat C, Ferrante M (2017) Phytoremediation of contaminated soils by heavy metals and PAHs. A brief review. Environ Technol Innov 8:309–326

Das P, Samantaray S, Rout GR (1997) Studies on cadmium toxicity in plants: a review. Environ Pollut 98:29–36

Das D, Chakraborty A, Bhar S, Sudarshan M, Santra SC (2013) Gamma irradiation in modulating cadmium bioremediation potential of Aspergillus sp. IOSR J Environ Sci Toxicol Food Technol 3:51–55

De Oliveira VH, Ullah I, Dunwell JM, Tibbett M (2020) Bioremediation potential of Cd by transgenic yeast expressing a metallothionein gene from Populus trichocarpa. Ecotoxicol Environ Saf 202:110917. https://doi.org/10.1016/j.ecoenv.2020.110917

Deshmukh R, Khardenavis AA, Purohit HJ (2016) Diverse metabolic capacities of fungi for bioremediation. Indian J Microbiol 56(3):247–264

Din G, Hassan A, Dunlap J (2022) Cadmium tolerance and bioremediation potential of filamentous fungus Penicillium chrysogenum FMS2 isolated from soil. Int J Environ Sci Technol 19:2761–2770. https://doi.org/10.1007/s13762-021-03211-7

Dongre NN, Suryakar AN, Patil AJ, Hundekari IA, Devar-navadagi BB (2013) Biochemical effects of lead exposure on battery manufacture workers with reference to blood pressure, calcium metabolism and bone mineral density. Indian J Clin Biochem 28(1):65–70

Duxbury T (1985) Ecological aspects of heavy metal responses in microorganisms. Adv Microb Ecol 8:185–235

Faroon O, Ashizawa A, Wright S, Tucker P, Jenkins K, Ingerman L, Rudisill C (2022) Toxicological profile for cadmium. ATSDR, Agency for Toxic Substances and Disease Registry, Atlanta, GA

Fay MJ, Alt LA, Ryba D, Salamah R, Peach R, Papaeliou A, Stubbs-Russell Z (2018) Cadmium nephrotoxicity is associated with altered microRNA expression in the rat renal cortex. Toxics 6(1):16–30

Fazli MM, Soleimani N, Mehrasbi M, Darabian S, Mohammadi J, Ramazani A (2015) Highly cadmium tolerant fungi: their tolerance and removal potential. J Environ Health Sci Eng 13:19. https://doi.org/10.1186/s40201-015-0176-0

Fornazier RF, Ferreira RR, Vitoria AP, Molina SMG, Lea PJ, Azevedo RA (2002) Effects of cadmium on antioxidant enzyme activities in sugar cane. Biol Plant 45:91–97

Fujiki K, Inamura H, Sugaya T, Matsuoka M (2019) Blockade of ALK4/5 signaling suppresses cadmium-and erastin-induced cell death in renal proximal tubular epithelial cells via distinct signaling mechanisms. Cell Death Differ 26(11):2371–2385

Garg N, Bhandari P (2014) Cadmium toxicity in crop plants and its alleviation by arbuscular mycorrhizal (AM) fungi: an overview. Plant Biosyst 148:609–621. https://doi.org/10.1080/11263504.2013.788096

Geng HX, Wang L (2019) Cadmium: toxic effects on placental and embryonic development. Environ Toxicol Pharmacol 67:102–107

Haider FU, Liqun C, Coulter JA, Cheema SA, Wu J, Zhang R, Wenjun M, Farooq M (2021) Cadmium toxicity in plants: impacts and remediation strategies. Ecotoxicol Environ Saf 211: 111887. https://doi.org/10.1016/j.ecoenv.2020.111887

Hassanin ES, El-Bassel H, El-Razek A (2017) Effect of Cadmium Exposure among workers in Battery Industries. Bull Natl Nutr Inst Arab Repub Egypt 49(1):1–20

Hashem A, Abd-Allah EF, Alqaravi AA, Egamberdieva D (2016) Bioremediation of adverse impact of cadmium toxicity on *Cassia italica* Mill by arbuscular mycorrhizal fungi. Saudi J Biol Sci 23:39–47

Hayat R, Din G, Farooqi A, Haleem A, Din S, Hasan F, Badshah M, Khan S, Shah AA (2023) Characterization of melanin pigment from *Aspergillus terreus* LCM8 and its role in cadmium remediation. Int J Env Sci Technol 20:3151–3160. https://doi.org/10.1007/s13762-022-04165-0

He Y, Li C, Sun Z, Zhang W, He J, Zhao Y, Xu Z, Zhao W (2023) Penicillium spp. XK10, fungi with potential to repair cadmium and antimony pollution. Appl Sci 13:1228. https://doi.org/10.3390/app13031228

Huang J, Fu Y, Liu Y (2014) Comparison of alkali-tolerant fungus Myrothecium sp. IMER1 and white-rot fungi for decolorization of textile dyes and dye effluents. J Bioremed Biodegr 5:1–5. https://doi.org/10.4172/2155-6199.1000221

Huang J, Liu Z, Li S, Xu B, Gong Y, Yang Y, Sun H (2016) Isolation and engineering of plant growth promoting rhizobacteria Pseudomonas aeruginosa for enhanced cadmium bioremediation. J Gen Appl Microbiol 62:258–265

Ilyas S, Kim MS, Lee JC (2018) Integration of microbial and chemical processing for a sustainable metallurgy. J Chem Technol Biotechnol 93:320–332. https://doi.org/10.1002/jctb.5402

Jebapriya GR, Gnanadoss JJ (2013) Bioremediation of textile dye using white-rot fungi: a review. Int J Curr Res Rev 5:1–13

Kapahi M, Sachdeva S (2019) Bioremediation options for heavy metal pollution. J Health Pollut 9: 191203. https://doi.org/10.5696/2156-9614-9.24.191203

Karahan F, Ozyigit II, Saracoglu IA, Yalcin IE, Hocaoglu-Ozyigit A, Ilcim A (2020) Heavy metal levels and mineral nutrient status in different parts of various medicinal plants collected from eastern Mediterranean region of Turkey. Biol Trace Elem Res 197:316–329

Kensa VM (2011) Bioremediation – an overview. J Ind Pollut Control 27:161–168

Kiran MG, Pakshirajan K, Das G (2018) Heavy metal removal from aqueous solution using sodium alginate immobilized sulfate reducing bacteria: mechanism and process optimization. J Environ Manag 218:486–496. https://doi.org/10.1016/j.jenvman.2018.03.020

Kmecick M, da Costa MCV, de Oliveira Ribeiro CA, Ortolani-Machado CF (2019) Morphological evidence of neurotoxic effects in chicken embryos after exposure to perfluorooctanoic acid (PFOA) and inorganic cadmium. Toxicology 427:152286

Kumari D, Qian XY, Pan X, Achal V, Li Q, Gadd GM (2016) Microbially-induced carbonate precipitation for immobilization of toxic metals. Adv Appl Microbiol 94:79–108. https://doi.org/10.1016/bs.aambs.2015.12.002

Kurniati E, Arfarita N, Imai T, Higuchi T, Kanno A, Yamamoto K, Sekine M (2014) Potential bioremediation of mercury-contaminated substrate using filamentous fungi isolated from forest soil. J Environ Sci 26:1223–1231. https://doi.org/10.1016/S1001-0742(13)60592-6

Lavryshyn YY, Gutyj BV (2019) Protein synthesize function of bulls liver at experimental chronic cadmium toxicity. Sci Messen LNU Vet Med Biotechnol Ser Vet Sci 21(94):92–96

Lee JY, Tokumoto M, Fujiwara Y, Hasegawa T, Seko Y, Shimada A, Satoh M (2016) Accumulation of p53 via down-regulation of UBE2D family genes is a critical pathway for cadmium-induced renal toxicity. Sci Rep 6(1):1–15

Liu J, Fu P, Wang L, Lin X, Enayatizamir N (2022) A fungus (*Trametes pubescens*) resists cadmium toxicity by rewiring nitrogen metabolism and enhancing energy metabolism. Front Microbiol 13. https://doi.org/10.3389/fmicb.2022.1040579

Lu CW, Ho HC, Yao CL, Tseng TY, Kao CM, Chen SC (2023) Bioremediation potential of cadmium by recombinant *Escherichia coli* surface expressing metallothionein MTT5 from *Tetrahymena thermophila*. Chemosphere 310:136850. https://doi.org/10.1016/j.chemosphere.2022.136850

Ma H, Wei M, Wang Z, Hou S, Li X, Xu H (2020) Bioremediation of cadmium polluted soil using a novel cadmium immobilizing plant growth promotion strain Bacillus sp. TZ5 loaded on biochar. J Hazard Mater 388:122065. https://doi.org/10.1016/j.jhazmat.2020.122065

Ma B, Song W, Zhang X, Chen M, Li J, Yang X, Zhang L (2023) Potential application of novel cadmium-tolerant bacteria in bioremediation of Cd-contaminated soil. Ecotoxicol Environ Saf 255:114766. https://doi.org/10.1016/j.ecoenv.2023.114766

Manzoor F, Karbassi A, Golzari A (2019) Removal of heavy metal contaminants from wastewater by using *Chlorella vulgaris* Beijerinck: a review. Curr Environ Manag 6(3):174–187

Meena V, Dotaniya ML, Saha JK, Meena BP, Das H, Patra AK (2020) Sustainable C and N management under metal-contaminated soils. In: Datta R, Meena RS, Pathan SI, Ceccherini MT (eds) Carbon and nitrogen cycling in soil. Springer, Singapore, pp 293–336

Minari GD, Saran LM, Constancio MTL, da Silva RC, Rosalen DL, De Melo WJ, Alves LCT (2020) Bioremediation potential of new cadmium, chromium, and nickel-resistant bacteria isolated from tropical agricultural soil. Ecotoxicol Environ Saf 204:111038. https://doi.org/10.1016/j.ecoenv.2020.111038

Mohsenzadeh F, Shahrokhi F (2014) Biological removing of cadmium from contaminated media by fungal biomass of Trichoderma species. J Environ Health Sci Eng 12:102. https://doi.org/10.1186/2052-336X-12-102

Moreno JL, Hernandez T, Garcia C (1999) Effects of cadmium containing sewage sludge compost on dynamics of organic matter and microbial activity in arid soils. Biol Fert Soils 28:230–237

Mumtaz S, Streten-Joyce C, Parry DL, Mc Guinness KA, Lu P, Gibb KS (2013) Fungi outcompete bacteria under increased uranium concentration in culture media. J Environ Radioact 120:39–44. https://doi.org/10.1016/j.jenvrad.2013.01.007

Nasiri K, Babaeinejad T, Ghanavati N, Mohsenifar K (2022) Arbuscular mycorrhizal fungi affecting the growth, nutrient uptake and phytoremediation potential of different plants in a cadmium-polluted soil. Biometals 35:1243–1125. https://doi.org/10.1007/s10534-022-00439-9

Nayak V, Pai PV, Pai A, Pai S, Sushma YD, Rao CV (2013) A comparative study of caffeine degradation by four different fungi. Biorem J 17:79–85. https://doi.org/10.1080/10889868.2012.751960

Nishijo M, Nambunmee K, Suvagandha D, Swaddiwudhipong W, Ru-angyuttikarn W, Nishino Y (2017) Gender-specific impact of cadmium exposure on bone metabolism in older people living in a cadmium-polluted area in Thailand. Int J Environ Res Public Health 14(4):401

Okoye CN, MacDonald-Jay N, Kamunde C (2019) Effects of bio-energetics, temperature and cadmium on liver mitochondria reactive oxygen species production and consumption. Aquat Toxicol 214:105264

Ozyigit AH, Genc BN (2020) Cadmium in plants, humans and the environment. Front Life Sci Relat Technol 1(1):12–21

Patel M, Patel K, Al-Keridis LA, Alshammari N, Badraoui R, Elasbali AM, Al-Soud WA, Hassan MI, Yadav DK, Adnan M (2022) Cadmium-tolerant plant growth-promoting bacteria Curtobacterium oceanosedimentum improves growth attributes and strengthens antioxidant system in chili (Capsicum frutescens). Sustainability 14:4335. https://doi.org/10.3390/su14074335

Person RJ, Tokar EJ, Xu Y, Orihuela R, Ngalame NNO, Waalkes MP (2013) Chronic cadmium exposure in vitro induces cancer cell characteristics in human lung cells. Toxicol Appl Pharmacol 273(2):281–288

Pinto AP, Mota AM, de Varennes A, Pinto FC (2004) Influence of organic matter on the uptake of cadmium, zinc, copper and iron by sorghum plants. Sci Total Environ 326:239–247

Pratinidhi SA, Sagare AA, Patil AJ (2018) Heavy metal levels in commonly used cosmetic products in Asia. MIMER Med J 2(2):31–36

Prokopowicz A, Sobczak A, Szuła-Chraplewska M, Ochota P, Kośmider L (2019) Exposure to cadmium and lead in cigarette smokers who switched to electronic cigarettes. Nicotine Tob Res 21(9):1198–1205

Rajput V, Minkina T, Semenkov I, Klink G, Tarigholizadeh S, Sushkova S (2021) Phylogenetic analysis of hyperaccumulator plant species for heavy metals and polycyclic aromatic hydrocarbons. Environ Geochem Health 43:1629–1654. https://doi.org/10.1007/s10653-020-00527-0

Ramos I, Esteban E, Lucena JJ, Garate A (2002) Cadmium uptake and subcellular distribution in plants of Lactuca sp. Cd-Mn interaction. Plant Sci 162:761–767

Reyes-Hinojosa D, Lozada-Pérez CA, Cuevas YZ, López-Reyes A, Martínez-Nava G, Fernández-Torres J, Martínez-Flores K (2019) Toxicity of cadmium in musculoskeletal diseases. Environ Toxicol Pharmacol 72:103219

Rivetta A, Negrini N, Cocucci M (1997) Involvement of Ca^{2+}-calmodulin in Cd toxicity during the early phases of radish (*Raphanus sativus* L.) seed germination. Plant Cell Environ 20:600–608

Rosales E, Pazos M, Angeles SM (2013) Feasibility of solid-state fermentation using spent fungi-substrate in the biodegradation of PAHs. Clean Soil Air Water 41:610–615. https://doi.org/10.1002/clen.201100305

Sandilo LM, Dalurzo HC, Gomez M, Romero-Puertas MC, del Rio LA (2001) Cadmium induced changes in the growth and oxidative metabolism of pea plants. J Exp Bot 52:2115–2126

Sanita di Toppi L, Gabbrielli R (1999) Response to cadmium in higher plants. Environ Exp Bot 41:105–130

Santoyo-Sánchez M, Thévenod F, Barbier O (2018) Cadmium handling, toxicity and molecular targets involved during pregnancy: lessons from experimental models. Metal Metab Anim 18:272

Sarma H, Nava AR, Prasad MNV (2019) Mechanistic understanding and future prospect of microbe-enhanced phytoremediation of polycyclic aromatic hydrocarbons in soil. Environ Toxicol Innovation 13:318–330. https://doi.org/10.1016/j.eti.2018.12.004

Sharma A, Shankhdhar D, Shankhdahr SC (2013) Enhancing grain iron content of rice by the application of plant growth promoting rhizobacteria. Plant Soil Environ 59(2):89–94

Silambarasan S, Abraham J (2013) Ecofriendly method for bioremediation of chlorpyrifos from agricultural soil by novel fungus Aspergillus terreus JAS1. Water Air Soil Pollut 224:1369. https://doi.org/10.1007/s11270-012-1369-0

Somashekaraiah BV, Padmaja K, Prasad ARK (1992) Phytotoxicity of cadmium ions on germinating seedlings of mung bean (*Phaseolus vulgaris*): involvement of lipid peroxides in chlorophyll degradation. Physiol Plant 85:85–89

Sreedevi PR, Suresh K, Jiang G (2022) Bacterial bioremediation of heavy metals in wastewater: a review of processes and applications. J Water Process Eng 48:102884. https://doi.org/10.1016/j.jwpe.2022.102884

Srivastava S, Agrawal SB, Mondal MK (2015) A review on progress of heavy metal removal using adsorbents of microbial and plant origin. Environ Sci Pollut Res 22:15386–15415. https://doi.org/10.1007/s11356-015-5278-9

Stohs SJ, Bagchi D (1995) Oxidative mechanisms in the toxicity of metal ions. Free Radic Biol Med 18:321–336

Sundblad BM, Ji J, Levänen B, Midander K, Julander A, Larsson K, Lindén A (2016) Extracellular cadmium in the bronchoalveolar space of long-term tobacco smokers with and without COPD and its association with inflammation. Int J Chron Obstruct Pulmon Dis 11:1005

Tao C, Pei Y, Zhang L, Zhang Y (2020) Microbial communities respond to microenvironments in lungs of mice under simulated expo-sure to cadmium aerosols. Sci Total Environ 710:136300

Tchounwou PB, Yedjou CG, Patlolla AK, Sutton DJ (2012) Heavy metal toxicity and the environment. Exp Suppl 101:133–164

Tester M, Leigh RA (2001) Partitioning of nutrient transport process in roots. J Exp Bot 52:445–457

Tulakdar D, Sharma R, Jaglan S, Vats R, Kumar R, Mahanashi MH, Umar A (2020) Identification and characterization of cadmium resistant fungus isolated from contaminated site and its potential for bioremediation. Environ Toxicol Innov 17:100604. https://doi.org/10.1016/j.eti.2020.100604

Vaid SK, Kumar B, Sharma A, Shukla AK, Srivastava PC (2014) Effect of Zn solubilizing bacteria on growth promotion and Zn nutrition of rice. J Soil Sci Plant Nutr 14:889–910

Wagner GJ (1993) Accumulation of cadmium in crop plants and its consequences to human health. Adv Agron 51:173–212

Wang XY, Yang H, Wang MG, Yang DB, Wang ZY, Wang L (2017) Trehalose protects against cadmium-induced cytotoxicity in primary rat proximal tubular cells via inhibiting apoptosis and restoring autophagic flux. Cell Death Dis 8(10):e3099–e3099

Wang LY, Fan RF, Yang DB, Zhang D, Wang L (2019) Puerarin reverses cadmium-induced lysosomal dysfunction in primary rat proximal tubular cells via inhibiting Nrf2 pathway. Biochem Pharmacol 162:132–141

Yang Y, Liang Y, Ghosh A, Song Y, Chen H, Tang M (2015) Assessment of arbuscular mycorrhizal fungi status and heavy metal accumulation characteristics of tree species in a lead–zinc mine area: potential applications for phytoremediation. Environ Sci Pollut Res Int 22:13179–13193. https://doi.org/10.1007/s11356-015-4521-8

Yaseen MA (2019) Effect of different cadmium levels on growth and biochemical parameters of Cyprinus carpio fingerlings reared in a close system. Zanco J Pure Appl Sci 31(4):139–152

Yongming S, Rongzhu L, Jie L, Yan X, Zhu Y, Schweigert M (2011) The occupational disease prevention and control act of the People's Republic of China: an awareness assessment among workers at foreign-invested enterprises. New Solut 21(1):103–116

Yu X, Xiao JT, Ding Z, Xiong F, Liu X, Tian J, Wu N (2023) Cadmium-absorptive *Bacillus vietnamensis* 151–6 reduces the grain cadmium accumulation in rice (*Oryza sativa* L.): potential for cadmium bioremediation. Ecotoxicol Environ Saf 254:114760. https://doi.org/10.1016/j.ecoenv.2023.114760

Yuan P, Wang J, Pan Y, Shen B, Wu C (2019) Review of biochar for the management of contaminated soil: preparation, application and prospect. Sci Total Environ 659:473–490. https://doi.org/10.1016/j.scitotenv.2018.12.400

Yuan B, Huang L, Liu X, Bai L, Liu H, Jiang H, Zhu P, Xiao Y, Geng J, Liu Q, Hao X (2022) Application of mixotrophic acidophiles for the bioremediation of cadmium-contaminated soils elevates cadmium removal, soil nutrient availability, and rice growth. Ecotoxicol Environ Saf. 236:113499. https://doi.org/10.1016/j.ecoenv.2022.113499

Zeng X, Tang J, Yin H, Liu X, Jiang P, Liu H (2010) Isolation, identification and cadmium adsorption of a high cadmium-resistant *Paecilomyces lilacinus*. Afr J Biotechnol 9(39): 6525–6533

Zheng Y, Yu S, Li Y, Peng J, Yu J, Chi R, Xiao C (2022) Efficient bioimmobilization of cadmium contamination in phosphate mining wastelands by the phosphate solubilizing fungus *Penicillium oxalicum* ZP6. Biochem Eng J 187:108667. https://doi.org/10.1016/j.bej.2022.108667

Microbial Tolerance Strategies Against Cadmium Toxicity

Gisela Adelina Rolón-Cárdenas and Alejandro Hernández-Morales

Abstract Cadmium (Cd) is a heavy metal found in the environment because it is released from natural and anthropogenic sources. Cd is a persistent, nonbiodegradable pollutant with great toxicity for plants, animals, and microorganisms. Although Cd toxicity, some microorganisms can inhabit different environments contaminated with toxic levels of Cd^{2+} because they have developed Cd tolerance and detoxification mechanisms. These mechanisms include the production of extracellular chelating compounds, the adsorption of Cd^{2+} in the cellular surface, the transport of Cd^{2+} from the cytoplasm using efflux pumps, and Cd intracellular complexation with glutathione, metallothionein, and phytochelatins. This chapter shows the state-of-the-art of Cd-tolerant microorganisms and describes the main mechanisms employed in Cd tolerance by bacteria and fungi.

Keywords Cd-tolerant bacteria · Cd-tolerant fungi · Cd-bioprecipitation · Cd efflux

1 Introduction

Cadmium (Cd) is a silvery-white, soft, and ductile metal listed in group 12 (IIB) of the periodic table of elements (Bashir et al. 2019; Morrow 2010). Cd is found naturally in the earth's crust in a concentration of about 0.1–0.2 mg/kg (Suhani et al. 2021) or 5 μg/L in water (Kubier et al. 2019). Cd presence in the environment can be from natural or anthropogenic sources. Primary natural sources of Cd are weathering of cadmium-rich rocks, forest fires, and volcanic eruptions (Suhani et al. 2021). While anthropogenic cadmium results from zinc refining, the industrial

G. A. Rolón-Cárdenas
Facultad de Estudios Profesionales Zona Huasteca, Universidad Autónoma de San Luis Potosí, Ciudad Valles, San Luis Potosi, Mexico

A. Hernández-Morales (✉)
Facultad de Estudios Profesionales Zona Huasteca, and Centro de Investigación y Estudios de Posgrado, Facultad de Ciencias Químicas, Universidad Autónoma de San Luis Potosi, Ciudad Valles, San Luis Potosi, Mexico
e-mail: alejandro.hernandez@uaslp.mx

A. K. Jha, N. Kumar (eds.), *Cadmium Toxicity Mitigation*,
https://doi.org/10.1007/978-3-031-47390-6_6

production of pigments for plastic and glass, nickel-cadmium batteries, television screens, lasers, cosmetics (Pinot et al. 2000), and the use of phosphate fertilizers in agricultural land (Samrane and Bouhaouss 2022).

The Agency for Toxic Substances and Disease Registry has estimated that human activities release 4,000–13,000 tons of Cd each year (ATSDR 2012). These activities increase Cd concentrations in the soil and water of different regions of the worldwide (Yuan et al. 2021). For instance, high Cd concentrations have been detected in the River Ganga basin, India (Maurya et al. 2019), in arable soils of Switzerland (Bigalke et al. 2017), and in the agricultural soil of Shanghai, China (Fei et al. 2019).

Cd is a persistent, nonbiodegradable pollutant in the environment and is present in different trophic levels in the food chain (Suhani et al. 2021), which represents a threat to food safety and a potential risk to human health (Suhani et al. 2021). Cadmium chronic exposure in humans has been linked to hepatotoxicity (Hyder et al. 2013), cardiovascular diseases (Tellez-Plaza et al. 2013), renal dysfunction (Orr and Bridges 2017), respiratory diseases (Bouley et al. 1977), affectations to the bone, and nervous system (Lim et al. 2016). In addition, Cd has been classified as a potential human and animal carcinogen by the World Health Organization (WHO) and the US Environmental Protection Agency (US EPA) (Hartwig 2013; IRIS-USEPA 1987).

Cadmium is also toxic to microorganisms at low concentrations (Ivanov et al. 2003; Choi 2009). For example, it has been observed that Cd affects the mycelial growth of *Trichoderma harzianum* (Oshiquiri et al. 2020), ectomycorrhizal (Mccreight and Schroeder 1982), and pathogenic soil fungi (Płaza et al. 1998), inhibits the sporulation and causes malformations in conidiophores in *Aspergillus carbonarius* and *Penicillium* sp. (Nazareth and Marbaniang 2008; Ramadan et al. 1988). On the other hand, Cd inhibits growth, induces genotoxic damage, and affects cell division in bacteria (Mitra and Bernstein 1978; Hossain et al. 2012; Trevors et al. 1986). Previous studies have shown that Cd causes alteration in cell morphology, reduces cell size, and damages the cell surface of *Escherichia coli* and *Pseudomonas putida* (Hossain et al. 2012; Higham et al. 1986). Cd also generates oxidative stress in the bacterial cell by interfering with the activity of antioxidant enzymes (Hossain et al. 2012; Salam et al. 2020), and reducing the intracellular levels of manganese and zinc (Begg et al. 2015).

2 Cd-Tolerant Microorganisms

Cadmium-tolerant microorganisms are bacteria and fungi capable of inhabiting different environments with toxic levels of Cd ions (Badu et al. 2015; Sher et al. 2020; Imran et al. 2020; Akbar et al. 2022). In previous studies, different Cd-tolerant fungi and bacteria have been isolated from sewage, mining sites, plants collected from contaminated sites, and industrial and agricultural soils (Table 1). These microorganisms have a high Cd tolerance (MIC range of 250–4000 μg/mL) and can also tolerate different heavy metals like lead (Pb), zinc (Zn), copper (Cu), cobalt

Table 1 Minimum inhibitory concentration (MIC) of bacterial and fungal strains against heavy metals

	Microorganism	Origin	MIC (μg/mL) Cd	Pb	Zn	Cu	Co	Ni	References
Bacteria	*Pseudomonas koreensis* AGB-1	Roots of *Miscanthus sinensis* growing in mine-tailing soil	2700	1800	3000	500	NR	NR	Badu et al. (2015)
	Enterobacter mori HZ21	Root nodules of *Robinia pseudoacacia* obtained from Pb/Zn mining area	2022	414	2615	12	NR	NR	Fan et al. (2018)
	Micrococcus luteus AS2	Industrial wastewater	337	1035	NR	NR	294	175	Sher et al. (2020)
	Acinetobacter johnsonii LWN6	Agricultural soils irrigated with domestic wastewater	674	1242	NR	NR	NR	117	Ajmal et al. (2021)
	Pseudomonas aeruginosa RF2	Arable fields cultivating Tulaipanji rice	600	200	1100	600	300	1400	Saha et al. (2022b)
	Enterobacter cloacae AS10	Rhizospheric soil of rice from a contaminated site	4000	3312	NR	NR	NR	NR	Ghosh et al. (2022)
Fungi	*Exophiala pisciphila* H93	*Arundinella bengalensis* of an old Pb-Zn mine smelting site	510	2300	3100	NR	NR	NR	Zhang et al. (2008)
	Aspergillus niger SF-5	Soils irrigated with untreated municipal/industrial effluent	1461	NR	NR	445	NR	117	Akhtar et al. (2013)
	Trichothecium sp.	Agricultural soil irrigated with industrial wastewater	400	NR	NR	1200	1200	1600	Imran et al. (2020)
	Trichoderma brevicompactum QYCD-6	Earthworm gut	250	1600	450	300	NR	NR	Zhang et al. (2020)
	Chaetomium sp.	Agricultural soil near waste dump site	400	2200	NR	NR	NR	NR	Paul et al. (2021)
	Penicillium oxalicum MS5	Industrial soil in Islamabad Pakistan	1000	NR	NR	600	NR	NR	Akbar et al. (2022)

NR, not reported

(Co), and nickel (Ni) (Table 1). The high tolerance to heavy metals in these microorganisms may be due to the prolonged exposure to metal ions at the site from which they were isolated (Pal et al. 2005). Therefore, high levels of heavy metals in the contaminated site inhibit the growth of sensible species and promote the survival of tolerant microorganisms (Diaz-Ravina and Baath 1996).

3 Microbial Mechanism of Cadmium Tolerance

Many microorganisms have developed different mechanisms to decrease the damage from non-essential metal ions like Cd (Loutet et al. 2015). These mechanisms include extracellular sequestration, chemical transformations, surface adsorption, intracellular complexation, and the expulsion of cations by Cd efflux systems (Fig. 1) (Silver 1992; Khan et al. 2022; Ron et al. 1992).

3.1 Chemical Transformation

Cadmium's toxicity for microorganisms depends mainly on their bioavailability, which is influenced by the concentration and speciation of ions (Pal et al. 2022). In solution, metals can be found as free ions (Cd^{2+}) or forming organic and inorganic complexes, depending on environmental factors such as pH, redox potential (Eh), temperature, concentration of anions, and organic matter content (Monteiro et al. 2012; Rieuwerts et al. 1998). Dissolved cadmium ions (Cd^{2+}) are the most toxic Cd species for microorganisms due to their easily penetrating microbial membranes affecting both the synthesis and the activity of the enzymes involved in essential biological functions (Braud et al. 2010; Bjerregaard et al. 2015).

In microorganisms, limiting Cd^{2+} uptake is the cell's first defense line, reducing the Cd bioavailability and toxicity (Kanwar et al. 2021).

Chemical transformations are one of the most essential mechanisms affecting the bioavailability of metal ions in the environment (Pal et al. 2022). Most of these transformations are mediated by the biological action of microorganisms, which result in the formation of organometallic compounds or precipitation of metal ions (Fashola et al. 2016; Raab and Feldmann 2003).

Biomethylation is an enzymatic process in which living organisms transfer one or more methyl groups ($–CH_3$) to metal atoms, forming methylated metal compounds (Fatoki 1997). Previously, As, Hg, Sn, Se, Te, Ge, and Pb methylated have been detected in soil and aquatic systems (Fatoki 1997). A previous study has also reported the biological methylation of Cd. Pongratz and Heumann (1999) reported that a polar marine bacteria methylates Cd^{2+} converting it into monomethyl cadmium ($MeCd^{+}$) under polar cold conditions. Biomethylation by the bacteria may result in a mechanism of Cd detoxification due to the addition of methyl groups

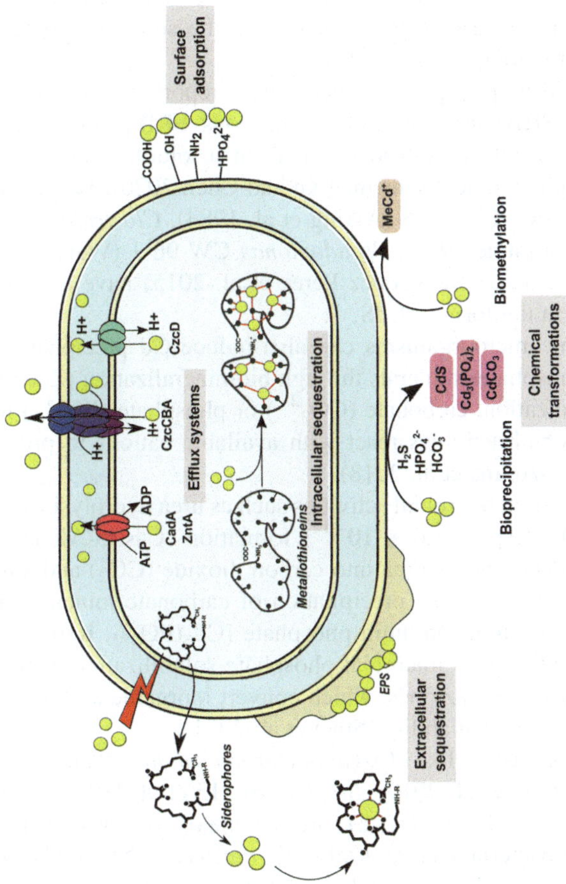

Fig. 1 Overview of microbial mechanisms involved in cadmium tolerance

reducing the water solubility and increasing the volatility of the metal ions, which changes its mobility and toxicity (Gadd 2004; Seshadri et al. 2015).

Precipitation is a chemical process in which dissolved metal ions react with inorganic anions to form insoluble compounds (Wang et al. 2005; Park et al. 2011). In solution, Cd can precipitate in the presence of hydroxide, phosphate, sulfide, or carbonate anions, depending on pH and metal concentration (Vallero 2014; Rumble 2017).

Microorganisms can also employ this process to reduce the bioavailability and toxicity of Cd ions (Xia et al. 2021). During bioprecipitation, metabolites produced by the microorganisms react with metal ions in the solution and form metal precipitates (Janyasuthwiong and Rene 2017).

The primary Cd-bioprecipitation mechanism reported is mediated by sulfate-reducing bacteria (SRB) under anaerobic conditions. SRB produce hydrogen sulfide (H_2S), which reacts with dissolved Cd to form insoluble cadmium sulfide (CdS, $Ksp = 8 \times 10^{-28}$) in anoxic sediments (Xu and Chen 2020). Some bacteria such as *Klebsiella aerogenes* NCTC 418 (Aiking et al. 1984), *Clostridium thernoaceticum* (Cunningham and Lundie 1993), *Pseudomonas* CW-96-1 (Wang et al. 1997), and *Desulfovibrio alaskensis* 6SR (López Pérez et al. 2015) have also been shown to bioprecipitate Cd in the form of CdS.

In nature, many microorganisms can also induce Cd precipitation by forming carbonates and phosphates minerals through biomineralization (Zheng et al. 2021). During biomineralization, carbonate (CO_3^{2-}) or phosphate (PO_4^{3-}) ions produced by different microbial activities react with available cations to produce insoluble minerals (Chaparro-Acuña et al. 2018).

In biomineralization, microbial activities such as urea hydrolysis induce cadmium carbonate ($CdCO_3$, $Ksp = 1.0 \times 10^{-12}$) formation (Castro-Alonso et al. 2019). Microbial urease decomposes urea into carbon dioxide (CO_2) and ammonia (NH_3) increases pH and promotes the precipitation of carbonate minerals (Kumari et al. 2014). On the other hand, cadmium phosphate [$Cd_3(PO_4)_2$, $Ksp = 2.53 \times 10^{-33}$] precipitation is mediated by microbial phosphate solubilization or by the action of phosphatase and phytase enzymes, which convert inorganic and organic phosphate sources into phosphate availability (Shan et al. 2021).

Many microorganisms like *Exiguobacterium undae* (Kumari et al. 2014), *Cupriavidus* sp. (Zhao et al. 2019), *Bacillus* sp. (Li et al. 2019; Zhao et al. 2017), and *Sporosarcina* sp. (Qiao et al. 2021) are capable of converting Cd^{2+} into $CdCO_3$. In addition, some bacteria such as *Klebsiella aerogenes* NCTC 418 (Aiking et al. 1984), *Citrobacter* sp. (Macaskie and Dean 1984), *Bacillus cereus* (Li et al. 2018), and *Burkholderia* sp. (Zeng et al. 2021) also precipitate Cd^{2+} ions in the form of $Cd_3(PO_4)_2$.

Free Cd ions can also be immobilized by co-precipitating with calcium carbonate and calcium phosphate minerals. For example, *Exiguobacterium undae* and *Enterobacter* sp. have been observed to immobilize free Cd^{2+} in soil co-precipitating with calcite and vaterite ($CaCO_3$) (Kumari et al. 2014; Peng et al. 2020). Similarly, Cd ions are co-precipitated with hydroxyapatite [$Ca_5(PO_4)_3(OH)$] by *Burkholderia* sp. (Zeng et al. 2021).

3.2 Extracellular Cd Sequestration

Extracellular Cd sequestration involves the production of chelating compounds such as siderophores, extracellular polymeric substances (EPS), or organic compounds produced non-specifically by microorganisms (Masood and Malik 2013; Wang et al. 2019).

The siderophores are organic iron-chelating compounds of low molecular mass (200–2000 Da) synthesized under iron deficiency by bacteria, fungi, and plants (Ghosh et al. 2020). Siderophores can be classified into catecholates, phenolates, hydroxamates, and carboxylates (Timofeeva et al. 2022), which can bind different metal cations in addition to iron (Schalk et al. 2011; Rajkumar et al. 2010). For example, pyoverdine siderophore produced by *Pseudomonas aeruginosa* can chelate Co, Cu, Ni, Pb, and Zn from the extracellular medium (Braud et al. 2010). Similarly, a siderophore extracted from *Aspergillus nidulans* possesses a high affinity for Cd ions (Khan et al. 2020).

Microbial siderophore production is essential in microbial tolerance to Cd due to its ability to complex heavy metal ions (Schalk et al. 2011). Previous studies have reported that Cd increases the production of siderophores in bacteria. In *Pseudomonas fulva* and *Streptomyces* sp., Cd stimulated the biosynthesis of phenolates and catecholates (Złoch et al. 2016; Thiem et al. 2018). Similarly, an increase in siderophore production was observed in *P. aeruginosa* (Sinha and Kumar-Mukherjee 2008), *P. veronii* (Ferreira et al. 2018), and *Enterobacter* sp. (Chen et al. 2016) after Cd exposure. The formation of Cd–siderophore complexes reduces the levels of extracellular Cd^{2+}, and limits their diffusion into the cell (Roskova et al. 2022).

Biofilm production has also been referred to as a mechanism of extracellular Cd sequestration by microorganisms. Biofilms are structures in which the extracellular polymeric substances (EPS) produced by bacterial cells link to neighboring cells forming a complex polymeric matrix in which bacteria remain immobilized (Yin et al. 2019b). EPS are different classes of polysaccharides, proteins, nucleic acids, glycoproteins, and phospholipids produced by microorganisms and secreted into the extracellular environment (Mohd-Nadzir et al. 2021; Czaczyk and Myszka 2007). The biofilm synthesized by microorganisms plays a vital role in protecting against stressful conditions, including toxic metals (Rizvi et al. 2019). Several studies have reported that some bacteria, such as *Pseudomonas* sp. EJ01 (Chien et al. 2013), *P. aeruginosa* JP-11 (Chakraborty and Das 2014), *Staphylococcus epidermidis* (Wu et al. 2015), and *Serratia marcescens* KMR-3 (Zhu et al. 2022) increased biofilm production in response to Cd stress.

Biofilm and EPS have a high capacity to remove metal ions. For example, biofilm produced by *Arthrobacter viscosus* and *Enterobacter* sp. removed Cd from aqueous solution (Braud et al. 2010; Dhanya et al. 2021). Similarly, Pb, Cd, Co, Ni, and Cu ions are removed by EPS extracted from *Paenibacillus jamilae* (Morillo-Pérez et al. 2008). The metal-binding capacity of biofilm is attributed to ionizable functional groups in the EPS matrix (Chakraborty and Das 2014; Naik et al. 2012). In the biofilm of *Serratia marcescens* KMR-3, the main groups involved in the Cd^{2+} adsorption are hydroxyl (–OH), carboxyl (–COOH), amine ($-NH_2$), carbonyl (C=O), and sulfhydryl (–SH) groups. On the other hand, amine ($-NH_2$), carboxyl (–COOH), and phosphoryl ($-PO_3$) groups of the EPS produced by *Pseudomonas veronni* are responsible for Cd^{2+} complexation (Ferreira et al. 2020). These groups form complexes with metal ions, making heavy metals inaccessible to microbes (Rizvi et al. 2019).

3.3 Surface Adsorption

The microbial envelope is the first component of microorganisms that interacts with the environment (Yun et al. 2011; Priyadarshini et al. 2021), and presents a multi-layer structure whose functions are maintaining the shape cell, controlling cellular permeability and protecting it from the environment (Brown et al. 2013; Holst et al. 2010). Depending on the microorganism, the cell envelope can be formed by an outer membrane and/or a cell wall in addition to the cytoplasmic membrane (Silhavy et al. 2010).

Gram-positive bacteria have a cell wall composed of a thick layer of peptidoglycan, which is formed by N-acetylglucosamine and N-acetylmuramic acid chains interconnected by peptide side chains (Priyadarshanee and Das 2021; Rohde 2019). On the other hand, in Gram-negative bacteria, the peptidoglycan wall is thinner and is surrounded by an external membrane made up of lipopolysaccharides (LPS), phospholipids, and membrane proteins (Dixit et al. 2015; Bertrand et al. 2015). The fungi also present a cell wall formed by chitin fibers, glucans, and different types of glycoproteins, depending on fungal species, morphotype, and growth stage (Garcia-Rubio et al. 2019; Gow et al. 2017).

In a contaminated environment, the microbial envelope is the first component of the cell that interacts with heavy metal ions, whereby Cd ions can be adsorbed in the cellular surface (Yun et al. 2011; Priyadarshini et al. 2021). The adsorption of Cd^{2+} to the cell surface has been reported previously in *Salmonella enterica*. This bacteria adsorbs 6 mM/g of Cd^{2+} on its surface when exposed to 1 mM of Cd for 8 days (Khan et al. 2016).

Cd ions can be adsorbed to the cell surface by electrostatic interaction with anionic functional groups presenting in the cell wall or cell membrane of bacteria and fungi (Chakravarty and Banerjee 2012; Kumar et al. 2021; Yin et al. 2019a). In previous studies, the analysis of microbial surfaces by Fourier transform infrared (FTIR) spectroscopy showed that the hydroxyl (–OH) and amine ($-NH_2$) are the main functional groups involved in Cd^{2+} binding of bacteria and fungi surface (Table 2). ,$^{2+}$ The phosphate groups in the cell surface of *Pseudomonas* sp., *P. aeruginosa*, *Salmonella enterica*, *Fusarium verticillioides*, and *Mucoromycote* sp., also participate in the Cd^{2+} adsorption (Table 2). The FTIR analysis also revealed that the functional groups carbonyl (–C=O) and carboxyl (–COOH) are responsible for Cd^{2+} binding in *Pseudomonas* sp., *Salmonella enterica*,

Table 2 Functional groups identified on the microbial surface by FTIR technique that are involved in Cd adsorption

	Microorganism	Functional groups	References
Bacteria	*Ochrobactrum intermedium* BB12	–OH, $–NH_2$, and –CH	Renu et al. (2022)
	Pseudomonas sp. 375	–OH, $–NH_2$, –CH, C=O, –CN, $–SO_3$, and P=O (phosphate)	Xu et al. (2020)
	Salmonella enterica 43C	–COOH, $–NH_2$, and P=O (phosphate)	Khan et al. (2016)
	Pseudomonas aeruginosa	–OH, $–NH_2$, –COOH, P=O (phosphate), and O=C–N (amides)	Durve and Chandra (2014)
Fungi	*Fusarium verticillioides*	–OH, $–NH_2$, C=O, and P=O (phosphate)	Hassan et al. (2021)
	Aspergillus spp.	$–NH_2$, O=C–O–C (esters), O=C–O–Ph (alkyl aril esters)	Hasgül et al. (2019)
	Paecilomyces lilacinus XLA	–OH, $–NH_2$, –COOH, and O=C–N (amides)	Xia et al. (2015)
	Mucoromycote sp. XLC	–OH, $–NH_2$, –CH, O=C–N (amides), and P=O (phosphate)	Xia et al. (2015)
	Trichoderma sp.	–OH, $–NH_2$, and C=O	Rahman et al. (2014)

P. aeruginosa, *Fusarium verticillioides, Paecilomyces lilacinus*, and *Trichoderma* sp. (Table 2). The adsorption of Cd ions to the cell surface prevents Cd entry inside the bacterial and fungal cells (Pal et al. 2022), and protects the microbial cell from metal stress (Yin et al. 2019a).

Microbial Cd adsorption capacity can be affected by the pH of the extracellular medium (Chaturvedi et al. 2015). For example, it is observed that an increase in pH increased Cd adsorption in *Bacillus cereus* (Li et al. 2018) and *Cupriavidus* sp. (Li et al. 2019), due to an increase in the deprotonation of metal-binding sites on the cell surface, which causes more negative charge and thereby an increase in electrostatic interactions between Cd ions and cells surface (Li et al. 2019).

3.4 Cadmium Efflux Mechanism

Metal ion efflux is one of the main mechanisms of metal homeostasis in bacteria (Chandrangsu et al. 2017). Efflux pumps are proteins localized in membranes that transport metal ions from the bacterial cell to the external environment (Amaral et al. 2014). In bacteria, the efflux of inorganic metal cations is facilitated mainly by three different types of efflux systems: (1) P-type adenosine triphosphatases (ATPases), (2) CBA (capsule biogenesis assembly) transporters, and (3) CDF (cation diffusion facilitator) transporters (Salam et al. 2020). These systems regulate the metal cations essential in bacterial metabolism and prevent the accumulation of toxic heavy metals inside the cell (Nies and Silver 1995).

3.4.1 P-Type ATPases

P-type ATPases are a superfamily of integral membrane proteins that transport metal ions from the cytoplasm to the periplasm using energy provided by ATP hydrolysis (Khan et al. 2022; Apell 2004). CadA protein system is one the best-characterized cadmium efflux system type ATPase in bacteria (Nies 2003), which also confers resistance to Zn^{2+} and Pb^{2+} through an energy-dependent efflux mechanism (Oger et al. 2003). This protein system is encoded by the *cadA* gene, which has been found on the plasmids pI258 and pLm74 of *Staphylococcus aureus* and *Listeria monocytogenes,* respectively (Yoon et al. 1991; Lebrun et al. 1994), as well as in the chromosome of Cd-tolerant bacteria such as *Cupriavidus nantongensis* X1T, *P. aeruginosa*, *Bacillus cereus* NWUAB01, and *Enterobacter* sp. DNB-S2 (Table 3).

The ZntA efflux system in *Escherichia coli* is a P-type ATPase transporter similar to the CadA system, which is regulated by ZntR, a transcriptional factor, and induced by metallic cations like Cd^{2+} (Sharma et al. 2000; Brocklehurst et al. 1999). For example, it has been observed that exposure to 4 mM $CdCl_2$ for 6 h increased the expression of the *zntA* gene in *Cellulomonas* sp. Y8 (Chen et al. 2021). ZntA system

Table 3 Genes of efflux system in bacteria Cd tolerant

Bacteria	Origin	MIC Cd (mM)	Cd-resistant genes detected	References
Pseudomonas qingdaonensis ZCR6	Rhizospheric soil of *Zea mays* from a contaminated site	5.0	*czcA, czcB, czcC*	Chlebek et al. (2022)
Cupriavidus metallidurans BS1	Soil of a gold–copper mine in China	2.5	*zntA, czcA, czcB, czcC, czcD*	Mazhar et al. (2020)
Cupriavidus nantongensis X1T	Activated sludge	2.7	*cadA, czcA, czcB, czcC, czcD*	Fang et al. (2022)
Pseudomonas aeruginosa	Rhizospheric soil of paddy fields	3.0	*cadA, czcA, czcB, czcC, czcD*	Saha et al. (2022a)
Burkholderia contaminans ZCC	Soil of a copper–gold mine in China	7.0	*cadA/zntA, czcA, czcB, czcC, czcD*	You et al. (2021)
Bacillus cereus NWUAB01	Soil of a gold mining area in South Africa	1.4	*cadA, czcD*	Ayangbenro and Babalola (2020)
Enterobacter sp. DNB-S2	Black soil in China	12.5	*cadA, czcA, czcD*	Sun et al. (2020)
Vibrio parahaemolyticus ST165	*Penaeus vannamei* from seawater farms.	7.0	*zntA, czcA, czcB, czcD*	Fang et al. (2021)
Citrobacter freundii SRS1	Soils irrigated with a tannery effluent	3.0	*zntA, czcA, czcD*	Uddin et al. (2023)
Caulobacter flavus RHGG3T	Rhizosphere soil of watermelon (*Citrullus lanatus*)	0.4	*zntA, czcA, czcB, czcC, czcD*	Yang et al. (2019)

is involved in the efflux of Cd^{2+} from the cytoplasm and is a vital mechanism in the cadmium resistance of bacteria (Liu et al. 2006). Orthologous of the *zntA* gene have also been found in the genome of *Cupriavidus metallidurans* BS1, *Burkholderia contaminans* ZCC, *Vibrio parahaemolyticus* ST165, *Citrobacter freundii* SRS1, and *Caulobacter flavus* RHGG3T (Table 3). The high Cd resistance in these bacteria could be attributed to an efflux of Cd ions by this P-type ATPase system.

3.4.2 CBA Transporters

CBA transporters are a protein complex formed by three components: an RND protein located in the cytoplasmic membrane, a membrane fusion protein (MFP), and an outer membrane channel protein (OMP) (Margaryan et al. 2021; Valencia et al. 2013). This efflux system acts as a chemiosmotic antiporter in Gram-negative bacteria and has been involved in metal resistance, nodulation, and cell division in *R. metallidurans*, *Mesorhizobium loti*, and *E. coli*, respectively (Liu et al. 2015).

The best-studied CBA transporter is the CzcCBA protein system, which is encoded by the *czcCBA* operon formed by three structural genes (*czcA*, *czcB*, and *czcC*) (Choudhury and Srivastava 2001). Genomic analysis revealed the presence of *czcCBA* operon in the genome of different bacteria with Cd resistance like *Pseudomonas qingdaonensis* ZCR6, *C. metallidurans* BS1, *C. nantongensis* X1T, *P. aeruginosa*, *Burkholderia contaminans* ZCC, and *Caulobacter flavus* RHGG3 (Table 3). The *czcCBA* operon encodes for three proteins: CzcA and CzcC are proteins located in the cytoplasmic and outer membrane, respectively (Yu et al. 2020). While CzcB is a membrane fusion protein located in the periplasm (Große et al. 1999). The three proteins form an efflux system that transports Cd^{2+} from the cytoplasm to the external environment across the periplasmic space, reducing the intracellular concentration of Cd^{2+} and protecting the bacterial cell against metal stress (Osman et al. 2019; Pal et al. 2022).

3.4.3 CDF Transporters

Cation diffusion facilitators (CDF) are a family of transmembrane proteins that transport heavy metal cations from the cytoplasm into the periplasm (Salam et al. 2020) through a chemiosmotic gradient formed by protons or potassium (Nies 2003; Legatzki et al. 2003).

Members of these transporters are found in both prokaryotes and eukaryotes and are involved in the export of divalent ions such as Zn^{2+}, Co^{2+}, Ni^{2+}, Mn^{2+}, Cd^{2+}, and Fe^{2+} (Russell and Soulimane 2012; Pal et al. 2022). Whereby CDF transporters are also involved in Cd tolerance in bacteria (Kolaj-Robin et al. 2015). CzcD protein, encoded by the *czcD* gene, is one the most characterized cadmium transporter type CDF in bacteria (Khan et al. 2022). Orthologs of *czcD* have been found in the genome of Cd-tolerant bacteria like *C. metallidurans* BS1, *C. nantongensis* X1T, *P. aeruginosa*, *Burkholderia contaminants* ZCC, *Bacillus cereus* NWUAB01,

Enterobacter sp. DNB S2, *Vibrio parahaemolyticus* ST165, *Citrobacter freundii* SRS1, *Halobacterium salinarum* AS1, and *Caulobacter flavus* RHGG3T (Table 3). CzcD transporter, located in the cytoplasmic membrane, reduces the accumulation of Cd^{2+} within the cytoplasm by efflux and confers a low degree of resistance to Cd compared to other efflux systems (Liu et al. 2021). Previous studies have reported that czcD is also involved in regulating CzcCBA efflux system (Anton et al. 1999). For instance, the deletion of the *czcD* gene resulted in a higher expression of *czcCBA* genes in *Cupriavidus metallidurans* CH34 (Anton et al. 1999).

3.5 Intracellular Cd Sequestration

Cadmium is a non-essential metal in microbial metabolism; thus, microorganisms do not have a specialized uptake system for its absorption (Jebril et al. 2022). Cd can enter the bacteria cell by non-specific transport systems or inducible transport systems that utilize the energy from ATP hydrolysis (Nies 1999).

Intracellular sequestration is a mechanism microorganisms use to reduce the toxicity of metal ions inside the cells (Worms et al. 2006). In this mechanism, the metal ions are complexed in the cell cytoplasm with several classes of organic ligands rich in cysteine residues such as glutathione (GSH), metallothioneins (MT), and phytochelatins (PC) (Worms et al. 2006; Igiri et al. 2018).

Glutathione (GSH) is a tripeptide of 307 Da, enzymatically synthesized by γ-glutamylcysteine synthetase and glutathione synthetase from the glutamic acid, cysteine, and glycine (Smirnova and Oktyabrsky 2005; Labarrere and Kassab 2022). While metallothionein (MT) are low molecular weight protein (3.5–14 kDa) with cysteine and other non-aromatic amino acids (Yu et al. 2022; Chatterjee et al. 2020). Previous studies have reported that different microorganisms synthesize GSH and MT in response to Cd stress. For example, it has been reported that *Escherichia coli* (Khan et al. 2015), *Rhizobium leguminosarum* (Figueira et al. 2005), *Acidithiobacillus ferrooxidans* (Zheng et al. 2015), *Hebeloma cylindrosporum* (Khullar and Reddy 2019), *Laccaria bicolor* (Khullar and Sudhakara Reddy 2019), and *Fusarium solani* (Kumar et al. 2019) produce GSH that sequester intracellular Cd by forming non-toxic GSH-Cd complex (Khullar and Reddy 2018). On the other hand, Cd exposure increases the synthesis of MT in *Pseudomonas putida* (Enshaei et al. 2010) and *Heliscus lugdunensis* (Jaeckel et al. 2005; Loebus et al. 2013), and increases the expression of MT genes in *Hebeloma mesophaeum* (Sácký et al. 2019), *Suillus himalayensis* (Kalsotra et al. 2018), and *Laccaria bicolor* (Reddy et al. 2014).

Phytochelatins (PC) are another ligand involved in Cd detoxification (Gutiérrez et al. 2019). These molecules are oligopeptides with a general structure $(\gamma\text{Glu-Cys})_n\text{Gly}$ ($n = 2$–11) synthesized from GSH (Bae et al. 2000). In some yeasts, it has been demonstrated that Cd induces the synthesis of PC. For example, 250 μM Cd^{2+} produces PC2 in *Saccharomyces cerevisiae* (Kneer et al. 1992). Similarly,

0.1 mM Cd^{2+} induces the synthesis of PC2 and PC3 in *S. pombe* (Mutoh and Hayashi 1988).

GSH, PC, and MT form stable complexes inside microbial cells with the Cd^{2+} protecting microbial metabolic processes (Naik et al. 2012; Khullar and Reddy 2018).

4 Conclusion

Cd is a heavy metal that is toxic to microorganisms at low concentrations. One of the key characteristics of microorganisms that inhabit contaminated sites is their ability to tolerate high levels of Cd. This characteristic results from one or several mechanisms produced in response to Cd stress. These mechanisms include the immobilization of ions in the cell surface (surface adsorption), the chelation by siderophores and EPS (extracellular sequestration), the transformation of Cd ions into less bioavailable species like carbonates, phosphates, and sulfide (chemical transformations), the expulsion of Cd ions from the cytoplasm through specialized protein systems (efflux system), and the complexation by cysteine-rich compounds such as metallothioneins, phytochelatins, and glutathione (intracellular sequestration). The Cd-tolerant bacterium and fungi could be a suitable alternative for bioremediation of Cd and other heavy metals from contaminated water and soil.

Acknowledgments This work was funded by grants from CONACYT, Programa Presupuestario F003 (Formerly Fondo Sectorial de Investigación para la Educación) CB2017-2018, A1-S-40454, to Alejandro Hernández-Morales.

References

Aiking H, Stijnman A, Van Garderen C et al (1984) Inorganic phosphate accumulation and cadmium detoxification in *Klebsiella aerogenes* NCTC 418 growing in continuous culture. Appl Environ Microbiol 47:374–377

Ajmal AW, Saroosh S, Mulk S et al (2021) Bacteria isolated from wastewater irrigated agricultural soils adapt to heavy metal toxicity while maintaining their plant growth promoting traits. Sustainability 13:7792

Akbar M, El-Sabrout AM, Shokralla S et al (2022) Preservation and recovery of metal-tolerant fungi from industrial soil and their application to improve germination and growth of wheat. Sustainability 14:5531

Akhtar S, Mahmood-Ul-Hassan M, Ahmad R et al (2013) Metal tolerance potential of filamentous fungi isolated from soils irrigated with untreated municipal effluent. Soil Environ 32:55–62

Amaral L, Martins A, Spengler G, Molnar J (2014) Efflux pumps of gram-negative bacteria: what they do, how they do it, with what and how to deal with them. Front Pharmacol 4

Anton A, Grosse C, Reissmann J et al (1999) CzcD is a heavy metal ion transporter involved in regulation of heavy metal resistance in *Ralstonia* sp. strain CH34. J Bacteriol 181:6876–6881

Apell H-J (2004) How do P-type ATPases transport ions? Bioelectrochemistry 63:149–156

ATSDR (2012) Toxicological profile for cadmium [Online]. Atlanta GA. https://www.atsdr.cdc.gov/toxprofiles/tp5-p.pdf. Accessed 24 July 2023

Ayangbenro AS, Babalola OO (2020) Genomic analysis of *Bacillus cereus* NWUAB01 and its heavy metal removal from polluted soil. Sci Rep 10:19660

Badu AG, Shea PJ, Sudhakar D et al (2015) Potential use of *Pseudomonas koreensis* AGB-1 in association with Miscanthus sinensis to remediate heavy metal(loid)-contaminated mining site soil. J Environ Manag 151:160–166

Bae W, Chen W, Mulchandani A, Mehra RK (2000) Enhanced bioaccumulation of heavy metals by bacterial cells displaying synthetic phytochelatins. Biotechnol Bioeng 70:518–524

Bashir S, Adeel M, Gulshan AB et al (2019) Effects of organic and inorganic passivators on the immobilization of cadmium in contaminated soils: a review. Environ Eng Sci 36:986–998

Begg SL, Eijkelkamp BA, Luo Z et al (2015) Dysregulation of transition metal ion homeostasis is the molecular basis for cadmium toxicity in *Streptococcus pneumoniae*. Nat Commun 6:6418

Bertrand J-C, Doumenq P, Guyoneaud R et al (2015) Applied microbial ecology and bioremediation. In: Bertrand J-C, Caumette P, Lebaron P et al (eds) Environmental microbiology: fundamentals and applications: microbial ecology. Springer Netherlands, Dordrecht

Bigalke M, Ulrich A, Rehmus A, Keller A (2017) Accumulation of cadmium and uranium in arable soils in Switzerland. Environ Pollut 221:85–93

Bjerregaard P, Andersen CBI, Andersen O (2015) Ecotoxicology of metals—sources, transport, and effects on the ecosystem. In: Nordberg GF, Fowler BA, Nordberg M (eds) Handbook on the toxicology of metals, 4th edn. Academic Press, San Diego

Bouley G, Dubreuil A, Despaux N, Boudene C (1977) Toxic effects of cadmium microparticles on the respiratory system: an experimental study on rats and mice. Scand J Work Environ Health 3: 116–121

Braud A, Geoffroy V, Hoegy F et al (2010) Presence of the siderophores pyoverdine and pyochelin in the extracellular medium reduces toxic metal accumulation in *Pseudomonas aeruginosa* and increases bacterial metal tolerance. Environ Microbiol Rep 2:419–425

Brocklehurst KR, Hobman JL, Lawley B et al (1999) ZntR is a Zn(II)-responsive MerR-like transcriptional regulator of zntA in *Escherichia coli*. Mol Microbiol 31:893–902

Brown S, Santa Maria JP, Jr. and Walker, S. (2013) Wall teichoic acids of gram-positive bacteria. Annu Rev Microbiol 67:313–336

Castro-Alonso MJ, Montañez-Hernandez LE, Sanchez-Muñoz MA et al (2019) Microbially induced calcium carbonate precipitation (MICP) and its potential in bioconcrete: microbiological and molecular concepts. Front Mater 6:126

Chakraborty J, Das S (2014) Characterization and cadmium-resistant gene expression of biofilm-forming marine bacterium *Pseudomonas aeruginosa* JP-11. Environ Sci Pollut Res 21:14188–14201

Chakravarty R, Banerjee PC (2012) Mechanism of cadmium binding on the cell wall of an acidophilic bacterium. Bioresour Technol 108:176–183

Chandrangsu P, Rensing C, Helmann JD (2017) Metal homeostasis and resistance in bacteria. Nat Rev Microbiol 15:338–350

Chaparro-Acuña SP, Becerra-Jiménez ML, Martínez-Zambrano JJ, Rojas-Sarmiento HA (2018) Soil bacteria that precipitate calcium carbonate: mechanism and applications of the process. Acta Agron 67:277–288

Chatterjee S, Kumari S, Rath S et al (2020) Diversity, structure and regulation of microbial metallothionein: metal resistance and possible applications in sequestration of toxic metals. Metallomics 12:1637–1655

Chaturvedi AD, Pal D, Penta S, Kumar A (2015) Ecotoxic heavy metals transformation by bacteria and fungi in aquatic ecosystem. World J Microbiol Biotechnol 31:1595–1603

Chen Y, Chao Y, Li Y et al (2016) Survival strategies of the plant-associated bacterium *Enterobacter* sp. strain EG16 under cadmium stress. Appl Environ Microbiol 82:1734–1744

Chen J, Wang L, Li W et al (2021) Genomic insights into cadmium resistance of a newly isolated, plasmid-free *Cellulomonas* sp. strain Y8. Front Microbiol 12:784575

Chien C-C, Lin B-C, Wu C-H (2013) Biofilm formation and heavy metal resistance by an environmental *Pseudomonas* sp. Biochem Eng J 78:132–137

Chlebek D, Płociniczak T, Gobetti S et al (2022) Analysis of the genome of the heavy metal resistant and hydrocarbon-degrading rhizospheric *Pseudomonas qingdaonensis* ZCR6 strain and assessment of its plant-growth-promoting traits. Int J Mol Sci 23:214

Choi J (2009) Adsorption, bioavailability, and toxicity of cadmium to soil microorganisms. Geomicrobiol J 26:248–255

Choudhury R, Srivastava S (2001) Zinc resistance mechanisms in bacteria. Curr Sci 2021:768–775

Cunningham DP, Lundie LL Jr (1993) Precipitation of cadmium by *Clostridium thermoaceticum.* Appl Environ Microbiol 59:7–14

Czaczyk K, Myszka K (2007) Biosynthesis of extracellular polymeric substances (EPS) and its role in microbial biofilm formation. Pol J Environ Stud 16

Dhanya BE, Athmika, Rekha PD (2021) Characterization of an exopolysaccharide produced by Enterobacter sp. YU16-RN5 and its potential to alleviate cadmium induced cytotoxicity in vitro. 3 Biotech 11:491

Diaz-Ravina M, Baath E (1996) Development of metal tolerance in soil bacterial communities exposed to experimentally increased metal levels. Appl Environ Microbiol 62:2970–2977

Dixit R, Wasiullah, Malaviya D et al (2015) Bioremediation of heavy metals from soil and aquatic environment: an overview of principles and criteria of fundamental processes. Sustainability 7: 2189

Durve A, Chandra N (2014) FT-IR analysis of bacterial biomass in response to heavy metal stress. Int J Biotechnol Photon 112:386–391

Enshaei M, Khanafari A, Akhavan SA (2010) Metallothionein induction in two species of *Pseudomonas* exposed to cadmium and copper contamination. Iran J Environ Health Sci Eng 7:287–298

Fan M, Liu Z, Nan L et al (2018) Isolation, characterization, and selection of heavy metal-resistant and plant growth-promoting endophytic bacteria from root nodules of *Robinia pseudoacacia* in a Pb/Zn mining area. Microbiol Res 217:51–59

Fang J, Jiao X, Cheng H et al (2021) Draft genome of a multidrug and multi-heavy metal resistant *Vibrio parahaemolyticus* ST165 strain of *Penaeus vannamei* from seawater farms in Zhejiang, China. J Glob Antimicrob Resist 26:323–325

Fang L, Zhu H, Geng Y et al (2022) Resistance properties and adaptation mechanism of cadmium in an enriched strain, *Cupriavidus nantongensis* X1T. J Hazard Mater 434:128935

Fashola MO, Ngole-Jeme VM, Babalola OO (2016) Heavy metal pollution from gold mines: environmental effects and bacterial strategies for resistance. Int J Environ Res Public Health 13:1047

Fatoki O (1997) Biomethylation in the natural environment: a review. South Afr J Anim Sci 93: 366–370

Fei X, Xiao R, Christakos G et al (2019) Comprehensive assessment and source apportionment of heavy metals in Shanghai agricultural soils with different fertility levels. Ecol Indic 106:105508

Ferreira ML, Ramirez SA, Vullo DL (2018) Chemical characterization and ligand behaviour of *Pseudomonas veronii* 2E siderophores. World J Microbiol Biotechnol 34:134

Ferreira ML, Gerbino E, Cavallero GJ et al (2020) Infrared spectroscopy with multivariate analysis to interrogate the interaction of whole cells and secreted soluble exopolymeric substances of *Pseudomonas veronii* 2E with Cd(II), Cu(II) and Zn(II). Spectrochim Acta A Mol Biomol Spectrosc 228:117820

Figueira EMDAP, Lima AIG, Pereira SIA (2005) Cadmium tolerance plasticity in *Rhizobium leguminosarum* bv. viciae: glutathione as a detoxifying agent. Can J Microbiol 51:7–14

Gadd GM (2004) Microbial influence on metal mobility and application for bioremediation. Geoderma 122:109–119

Garcia-Rubio R, De Oliveira HC, Rivera J, Trevijano-Contador N (2019) The fungal cell wall: *Candida, Cryptococcus*, and *Aspergillus* species. Front Microbiol 10:2993

Ghosh SK, Bera T, Chakrabarty AM (2020) Microbial siderophore – a boon to agricultural sciences. Biol Control 144:104214

Ghosh A, Pramanik K, Bhattacharya S et al (2022) A potent cadmium bioaccumulating *Enterobacter cloacae* strain displays phytobeneficial property in Cd-exposed rice seedlings. Curr Res Microb Sci 3:100101

Gow NAR, Latge JP, Munro CA (2017) The fungal cell wall: structure, biosynthesis, and function. Microbiol Spectr 5

Große C, Grass G, Anton A et al (1999) Transcriptional organization of the czc heavy-metal homeostasis determinant from *Alcaligenes eutrophus*. J Bacteriol 181:2385–2393

Gutiérrez J-C, De Francisco P, Amaro F et al (2019) Structural and functional diversity of microbial metallothionein genes. In: Das S, Dash HR (eds) Microbial diversity in the genomic era. Academic Press

Hartwig A (2013) Cadmium and cancer. In: Sigel A, Sigel H, Sigel RKO (eds) Cadmium: from toxicity to essentiality. Springer Netherlands, Dordrecht

Hasgül E, Malkoç S, Güven A et al (2019) Biosorption of cadmium and copper by *Aspergillus* spp. isolated from industrial ceramic waste sludge. Biyolojik Çeşitlilik ve Koruma 12:44–56

Hassan E, Nafady N, Hassan S et al (2021) Cadmium biosorption potential and kinetic behavior of endophytic *Fusarium verticillioides* and its green synthesized silver nanoparticles (GSNPs). Glob Nest J 23:449–457

Higham DP, Sadler PJ, Scawen MD (1986) Effect of cadmium on the morphology, membrane integrity and permeability of *Pseudomonas putida*. J Gen Microbiol 132:1475–1482

Holst O, Moran AP, Brennan PJ (2010) Overview of the glycosylated components of the bacterial cell envelope. In: Holst O, Brennan PJ, Itzstein MV et al (eds) Microbial glycobiology. Academic Press, San Diego

Hossain ST, Mallick I, Mukherjee SK (2012) Cadmium toxicity in *Escherichia coli*: cell morphology, Z-ring formation and intracellular oxidative balance. Ecotoxicol Environ Saf 86:54–59

Hyder O, Chung M, Cosgrove D et al (2013) Cadmium exposure and liver disease among US adults. J Gastrointest Surg 7:1265–1273

Igiri BE, Okoduwa SIR, Idoko GO et al (2018) Toxicity and bioremediation of heavy metals contaminated ecosystem from tannery wastewater: a review. J Toxicol 2018:2568038

Imran M, Ahmad I, Barasubiye T et al (2020) Heavy metal tolerance among free-living fungi isolated from soil receiving long term application of wastewater. J Pure Appl Microbiol 14:157–170

IRIS-USEPA (1987) Cadmium [Online]. U.S. Environmental Protection Agency. https://iris.epa.gov/ChemicalLanding/&substance_nmbr=141

Ivanov VB, Bystrova EI, Seregin IV (2003) Comparative impacts of heavy metals on root growth as related to their specificity and selectivity. Russ J Plant Physiol 50:398–406

Jaeckel P, Krauss G, Menge S et al (2005) Cadmium induces a novel metallothionein and phytochelatin 2 in an aquatic fungus. Biochem Biophys Res Commun 333:150–155

Janyasuthwiong S, Rene E (2017) Bioprecipitation—a promising technique for heavy metal removal and recovery from contaminated wastewater streams. MOJ Civil Eng 2:191–193

Jebril N, Boden R, Braungardt C (2022) Cadmium resistant bacteria mediated cadmium removal: a systematic review on resistance, mechanism and bioremediation approaches. In: IOP Conference Series: Earth and Environmental Science. IOP Publishing, p 012006

Kalsotra T, Khullar S, Agnihotri R, Reddy MS (2018) Metal induction of two metallothionein genes in the ectomycorrhizal fungus *Suillus himalayensis* and their role in metal tolerance. Microbiology 164:868–876

Kanwar P, Goswami N, Mishra T (2021) Heavy metal tolerance mechanisms in microorganisms - a review. Pollut Res 40:562–571

Khan Z, Atif Nisar M, Zajif Hussain S et al (2015) Cadmium resistance mechanism in *Escherichia coli* P4 and its potential use to bioremediate environmental cadmium. Appl Microbiol Biotechnol 99:10745–10757

Khan Z, Rehman A, Hussain SZ et al (2016) Cadmium resistance and uptake by bacterium, *Salmonella enterica* 43C, isolated from industrial effluent. AMB Express 6:54

Khan A, Gupta A, Singh P et al (2020) Siderophore-assisted cadmium hyperaccumulation in *Bacillus subtilis*. Int Microbiol 23:277–286

Khan Z, Elahi A, Bukhari DA, Rehman A (2022) Cadmium sources, toxicity, resistance and removal by microorganisms: a potential strategy for cadmium eradication. J Saudi Chem Soc 26:101569

Khullar S, Reddy MS (2018) Ectomycorrhizal fungi and its role in metal homeostasis through metallothionein and glutathione mechanisms. Curr Biotechnol 7:231–241

Khullar S, Reddy MS (2019) Cadmium induced glutathione bioaccumulation mediated by γ-glutamylcysteine synthetase in ectomycorrhizal fungus *Hebeloma cylindrosporum*. Biometals 32:101–110

Khullar S, Sudhakara Reddy M (2019) Cadmium and arsenic responses in the ectomycorrhizal fungus *Laccaria bicolor*: glutathione metabolism and its role in metal(loid) homeostasis. Environ Microbiol Rep 11:53–61

Kneer R, Kutchan TM, Hochberger A, Zenk MH (1992) *Saccharomyces cerevisiae* and *Neurospora crassa* contain heavy metal sequestering phytochelatin. Arch Microbiol 157:305–310

Kolaj-Robin O, Russell D, Hayes KA et al (2015) Cation diffusion facilitator family: structure and function. FEBS Lett 589:1283–1295

Kubier A, Wilkin RT, Pichler T (2019) Cadmium in soils and groundwater: a review. Appl Geochem 108:104388

Kumar V, Singh S, Singh G, Dwivedi SK (2019) Exploring the cadmium tolerance and removal capability of a filamentous fungus *Fusarium solani*. Geomicrobiol J 36:782–791

Kumar A, Subrahmanyam G, Mondal R et al (2021) Bio-remediation approaches for alleviation of cadmium contamination in natural resources. Chemosphere 268:128855

Kumari D, Pan X, Lee D-J, Achal V (2014) Immobilization of cadmium in soil by microbially induced carbonate precipitation with *Exiguobacterium undae* at low temperature. Int Biodeterior Biodegrad 94:98–102

Labarrere CA, Kassab GS (2022) Glutathione: a Samsonian life-sustaining small molecule that protects against oxidative stress, ageing and damaging inflammation. Front Nutr 9:1007816

Lebrun M, Audurier A, Cossart P (1994) Plasmid-borne cadmium resistance genes in *Listeria monocytogenes* are similar to cadA and cadC of *Staphylococcus aureus* and are induced by cadmium. J Bacteriol 176:3040–3048

Legatzki A, Grass G, Anton A et al (2003) Interplay of the Czc system and two P-type ATPases in conferring metal resistance to *Ralstonia metallidurans*. J Bacteriol 185:4354–4361

Li F, Wang W, Li C et al (2018) Self-mediated pH changes in culture medium affecting biosorption and biomineralization of Cd^{2+} by *Bacillus cereus* Cd01. J Hazard Mater 358:178–186

Li F, Zheng Y, Tian J et al (2019) *Cupriavidus* sp. strain Cd02-mediated pH increase favoring bioprecipitation of Cd^{2+} in medium and reduction of cadmium bioavailability in paddy soil. Ecotoxicol Environ Saf 184:109655

Lim H-S, Lee H-H, Kim T-H, Lee B-R (2016) Relationship between heavy metal exposure and bone mineral density in Korean adult. J Bone Metab 23:223–231

Liu J, Dutta SJ, Stemmler AJ, Mitra B (2006) Metal-binding affinity of the transmembrane site in ZntA: implications for metal selectivity. Biochemistry 45:763–772

Liu P, Chen X, Huang Q, Chen W (2015) The Role of CzcRS Two-component systems in the heavy metal resistance of Pseudomonas putida X4. Int J Mol Sci 16:17005

Liu H, Zhang Y, Wang Y et al (2021) The connection between Czc and Cad systems involved in cadmium resistance in *Pseudomonas putida*. Int J Mol Sci 22:9697

Loebus J, Leitenmaier B, Meissner D et al (2013) The major function of a metallothionein from the aquatic fungus *Heliscus lugdunensis* is cadmium detoxification. J Inorg Biochem 127:253–260

López Pérez PA, Aguilar López R, Neria González MI (2015) Cadmium removal at high concentration in aqueous medium: mediated by *Desulfovibrio alaskensis*. Int J Environ Sci Technol 12: 1975–1986

Loutet SA, Chan ACK, Kobylarz MJ et al (2015) The fate of intracellular metal ions in microbes. In: Nriagu JO, Skaar EP (eds) Trace metals and infectious diseases. MIT Press, Cambridge, MA
Macaskie LE, Dean ACR (1984) Cadmium Accumulation by a Citrobacter sp. J Gen Microbiol 130: 53–62
Margaryan A, Panosyan H, Birkeland N-K (2021) Heavy metal resistance in prokaryotes: mechanism and application. In: Egamberdieva D, Birkeland N-K, Li W-J et al (eds) Microbial communities and their interactions in the extreme environment. Springer Singapore, Singapore
Masood F, Malik A (2013) Current aspects of metal resistant bacteria in bioremediation: from genes to ecosystem. In: Malik A, Grohmann E, Alves M (eds) Management of microbial resources in the environment. Springer Netherlands, Dordrecht
Maurya PK, Malik DS, Yadav KK et al (2019) Bioaccumulation and potential sources of heavy metal contamination in fish species in river Ganga basin: possible human health risks evaluation. Toxicol Rep 6:472–481
Mazhar SH, Herzberg M, Ben Fekih I et al (2020) Comparative insights into the complete genome sequence of highly metal resistant *Cupriavidus metallidurans* strain BS1 isolated from a gold–copper mine. Front Microbiol 11
Mccreight JD, Schroeder DB (1982) Inhibition of growth of nine ectomycorrhizal fungi by cadmium, lead, and nickel *in vitro*. Environ Exp Bot 22:1–7
Mitra RS, Bernstein IA (1978) Single-strand breakage in DNA of *Escherichia coli* exposed to Cd^{2+}. J Bacteriol 133:75–80
Mohd-Nadzir M, Nurhayati RW, Idris FN, Nguyen MH (2021) Biomedical applications of bacterial exopolysaccharides: a review. Polymers 13:530
Monteiro CM, Castro PML, Malcata FX (2012) Metal uptake by microalgae: underlying mechanisms and practical applications. Biotechnol Prog 28:299–311
Morillo-Pérez JA, García-Ribera R, Quesada T et al (2008) Biosorption of heavy metals by the exopolysaccharide produced by *Paenibacillus jamilae*. World J Microbiol Biotechnol 24:2699–2704
Morrow H (2010) Cadmium and cadmium alloys. In: Kirk-Othmer encyclopedia of chemical technology. Wiley
Mutoh N, Hayashi Y (1988) Isolation of mutants of *Schizosaccharomyces pombe* unable to synthesize cadystin, small cadmium-binding peptides. Biochem Biophys Res Commun 151: 32–39
Naik MM, Pandey A, Dubey SK (2012) Biological characterization of lead-enhanced exopolysaccharide produced by a lead resistant *Enterobacter cloacae* strain P2B. Biodegradation 23:775–783
Nazareth S, Marbaniang T (2008) Effect of heavy metals on cultural and morphological growth characteristics of halotolerant *Penicillium* morphotypes. J Basic Microbiol 48:363–369
Nies DH (1999) Microbial heavy-metal resistance. Appl Microbiol Biotechnol 51:730
Nies DH (2003) Efflux-mediated heavy metal resistance in prokaryotes. FEMS Microbiol Rev 27: 313–339
Nies DH, Silver S (1995) Ion efflux systems involved in bacterial metal resistances. J Ind Microbiol 14:186–199
Oger C, Mahillon J, Petit F (2003) Distribution and diversity of a cadmium resistance (*cadA*) determinant and occurrence of IS257 insertion sequences in Staphylococcal bacteria isolated from a contaminated estuary (Seine, France). FEMS Microbiol Ecol 43:173–183
Orr S, Bridges C (2017) Chronic kidney disease and exposure to nephrotoxic metals. Int J Mol Sci 18:1039
Oshiquiri LH, Dos Santos KRA, Ferreira Junior SA et al (2020) *Trichoderma harzianum* transcriptome in response to cadmium exposure. Fungal Genet Biol 134:103281
Osman GE, Abulreesh HH, Elbanna K et al (2019) Recent progress in metal-microbe interactions: prospects in bioremediation. J Pure Appl Microbiol 13:13
Pal A, Dutta S, Mukherjee PK, Paul AK (2005) Occurrence of heavy metal-resistance in microflora from serpentine soil of Andaman. J Basic Microbiol 45:207–218

Pal A, Bhattacharjee S, Saha J et al (2022) Bacterial survival strategies and responses under heavy metal stress: a comprehensive overview. Crit Rev Microbiol 48:327–355

Park JH, Lamb D, Paneerselvam P et al (2011) Role of organic amendments on enhanced bioremediation of heavy metal(loid) contaminated soils. J Hazard Mater 185:549–574

Paul A, Mondal B, Ghosh R, Bhattacharyya S (2021) Heavy metal tolerance of filamentous fungal strains isolated from agricultural field near waste dump site of Dhapa, Kolkata. J Mycopathol Res 59:23–30

Peng D, Qiao S, Luo Y et al (2020) Performance of microbial induced carbonate precipitation for immobilizing Cd in water and soil. J Hazard Mater 400:123116

Pinot F, Kreps SE, Bachelet M et al (2000) Cadmium in the environment: sources, mechanisms of biotoxicity, and biomarkers. Rev Environ Health 15:299–324

Płaza G, Łukasik W, Ulfig K (1998) Effect of cadmium on growth of potentially pathogenic soil fungi. Mycopathologia 141:93–100

Pongratz R, Heumann KG (1999) Production of methylated mercury, lead, and cadmium by marine bacteria as a significant natural source for atmospheric heavy metals in polar regions. Chemosphere 39:89–102

Priyadarshanee M, Das S (2021) Biosorption and removal of toxic heavy metals by metal tolerating bacteria for bioremediation of metal contamination: a comprehensive review. J Environ Chem Eng 9:104686

Priyadarshini E, Priyadarshini SS, Cousins BG, Pradhan N (2021) Metal-fungus interaction: review on cellular processes underlying heavy metal detoxification and synthesis of metal nanoparticles. Chemosphere 274:129976

Qiao S, Zeng G, Wang X et al (2021) Multiple heavy metals immobilization based on microbially induced carbonate precipitation by ureolytic bacteria and the precipitation patterns exploration. Chemosphere 274:129661

Raab A, Feldmann J (2003) Microbial transformation of metals and metalloids. Sci Prog 86:179–202

Rahman NNNA, Shahadat M, Won CA, Omar FM (2014) FTIR study and bioadsorption kinetics of bioadsorbent for the analysis of metal pollutants. RSC Adv 4:58156–58163

Rajkumar M, Ae N, Prasad MNV, Freitas H (2010) Potential of siderophore-producing bacteria for improving heavy metal phytoextraction. Trends Biotechnol 28:142–149

Ramadan SE, Razak AA, Soliman HG (1988) Influence of cadmium on certain biological activities in a cadmium-tolerant fungi. Biol Trace Elem Res 18:179–190

Reddy MS, Prasanna L, Marmeisse R, Fraissinet-Tachet L (2014) Differential expression of metallothioneins in response to heavy metals and their involvement in metal tolerance in the symbiotic basidiomycete *Laccaria bicolor*. Microbiology 160:2235–2242

Renu S, Sarim KM, Singh DP et al (2022) Deciphering cadmium (Cd) tolerance in newly isolated bacterial strain, *Ochrobactrum intermedium* BB12, and its role in alleviation of Cd stress in spinach plant (*Spinacia oleracea* L.). Front Microbiol 12

Rieuwerts JS, Thornton I, Farago ME, Ashmore MR (1998) Factors influencing metal bioavailability in soils: preliminary investigations for the development of a critical loads approach for metals. Chem Speciat Bioavailab 10:61–75

Rizvi A, Ahmed B, Zaidi A, Khan MS (2019) Bioreduction of toxicity influenced by bioactive molecules secreted under metal stress by *Azotobacter chroococcum*. Ecotoxicology 28:302–322

Rohde M (2019) The gram-positive bacterial cell wall. Microbiol Spectr 7. https://doi.org/10.1128/microbiolspec.gpp3-0044-2018

Ron EZ, Minz D, Finkelstein NP, Rosenberg E (1992) Interactions of bacteria with cadmium. Biodegradation 3:161–170

Roskova Z, Skarohlid R, Mcgachy L (2022) Siderophores: an alternative bioremediation strategy? Sci Total Environ 819:153144

Rumble J (2017) CRC handbook of chemistry and physics. Routledge

Russell D, Soulimane T (2012) Evidence for zinc and cadmium binding in a CDF transporter lacking the cytoplasmic domain. FEBS Lett 586:4332–4338

Sácký J, Beneš V, Borovička J et al (2019) Different cadmium tolerance of two isolates of *Hebeloma mesophaeum* showing different basal expression levels of metallothionein (HmMT3) gene. Fungal Biol 123:247–254

Saha J, Dey S, Pal A (2022a) Whole genome sequencing and comparative genomic analyses of *Pseudomonas aeruginosa* strain isolated from arable soil reveal novel insights into heavy metal resistance and codon biology. Curr Genet 68:481–503

Saha J, Sarkar M, Mandal P, Pal A (2022b) Comparative study of heavy metal uptake and analysis of plant growth promotion potential of multiple heavy metal-resistant bacteria isolated from arable land. Curr Microbiol 79:7

Salam LB, Obayori OS, Ilori MO, Amund OO (2020) Effects of cadmium perturbation on the microbial community structure and heavy metal resistome of a tropical agricultural soil. Bioresour Bioprocess 7:25

Samrane K, Bouhaouss A (2022) Cadmium in phosphorous fertilizers: balance and trends. Rasayan J Chem 15:2103–2117

Schalk IJ, Hannauer M, Braud A (2011) New roles for bacterial siderophores in metal transport and tolerance. Environ Microbiol 13:2844–2854

Seshadri B, Bolan NS, Naidu R (2015) Rhizosphere-induced heavy metal(loid) transformation in relation to bioavailability and remediation. J Soil Sci Plant Nutr 15:524–548

Shan B, Hao R, Xu H et al (2021) A review on mechanism of biomineralization using microbial-induced precipitation for immobilizing lead ions. Environ Sci Pollut Res 28:30486–30498

Sharma R, Rensing C, Rosen BP, Mitra B (2000) The ATP hydrolytic activity of purified ZntA, a Pb(II)/Cd(II)/Zn(II)-translocating ATPase from *Escherichia coli*. J Biol Chem 275:3873–3878

Sher S, Hussain SZ, Rehman A (2020) Phenotypic and genomic analysis of multiple heavy metal–resistant *Micrococcus luteus* strain AS2 isolated from industrial waste water and its potential use in arsenic bioremediation. Appl Microbiol Biotechnol 104:2243–2254

Silhavy TJ, Kahne D, Walker S (2010) The bacterial cell envelope. Cold Spring Harb Perspect Biol 2:a000414

Silver S (1992) Plasmid-determined metal resistance mechanisms: range and overview. Plasmid 27: 1–3

Sinha S, Kumar-Mukherjee S (2008) Cadmium–induced siderophore production by a high Cd-resistant bacterial strain relieved Cd toxicity in plants through root colonization. Curr Microbiol 56:55–60

Smirnova GV, Oktyabrsky ON (2005) Glutathione in bacteria. Biochemistry (Mosc) 70:1199–1211

Suhani I, Sahab S, Srivastava V, Singh RP (2021) Impact of cadmium pollution on food safety and human health. Curr Opin Toxicol 27:1–7

Sun R, Wang L, Huang R et al (2020) Cadmium resistance mechanisms of a functional strain *Enterobacter* sp. DNB-S2, isolated from black soil in Northeast China. Environ Pollut 263: 114612

Tellez-Plaza M, Guallar E, Howard BV et al (2013) Cadmium exposure and incident cardiovascular disease. Epidemiology 24:421–429

Thiem D, Złoch M, Gadzała-Kopciuch R et al (2018) Cadmium-induced changes in the production of siderophores by a plant growth promoting strain of *Pseudomonas fulva*. J Basic Microbiol 58: 623–632

Timofeeva AM, Galyamova MR, Sedykh SE (2022) Bacterial siderophores: classification, biosynthesis, perspectives of use in agriculture. Plants (Basel) 11:3065

Trevors JT, Stratton GW, Gadd GM (1986) Cadmium transport, resistance, and toxicity in bacteria, algae, and fungi. Can J Microbiol 32:447–464

Uddin MJ, Haque F, Jabeen I, Shuvo SR (2023) Characterization and whole-genome sequencing of an extreme arsenic-tolerant *Citrobacter freundii* SRS1 strain isolated from Savar area in Bangladesh. Can J Microbiol 69:44–52

Valencia EY, Braz VS, Guzzo C, Marques MV (2013) Two RND proteins involved in heavy metal efflux in *Caulobacter crescentus* belong to separate clusters within proteobacteria. BMC Microbiol 13:79

Vallero D (2014) Chapter 23 - Metal and metalloid cycles. In: Vallero D (ed) Fundamentals of air pollution, 5th edn. Academic Press, Boston

Wang CL, Michels PC, Dawson SC et al (1997) Cadmium removal by a new strain of *Pseudomonas aeruginosa* in aerobic culture. Appl Environ Microbiol 63:4075–4078

Wang LK, Vaccari DA, Li Y, Shammas NK (2005) Chemical precipitation. In: Wang LK, Hung Y-T, Shammas NK (eds) Physicochemical treatment processes. Humana Press, Totowa, NJ

Wang X, Zhang X, Liu X et al (2019) Physiological, biochemical and proteomic insight into integrated strategies of an endophytic bacterium *Burkholderia cenocepacia* strain YG-3 response to cadmium stress. Metallomics 11:1252–1264

Worms I, Simon DF, Hassler CS, Wilkinson KJ (2006) Bioavailability of trace metals to aquatic microorganisms: importance of chemical, biological and physical processes on biouptake. Biochimie 88:1721–1731

Wu X, Santos RR, Fink-Gremmels J (2015) Cadmium modulates biofilm formation by *Staphylococcus epidermidis*. Int J Environ Res Public Health 12:2878–2894

Xia L, Xu X, Zhu W et al (2015) A comparative study on the biosorption of Cd^{2+} onto *Paecilomyces lilacinus* XLA and *Mucoromycote* sp. XLC. Int J Mol Sci 16:15670–15687

Xia X, Wu S, Zhou Z, Wang G (2021) Microbial Cd(II) and Cr(VI) resistance mechanisms and application in bioremediation. J Hazard Mater 401:123685

Xu Y-N, Chen Y (2020) Advances in heavy metal removal by sulfate-reducing bacteria. Water Sci Technol 81:1797–1827

Xu S, Xing Y, Liu S et al (2020) Characterization of Cd^{2+} biosorption by *Pseudomonas* sp. strain 375, a novel biosorbent isolated from soil polluted with heavy metals in Southern China. Chemosphere 240:124893

Yang E, Sun L, Ding X et al (2019) Complete genome sequence of *Caulobacter flavus* RHGG3T, a type species of the genus *Caulobacter* with plant growth-promoting traits and heavy metal resistance. 3 Biotech 9:42

Yin K, Wang Q, Lv M, Chen L (2019a) Microorganism remediation strategies towards heavy metals. Chem Eng J 360:1553–1563

Yin W, Wang Y, Liu L, He J (2019b) Biofilms: the microbial "protective clothing" in extreme environments. Int J Mol Sci 20:3423

Yoon KP, Misra TK, Silver S (1991) Regulation of the cadA cadmium resistance determinant of *Staphylococcus aureus* plasmid pI258. J Bacteriol 173:7643–7649

You L-X, Zhang R-R, Dai J-X et al (2021) Potential of cadmium resistant *Burkholderia contaminans* strain ZCC in promoting growth of soy beans in the presence of cadmium. Ecotoxicol Environ Saf 211:111914

Yu X, Ding Z, Ji Y et al (2020) An operon consisting of a P-type ATPase gene and a transcriptional regulator gene responsible for cadmium resistances in *Bacillus vietamensis* 151–6 and *Bacillus marisflavi* 151–25. BMC Microbiol 20:18

Yu Y, Shi K, Li X et al (2022) Reducing cadmium in rice using metallothionein surface-engineered bacteria WH16-1-MT. Environ Res 203:111801

Yuan X, Xue N, Han Z (2021) A meta-analysis of heavy metals pollution in farmland and urban soils in China over the past 20 years. J Environ Sci 101:217–226

Yun YS, Vijayaraghavan K, Won SW (2011) Bacterial biosorption and biosorbents. In: Kotrba P, Mackova M, Macek T (eds) Microbial biosorption of metals. Springer, Dordrecht, The Netherlands

Zeng G, Qiao S, Wang X et al (2021) Immobilization of cadmium by *Burkholderia* sp. QY14 through modified microbially induced phosphate precipitation. J Hazard Mater 412:125156

Zhang Y, Zhang Y, Liu M et al (2008) Dark septate endophyte (DSE) fungi isolated from metal polluted soils: their taxonomic position, tolerance, and accumulation of heavy metals *in vitro*. J Microbiol 46:624–632

Zhang D, Yin C, Abbas N et al (2020) Multiple heavy metal tolerance and removal by an earthworm gut fungus *Trichoderma brevicompactum* QYCD-6. Sci Rep 10:6940

Zhao Y, Yao J, Yuan Z et al (2017) Bioremediation of Cd by strain GZ-22 isolated from mine soil based on biosorption and microbially induced carbonate precipitation. Environ Sci Pollut Res 24:372–380

Zhao X, Wang M, Wang H et al (2019) Study on the remediation of Cd pollution by the biomineralization of urease-producing bacteria. Int J Environ Res Public Health 16:268

Zheng C, Chen M, Tao Z et al (2015) Differential expression of sulfur assimilation pathway genes in *Acidithiobacillus ferrooxidans* under Cd^{2+} stress: evidence from transcriptional, enzymatic, and metabolic profiles. Extremophiles 19:429–436

Zheng Y, Xiao C, Chi R (2021) Remediation of soil cadmium pollution by biomineralization using microbial-induced precipitation: a review. World J Microbiol Biotechnol 37:208

Zhu G, Xie L, Tan W et al (2022) Cd^{2+} tolerance and removal mechanisms of *Serratia marcescens* KMR-3. J Biotechnol 359:65–74

Złoch M, Thiem D, Gadzała-Kopciuch R, Hrynkiewicz K (2016) Synthesis of siderophores by plant-associated metallotolerant bacteria under exposure to Cd^{2+}. Chemosphere 156:312–325

Cadmium Toxicity and Role of Plant Growth Promoting Bacteria in Phytoremediation

Harsh Kumar, Shumailah Ishtiyaq, Vinamrata Ponia, Paulo J. C. Favas, Rohan J. D'Souza, Mayank Varun, and Manoj S. Paul

Abstract Heavy metal contamination has become a household word in recent decades as more and more information about the persistence of such pollutants in soil and water, as well as their toxicity to all life forms is being gained. Anthropogenic practices ranging from agriculture to space science are the main culprits. Lead, nickel, zinc, and copper are the foremost pollutants while cadmium is among the most toxic even at very low concentrations. Cadmium is a rare metal but has been identified as a Group I human carcinogen and a threat to the environment at large. It damages almost every organ system in the long run. Among the various technologies being used to remediate polluted soil and water, phytoremediation is a green, clean method which utilizes plants to accumulate and degrade harmful pollutants from the growth matrix. It is both cost-effective and less invasive. The potential of suitable plants can be further enhanced by addition of amendments like Plant Growth Promoting Bacteria (PGPB). These soil bacteria are found in and around the roots of plants and secrete chemicals which promote plant growth and affect bioavailability of substances like minerals, salts and heavy metals in soil. The chapter discusses all these aspects related to phytoremediation in some detail. A small case study which evaluates the phytoremedial potential of *Chenopodium album*, and its

H. Kumar
Department of Microbiology, School of Life Sciences, Dr. B.R. Ambedkar University, Agra, India

S. Ishtiyaq · R. J. D'Souza · M. S. Paul
Department of Botany, St. John's College, Agra, India

V. Ponia
Department of Botany, R.B.S. College, Agra, India

P. J. C. Favas
School of Life Sciences and the Environment, University of Tras-os-Montes e Alto Douro, Vila Real, Portugal

Faculty of Sciences and Technology, MARE – Marine and Environmental Sciences Centre, University of Coimbra, Coimbra, Portugal

M. Varun (✉)
Department of Botany, Hislop College, Nagpur, India

A. K. Jha, N. Kumar (eds.), *Cadmium Toxicity Mitigation*,
https://doi.org/10.1007/978-3-031-47390-6_7

enhancement in the above context is also described. The two PGPBs used here are *Pseudomonas fluorescens* and *Bradyrhizobium japonicum.* The test plant has a good potential for accumulating Cd, especially in synergy with the two bacterial strains. The Cd accumulation patterns along with the physiological and biochemical parameters analysed suggest that plants treated with a combination of the PGPBs performed better than those treated with either one individually. PGPB treatments worked better than non-PGPB treatments as hypothesized.

Keywords Heavy metals · Cadmium · Soil pollution · Phytotoxicity · Antioxidative defence · Rhizosphere

1 Introduction

Heavy metal pollution of the environment has been an issue of great concern globally for the past few decades now. Heavy metals distributed into the environment are pervasive and because of their hazardous nature pose a serious threat to humans, animals as well as plants, beyond certain limits. Heavy metals, sometimes referred to as 'trace metals' (found in low quantity, usually <1 ppm), in particular, locations such as soil, groundwater, plants and animal tissues (Phillips et al. 1981) are not considered harmful pollutants if found in low concentrations. As heavy metals are also found in bedrock, they may be naturally present in the earth's crust. Some like manganese, copper and zinc are essential micronutrients for living organisms (Marschner 1995). Average concentrations of heavy metals in soil generally range from <1 to >1000 mg kg^{-1} of soil (Adriano 2003).

Anthropogenic practices such as mining, waste sludge applications, improper waste management and lowland operations are the main reasons for severe metal emissions (Ha et al. 2014). Industries like metallurgy, mining, smelting as well as appliance, vehicle, and battery manufacturing also add to heavy metal load of the environment, more so in developing countries (Toth et al. 2016; Zhao et al. 2019; Ahmad et al. 2022). Anthropogenic addition is the key reason behind heavy metal contamination of the environment (Table 1). Lead, nickel, zinc and copper are the foremost pollutants while cadmium is among the most toxic even at very low concentrations.

Some metals are necessary for the biological functioning of plants and animals, but at higher concentrations, they disrupt their metabolic and physiological processes. Other toxic heavy metals such as cadmium, arsenic, chromium, uranium, lead and mercury inhibit photosynthesis and activity of essential enzymes as well as reduce mineral nutrition in plants. This results in reduced growth (Kabata-Pendias and Pendias 2001; Nematian and Kazemeini 2013).

In humans and animals, heavy metals can accumulate in vital organs for years after being absorbed posing severe health concerns (Tak et al. 2013, Witkowska et al. 2021). Heavy metals may be deleterious at minimal levels and pose the risk of cancer (Dixit et al. 2015). Cadmium is one of the most widespread hazardous heavy metals released from mining/industrial activity and poses a serious threat to human

Table 1 Rough estimation of yearly emission of heavy metals in the atmosphere by natural and anthropogenic resources (modified from Valavanidis and Vlachogianni 2010)

Heavy metals	Sources	Emission by natural resources (thousand tonnes year^{-1})	Emission by anthropogenic resources (thousand tonnes year^{-1})	References
Cd	Paints, electroplating, plastic combustion, chemical fertilizers	1.0	7.5	Pulford and Watson (2003)
As	Wood preservatives and pesticides	7.8	24	Thangavel and Subbhuraam (2004)
Pb	Burning of leaded fuel, manufacturing of batteries, pesticides	19	450	Wuana and Okieimen (2011)
Ni	Industrial waste, room appliances, medical equipment, automobile batteries, and steel	26	47	Tariq et al. (2006)
Cu	Cu-based pesticides, fertilizers	19	56	Khan et al. (2007)

health. Most of the Cd present in the human body enters through contaminated food and/or water. Thus, the potential transfer of Cd from soil through the food chain is a valid and considerable risk (Huang et al. 2004).

2 Heavy Metal Toxicity in the Soil

Soil pollution by toxic metals often results in imbalanced micro-flora and fauna and consequently affects the meso-fauna and plants. In general, soils with high metal exposure have lower plant cover and growth. This increases surface erosion, resulting in metal leakage into runoff water which subsequently absorbs into surrounding soil and surface waters (Lasat 2002). The most hazardous heavy metal ranges and regulatory guidelines are listed in Table 2, where the limits of their tolerance are also mentioned.

2.1 Sources of Cadmium

Cadmium is a silver-white, group IIb metal of the periodic table with relatively low melting and boiling points. It is a rare element and is found at a low level in the natural environment. It is the seventh most toxic heavy metal in the world ranking as

Table 2 Soil concentration ranges and regulatory guidelines for heavy metals

Metals	Soil concentration range[a] (mg kg^{-1})	Regulatory limit[b] (mg kg^{-1})
As	0.10–02	20
Pb	1–6900	600
Hg	<0.01–800	270
Cr	0.05–3950	100
Cd	0.10–345	100
Zn	150–5000	1500
Ni	10–1000	35–40

[a]Riley et al. (1992)
[b]Non-residential direct contact soil clean-up criteria (NJDEP 1996)

per ATSDR (2019) and categorized as a Group I human carcinogen by IARC. It has also been classified as a 'priority substance' according to the EU Water Frame Directive 2000/60/EC, highlighting it as a 'significant risk to the aquatic environment or others by its intermediation'.

Cadmium is released into the environment through power stations, photography, electroplating, heating systems and paint, dye, glass and metal-working industries. It was first used extensively in painting pigments, glass production, electroplating, photography and dyeing. The heat, light, weathering and alkali resistance of Cd, as well as its colour, promoted its use as a pigment in various plastics, paints, textiles, rubber, glass and enamels. In the 1970s, Cd was used as a plating agent, pigment, stabilizer and in batteries. In batteries, Cd is complexed with Ni, while in alloys, it is combined with metals like Cu and Zn (Nriagu et al. 1982). Other important sources of Cd, which particularly affect agricultural soils, are phosphate fertilizers, sewage sludge, and other organic waste (Adriano 1992).

The absorption of Cd in soil and water has triggered significant environmental and health problems (Salt et al. 1995). It is poisonous at extremely low levels, and it has no known useful function in higher organisms. After World War II, in the 1960s and 1970s, Japanese people suffered from different pollution-related health issues. Itai-itai (which affects the skeleton system) was one of these conditions caused by chronic Cd-contaminated rice fields (Kaji 2012).

Short-term intake of Cd may cause serious illnesses such as damage to respiratory organs. Higher doses can cause stomach pain, diarrhoea, nausea as well as Cd deposition on organs such as lungs and bones, softening of bones, bone deformity, osteoporosis, hypercalciuria, and fractures. Cd concentration increases 3000-fold as it attaches to the cysteine-rich protein (metallothionein) forming the cysteine–metallothionein complex. This cysteine–metallothione combination causes hepatotoxicity and circulates toward the kidney where it may cause nephrotoxicity after accumulating in renal tissue (Sabolic et al. 2010). It also affects the growth and productivity of plants negatively which is discussed in some detail later.

3 Remediation of Heavy Metal Contaminated Soil

3.1 Conventional Remediation Methods

A number of physical and chemical techniques have been developed and used to extract/remove heavy metals from soil, water and sediments. Among these, excavation, chemical precipitation, ion-exchange, chemical oxidation/reduction, adsorption, coagulation-flocculation, reverse osmosis, floatation and solvent extraction are some of the most used techniques (Volesky and Holan 1995). However, these conventional approaches pose some constraints. They are economically viable only if the metal contamination is in the moderate to high bracket (Addour et al. 1999; Cossich et al. 2002). Additionally, most of these techniques include complications such as unpredictable removal of metal and a highly toxic sludge by-product. This is difficult to dewater and demands extra prudent disposal (Xia and Liyuan 2002). The removal of heavy metal ions by such techniques is also energy intensive (Xia and Liyuan 2002).

3.2 Use of Plants in Remediation

Phytoremediation is a technique which utilizes plants to accumulate and degrade harmful pollutants from both soil and water. It is a blanket term for plant-based technologies with different mechanisms of action (Kavamura and Esposito 2010). It is cost-effective and less invasive. It also happens to be aesthetically pleasing and suitable for developing countries (Pivertz 2001).

Phytoextraction: This involves the use of plants with pollutants (heavy metals and organic compounds) accumulating characteristics to concentrate them in the above-ground parts like shoots, leaves and fruits which can later be harvested and disposed of safely (Sekara et al. 2005).

Rhizofiltration: It is defined as the use of plants, both terrestrial and aquatic, to accumulate contaminants from polluted water in their roots (Mesjasz et al. 2004).

Phytostabilization: This involves the use of plants to reduce the mobility of contaminants in the soil and/or limit their availability to plant roots. This prevents/slows down the leaching of the contaminants into groundwater and their subsequent entry into the food chain (Erakhrumen 2007).

Phytovolatilization: The use of plants to uptake pollutants from soil/water and transform them into volatile compounds that can be released into the atmosphere through the stomata is known as phytovolatilization (Karami and Shamsuddin 2010).

The success of phytoremediation may not only depend on the plant itself but also on the interaction of the plant roots with bacteria and the concentrations of heavy metals in the soil (Wang et al. 2010). This includes the bioremediation process of utilizing microorganisms and their enzymes to break down toxic pollutants into less

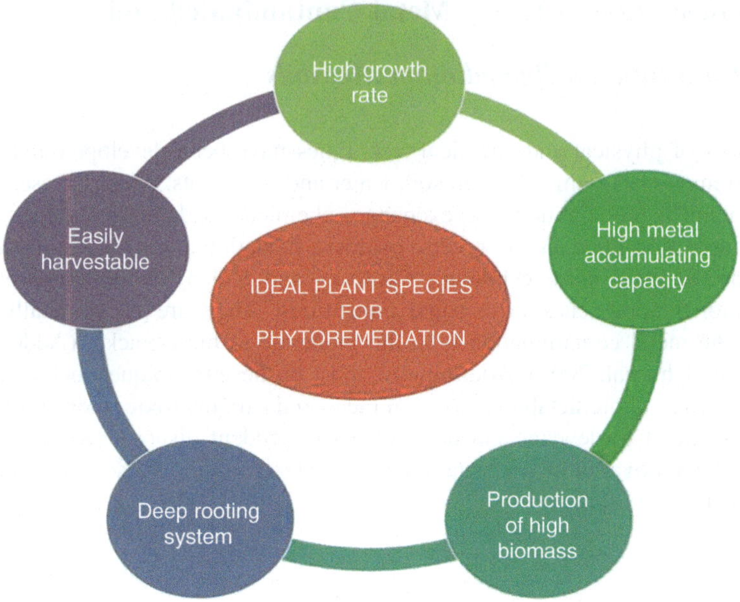

Fig. 1 Ideal characteristics of a plant for the purpose of phytoremediation

toxic and even more bioavailable forms. The characteristics that make a plant species ideal for phytoremediation are summed up in Fig. 1. According to the USEPA (2000), an ideal plant for phytoremediation should be one with either low biomass and very high metal uptake or high biomass with enhanced metal uptake potential.

4 Phytoremediation of Cadmium: Accumulation and Toxicity

Plants with the potential to phytoextract contaminants accumulate higher metal/contaminant concentrations in their tissues as compared to other surrounding plants (Ghosh and Singh 2005). Accumulation of Cd in a plant depends on the soil pH and redox potential as well as the chemistry of its roots. Roots are known to alter their physiology under heavy metal contamination. This leads to physical and chemical reactions which limit the bioavailability of the metal and its uptake (Dong et al. 2007). These are a part of the stress tolerance and defence mechanism of the plant. Many plants have been found to accumulate Cd viz.—*Thlaspi caerulescens*, *Arabidopsis hallerii*, *Solanum lycopersicum*, *Calotropis procera, Poa annua, Chenopodium murale, Abutilon indicum* and *Vetiver* grass (Baker et al. 2000; Chen et al. 2000; Cosio et al. 2004; Varun et al. 2012; D'Souza et al. 2013) but limited attempts have been made to enhance this natural potential with agronomic amendments.

Sequestration of Cd at elevated concentrations not only results in toxicity symptoms but also causes structural and ultrastructural changes. The effect of higher levels of cadmium on plant species has been widely studied with ecological and environmental implications that rely on the presence of Cd in soil or water (Mohanpuria et al. 2007). Plants that are able to accumulate Cd to appreciable levels have the ability to tolerate Cd in their systems and alleviate any contrary effects by metabolic pathways or isolation in vacuoles.

Cadmium toxicity symptoms in plants include chlorosis, leaf rolling, root tip burning as well as reduced stomatal opening, transpiration, photosynthesis, stunted growth and finally death (Kahle 1993; Sanita and Gabbrielli 1999). Cd is known to inhibit important root enzymes like iron reductase and nitrate reductase leading to Fe (II) deficiency and reduced absorption of nitrates. Thus, it seriously affects key physiological processes of plants including photosynthesis and nitrogen fixation (Alcantara et al. 1994; Hernández and Cooke 1997). It affects the uptake, translocation and use of water and minerals since it reduces the population of soil microbes and eventually the availability of minerals from the soil (Duxbury 1985; Benavides et al. 2005).

One of the major reasons for Cd toxicity is the generation of free radicals like ROS (reactive oxygen species) responsible for oxidative damage to the cell. Cadmium toxicity disturbs redox equilibrium of the cell and decreases the production of enzymatic and non-enzymatic antioxidants (Somashekaraiah et al. 1992; Fornazier et al. 2002; Cho and Seo 2004).

Plants accelerate the transcription of genes responsible for metal assimilation pathways as a tolerance and defence mechanism against metals in the soil. Many gene families responsible directly or indirectly for heavy metal transport and accumulation have been identified in *Arabidopsis thaliana*. Genes of families like IRT (Iron Regulated Transporter) and NRAMP (Natural Resistance-Associated Macrophage Protein) facilitate heavy metal uptake from soil to root (Sasaki et al. 2012) while those of families like HMA (Heavy metal ATPase) are involved in root to shoot transfer in xylem (Takahashi et al. 2012).

Despite all the advantages of phytoremediation, the use of plants alone for the purpose of heavy metal remediation has many curbs and constraints because of their toxic symptoms on plants. This problem of heavy metal toxicity could be alleviated by exploring associations of Plant Growth Promoting Rhizobacteria or Bacteria (PGPB) with plants. PGPB are free-living soil bacteria that live in conjunction with the roots of many different plants (Kloepper and Schroth 1978). They produce and secrete numerous regulatory chemicals that help in plant growth promotion and development. PGPB form this association by colonizing the roots of the plant and also act like a biocontrol agent and provide protection from many plant pathogens and diseases. This association not only improves the growth of the plant but could also help in facilitating the sequestration of toxic heavy metals from contaminated soil.

5 Plant Growth Promoting Bacteria (PGPB) and Plant Growth Promotion

The narrow soil zone surrounding a plant's roots is referred to as its rhizosphere. It is a highly complex region inhabited by a wide range of microbes, insects and nematodes (Venturi and Keel 2016). The presence of high concentrations of nutritive molecules such as proteins, sugar and organic acids is the primary cause for the presence of PGPBs in the rhizosphere. These compounds are significant components of root exudates, which influence bacterial concentrations in the rhizosphere (Beneduzi et al. 2012; Goswami et al. 2016).

PGPB may be classified into symbiotic or intracellular rhizobacteria which live within plant root cells, e.g. nodule bacteria, and free-living or extracellular rhizobacteria that live outside plant cells (known as PGPB, e.g. *Pseudomonas,* and *Azotobacter*) (Khan 2005; Babalola and Akindolire 2011). They can also be classified on the basis of their activities as (1) biofertilizers (which increases the bioavailability of the nutrients in the soil); (2) phytostimulators (promoting plant growth through phytohormones); (3) rhizoremediators (break down of organic contaminants); and (4) biopesticides (control diseases by synthesis of antibiotics and antifungal metabolites) (Somers et al. 2004).

PGPB can benefit plants directly by making essential nutrients such as nitrogen (N), phosphorus (P), sulphur (S), magnesium (Mg) and calcium (Ca) available in soils to promote plant growth (Fig. 2). They also have indirect benefits like suppression of pathogens by production of siderophores, bacterial and fungal antagonistic

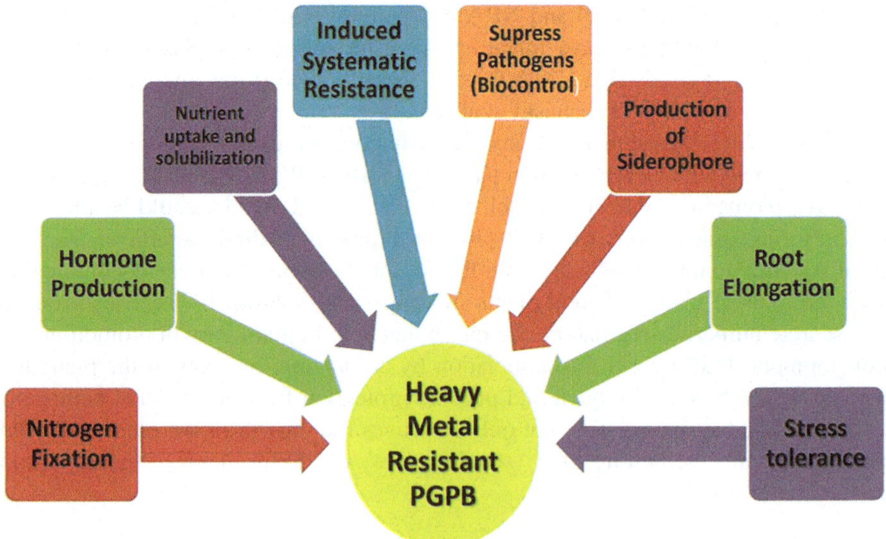

Fig. 2 Different mechanisms of PGPB for phytoremediation

substances and phytohormone production (IAA, ethylene etc.) (Glick 2003; Nehra and Choudhary 2015; Venturi and Keel 2016; Kumar et al. 2022).

Many PGPB naturally tolerate toxic metals and are thus used for phytoremediation to replace organic compounds and metal pollutants in crops. Various bacterial strains and their use in phytoextraction of heavy metals are listed in Table 3. Bacterial 1-aminocyclopropane1-carboxylate (ACC) a deaminase enzyme synthesized by PGPB is found to regulate the toxic symptoms of ethylene (Babalola et al. 2003). In addition, rhizobacterial strains can solubilize or mineralize organic and inorganic phosphorus (Glick et al. 1998) and thus strengthen the tolerance potential of the plant toward environmental stresses like drought and heavy metals.

PGPBs develop several intrinsic and extrinsic mechanisms for survival under heavy metal-stressed environments. They have the ability to immobilize, mobilize or transform metals and thus make them inactive or less toxic. The different mechanisms employed by PGPB for the bioremediation of heavy metals are depicted in Fig. 3.

6 Case Study

A combination of both plants and microorganisms could increase the overall efficiency of phytoremediation efforts and have been successfully incorporated in various experiments. However, the success of this approach depends upon the concentration of heavy metals in soil and the type of microbes and plants involved. A PGPB-enhanced phytoremediation experiment was set up to evaluate the tolerance of *Chenopodium album* L. toward Cd stress and accumulation patterns along with any PGPB-related enhancement of this potential. The enhancement would be quantified in terms of increase in biomass and/or bioavailability of Cd in soil.

6.1 Test Plant and PGPB

Chenopodium album L. or white goosefoot belongs to the Chenopodiaceae family. It is a fast-growing weed tolerant to xeric, cold and saline environments. Two PGPB strains were selected for the study *Pseudomonas fluorescens*—NCIM2100 (PGPB1) and *Bradyrhizobium japonicum*—NCIM5350 (PGPB2) based on their reported Cd resistance, mutual compatibility and plant growth promotion (Al-Dhabi et al. 2019; Alsohim 2020; Zefa et al. 2020). The cultures were obtained from the National Chemical Laboratory, Pune, India, as freeze-dried cultures and stored at −20 °C. Each culture was revived for use as per the protocol provided by the laboratory. Before application, the PGPB strains were tested for compatibility by cross streaking. A strain was streaked to one side of petri-plate while the other was streaked

Table 3 PGP activities of bacterial strains and their role in phytoremediation

Metals accumulation	Bacterial strains	PGP activities	Plants	References
As	*Actinomyces* sp. MC1, *Pseudomonas aeruginosa* strain Jogii, *Azotobacter* sp. Jogii, *Azotobacter brasilense*, *Bacillus subtilis*	Siderophore, indole acetic acid, phosphate solubilization	*Eichhornia crassipes*	Parvinder et al. (2017)
Cd	*Bacillus* sp., *Klebsiella* sp., *Leifsonia* sp., *Enterobacter* sp.	Indole acetic acid, phosphate solubilization	*Zea mays*	Ahmad et al. (2015)
	Kocuria flav AB402, *Bacillus vietnamensis* AB403	Indole acetic acid, siderophore, EPS	*Oryza sativa*	Ivy et al. (2018)
	Arthrobacter sp. SrN1, *Bacillus altitudinis* SrN9, *Bacillus megatherium*	ACCD, indole acetic acid, Siderophores	*Brassica napus, Sedum alfredii*	Pan et al. (2016)
	Chryseobacterium humi ECP37, *Pseudomonas reactants* EDP28, *Pseudomonas fluorescens* S3X	Phosphate solubilization, ACCD, siderophore, HCN, NH_3	*Zea mays*	Moreira et al. (2016)
	Bacillus safensis, Kocuria rosea		*Helianthus annuus*	Mohammadzadeh et al. 2017
	Kocuria flav, Bacillus vietnamensis	Indole acetic acid, siderophore, EPS	*Oryza sativa*	Ivy et al. (2018)
	Serratia marcescens	Phosphate solubilization, ACCD, indole acetic acid, siderophore	*Pteris vittata*	Wang et al. (2017a, b)
	Brevibacterium casei MH8	ACCD, indole acetic acid, HCN	*Sinapis alba*	Plociniczak et al. (2016)
	Pseudomonas sp., *Azotobacter* sp., *Paenibacillus* sp.	Siderophores	*Pennisetum glaucum, Sorghum bicolor*	Vartika et al. (2016)
	Variovorax paradoxus 5C-2, *Mesorhizobium* loti Le2	ACCD	*Lotus edulis*	Safronova et al. (2012)
	Pseudomonas moraviensis, Bacillus cereus	Indole acetic acid	*Cenchrus ciliaris*	Tamoor et al. (2017)
	Pseudomonas sp. 228, *Serratia*	ACCD, indole acetic acid, N	*Helianthis tuberosus*	Blanca et al. (2017)

(continued)

Table 3 (continued)

Metals accumulation	Bacterial strains	PGP activities	Plants	References
	sp. 246, *Pseudomonas* sp. 262	fixation, phosphate solubilization		
	Burkholderia cepacia		*Sedum alfredii*	Li et al. (2007)
Co	*Kocuria flav* AB402, *Bacillus vietnamensis* AB403	Indole acetic acid, siderophore, EPS	*Oryza sativa*	Ivy et al. (2018)
	Planomicrobium chinense, *Bacillus cereus*, *Pseudomonas fluorescens*		*Zea mays*	Khan and Bano (2016)
Cr	*Kocuria flav* AB402, *Bacillus vietnamensis* AB403	EPS, indole acetic acid, siderophore	*Oryza sativa*	Ivy et al. (2018)
	Serratia marcescens PRE01	ACCD, indole acetic acid, siderophore, phosphate solubilization	*Pteris vittata*	Wang et al. (2017a, b)
Cu	*Azotobacter vinelandii* GZC24, *Microbacterium lactium* YJ7, *Arthrobacter* sp. MT16, *Pseudomonas chlororaphis* SZY6, *Microbacterium* sp. JYC17	HCN, NH_3, phosphate solubilization, ACCD, siderophore,	*Brassica napus*	He et al. (2010)
	Achromobacter xylosoxidans Ax10	Phosphate solubilisation, ACCD, indole acetic acid	*Brassica juncea*	Ma et al. (2009)
	Kluyvera ascorbata SUD165	ACCD, Siderophores	*Brassica campestris*	Burd et al. (1998)
	Klebsiella Mc173, *Escherichia* N16, *Enterobacter* N9, *Serratia* K120	Phosphate solubilization, indole acetic acid, ACCD, siderophores	*Helianthus annuus*	Carlos et al. (2016)
	Bacillus weihenstephanensis SM3	Indole acetic acid, phosphate solubilization	*Helianthus annuus*	Rajkumar et al. (2008)
	Kocuria flav, *Bacillus vietnamensis*	Indole acetic acid, siderophore, EPS	*Oryza sativa*	Ivy et al. (2018)

(continued)

Table 3 (continued)

Metals accumulation	Bacterial strains	PGP activities	Plants	References
	Brevibacterium casei MH8	ACCD, indole acetic acid, HCN	*Sinapis alba*	Ploсiniczak et al. (2016)
Fe	*Azotobacter* sp., *Paenibacillus* sp., *Pseudomonas* sp.,	Siderophores production	*Sorghum bicolor, Pennisetum glaucum*	Vartika et al. (2016)
	Variovorax paradoxus 5C-2, *Mesorhizobium* loti Le2	ACCD	*Lotus edulis*	Safronova et al. (2012)
Mn	*Planomicrobium chinense*, *Bacillus cereus*, *Pseudomonas fluorescens*		*Zea mays*	Khan and Bano (2016)
	Bacillus cereus, Pseudomonas moraviensis	Indole acetic acid	*Cenchrus ciliaris*	Tamoor et al. (2017)
Ni	*Microbacterium, Arthrobacter*	Indole acetic acid, ACCD, siderophores	*Noccaea caerulescens*	Visioli et al. (2015)
	Psychrobacter sp. SRA1/SRA2	ACCD, indole acetic acid, phosphate solubilization	*Brassica juncea*	Ma et al. (2009)
	Arthrobacter nicotinovorans	Indole acetic acid, siderophore, P solubilization	*Alyssum pintodasilvae*	Cabello-Conejo et al. (2014)
	Bacillus megaterium, Pseudomonas sp.	Phosphate solubilization, ACCD, indole acetic acid, siderophore	*Brassica juncea*	Rajkumar and Freitas (2008b)
	Variovorax paradoxus NB24	ACCD, indole acetic acid, siderophore	*Bornmuellera tymphaea*	Durand et al. (2016)
	Klebsiella Mc173, *Escherichia* N16, *Enterobacter* N9, *Serratia* K120	Phosphate solubilization, ACCD, indole acetic acid, siderophores	*Helianthus annuus*	Carlos et al. (2016)
	Planomicrobium chinense, *Bacillus cereus*, *Pseudomonas fluorescens*		*Zea mays*	Khan and Bano (2016)
	Kocuria flav AB402, *Bacillus vietnamensis* AB403	Indole acetic acid,	*Oryza sativa*	Ivy et al. (2018)

(continued)

Table 3 (continued)

Metals accumulation	Bacterial strains	PGP activities	Plants	References
		siderophore, EPS		
Pb	*Streptomyces pactum* Act12	Siderophore	*Sorghum bicolor*	Amjad et al. (2017)
	Pseudomonas fluorescens, *Bacillus safensis*	Indole acetic acid, siderophore, ACCD	*Helianthus annuus*	Seyed et al. (2018)
	Klebsiella Mc173, *Enterobacter* N9, *Escherichia* N16, *Serratia* K120	ACCD, indole acetic acid, siderophores, phosphate solubilization	*Helianthus annuus*	Carlos et al. (2016)
	Pseudomonas fluorescens, *Bacillus cereus*, *Planomicrobium chinense*		*Zea mays*	Khan and Bano (2016)
Zn	*Klebsiella* Mc173, *Escherichia* N16, *Serratia* K120, *Enterobacter* N9	Siderophores, phosphate solubilization, ACCD, indole acetic acid	*Helianthus annuus*	Carlos et al. (2016)
	Pseudomonas sp. PsM6, *Pseudomonas jessenii* PjM15	ACCD, indole acetic acid, siderophore, phosphate solubilization	*Ricinus communis*	Rajkumar and Freitas (2008a)
	Klebsiella oxytoca	Indole acetic acid, HCN, ACCD, ammonia	*Helianthus annuus*	Arunakumara et al. (2015)
	Bacillus weihenstephanensis SM3	Indole acetic acid, phosphate solubilization	*Helianthus annuus*	Rajkumar et al. (2008)
	Streptomyces pactum	Siderophore production	*Sorghum bicolor*	Amjad et al. (2017)
	Pseudomonas fluorescens, *Bacillus safensis*	ACCD, indole acetic acid, siderophore	*Helianthus annuus*	Seyed et al. (2018)
	Pseudomonas sp. 228, *Serratia* sp. 246, *Pseudomonas* sp. 262	Indole acetic acid, N fixation, ACCD, phosphate solubilization	*Helianthis tuberosus*	Blanca et al. (2017)
			Oryza sativa	Ivy et al. (2018)

(continued)

Table 3 (continued)

Metals accumulation	Bacterial strains	PGP activities	Plants	References
	Kocuria flav AB402, *Bacillus vietnamensis* AB403	EPS, indole acetic acid, siderophore		
	Burkholderia cepacia		*Sedum alfredii*	Li et al. (2007)
	Brevibacterium casei MH8	ACCD, indole acetic acid, HCN	*Sinapis alba*	Plociniczak et al. (2016)

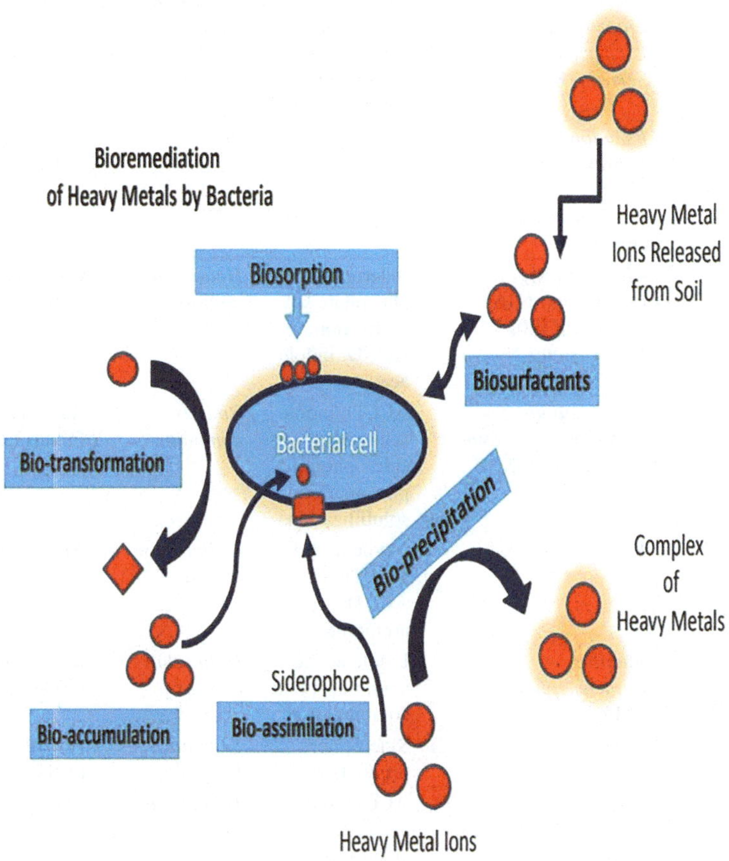

Fig. 3 Mechanism of heavy metal remediation by bacteria

perpendicularly up to the first. After 3 days of incubation, the cultures were observed and visually identified as compatible.

6.2 Experimental Design

Seeds of *Chenopodium album* were sterilized using a 0.1% $HgCl_2$ solution followed by soaking in double distilled water for 3 h. They were then germinated in sterile soil. After the first five true leaves emerged, the seedlings were first soaked in PGPB suspensions for 2 h followed by transplantation in plastic pots as per the treatment plan. These were placed in a greenhouse with a 16/8-h light/dark cycle, day/night temperatures of 27/22 °C and 70–80% relative humidity. After 2 weeks, 5 ml inoculum (10^6–10^8 CFU ml^{-1}) of respective PGPB was injected into the rhizosphere of each plant.

Each pot could hold 4 kg of autoclaved and air-dried soil. Plastic was the material of choice to minimize leaching of Cd. The pots were artificially spiked with cadmium in aqueous solutions using $CdCl_2 \cdot 5H_2O$. The soil used was a sandy loam made of 60–80% sand, 10–24% silt, and 8–16% clay from the botanical garden of St. John's College, Agra. It had a moderate water holding capacity with pH 7.26 ± 0.34, Moisture content 8.52 ± 0.44% and Electrical conductivity 1.14 ± 0.17 dSm^{-1}. The Organic carbon content was 1.16 ± 0.4%, available P_2O_5 7.3 ± 0.2 mg kg^{-1}, available Nitrogen 137.24 ± 1.3 mg kg^{-1} and available K_2O 258 ± 5.3 mg kg^{-1} Sodium (Na), Calcium-Magnesium (Ca + Mg) and Potassium (K) content of soil was 743 ± 4, 4.8 ± 0.4, and 27 ± 3 meq l^{-1}, respectively.

Plant treatments were: T0 (Control—Cd0); T1 (Cd 25 mg kg^{-1}); T2 (Cd 50 mg kg^{-1}); T3 (Cd 100 mg kg^{-1}); T4 (Cd 25 mg kg^{-1} + PGPB1); T5 (Cd 25 mg kg^{-1} + PGPB2) and T6 (Cd 25 mg kg^{-1} + PGPB1 + PGPB2). All treatments were applied in triplicate and maintained in a randomized block design. Watering was done using distilled water as needed to reduce the leaching of Cd salt. A plastic tray was kept under each pot to collect leachates, if any, which were collected and added to the same pot at the next watering. At 120 DAT (days after treatment), the plants were harvested and washed in deionized water individually. Each was then divided into root and shoot and growth parameters (root/shoot length, biomass) were noted. These were then dried, digested and analyzed for Cd content by AAS. Chlorophyll and proline contents in fresh leaves were measured on 30, 60, 90 and 120 DAT. The antioxidative enzyme activity of plants was measured at the end of the study.

Chlorophyll estimation was performed using acetone (80% v/v) (Lichtenthaler 1987). Estimation of proline content was done according to Bates et al. (1973). For antioxidative enzyme activity fresh leaf samples were taken for analysis. Catalase (CAT), Ascorbate peroxidase (APX) and Superoxide dismutase (SOD) activity was estimated according to Aebi (1984), Nakano and Asada (1981) and Dhindsa et al. (1981), respectively.

6.3 Results

6.3.1 Cd Uptake and Growth Response of *Chenopodium album*

Accumulation patterns of Cd in *Chenopodium album* indicate greater allocation of the metal to the roots (Table 4). At the lowest dose root uptake was almost twice that of the shoot (6.31/3.32 mg Cd kg^{-1}) at 120 DAT. Interestingly, the uptake was almost equal at Cd 50 but followed the pattern of lower dose when Cd concentration was further increased to 100 mg kg^{-1} soil. The phytoremedial potential of the test plant thus, points towards phytostabilization being the mechanism behind it.

The PGPB treatments enhanced Cd uptake when compared to their parallel treatment (Cd 25) without any bacterial inoculation. *Chenopodium album* showed good potential for accumulating Cd, especially in synergy with *Pseudomonas fluorescens* (PGPB1) and *Bradyrhizobium japonicum* (PGPB2).

Treatments with Cd significantly affected the root/shoot length and biomass of *C. album*. Symptoms of phytotoxicity like rolling of leaves, stunted growth and chlorosis were observed. Plants treated with Cd alone suffered more as quantified by lesser shoot/root length and biomass. PGPB inoculations protected the plants from Cd phytotoxicity and promoted growth significantly. Application of PGPB1 + PGPB2 was the most effective in this regard.

6.3.2 Chlorophyll and Proline Content

Chlorophyll content of a leaf indicates the photosynthetic capability of its tissues. Data in Fig. 4 indicates a marked reduction in total chlorophyll content of *C. album* leaves with an increase in Cd content. Combined inoculation with the two PGPB strains was most protective against the negative effects of Cd. The synergistic impact of both PGPB strains under Cd contamination resulted in the highest increase of chlorophyll content by 21% when compared to Cd treatment alone. Similarly, accumulation of the stress indicating osmolyte, proline, was found to be notably

Table 4 Metal uptake, partition and effect on root length, shoot length and biomass in *Chenopodium album*

Treatments	Cd in root (mg kg^{-1})	Cd in shoot (mg kg^{-1})	Shoot length (cm)	Root length (cm)	Biomass (g)
T0—Control	ND	ND	43.5 ± 1.7	18.7 ± 0.7	35.5 ± 1.1
T1—$Cd_{25\ ppm}$	6.31 ± 0.5	3.32 ± 0.4	40.1 ± 2.5	16.3 ± 1.1	34.2 ± 1.5
T2—$Cd_{50\ ppm}$	11.2 ± 0.8	10.24 ± 0.6	38.4 ± 2.0	14.1 ± 0.6	33.3 ± 1.7
T3—$Cd_{100\ ppm}$	18.62 ± 0.4	9.12 ± 0.6	35.2 ± 1.5	12.8 ± 1.1	31.2 ± 1.2
T4—$Cd_{25\ ppm}$ + PGPB1	7.25 ± 0.6	4.87 ± 0.8	45.3 ± 2.8	20.2 ± 0.8	38.4 ± 1.3
T5—$Cd_{25\ ppm}$ + PGPB2	6.92 ± 0.5	4.32 ± 0.7	44.8 ± 2.7	19.6 ± 0.5	37.8 ± 1.5
T6—$Cd_{25\ ppm}$ + PGPB1 + PGPB2	6.84 ± 0.4	5.31 ± 0.5	46.1 ± 2.2	21.1 ± 0.6	40.2 ± 1.7

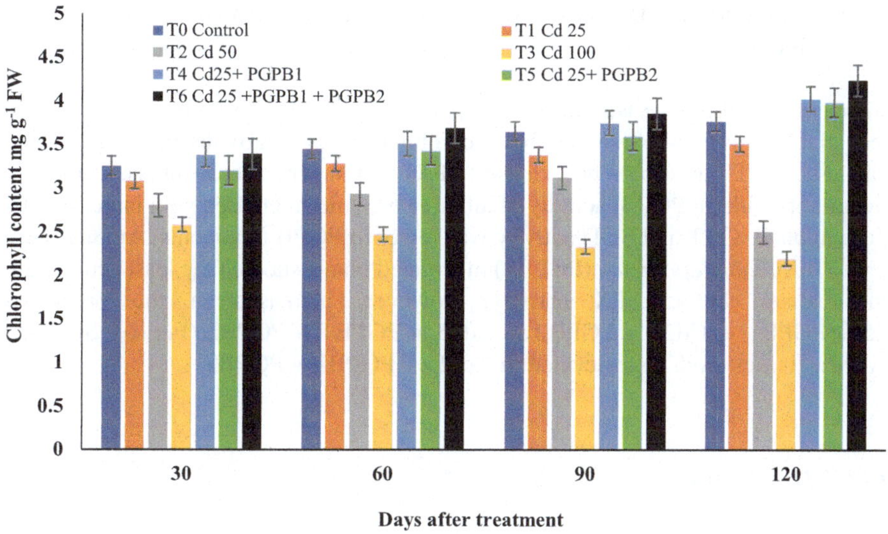

Fig. 4 Chlorophyll content (mg g^{-1} fresh weight) in *C. album*

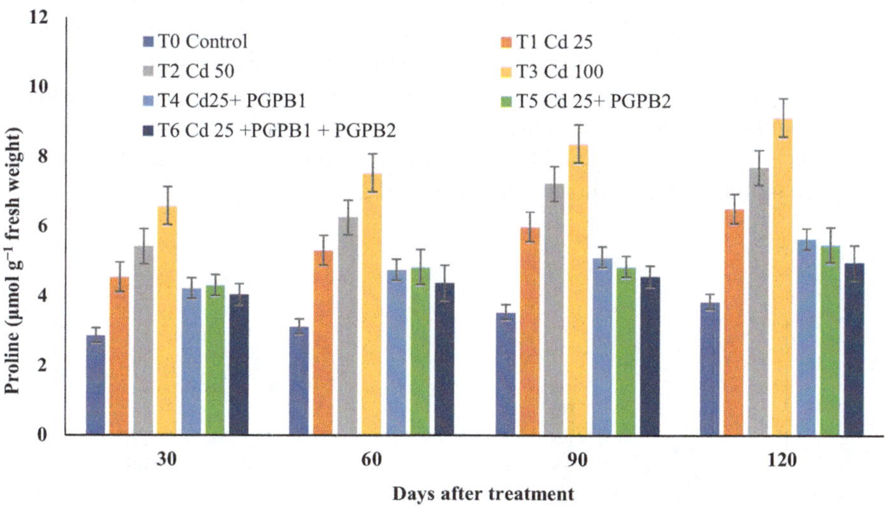

Fig. 5 Proline content in *C. album* leaves

affected under metal stress. An increase of up to 41% (Cd25 ppm), 50% (Cd50 ppm) and 58% (Cd100 ppm) was observed (Fig. 5). Proline content showed a gradual decrease at each testing date in plants inoculated with PGPB. The lowest range of proline (4.05–4.96 μmol g^{-1} fw) was observed in Cd25+ PGPB1 + PGPB2.

6.3.3 Antioxidative Defence System with Plant Growth Promoting Bacteria

Increased metal uptake was strongly dependent on the antioxidative defence of the test plant. This was expressed by a combination of antioxidative enzymes like catalase (CAT), ascorbate peroxidase (APX) and superoxide dismutase (SOD). In treatments without PGPB, activity of all three enzymes increased with increasing Cd dose applied. CAT and SOD activity was lower in PGPB treatments. Among these, *Bradyrhizobium japonicum* (PGPB2) inoculated plants showed the least activity and thus lesser stress. APX activity, however, increased in the order Cd 25 + PGPB1 < Cd25 + PGPB2 < Cd25 + PGPB1 + PGPB2 (Fig. 6). SOD and APX activities were also elevated in Cd 25 + PGPB1 + PGPB2.

6.4 Discussion

Although, *C. album* indicated a reduction in growth with increased Cd concentrations, it was able to survive the Cd dosage regime of the study even without the PGPB amendments. It preferentially stored Cd in the roots. Such strategies are known in halophytes which can tolerate environmental stresses including toxic metal ions and could prove better for the extraction of heavy metals from soils when compared to glycophytes (Ishtiyaq et al. 2023). Another species of the genus *Chenopodium, C. botrys* may remove up to 180 g Cd ha^{-1}. This is six times more than the extraction potential of *Noccaea caerulescens,* a known Cd hyperaccumulator (Mazharia and Homaeed 2012).

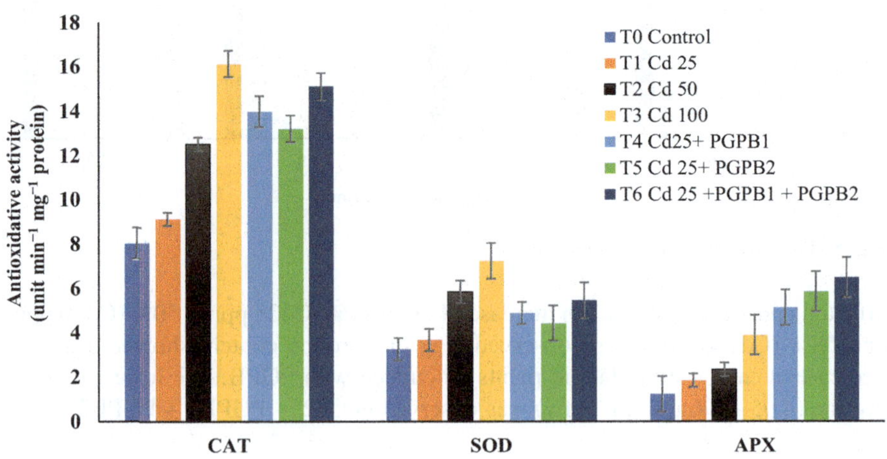

Fig. 6 Antioxidative activity in *C. album*

The data clearly indicates that *Pseudomonas fluorescens* (PGPB1) and *Bradyrhizobium japonicum* (PGPB2) treatments gave better results in terms of metal uptake. Further, they also increased the root/shoot length and biomass of *C. album* as well as its vigour thereby enhancing its Cd tolerance. It showed maximum absorption in the treatment (Cd25+ PGPB1 + PGPB2) with the values of 6.84 mg Cd kg^{-1} in roots and 5.31 mg Cd kg^{-1} in shoots. It could be due to increased Cd adsorption on the root surface, facilitated by comparatively less mobility of metals in the root zone. These microorganisms can directly enhance phytoremediation by changing the bioavailability of heavy metals in soil as well as the pH of soil. Besides, they facilitate survival and growth under metal stress (Kong and Glick 2017). Siderophores from organic complexes with trace elements (Cd, Cu and Pb) alleviate stress in plants (Rajkumar et al. 2010). Other studies also report 4–25% (root) and 2–55% (shoot) increases in *Sorghum* and *Oryza sativa*, respectively, after inoculation with various strains of *Streptomyces* (Gopalakrishnan et al. 2013). Chauhan and Rai (2009) confirmed the role of *Trichoderma harzianum* and *Pseudomonas fluorescens* inoculation on the uptake of Cd and Zn (individually) by Indian mustard (*Brassica juncea*) from the soil. Whiting et al. (2001) reported a doubled accumulation of Zn in *Thlaspi caerulescens* shoots in autoclaved soil when its surface-sterilized seeds were inoculated with a mixture of rhizosphere bacteria before sowing.

Chellaiah (2018) demonstrated Cd resistance of *Pseudomonas* up to 500 ppm and its plant growth promoting traits up to 200 ppm Cd in soil. Egamberdieva et al. (2017) found that *Bradyrhizobium* and *Pseudomonas* can together improve the tolerance of a plant by altering the root architecture, increasing nodulation and uptake of nitrogen and phosphorus from soil.

Cd phytotoxicity commonly presents as reduced in plant height, pigment, protein, inhibited photosynthesis and production of stress markers such as proline. An observed decrease in chlorophyll content may be attributed to a slow or inhibited biosynthesis or accelerated degradation (Gopal and Rizvi 2008). Reduction in net photosynthetic rate due to increasing Cd concentration has been recorded in *A. halimus* (Mesnoua et al. 2016; Ishtiyaq et al. 2023). Results of this study indicate that *C. album* was tolerant to moderate concentrations of Cd. This may very well be because of its ability to compartmentalize the metal in its root tissues while simultaneously increasing the antioxidant enzymatic defence in its leaves. Proline is an osmolyte that contributes widely to higher plants and usually accumulates in large amounts in response to environmental stresses and known for accumulating inside the leaves in several plant species when subjected to heavy metals, osmotic stresses, low temperature or even starvation (Tully and Hanson 1979). *C. album* showed highest production of proline in response to a higher dose of Cd. Inoculation with PGPB reduced proline content indicating improved physiology, reduced oxidative stress and its adaptation to Cd stress. Elevation in proline content has also been reported in *Atriplex halimus* grown in Cd-contaminated soil (Ishtiyaq et al. 2023).

In plants, antioxidant enzymes are the chief defence against oxidative stress caused by heavy metals (Weckx and Clijsters 1996). Catalase, superoxide dismutase and ascorbate peroxidase activity increase under such conditions (Gill and Tuteja

2010). Thus, fluctuations in their activity are often used to understand metal tolerance in plants (Ishtiyaq et al. 2020). Cd is thought to induce oxidative damage to plants by producing ROS, which destroys biomolecules like membrane lipids and proteins. Previous studies indicate that CAT, APX and SOD activity increase against lower metal concentrations and decrease under high metal concentrations as these disturb plant defence mechanisms (Pinto et al. 2009; Ishtiyaq et al. 2020). Previous studies include observations on increased CAT and SOD activity in *A. halimus* leaves due to Cd and Pb stress (El-Bakatoushi et al. 2015; Mesnoua et al. 2016). Karthikeyan et al. (2007) reported an increase in the activity of antioxidant enzymes such as GPX, CAT and SOD due to the treatment with diazotrophic bacteria such as *Azospirillum* and *Azotobacter*. Kumar et al. (2022) also reported a synergistic association of *Pseudomonas fluorescens, Bradyrhizobium japonicum* and organic matter against Cd toxicity which improved plant growth, antioxidative defence and metal uptake in *Atriplex lentiformis.*

7 Conclusion

Phytoremediation as a clean-up strategy is beneficial as it preserves soil structure and microbial communities and is low-cost and eco-friendly while being aesthetically pleasing. Studies investigating phytoremediation using specific plants and individual contaminants have provided encouraging results. Thus, plants with good Cd tolerance and accumulation along with quick growth and good biomass, like *Chenopodium album*, can potentially be used to phytostabilize Cd. The synergy between PGPBs and test plant not only improved the plant's vigour and biomass but also its Cd tolerance and uptake. Thus, inoculating plants/ rhizosphere soils with selected metal-tolerant bacteria can be promoted as a sustainable and economical method to boost the overall phytoremedial potential of plants being utilized.

References

Addour L, Belhocine D, Boudries N, Comeau Y, Pauss A, Mameri N (1999) Zinc uptake by *Streptomyces rimosus* biomass using a packed-bed column. J Chem Technol Biotechnol 74: 1089–1095

Adriano DC (1992) Biogeochemistry of trace metals. Lewis Publishers, Boca Raton, FL

Adriano DC (2003) Trace elements in terrestrial environments: biogeochemistry, bioavailability and risks of metals, vol 2. Springer, New York, pp 1–6

Aebi H (1984) Catalase in vitro. Methods Enzymol 105:121–126

Agency for Toxic Substances and Disease Registry (ATSDR) (2019) Substance priority list. U.-S. Department of Health and Human Services, Public Health Service, Atlanta, GA

Ahmad I, Javed AM, Naeem AH, Ghafoor U, Shahid M (2015) Differential effects of plant growth-promoting rhizobacteria on maize growth and cadmium uptake. J Plant Growth Regul 35:303–315

Ahmad B, Zaid A, Zulfqar F, Bovand F, Dar TA (2022) Nanotechnology: a novel and sustainable approach towards heavy metal stress alleviation in plants. Nanotechnol Environ Eng 8:27–40

Alcantara E, Romera FJ, Cañete M, De La GMD (1994) Effects of heavy metals on both induction and function of root Fe(III) reductase in Fe-deficient cucumber (*Cucumis sativus* L.) plants. J Exp Bot 45:1893–1898

Al-Dhabi NA, Esmail GA, Mohammed GAK, Valan AM (2019) Optimizing the management of cadmium bioremediation capacity of metal-resistant *pseudomonas* sp. strain Al-Dhabi-126 isolated from the industrial City of Saudi Arabian environment. Int J Environ Res Public Health 16(4788)

Alsohim AS (2020) Inluence of *Pseudomonas fluorescens* mutants produced by transposon mutagenesis on *in vitro* and *in vivo* biocontrol and plant growth promotion. Egypt J Biol Pest Control 30:19

Amjad A, Di G, Amanullah M, Fang M, Ronghua L, Feng S, Ping W, Zengqiang Z (2017) *Streptomyces pactum* assisted phytoremediation in Zn/Pb smelter contaminated soil of Feng County and its impact on enzymatic activities. Sci Rep 7:46087

Arunakumara KKIU, Walpola BC, Yoon MH (2015) Bioaugmentation-assisted phytoextraction of Co, Pb and Zn: an assessment with a phosphate-solubilizing bacterium isolated from metal-contaminated mines of Boryeong area in South Korea. Biotechnol Agron Soc Environ 19:143–152

Babalola OO, Akindolire AM (2011) Identification of native rhizobacteria peculiar to selected food crops in Mmabatho municipality of South Africa. Biol Agric Hort 27(3–4):294–309

Babalola OO, Osir EO, Sanni AI, Odhiambo GD, Bulimo WD (2003) Amplification of 1 aminocyclopropane-1-carboxylic (ACC) deaminase from plant growth promoting bacteria in Strigainfested soil. Afr J Biotechnol 2:157–160

Baker AJM, McGrath SP, Reeves RD, Smith JAC (2000) Metal hyperaccumulator plants: a review of the ecology and physiology of a biological resource for phytoremediation of metal-polluted soils. In: Terry N, Bañuelos G (eds) Phytoremediation of contaminated soil and water. CRC Press, Boca Raton, FL, pp 85–108

Bates LS, Waldren RP, Teare ID (1973) Reclassification of Brevibacterium halotolerans DSM8802 as *Bacillus halotolerans* comb. nov. based on microbial and biochemical characterization and multiple gene sequence, rapid determination of free proline for water-stress studies. Plant Soil 39:205–207

Benavides MP, Gallegos SM, Tomaro M (2005) Cadmium toxicity in plants. Braz J Plant Physiol 17(1):21–34

Beneduzi A, Ambrosini A, Passaglia LM (2012) Plant growth-promoting rhizobacteria (PGPR): their potential as antagonists and biocontrol agents. Genet Mol Biol 35:1044–1051

Blanca M, Thijs S, Lobo MC, Weyens N, Ameloot M, Vangronsveld J, Perez- Sanz A (2017) Cultivar and metal-specific effects of endophytic bacteria in *Helianthus tuberosus* exposed to Cd and Zn. Int J Mol Sci 18:2026

Burd GI, Dixon DG, Glick BR (1998) A plant growth-promoting bacterium that decreases nickel toxicity in seedlings. Appl Environ Microbiol 64(10):3663–3668

Cabello-Conejo MI, Becerra-Castro C, Prieto-Fernández A, Monterroso C, Saavedra-Ferro A, Mench M, Kidd PS (2014) Rhizobacterial inoculants can improve nickel phytoextraction by the hyperaccumulator *Alyssum pintodasilvae*. Plant Soil 379(1–2):35–50

Carlos MHJ, Stefani PVY, Janette AM, Melani MSS, Gabriela PO (2016) Assessing the effects of heavy metals in ACC deaminase and IAA production on plant growth-promoting bacteria. Microbial Res 188–189:53–61

Chauhan JS, Rai JPN (2009) Phytoextraction of soil cadmium and zinc by microbes-inoculated Indian mustard (*Brassica juncea*). J Plant Interact 4:279–287

Chellaiah E (2018) Cadmium (heavy metals) bioremediation by *Pseudomonas aeruginosa*: a minireview. Appl Water Sci 8:154

Chen HM, Zeng CR, Tu C, Shen ZG (2000) Chemical methods and phytoremediation of soil contaminated with heavy metals. Chemosphere 41:229–234

Cho U, Seo N (2004) Oxidative stress in *Arabidopsis thaliana* exposed to cadmium is due to hydrogen peroxide accumulation. Plant Sci 168:113–120

Cosio C, Martinoia E, Keller C (2004) Hyperaccumulation of cadmium and zinc in *Thlaspi caerulescens* and *Arabidopsis halleri* at the leaf cellular level. Plant Physiol 134:716–725

Cossich ES, Tavares CRG, Ravagnani TMK (2002) Biosorption of Chromium(III) by *Sargassum* sp. Biomass Electron J Biotechnol 5:133–141

D'Souza R, Varun M, Pratas J, Paul MS (2013) Spatial distribution of heavy metals in soil and flora associated with the glass industry in North Central India: implications for phytoremediation. Soil Sediment Contam 22:1–20

Dhindsa RS, Plumb-Dhindsa P, Thorpe TA (1981) Leaf senescence: correlated with increased levels of membrane permeability and lipid peroxidation, and decreased levels of superoxide dismutase and catalase. J Exp Bot 32:93–101

Dixit R, Malaviya D, Pandiyan K, Singh UB, Sahu A, Shukla R, Singh BP, Rai JP, Sharma PK, Lade H (2015) Bioremediation of heavy metals from soil and aquatic environment: an overview of principles and criteria of fundamental processes. Sustainability 7:2189–2212

Dong J, Mao WH, Zhang GP, Wu FB, Cai Y (2007) Root excretion and plant tolerance to cadmium toxicity—a review. Plant Soil Environ 53:193–200

Durand A, Piutti S, Rue M, Morel JL, Echevarria G, Benizri E (2016) Improving nickel phytoextraction by co-cropping hyperaccumulator plants inoculated by plant growth promoting rhizobacteria. Plant Soil 399(1–2):179–192

Duxbury T (1985) Ecological aspects of heavy metal responses in microorganisms. Adv Microb Ecol 8:185–235

Egamberdieva D, Stephan W, Dilfuza J, Leena AR, Hong L (2017) Coordination between *Bradyrhizobium* and *Pseudomonas* alleviates salt stress in soybean through altering root system architecture. J Plant Interact 12:100–107

El-Bakatoushi R, Alframawy AM, Tammam A, Youssef D, El-Sadek L (2015) Molecular and physiological mechanisms of heavy metal tolerance in *Atriplex halimus*. Int J Phytoremediation 17:789–800

Erakhrumen AA (2007) Phytoremediation: an environmentally sound technology for pollution prevention, control and remediation in developing countries. Educ Res Rev 2:151–156

Fornazier RF, Ferreira RR, Vitória AP, Molina SMG, Lea PJ, Azevedo RA (2002) Effects of cadmium on antioxidant enzyme activities in sugar cane. Biol Plant 45:91–97

Ghosh M, Singh SP (2005) A review on phytoremediation of heavy metals and utilization of its byproducts. Appl Ecol Environ Res 3:1–18

Gill SS, Tuteja N (2010) Reactive oxygen species and antioxidant machinery in abiotic stress tolerance in crop plants. Plant Physiol Biochem 48:909–930

Glick BR (2003) Phytoremediation: synergistic use of plants and bacteria to clean up the environment. Biotechnol Adv 21:383–393

Glick BR, Penrose DM, Li J (1998) A model for the lowering of plant ethylene concentrations by plant growth promoting bacteria. J Theor Biol 190:63–68

Gopal R, Rizvi AH (2008) Excess lead alters growth, metabolism, and translocation of certain nutrients in radish. Chemosphere 70:1539–1544

Gopalakrishnan S, Srinivas V, Vidya MS, Rathore R (2013) Plant growth-promoting activities of Streptomyces spp. in sorghum and rice. SpringerPlus 2:574

Goswami D, Thakker JN, Dhandhukia PC (2016) Portraying mechanics of plant growth promoting rhizobacteria (PGPR): a review. Cogent Food Agric 2:1127500

Ha H, Olson J, Bian LA (2014) Analysis of heavy metals sources in soil using kriging interpolation on principal components. Environ Sci Technol 48:4999–5007

He LY, Zhang YF, Ma HY, Su LN, Chen ZJ, Wang QY, Qian M, Sheng XF (2010) Characterization of copper-resistant bacteria and assessment of bacterial communities in rhizosphere soils of copper-tolerant plants. Appl Soil Ecol 44(1):49–55

Hernández LE, Cooke DT (1997) Modification of the roots plasma membrane lipid composition of cadmium-treated *Pisum sativum*. J Exp Bot 48:1375–1381

Huang XD, El-Alawi Y, Penrose DM, Glick BR, Greenberg BM (2004) Responses of three grass species to creosote during phytoremediation. Environ Pollut 130:453–463

Ishtiyaq S, Kumar H, Clement O, Varun M, Paul MS (2020) Role of secondary metabolites in salt and heavy metal stress mitigation by halophytic plants: an overview. In: Mirza H, Prasad MNV (eds) Handbook of bioremediation. Academic Press, Cambridge, MA, pp 307–321

Ishtiyaq S, Kumar H, D'Souza RJ, Varun M, Favas PJC, Paul MS (2023) Physiological responses and adaptations of the halophyte *Atriplex halimus* to soil contaminated with Cd, Ni, and NaCl. Soil Syst 7:46

Ivy M, Chandrima B, Shayantan M, Dhritiman D, Somesh CS, Ujjal KM, Abhrajyoti G (2018) Effective rhizoinoculation and biofilm formation by arsenic immobilizing halophilic plant growth promoting bacteria (PGPB) isolated from mangrove rhizosphere: a step towards arsenic rhizoremediation. Sci Total Environ 610–611:1239–1250

Kabata-Pendias A, Pendias H (2001) Trace metals in soils and plants, 2nd edn. CRC Press, Boca Raton, FL

Kahle H (1993) Response of roots of trees to heavy metals. Environ Exp Bot 33:99–119

Kaji M (2012) Role of experts and public participation in pollution control: the case of Itai-itai disease in Japan. Ethics Sci Environ 12:99–111

Karami A, Shamsuddin ZH (2010) Phytoremediation of heavy metals with several efficiency enhancer methods. Afr J Biotechnol 9:3689–3698

Karthikeyan B, Jaleel CA, Gopi R, Deiveekasundaram M (2007) Alterations in seedling vigour and antioxidant enzyme activities in *Catharanthus roseus* under seed priming with native diazotrophs. J Zhejiang Univ Sci 8:453–457

Kavamura VN, Esposito E (2010) Biotechnological strategies applied to the decontamination of soils polluted with heavy metals. Biotechnol Adv 28:61–69

Khan AG (2005) Role of soil microbes in the rhizospheres of plant growing on trace metal contaminated soils in phytoremediation. J Trace Elem Med Biol 18:355–364

Khan N, Bano A (2016) Modulation of phytoremediation and plant growth by the treatment with PGPR, Ag nanoparticle and untreated municipal wastewater. Int J Phytoremediation 18:1258–1269

Khan NA, Samiullah SS, Naza R (2007) Activities of antioxidative enzymes, Sulphur assimilation, photosynthetic activity, and growth of wheat (*Triticum aestivum*) cultivars differing in yield potential under cadmium stress. J Agron Crop Sci 193:435–444

Kloepper JW, Schroth MN (1978) Plant growth-promoting rhizobacteria on radishes. In: Proceedings of the 4th international conference on plant pathogenic bacteria, vol 2. Station de Pathologie Végétale et de Phytobactériologie, INRA, Angers, France, pp 879–882

Kong Z, Glick BR (2017) The role of plant growth-promoting bacteria in metal phytoremediation. Adv Microb Physiol 71:97–132

Kumar H, Ishtiyaq S, Varun M, Favas PJC, Ogunkunle CO, Paul MS (2022) Bioremediation: plants and microbes for restoration of heavy metal contaminated soils. In: Puthur JT, Dhankher OP (eds) Bioenergy crops: a sustainable means of phytoremediation. CRC Press, pp 37–70

Lasat MM (2002) Phytoextraction of toxic metals: a review of biological mechanisms. J Environ Qual 31:109–120

Li WC, Ye ZH, Wong MH (2007) Effects of bacteria on enhanced metal uptake of the Cd/Zn-hyperaccumulating plant, *Sedum alfredii*. J Exp Bot 58:4173–4182

Lichtenthaler HK (1987) Chlorophylls and carotenoids: pigments of photosynthetic biomembranes. In: Methods in enzymology, vol 148. Academic Press, pp 350–382

Ma Y, Rajkumar M, Freitas H (2009) Inoculation of plant growth promoting bacterium *Achromobacter xylosoxidans* strain Ax10 for the improvement of copper phytoextraction by *Brassica juncea*. J Environ Manag 90(2):831–837

Marschner H (1995) Mineral nutrition of higher plants. Academic, London, pp 24–28

Mazharia M, Homaeed M (2012) Annual halophyte *Chenopodium botrys* can phytoextract cadmium from contaminated soils. J Basic Appl Sci Res 2:1415–1422

Mesjasz PJ, Nakonieczny M, Migula P, Augustyniak M, Tarnawska M, Reimold WU, Koeberl C, Przybylowicz W, Glowacka E (2004) Uptake of cadmium, lead, nickel and zinc from soil and water solutions by the nickel hyper accumulator *Berkheya coddii*. Acta Biol Cracov Ser Bot 46: 75–85

Mesnoua M, Mateos-Naranjo E, Barcia-Piedras JM, Pérez-Romero JA, Lotmani B, Redondo-Gómez S (2016) Physiological and biochemical mechanisms preventing Cd-toxicity in the hyperaccumulator *Atriplex halimus* L. Plant Physiol Biochem 106:30–38

Mohammadzadeh A, Tavakoli M, Motesharezadeh B, Reza CM (2017) Effects of plant growth-promoting bacteria on the phytoremediation of cadmium-contaminated soil by sunflower. Arch Agron Soil Sci 63:807–816

Mohanpuria P, Rana NK, Yadav SK (2007) Cadmium-induced oxidative stress influence on glutathione metabolic genes of *Camellia sinensis* (L.) O. Kuntze. Environ Toxicol 22:368–374

Moreira H, Pereira SIA, Marques APGC, Rangel AOSS, Castro PML (2016) Selection of metal resistant plant growth promoting rhizobacteria for the growth and metal accumulation of energy maize in a mine soil – effect of the inoculum size. Geoderma 278:1–11

Nakano Y, Asada K (1981) Hydrogen peroxide is scavenged by ascorbate specific peroxidase in spinach chloroplasts. Plant Cell Physiol 22:867–880

Nehra V, Choudhary M (2015) A review on plant growth-promoting rhizobacteria acting as bioinoculants and their biological approach towards the production of sustainable agriculture. J Appl Nat Sci 7:540–556

Nematian MA, Kazemeini F (2013) Accumulation of Pb, Zn, Cu and Fe in plants and hyperaccumulator choice in Galali iron mine area, Iran. Int J Agric Crop Sci 5(4):426–432

NJDEP (1996) Soil cleanup criteria. Department of Environmental Protection. Proposed cleanup standards for contaminated sites, NJAC 7, 26D, New Jersey

Nriagu JO, Wong HKT, Coker RD (1982) Deposition and chemistry of pollutant metals in lake around the smelters of Sudbury, Ontario. Environ Sci Technol 16:551–560

Pan F, Meng Q, Luo S, Shen J, Chen B, Yasmin KK, Japenga J, Ma X, Yang X, Feng Y (2016) Enhanced Cd extraction of oilseed rape (*Brassica napus*) by plant growth promoting bacteria isolated from Cd hyperaccumulator *Sedum alfredii* hance. Int J Phytoremediation 19:281–289

Parvinder K, Singh S, Kumar V, Singh N, Singh J (2017) Effect of rhizobacteria on arsenic uptake by macrophyte *Eichhornia crassipes* (Mart.) Solms. Int J Phytoremediation 28:114–120

Phillips DA, Srimathi K, Ann Suji H (1981) Chemistry and biochemistry of trace metals in biological systems. Applied Science Publishers, London

Pinto AP, Alves AS, Candeias AJ, Cardoso AI, de Varennes A, Martins LL, Mourato MP, Gonçalves ML, Mota AM (2009) Cadmium accumulation and antioxidative defences in *Brassica juncea* L. Czern, *Nicotiana tobacum* L. and *Solanum nigrum* L. Int J Environ Res 89:661–676

Pivertz EB (2001) Phytoremediation of contaminated soil and ground water at hazardous waste sites. Environmental Research Services Corporation. EPA/540/S- 01/500

Plociniczak T, Sinkkonen A, Romantschuk M, Sułowicz S, Piotrowska-Seget Z (2016) Rhizospheric bacterial strain *Brevibacterium casei* MH8a colonizes plant tissues and enhances Cd, Zn, Cu phytoextraction by white mustard. Front Plant Sci 7:101

Pulford I, Watson C (2003) Phytoremediation of heavy metal-contaminated land by trees - a review. Environ Int 29:529–540

Rajkumar M, Freitas H (2008a) Influence of metal resistant-plant growth-promoting bacteria on the growth of *Ricinus communis* in soil contaminated with heavy metals. Chemosphere 71(5): 834–842

Rajkumar M, Freitas H (2008b) Effects of inoculation of plant-growth promoting bacteria on Ni uptake by Indian mustard. Bioresour Technol 99(9):3491–3498

Rajkumar M, Ma Y, Freitas H (2008) Characterization of metal-resistant plant-growth promoting *Bacillus weihenstephanensis* isolated from serpentine soil in Portugal. J Basic Microbiol 48(6): 500–508

Rajkumar M, Ae N, Prasad MNV, Freitas H (2010) Potential of siderophore-producing bacteria for improving heavy metal phytoextraction. Trends Biotechnol 28:142–149

Riley RG, Zachara JM, Wobber FJ (1992) Chemical contaminants on DOE lands and selection of contaminated mixtures for subsurface science research. US-DOE Office, Energy Res Suburb Science Programme, Washington, DC

Sabolic I, Breljak D, Skarica M, Herak-Kramberger CM (2010) Role of metallothionein in cadmium traffic and toxicity in kidneys and other mammalian organs. Biometals 23(5):897–926

Safronova VI, Piluzza G, Zinovkina NY, Kimeklis AK, Belimov AA, Bullitta S (2012) Relationships between pasture legumes, rhizobacteria and nodule bacteria in heavy metal polluted mine waste of SW Sardinia. Symbiosis 58:149–159

Salt DE, Blaylock M, Kuma RN, Dushenkov V, Ensley BD, Chet I, Raskin I (1995) Phytoremediation—a novel strategy for the removal of toxic metals from the environment using plants. Bioresour Technol 134:68–474

Sanita TL, Gabbrielli R (1999) Response to cadmium in higher plants. J Environ Exp Bot 41:105–130

Sasaki A, Yamaji N, Yokosho K, Ma JF (2012) Nramp5 is a major transporter responsible for manganese and cadmium uptake in rice. Plant Cell 24:2155–2167

Sekara A, Poniedzialeek M, Ciura J, Jedrszczyk E (2005) Cadmium and lead accumulation and distribution in the organs of nine crops: implications for phytoremediation. Pol J Environ Stud 14:509–516

Seyed MM, Babak M, Hossein MH, Hoseinali A, Ali AZ (2018) Root-induced changes of Zn and Pb dynamics in the rhizosphere of sunflower with different plant growth promoting treatments in a heavily contaminated soil. Ecotoxicol Environ Saf 147:206–216

Somashekaraiah BV, Padmaja K, Prasad ARK (1992) Phytotoxicity of cadmium ions on germinating seedlings of mung bean (*Phaseolus vulgaris*): involvement of lipid peroxides in chlorophyll degradation. Plant Physiol 85:85–89

Somers E, Vanderleyden J, Srinivasan M (2004) Rhizosphere bacterial signalling: a love parade beneath our feet. Crit Rev Microbiol 30:205–240

Tak HI, Ahmad F, Babalola OO (2013) Advances in the application of plant growth-promoting rhizobacteria in phytoremediation of heavy metals. Rev Environ Contam Toxicol 223:33–47

Takahashi R, Shimaru Y, Shimo H, Ogo Y, Senoura T, Nishizawa NK, Nakanishi H (2012) The OsHMA2 transporter is involved in root-to-shoot translocation of Zn and Cd in rice plant. Cell Environ 35:1948–1957

Tamoor UH, Bano A, Naz I (2017) Alleviation of heavy metals toxicity by the application of plant growth promoting rhizobacteria and effects on wheat grown in saline sodic field. Int J Phytoremediation 19:522–529

Tariq M, Ali M, Shah Z (2006) Characteristics of industrial effluents and their possible impacts on quality of underground water. Soil Environ 25:64–69

Thangavel P, Subbhuraam C (2004) Phytoextraction: role of hyperaccumulators in metal contaminated soils. Proc Ind Nat Sci Acad 70:109–130

Toth G, Hermann T, Da Silva MR, Montanarella L (2016) Heavy metals in agricultural soils of the European Union with implications for food safety. Environ Pollut 88:299–309

Tully RE, Hanson AD (1979) Amino acid translocated from turgid and water stressed barley leaves. Plant Physiol 64:460–466

USEPA. United States Environmental Protection Agency. (2000, July). A guide to developing and documenting cost estimates during the feasibility study (EPA 540-R-00-002). Available: http://www.epa.gov/superfund/resources/remedy/costest.html

Valavanidis A, Vlachogianni T (2010) Metal pollution in ecosystems: ecotoxicology studies and risk assessment in the marine environment. Ecotoxicol Environ Saf 1:14

Vartika M, Antriksh G, Parvinder G, Simranjeet S, Nasib S, Praveen G, Joginder S (2016) Synergistic effects of *Arbuscular mycorrhizal* fungi and plant growth promoting rhizobacteria in bioremediation of iron contaminated soils. Int J Phytoremediation 18:697–703

Varun M, D'Souza R, Pratas J, Paul MS (2012) Metal contamination of soils and plants associated with the glass industry in North Central India: prospects of phytoremediation. Environ Sci Pollut Res 19:269–281

Venturi V, Keel C (2016) Review special issue: unravelling the secrets of the rhizosphere signaling in the rhizosphere. Trends Plant Sci 21(3):187–198

Visioli G, Vamerali T, Mattarozi M, Dramis L, Anna MS (2015) Combined endophytic inoculants enhance nickel phytoextraction from serpentine soil in the hyperaccumulator *Noccaea caerulescens*. Front Plant Sci 14(6):638

Volesky B, Holan ZR (1995) Biosorption of heavy metals. Biotechnol Prog 11:235–250

Wang S, Teng S, Fan M (2010) Interaction between heavy metals and aerobic granular sludge. In: Sarkar SK (ed) Environmental management. Sciyo, Croatia, pp 173–188

Wang J, Zhou C, Xiao X, Xie Y, Zhu L, Ma Z (2017a) Enhanced iron and selenium uptake in plants by volatile emissions of *Bacillus amyloliquefaciens* (BF06). Appl Sci 7:85

Wang L, Lin H, Dong Y, He Y, Liu C (2017b) Isolation of vanadium-resistance endophytic bacterium PRE01 from *Pteris vittata* in stone coal smelting district and characterization for potential use in phytoremediation. J Hazard Mater 341:1–9

Weckx JEJ, Clijsters HMM (1996) Oxidative damage and defense mechanisms in primary leaves of *Phaseolus vulgaris* as a result of root assimilation of toxic amounts of copper. Physiol Plant 96: 506–512

Whiting SN, De Souza M, Terry N (2001) Rhizosphere bacteria mobilize Zn for hyperaccumulator by *Thlaspi caerulescens*. Environ Sci Technol 35:3144–3150

Witkowska D, Słowik J, Chilicka K (2021) Heavy metals and human health: possible exposure pathways and the competition for protein binding sites. Molecules 26(19):6060

Wuana RA, Okieimen FE (2011) Heavy metals in contaminated soils: a review of sources, chemistry, risks and best available strategies for remediation. ISRN Ecol 2011:1–20

Xia Y, Liyuan C (2002) Study of gelatinous supports for immobilizing inactivated cells of *Rhizopus oligosporus* to prepare biosorbent for Lead ions. Int J Environ Stud 5:1–6

Zefa DM, Fantin LH, Koltun A, Andre LMO, Maria PBAN, Marcelo CG, Leandro SAG (2020) Effects of plant growth-promoting rhizobacteria on co-inoculation with *Bradyrhizobium* in soybean crop: a meta-analysis of studies from 1987 to 2018. Peer J 8:7905

Zhao K, Fu W, Qiu Q, Ye Z, Li Y, Tunney H (2019) Spatial patterns of potentially hazardous metals in paddy soils in a typical electrical waste dismantling area and their pollution characteristics. Geoderma 337:453–462

Antioxidant Defense: A Key Mechanism of Cadmium Tolerance

Erna Karalija and Adisa Parić

Abstract Cadmium (Cd) is a persistent soil pollutant, with high toxicity to plants and humans. Its availability to plants is affected by several factors, including pH of the soil, while its accumulation in the plants is related to the plant's tolerance, genotype, environmental factors, etc. As a response to oxidative stress induced by heavy metals plants activate antioxidant response, which consists out of enzymatic and nonenzymatic components. Enzymatic antioxidant system in plants includes enzymes such as superoxide dismutase, peroxidases, and catalases, while nonenzymatic antioxidant system includes molecules with ability to scavenge radicals. Antioxidant molecules mainly responsible for nonenzymatic antioxidant response in plants are ascorbate, tocopherol, carotenoids, and flavonoids. Effective antioxidant system is related to plants' tolerance to Cd and numerous research have demonstrated that increased levels of antioxidant enzymes and molecules can alleviate Cd stress and increase Cd accumulation levels in plants.

Keywords Oxidative stress · Enzymatic activity · Nonenzymatic antioxidants · Plant tolerance

1 Introduction

Cadmium (Cd) belongs to the group of trace metals together with Al, Cu, As, Cr, Fe, Pb, Hg, Ni, and Zn (Karimpour et al. 2018). It is widely distributed in soil, water, and atmosphere. After mercury and lead, Cd is the third largest pollutant that poses the greatest threat to the environment. Therefore, its bioaccumulation in the food chain can be very dangerous and can lead to serious problems in all living organisms. The main reason for food contamination with Cd is its efficient transfer from the soil to the plant, and the long biological half-life of Cd, which is up to 30 years, cumulatively increasing its level in the organism with age. Bioaccumulation effect

E. Karalija (✉) · A. Parić
Faculty of Science, University of Sarajevo, Sarajevo, Bosnia and Herzegovina
e-mail: erna.k@pmf.unsa.ba

A. K. Jha, N. Kumar (eds.), *Cadmium Toxicity Mitigation*,
https://doi.org/10.1007/978-3-031-47390-6_8

additionally increases health risks even at very low levels of Cd under chronic exposure (Ismael et al. 2019). An additional problem is the fact that Cd can be present in plants in concentrations that are not phytotoxic but are very toxic to humans and animals in ingested (Peijnenburg et al. 2000). So even though the plant does not show any symptoms of poisoning, the concentration of Cd in its edible parts can be higher than the permissible levels for humans.

Plants absorb cadmium from the soil and water through the roots, depending on its availability and concentration, and a small amount can be taken even from the atmosphere (Clemens 2006). Its availability can be affected by a number of factors such as pH and characteristics of the soil, the rhizosphere, and the presence of organic acids (Benavides et al. 2005), while its accumulation in plants can be affected by the plant species, plant genotype, environmental factors, and the presence of other minerals and nutrients (Volpe et al. 2015). Cadmium negatively affects plant growth, both morphologically and physiologically (Shanying et al. 2017). However, the threshold of phytotoxic concentration of Cd can vary depending on the species, ecotype, cultivar, etc. (Sparks 2012), but also on the period of application of toxic metals.

Cadmium cannot be degraded, which combined with its high mobility and no persistence increases the risk of exposure leading to oxidative stress and physiological damage from cellular level to the whole organism (Cuypers et al. 2010). Cadmium, as a bivalent cation, is not responsible for direct generation of reactive oxygen species (ROS) but an increase in the production of ROS has been recorded in many studies of Cd toxicity in plants as well as other organisms (Belyaeva et al. 2004; Pathak and Khandelwal 2006; Zhou et al. 2009; Karalija and Selović 2018; Karalija et al. 2022; Subašić et al. 2022). In the state of oxidative stress plants activate different mechanisms to mitigate this stress including enzymatic and nonenzymatic molecules that are capable of scavenging oxidative stress-inducing radicals. More and more data are unravelling the mechanisms of how Cd induces oxidative damage and how this damage can be mitigated in plants.

2 ROS Production in Plants Under Cd Stress

Since Cd is unable to catalyze redox reaction, it is assumed that Cd can increase ROS production indirectly by increasing the free Fe concentration through the replacement of different proteins increasing the amount of free redox-active metals in the cell (Cuypers et al. 2010). The production of $^{\bullet}OH$ (hydroxyl) radicals is then enabled directly through Fenton reaction by free redox-active metals as shown in Eq. (1) with Cu as an example of redox-active metal.

$$\begin{aligned} &Cu^{2+} + O_2^{\cdot -} \rightarrow Cu^{+} + O_2 \\ &\textbf{Fenton reaction}\ Cu^{+} + H_2O_2 \rightarrow Cu^{2+} + {}^{\circ}OH + OH^{-} \end{aligned} \tag{1}$$

Normal production of ROS in plant cells serves as a signalling system related to plant cell growth and developmental processes, some stress signalling (biotic and abiotic) and for processes of programmed cell death (Hasanuzzaman et al. 2020). To maintain the balance between production of ROS for signalling purposes and possible damage beyond physiological functions, plants activate antioxidative mechanisms in the cell to keep the ROS production under control (Jaskulak et al. 2018). The main source of oxidative reactive molecules, reactive oxygen species, are products of one-electron oxygen reduction or a result of oxygen photochemical excitation and include singlet oxygen, superoxide anion radicals, hydrogen peroxide, and hydroxyl radicals (Hasanuzzaman et al. 2020).

In plants, ROS production is a consequence of aerobic metabolic reactions (Buchanan et al. 2015). The major sites of ROS production are the electron transport chain of chloroplasts and mitochondria. In chloroplasts, ROS is produced in three places: singlet oxygen ($^1O_2$) is produced in photosystem II (PSII) by direct reduction of oxygen, and superoxide anion radicals ($O_2^{\bullet -}$) are mainly produced in photosystem I (PSI) by reduction of molecular oxygen (Mehler reaction), while large amounts of H_2O_2 are produced during photorespiration. Also, cytochrome b6f (Cyt b6f) can produce negligible amounts of $O_2^{\bullet -}$ or hydrogen peroxide (H_2O_2). Mitochondria produce $O_2^{\bullet -}$ in complexes I (NADH dehydrogenase) and III (cytochrome bc1 complex). Peroxisomes produce $O_2^{\bullet -}$ and H_2O_2, while at NADPH oxidases on the plasma membrane $O_2^{\bullet -}$ is produced. Some radicals, such as H_2O_2, are quite stable and can travel relatively long distances, while most other ROS have very short half-life and therefore can only travel short distances. In mitochondria, different amounts of ROS produced depending on whether the plant is under optimal or stressed conditions. ROS production in mitochondria happens as part of the physiological activity of respiratory chain at complex I (NADH dehydrogenase) and complex III (cytochrome bc1 complex) in the inner mitochondrial membrane (Buchanan et al. 2015). Besides mitochondria, ROS in plants can be formed at cell membranes, chloroplasts, peroxisomes, and apoplasts (Asada 2006). In cases where ROS molecules have an unpaired electron (free radicals) hydrogen peroxide can be formed as shown in Eq. (2) (Sharkey 2005):

$$O_2^{-} + O_2^{-} + 2H_2O \rightarrow O_2 + 2H_2O_2 \tag{2}$$

In optimal conditions ROS production is low and plants can easily scavenge produced free radicals while in stress conditions, such as exposure to heavy metals, production of ROS can be substantially increased (Borisova et al. 2016). An increase in ROS has multiple damaging effects such as membrane lipid peroxidation, mtDNA cleavage, and impaired ATP generation leading to the damage of mitochondria and cell apoptosis (Takahashi et al. 2004). In plants oxidative stress affects activity of δ-aminolevulinic acid dehydratase (ALA-D) activity, S-adenosyl-l-methionine

(SAM), thiobarbituric acid reactive substances (TBARS), GSH levels, δ-aminolevulinic acid (ALA) and porphobilinogen (PBG), superoxide dismutase (SOD), guaiacol peroxidase (GPOX), proteins, and gene expression changes (Foyer and Noctor 2005). Excessive production of ROS usually accompanying stress-related responses result in oxidative damage of nucleic acids, lipids, and proteins in cells. Oxidative-related changes in proteins are normal changes at the level of their structure and include processes of carbonylation, fragmentation of polypeptide chain, protein aggregation due to the formation of S-S connections as well as oxidation of thiol groups contributing to the formation of disulfide bridges (Hasanuzzaman et al. 2020).

ROS production on plasma membrane is normally on the side facing apoplast and it is considered that ROS is a result of reduction of external O_2 by xanthine oxidase and NADPH oxidases (NOXs) through cytoplasmic NADPH as the electron source (Heyno et al. 2008). After short-term exposure to Cd^{2+} ions induction of ROS caused by the failure of the cellular antioxidant system, cell death, and growth defects. Increased production of oxygen radicals at the plasma membrane requires signalling through kinases, Ca^{2+} fluxes, and de novo synthesis of NOX (Rodríguez-Serrano et al. 2006).

3 Antioxidant System of Plants

The impact of oxidative stress is dependent on the type of oxidant, site of its production/accumulation, the levels of antioxidants as well as the cells' ability to repair systems. Main components of plant defense systems against oxidative stress (excessive production of ROS) include enzymatic and nonenzymatic antioxidants (Fig. 1). Enzymatic components of antioxidant plant defense systems can directly remove ROS through oxidation reactions and reduction of ascorbic acid and glutathione (Hasanuzzaman et al. 2020).

(a) *Antioxidant enzymes*

In the enzymatic antioxidant system of plant defense against oxidative stress important role play enzymes containing Fe in redox-active form including catalase (CAT), Class I and Class II peroxidases (POD), and NADPH oxidase participating in the production and scavenging of ROS (Tchounwou et al. 2012). Basic enzymatic protection system in plant cells consists of enzymes (superoxide dismutase, catalase, or/and ascorbate peroxidase) and different metabolites participating in scavenging reactions such as glutathione (GSH) and ascorbate (AsA). Additionally, GSH can further participate in neutralization of oxidative stress-inducing molecules, like Cd, through their function as a precursor for synthesis of phytochelatins (PCs), which can chelate heavy metals like Cd (Jozefczak et al. 2015).

Superoxide dismutase (SOD) can catalyze reactions of dismutation of two superoxide anion radicals producing hydrogen peroxide and oxygen (Eq. 3). In the first

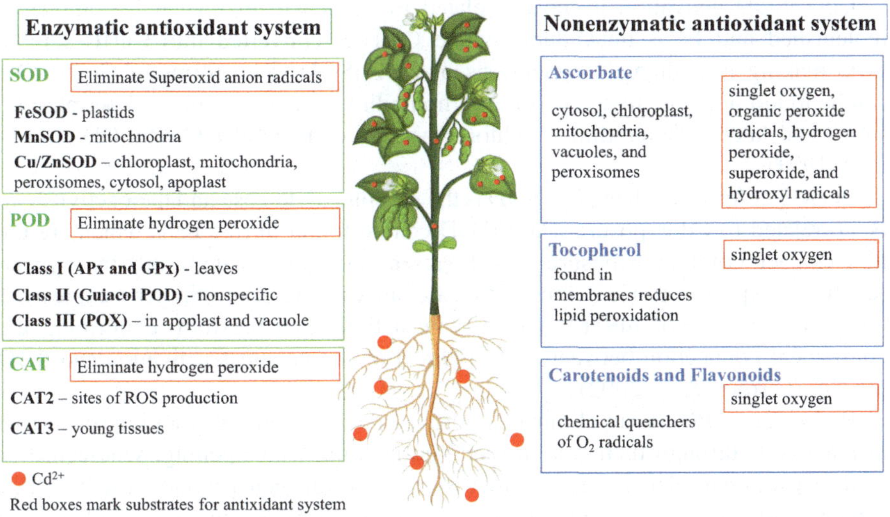

Fig. 1 Overview of enzymatic and nonenzymatic antioxidant systems included in ROS detoxification under Cd stress

step of the reaction metal enzyme in the catalytic center of the enzyme is being reduced followed by oxidation.

$$O_2 + O_2 + 2H^+ \rightarrow H_2O_2 + O_2 \qquad (3)$$

SODs are differentiated based on their cofactor type and include FeSOD (iron superoxide dismutase) mainly found in plastids, MnSOD (manganese-superoxide dismutase) found in mitochondria and Cu/ZnSOD (copper-zinc superoxide dismutase) present in peroxisomes, chloroplasts, mitochondria as well as in apoplast and cytosol (Zhang et al. 2009). Comparison based on amino acid composition suggests that Mn and FeSOD are more ancient and probably are derived from ancestral enzymes while Cu/ZnSOD probably evolved specifically in eukaryotes (Alscher et al. 2002). FeSOD can be found in eukaryotes as well as prokaryotes (but absent in animals), with chloroplast constricted location, and can be inactivated by hydrogen peroxide. FeSOD can be classified into two groups, first consisting of a homodimer formed from two identical protein subunits and an iron atom in the center (usually found in bacteria) and second group of four equal subunits forming tetramer containing double the iron in the active center compared to the first group (found in most higher plants) (Alscher et al. 2002).

MnSOD activity is connected to the levels of O_2 in the environment, increased O_2 availability decreases the availability of FE shifting SOD activity from FeSOD to MnSOD, which are active in mitochondria and peroxisomes. Reduction of oxygen radicals is performed by the formation of hydrogen peroxide through reaction of a donated electron by Mn center and superoxide radical reducing the radical which in turn reacts with proton forming hydrogen peroxide (Alscher et al. 2002).

Cu/ZnSOD activation is mostly related to cases where oxygen is completely replenished and Fe is unavailable, while Cu(I) is converted into Cu(II) soluble form making it available for synthesis of Cu/ZnSOD. Two forms of Cu/ZnSOD can be found in plants first consisting of homodimeric cytoplasmic and periplasmic forms and later of heterodimeric chloroplast and extracellular Cu/ZnSOD (Bordo et al. 1994).

In ROS defense mechanisms, SOD is the first line of defense and their activity can be correlated to Cd exposure. Cu/ZnSOD can be found in root cells which are the first contact point in exposure to Cd present in contaminated soil. In soybean seedlings, exposure to Cd^{2+} ions induced changes in Cu/ZnSOD gene expression levels at the level of mRNA and enzyme activities in the root, confirming the consequent connection between Cd^{2+} indirect oxidative stress activation and SOD activity (Ferreira et al. 2002).

Peroxidases (POD) can eliminate hydrogen peroxide and play an important role in plant cells through their role in seed germination, fruit ripening, senescence as well as pathogen-related defense processes. Based on their primary structure they can be divided into three main groups: Class I—intracellular enzymes in plants and yeast, Class II—extracellular peroxidases found in fungi, and Class III—classical peroxidases of higher plants, marked as secretory peroxidases (Jouili et al. 2011). Class I peroxidases play a major role in oxidative stress in H_2O_2 detoxification. They include cytochrome c peroxidase, ascorbate peroxidase, and catalase peroxidase. Cytochrome c peroxidase (CCP) reduces hydrogen peroxide to water using reducing equivalents of cytochrome c. Ascorbate peroxidases (APx) use ascorbate as a reducing equivalent to detoxify hydrogen peroxide. They also participate in the photo-protection of chloroplasts and cytosol in higher plants. Catalase peroxidases (CPs) exhibit catalase and peroxidase enzyme activity and are predominantly present in bacteria (Pandey et al. 2017). Class II peroxidases are often called as guaiacol peroxidases due to their affinity to guaiacol, universal substrate for peroxidases (Hiraga et al. 2001), but in general they can catalyze several substrates and are called nonspecific or phenol peroxidases dependent (Otter and Polle 1994) with general pattern of catalyzed reaction shown in Eq. (4).

$$\begin{aligned} &\text{Peroxidase} + H_2O_2 \rightarrow \text{Compound I} + H_2O \\ &\text{Compound I} + AH_2 \rightarrow \text{Compound II} + AH \\ &\text{Compound II} + AH \rightarrow \text{Peroxidase} + AH + H_2O \end{aligned} \tag{4}$$

In the peroxidase cycle, peroxidase is converted to compound I through catalyzation of hydrogen peroxide, with successive catalyzation of dehydrogenation reaction of electron donor molecules (phenolic compounds, lignin precursors, secondary metabolites, auxins) by Compounds I and II (Jouili et al. 2011).

Guaiacol peroxidase is usually found in intercellular spaces, cell walls, vacuoles, and cytosol with its main function in processes of lignification, ethylene biosynthesis, and protection against pathogens (Batish et al. 2006; Chen et al. 2008).

Class III peroxidases (POX) are mostly found in apoplast and vacuoles (Andrews et al. 2002), but considering they are secretory type of enzymes they can be found in the endoplasmic reticulum, the Golgi apparatus, and transport vesicles due to their synthesis, packaging, and transport processes in the cell compartments (Mäder 1992).

Ascorbate peroxidases (APx) and glutathione peroxidases (GPx), included in ascorbate-glutathione cycle in which two ascorbates can reduce hydrogen peroxide to water forming monodehydroascorbate (MDHA) (Gill and Tuteja 2010). Other types of peroxidases, glutathione peroxidases, have a role in the protection of cellular structures against oxidation and this enzyme can tolerate high temperatures (up to 100 °C) suggesting its role in heat stress. Other antioxidant enzymes, glutathione reductase (GR), dehydroascorbate reductase (DHAR), and monodehydroascorbate reductase (MDAR) are also involved in the ascorbate-glutathione cycle, although they are not detoxify ROS directly.

Catalases (CAT) facilitate dismutation reaction of hydrogen peroxide to oxygen and water (Eq. 5) and these enzymes can be found in peroxisomes, glyoxysomes, cytosol, and mitochondria. All eucaryotic CATs have the same tetrameric structure with a catalytic center composed out of NADPH molecules and heme groups. The reaction of CAT with hydrogen peroxide is up to 10,000 faster than those with peroxidases. Three main isomeric forms of catalases are differentiated: CAT1, CAT2, and CAT3. CAT2 is mainly found at the sites of increased ROS production while CAT3 is mainly associated with young tissues of plants and seeds where H_2O_2 is produced by degradation of fatty acids. Catalases mainly come into play when there is a significant increase in ROS production while peroxidase activity is more associated with lower ROS production (Azevedo et al. 2005).

$$2H_2 + O_2 \rightarrow 2H_2O + O \qquad (5)$$

δ-Aminolevulinic acid dehydratase (ALA-D) and δ-aminolevulinic acid (ALA) are part of the enzymatic response against Cd stress (among other stressors). The formation of ALA follows one of two possible pathways: (i) enzymatic condensation of glycine and succinyl-CoA by ALA synthase (ALA-S) or (ii) C5 pathway where glutamate is transformed to ALA. In plants, the dominant way of ALA production is C5 pathway. ALA can induce production of superoxide radical ($O_2^{\bullet-}$), hydrogen peroxide (H_2O_2), and hydroxyl radical ($HO^{\bullet}$) through enolization and metal-catalyzed aerobic oxidation leading to oxidative stress damage of cell structures. A strong correlation between Cd^{2+}-induced stress and increase in ALA-induced production of ROS has been confirmed in soybean plants (Noriega et al. 2007).

Reduced glutathione (GSH) is one of the most abundant antioxidants with an essential role in redox equilibrium preservation in plant cells. It is considered a key signalling molecule connected to the activity of phenylalanine ammonia lyase (PAL), especially in Cd-tolerant plants (Mohanpuria et al. 2007; DalCorso et al. 2010). It was recorded that levels of Glutathione reductase (GR) are increased under Cd stress leading to the reduction of oxidized glutathione (GSSG) to reduced glutathione (GSH), contributing to the Cd tolerance (Ammar et al. 2008).

The malondialdehyde (MDA) and thiobarbituric acid reactive substances (TBARS) are two assays often used for tracking levels of peroxidation of lipids in plants. In plants, MDA is a product derived because of auto-oxidation and enzymatic degradation of polyunsaturated fatty acids, and it can react with thiobarbituric acid (TBA) via an acid-catalyzed nucleophilic addition reaction (Hodges et al. 1999).

(b) *Nonenzymatic Antioxidants*

Besides enzymatic components of antioxidant plant defense system, nonenzymatic proteins and small molecules also play important roles in ROS scavenging and detoxification. One of the examples is ascorbate (AsA), present in cytosol, chloroplast, mitochondria, vacuoles, and peroxisomes, that can react with singlet oxygen, organic peroxide radicals, hydrogen peroxide, superoxide, and hydroxyl radicals. AsA together with Class III POD enzymes is a key player in redox system where hydrogen peroxide is a sensor of redox balance, as a signalling molecule in the process of lignification as well as a substrate for lignifying PODs (Loix et al. 2017).

Ascorbate is an anionic stabilized form of ascorbic acid (vitamin C) that is formed by deproteinization of the hydroxyl group at C3 of ascorbic acid molecule (Akram et al. 2017). Ascorbate anion (AH^-), water soluble, is an important antioxidant molecule vital for the downstream signalization of environmental stressors in plant cells (Shapiguzov et al. 2012). The reservoir of ascorbate seems to be cytoplasm with some transport to apoplast (5%) where the latter (apoplastic fraction) is involved in oxidative stress signalling (Gallie 2013).

a-Tocopherol (vitamin E) is another antioxidant usually found in membranes and can reduce the levels of lipid peroxidation and damages by singlet oxygen through the formation of tocopherol radicals that are scavenged further by glutathione and ascorbate (Hasanuzzaman et al. 2020). Tocopherol synthesis is highly regulated by environmental stress with close interplay with stress hormones such as abscisic acid, jasmonic acid, and salicylic acid (Szarka et al. 2012).

Carotenoids can scavenge triplet excited state molecules (like chlorophyll) and singlet oxygen through quenching of triplet chlorophyll preventing the overproduction of ROS. β-Carotene can act as chemical quenchers of O_2 by formation of β-carotene endoperoxide, a specific product that can be used as a marker of oxygen stress in plants (Ramel et al. 2012).

Other nonenzymatic antioxidants include flavonoids, which inhibit lipoxygenase activity, polyamines, proline, mannitol, sorbitol, and decoupling protein (UCP) found at the inner membrane of mitochondria with the ability to scatter proton electrochemical gradient thus reducing ROS production (Tchounwou et al. 2012).

4 Overcoming Cd-Induced Oxidative Stress Through Employment of Enzymatic and Nonenzymatic Antioxidants

Cadmium toxicity to plants is greatly dependent on the soil pH properties and its mobility. Acidic soils in general facilitate Cd absorption and mobility (maximum at pH 2) while alkaline pH makes it less accessible to plants (minimum mobility at pH 7) (Kabata-Pendias and Pendias 2001). The main toxic effects of Cd on plants include a significant reduction of biomass, root structures, photosynthesis, and mineral deficiencies (Hédiji et al. 2015), as well as impaired water uptake and water relations (Srivastava et al. 2014). Effects on photosynthesis are mainly the result of Cd-related inhibition donor as well as acceptor sides of PSII by replacing Ca on oxygen evolving complex (OEC) and reducing the rate of electron transport from primary quinone acceptor (QA) to secondary quinone acceptor (QB), respectively (Mahmoud et al. 2021). Antioxidant response of SOD to Cd in chloroplasts is based on the activity of two SOD forms Cu/ZnSOD and FeSOD, the former being attached to thylakoid membrane close to PSI and the latter attached to the stromal side of thylakoid close to PSII (Gallie and Chen 2019) (Fig. 2). Recently, it has been confirmed that exposure to Cd increases expression of FeSOD suggesting that this SOD has a key role in mitigation of chloroplast and photosynthesis-related damage caused by Cd exposure (Smeets et al. 2008; Luo et al. 2011; Hossain et al. 2012). Different plants react differently to Cd stress, some plants are more tolerant while others are more sensitive, some activate enzymatic antioxidant systems while others activate nonenzymatic antioxidant systems (Tables 1 and 2).

The main reason for mineral deficiencies lies in Cd effects on cell wall structure and its competition with other divalent cations such as Fe, Zn, Mg, and Ca absorption and inducing chlorosis (Xu et al. 2015). Across the cell membrane Cd is usually transported using ZIP family of metal transporters and it competes with other metals using the same transporters (Karalija et al. 2022).

Induction of peroxidase activity seems to be a general response of higher plants to heavy metal exposure, with a significant correlation of the enzymatic activity with Cd stress (Khan 2007). Additionally, there is research that confirms Cd effect on peroxidase activity, especially of different electrophoretic profiles of peroxidases as demonstrated in bean leaves where several isoforms showed increased activity (Smeets et al. 2005).

Application of AsA can alleviate Cd stress as recorded in oilseed rape seedlings where it was recorded that AsA mediates Cd-induced oxidative stress through nonenzymatic scavenging of radicals leading to decrease in MDA production. Cd-induced oxidative stress levels were dependent on the AsA–GSH–NADPH cycle (Akram et al. 2017) (Fig. 2). Increase of Cd tolerance has been recorded also in *Arabidopsis* with enhanced synthesis of tocopherol, another nonenzymatic antioxidant (vitamin E), is directly correlated to increased ability to tolerate Cd^{2+} (Collin et al. 2008).

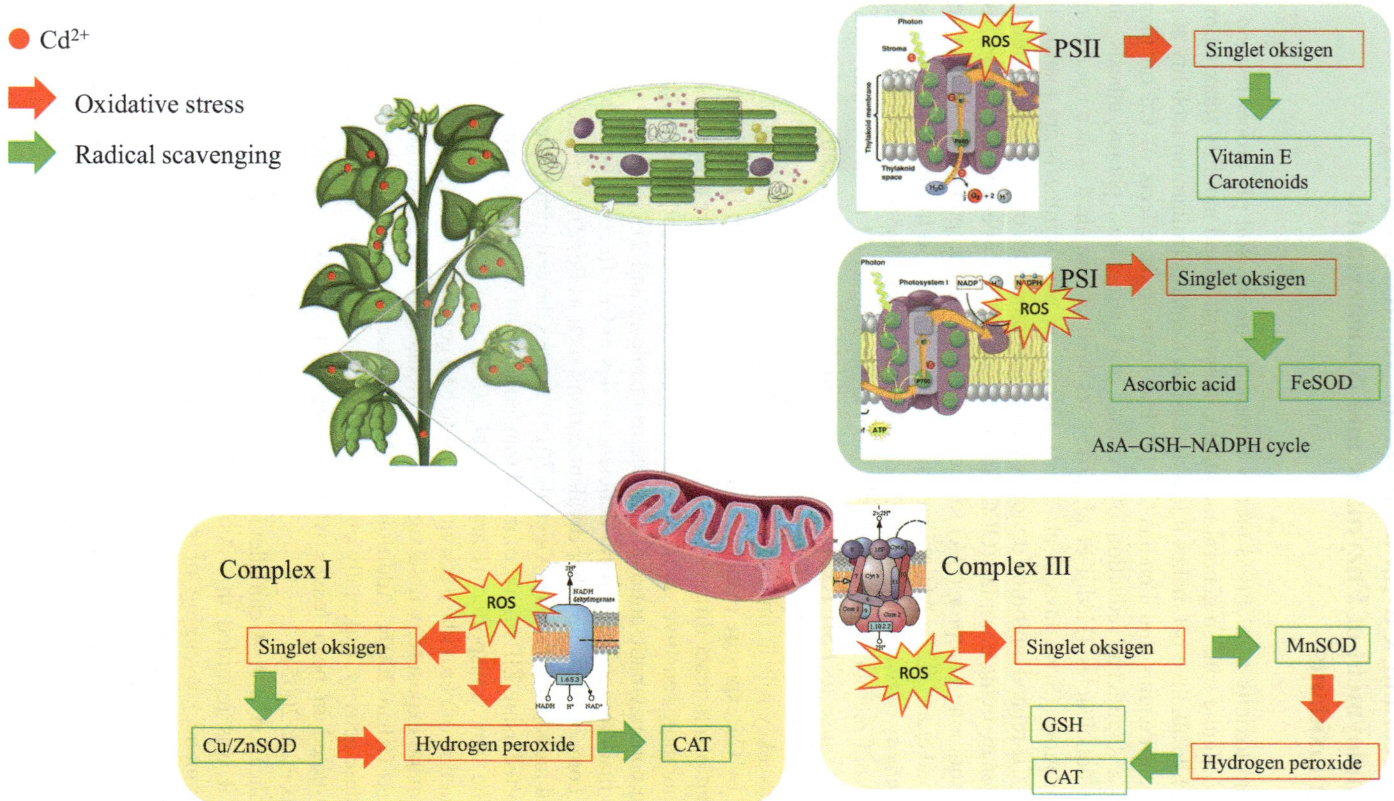

Fig. 2 Overview of ROS production and its detoxification in chloroplasts and mitochondria under Cd stress

Table 1 Examples of enzymatic system activation under Cd stress in different plant species

Plant species	Exposure time	Cadmium form and concentration	Antioxidant enzymes (AOE)	AOE concentration
Tomato (*Solanum lycopersicum*) Baruah et al. (2019)	7 days	20 $CdSO_4$ (ppm)	CAT	18 U mg^{-1} protein
		65 $CdSO_4$ (ppm)		25 U mg^{-1} protein
		110 $CdSO_4$ (ppm)		42 U mg^{-1} protein
		175 $CdSO_4$ (ppm)		55 U mg^{-1} protein
		220 $CdSO_4$ (ppm)		60 U mg^{-1} protein
Wheat (*Triticum aestivum*) Baruah et al. (2019)	7 days	20 $CdSO_4$ (ppm)	CAT	10 U mg^{-1} protein
		65 $CdSO_4$ (ppm)		12 U mg^{-1} protein
		110 $CdSO_4$ (ppm)		15 U mg^{-1} protein
		175 $CdSO_4$ (ppm)		22 U mg^{-1} protein
		220 $CdSO_4$ (ppm)		32 U mg^{-1} protein
Pea (*Pisum sativum*) Baruah et al. (2019)	7 days	20 $CdSO_4$ (ppm)	CAT	10 U mg^{-1} protein
		65 $CdSO_4$ (ppm)		25 U mg^{-1} protein
		110 $CdSO_4$ (ppm)		24 U mg^{-1} protein
		175 $CdSO_4$ (ppm)		10 U mg^{-1} protein
		220 $CdSO_4$ (ppm)		15 U mg^{-1} protein
Bidens pilosa Dai et al. (2021)	40 days	37 $CdCl_2$ (mg/kg)	SOD	200.5 U mg^{-1} protein
			GPX	330 U mg^{-1} protein
			GR	16.3 U mg^{-1} protein
JS-2002 (*Sorghum*) Jawad Hassan et al. (2020)	55 days	5 $CdCl_2$ (μM)	SOD	5 U min^{-1} g^{-1} protein
		25 $CdCl_2$ (μM)		4.8 U min^{-1} g^{-1} protein
		50 $CdCl_2$ (μM)		4 U min^{-1} g^{-1} protein
		100 $CdCl_2$ (μM)		3 U min^{-1} g^{-1} protein
		5 $CdCl_2$ (μM)	CAT	

(continued)

Table 1 (continued)

Plant species	Exposure time	Cadmium form and concentration	Antioxidant enzymes (AOE)	AOE concentration
				70 U min^{-1} g^{-1} protein
		25 $CdCl_2$ (μM)		80 U min^{-1} g^{-1} protein
		50 $CdCl_2$ (μM)		50 U min^{-1} g^{-1} protein
		100 $CdCl_2$ (μM)		30 U min^{-1} g^{-1} protein
		5 $CdCl_2$ (μM)	POD	110 U min^{-1} g^{-1} protein
		25 $CdCl_2$ (μM)		140 U min^{-1} g^{-1} protein
		50 $CdCl_2$ (μM)		90 U min^{-1} g^{-1} protein
		100 $CdCl_2$ (μM)		60 U min^{-1} g^{-1} protein
Chakwal *Sorghum* Jawad Hassan et al. (2020)	55 days after sowing	5 $CdCl_2$ (μM)	SOD	4 U min^{-1} g^{-1}
		25 $CdCl_2$ (μM)		2 U min^{-1} g^{-1}
		50 $CdCl_2$ (μM)		3 U min^{-1} g^{-1}
		100 $CdCl_2$ (μM)		2.8 U min^{-1} g^{-1}
		5 $CdCl_2$ (μM)	POD	70 U min^{-1} g^{-1}
		25 $CdCl_2$ (μM)		80 U min^{-1} g^{-1}
		50 $CdCl_2$ (μM)		50 U min^{-1} g^{-1}
		100 $CdCl_2$ (μM)		30 U min^{-1} g^{-1}
		5 $CdCl_2$ (μM)	CAT	110 U min^{-1} g^{-1}
		25 $CdCl_2$ (μM)		150 U min^{-1} g^{-1}
		50 $CdCl_2$ (μM)		80 U min^{-1} g^{-1}
		100 $CdCl_2$ (μM)		60 U min^{-1} g^{-1}
Lolium perenne Jiang et al. (2022)	30 days	3 $CdCl_2$ (ppm)	POD	
				1000 μmol g^{-1}
			SOD	850 μmol g^{-1}
			CAT	900 μmol g^{-1}
Polygonatum sibiricum	90 days	0.37 $CdCl_2$ (mg/kg)	SOD	300 U/g protein
			POD	1 μg/g·min

(continued)

Table 1 (continued)

Plant species	Exposure time	Cadmium form and concentration	Antioxidant enzymes (AOE)	AOE concentration
			CAT	4 μg/g·min
Cucurbita pepo Labidi et al. (2021)	10 days	100 μM $CdSO_4$	CAT	3 units per mg protein
			SOD	900 units per mg protein
Solanum nigrum Dai et al. (2022)	90 days	2 mg/kg $CdCl_2$	POD	40 U mg/g protein
		5 mg/kg $CdCl_2$		40 U mg/g protein
		10 mg/kg $CdCl_2$		40 U mg/g protein
		20 mg/kg $CdCl_2$		30 U mg/g protein
		30 mg/kg $CdCl_2$		20 U mg/g protein
		50 mg/kg $CdCl_2$		15 U mg/g protein
	90 days	2 mg/kg $CdCl_2$	SOD	260 U mg/g protein
		5 mg/kg $CdCl_2$		260 U mg/g protein
		10 mg/kg $CdCl_2$		260 U mg/g protein
		20 mg/kg $CdCl_2$		220 U mg/g protein
		30 mg/kg $CdCl_2$		200 U mg/g protein
		50 mg/kg $CdCl_2$		170 U mg/g protein
	90 days	2 mg/kg $CdCl_2$	GPX	270 U mg/g protein
		5 mg/kg $CdCl_2$		270 U mg/g protein
		10 mg/kg $CdCl_2$		270 U mg/g protein
		20 mg/kg $CdCl_2$		160 U mg/g protein
		30 mg/kg $CdCl_2$		140 U mg/g protein
		50 mg/kg $CdCl_2$		110 U mg/g protein
Centella asiatica Biswas et al. (2020)	30 days	50 mg/kg $Cd(NO_3)_2$	SOD	580 U/mg protein
		100 mg/kg $Cd(NO_3)_2$		300 U mg/g protein

(continued)

Table 1 (continued)

Plant species	Exposure time	Cadmium form and concentration	Antioxidant enzymes (AOE)	AOE concentration
		150 mg/kg $Cd(NO_3)_2$		590 U mg/g protein
		200 mg/kg $Cd(NO_3)_2$		600 U mg/g protein
		50 mg/kg $Cd(NO_3)_2$	AsPOX	4.5 U mg/g protein
		100 mg/kg $Cd(NO_3)_2$		5.8 U mg/g protein
		150 mg/kg $Cd(NO_3)_2$		5.9 U mg/g protein
		200 mg/kg $Cd(NO_3)_2$		6.8 U mg/g protein
		50 mg/kg $Cd(NO_3)_2$	GPx	1.2 U mg/g protein
		100 mg/kg $Cd(NO_3)_2$		0.8 U mg/g protein
		150 mg/kg $Cd(NO_3)_2$		1.8 U mg/g protein
		200 mg/kg $Cd(NO_3)_2$		2.7 U mg/g protein

Table 2 Examples of nonenzymatic system activation under Cd stress in different plant species

Plant species	Cadmium type/ concentration	Exposure time (duration time)	Nonenzymatic antioxidants (NEA)	NEA concentration
Cucurbita pepo Labidi et al. (2021)	100 $CdSO_4$	10 days	Malondialdehyde	12.13 nmol/g FW
			Proline	0.57 μmol/g FW
JS-2002 (*Sorghum*) Jawad Hassan et al. (2020)	5 μM $CdCl_2$	55 days	MDA	0.6 mmol g^{-1} FW
	25 μM $CdCl_2$			1.0 mmol g^{-1} FW
	50 μM $CdCl_2$			0.8 mmol g^{-1} FW
	100 μM $CdCl_2$			0.7 mmol g^{-1} FW
Chakwal *Sorghum* Jawad Hassan et al. (2020)	5 μM $CdCl_2$	55 days	MDA	0.7 mmol g^{-1} FW
	25 μM $CdCl_2$			1.2 mmol g^{-1} FW
	50 μM $CdCl_2$			1.0 mmol g^{-1} FW
	100 μM $CdCl_2$			0.8 mmol g^{-1} FW
Solanum nigrum Dai et al. (2022)	2 mg/kg $CdCl_2$	90 days	AsA(ascorbate) HZ leaf (mg/g)	10 mg/g
	5 mg/kg $CdCl_2$			10 mg/g
	10 mg/kg $CdCl_2$			10 mg/g
	20 mg/kg $CdCl_2$			8 mg/g
	30 mg/kg $CdCl_2$			6 mg/g
	50 mg/kg $CdCl_2$			3 mg/g
Solanum nigrum Dai et al. (2022)	2 mg/kg $CdCl_2$	90 days	GSH (glutathione)HZ leaf (mg/g)	18 mg/g
	5 mg/kg $CdCl_2$			18 mg/g
	10 mg/kg $CdCl_2$			18 mg/g
	20 mg/kg $CdCl_2$			13 mg/g
	30 mg/kg $CdCl_2$			11 mg/g
	50 mg/kg $CdCl_2$			5 mg/g
Solanum nigrum Dai et al. (2022)	2 mg/kg $CdCl_2$	90 days	MDA(malondialdehyde) HZ leaf (mmol/g)	0.20 mmol/g
	5 mg/kg $CdCl_2$			0.20 mmol/g
	10 mg/kg $CdCl_2$			0.20 mmol/g

(continued)

Table 2 (continued)

Plant species	Cadmium type/ concentration	Exposure time (duration time)	Nonenzymatic antioxidants (NEA)	NEA concentration
	20 mg/kg $CdCl_2$			0.40 mmol/g
	30 mg/kg $CdCl_2$			0.60 mmol/g
	50 mg/kg $CdCl_2$			0.80 mmol/g
Salvia sclarea Dobrikova et al. (2021)	100 μM $CdSO_4$	8 days	MDA	140% of control
			Anthocyanins	0.478 mg g^{-1} FW
			Total phenolic content	8.56 mg g^{-1} FW
Lolium perenne Jiang et al. (2022)	3 mg/kg $CdCl_2$	30th day after germination	MDA	14 nmol/g
Centella asiatica Biswas et al. (2020)	50 mg/kg Cd $(NO_3)_2$	30 days	Total phenolics (μg of gallic acid equivalent/gDW)	165.4 μgGA/gDW
	100 mg/kg Cd $(NO_3)_2$			218.11 μgGA/gDW
	150 mg/kg Cd $(NO_3)_2$			177.6 μgGA/gDW
	200 mg/kg Cd $(NO_3)_2$			194.03 μgGA/gDW
Centella asiatica Biswas et al. (2020)	50 mg/kg Cd $(NO_3)_2$		Total flavonoids (μg of rutin equivalent/gDW	154.3 μgR/gDW
	100 mg/kg Cd $(NO_3)_2$			146.97 μgR/gDW
	150 mg/kg Cd $(NO_3)_2$			174.50 μgR/gDW
	200 mg/kg Cd $(NO_3)_2$			207.86 μgR/gDW

References

Akram NA, Shafiq F, Ashraf M (2017) Ascorbic acid-a potential oxidant scavenger and its role in plant development and abiotic stress tolerance. Front Plant Sci 8:613

Alscher RG, Erturk N, Heath LS (2002) Role of superoxide dismutases (SODs) in controlling oxidative stress in plants. J Exp Bot 53(372):1331–1341

Ammar B, Mediouni WC, Tray B, Ghorbel MH, Jemal F (2008) Glutathione and phytochelatin contents in tomato plants exposed to cadmium. Biol Plant 52:314

Andrews J, Adams SR, Burton KS, Edmondson RN (2002) Partial purification of tomato fruit peroxidase and its effect on the mechanical properties of tomato fruit skin. J Exp Bot 53(379): 2393–2399

Asada K (2006) Production and scavenging of reactive oxygen species in chloroplasts and their functions. Plant Physiol 141:391–396

Azevedo H, Gomes C, Pinto G, Santos C (2005) Cadmium effects in sunflower: membrane permeability and changes in catalase and peroxidase activity in leaves and calluses. J Plant Nutr 28:2233–2241

Baruah N, Subham CM, Farooq M, Gogoi N (2019) Influence of heavy metals on seed germination and seedling growth of wheat, pea, and tomato. Water Air Soil Pollut 230:273–288

Batish DR, Singh HP, Setia N, Kaur S, Kohli RK (2006) 2-Benzoxazolinone (BOA) induced oxidative stress, lipid peroxidation and changes in some antioxidant enzyme activities in mung bean (*Phaseolus aureus*). Plant Physiol Biochem 44:819–827

Belyaeva EA, Glazunov VV, Korotkov SM (2004) Cd2? Versus Ca2+-produced mitochondrial membrane permeabilization: a proposed direct participation of respiratory complexes I and III. Chem Biol Interact 150:253–270

Benavides MP, Gallego SM, Tomaro ML (2005) Cadmium toxicity in plants. Braz J Plant Physiol 17:21–34

Biswas T, Parveen O, Pandey VP, Mathur A, Dwivedi UN (2020) Heavy metal accumulation efficiency, growth and centelloside production in the medicinal herb *Centella asiatica* (L.) urban under different soil concentrations of cadmium and lead. Ind Crop Prod 157:112948

Bordo D, Djinovic K, Bolognesi M (1994) Conserved patterns in the Cu, Zn superoxide dismutase family. J Mol Biol 238(3):366–386

Borisova G, Chukina N, Maleva M, Kumar A, Prasad MNV (2016) Thiols as biomarkers of heavy metal tolerance in the aquatic macrophytes of Middle Urals, Russia. Int J Phytoremediation 18(10):1037–1045

Buchanan BB, Gruissem W, Jones, R.L. (Eds.). (2015) Biochemistry and molecular biology of plants, 2nd edn Wiley-Blackwell

Chen L, Guo Y, Yang L, Wang Q (2008) Synergistic defensive mechanism of phytochelatins and antioxidative enzymes in *Brassica chinensis* L. against Cd stress. Chin Sci Bull 53:1503–1511

Clemens S (2006) Evolution and function of phytochelatin synthases. J Plant Physiol 163:319–332

Collin VC, Eymery F, Genty B, Rey P, Havaux M (2008) Vitamin E is essential for the tolerance of Arabidopsis thaliana to metal-induced oxidative stress. Plant Cell Environ 31:244–257

Cuypers A, Plusquin M, Remans T, Jozefczak M, Keunen E, Gielen H, Opdenakker K, Nair AR, Munters E, Artois TJ, Nawrot T (2010) Cadmium stress: an oxidative challenge. Biometals 23: 927–940

Dai H, Wei S, Pogrzeba M, Krzyżak J, Rusinowski S, Zhang Q (2021) The cadmium accumulation differences of two Bidens pilosa L. ecotypes from clean farmlands and the changes of some physiology and biochemistry indices. Ecotoxicol Environ Saf 209:111847

Dai H, Wei S, Twardowska I, Hou N, Zhang Q (2022) Cosmopolitan cadmium hyperaccumulator *Solanum nigrum*: exploring cadmium uptake, transport and physiological mechanisms of accumulation in different ecotypes as a way of enhancing its hyperaccumulative capacity. J Environ Manag 320:115878

DalCorso G, Farinati S, Furini A (2010) Regulatory networks of cadmium stress in plants. Plant Signal Behav 5:663–667

Dobrikova AG, Apostolova EL, Hanć A, Yotsova E, Borisova P, Sperdouli I, Adamakis IS, Moustakas M (2021) Cadmium toxicity in *Salvia sclarea* L.: an integrative response of element uptake, oxidative stress markers, leaf structure and photosynthesis. Ecotoxicol Environ Saf 209(11851):111851

Ferreira RR, Fornazier RF, Vitória AP, Lea PJ, Azevedo RA (2002) Changes in antioxidant enzyme activities in soybean under cadmium stress. J Plant Nutr 25(2):327–342

Foyer CH, Noctor G (2005) Oxidant and antioxidant signalling in plants: a re-evaluation of the concept of oxidative stress in a physiological context. Plant Cell Environ 28:1056–1071

Gallie DR (2013) Increasing vitamin C content in plant foods to improve their nutritional value-successes and challenges. Nutrients 5:3424–3446

Gallie DR, Chen Z (2019) Chloroplast-localized iron superoxide dismutases FSD2 and FSD3 are functionally distinct in *Arabidopsis*. PLoS One 14:e0220078

Gill SS, Tuteja N (2010) Reactive oxygen species and antioxidant machinery in abiotic stress tolerance in crop plants. Plant Physiol Biochem 48:909–930

Hasanuzzaman M, Bhuyan MB, Zulfiqar F, Raza A, Mohsin SM, Mahmud JA, Fujita M, Fotopoulos V (2020) Reactive oxygen species and antioxidant defense in plants under abiotic stress: Revisiting the crucial role of a universal defense regulator. Antioxidants 9(8):681

Hédiji H, Djebali W, Belkadhi A, Cabasson C, Moing A, Rolin D, Brouquisse R, Gallusci P, Chaïbi W (2015) Impact of long-term cadmium exposure on mineral content of *Solanum lycopersicum* plants: consequences on fruit production. S Afr J Bot 97:176–181

Heyno E, Klose C, Krieger-Liszkay A (2008) Origin of cadmium-induced reactive oxygen species production: mitochondrial electron transfer versus plasma membrane NADPH oxidase. New Phytol 179(3):687–699

Hiraga S, Sazaki K, Ito H, Ohashi Y, Matsui H (2001) A large family of class III plant peroxidases. Plant Cell Physiol 42(5):462–468

Hodges D, DeLong J, Forney C, Prange RK (1999) Improving the thiobarbituric acid-reactive substances assay for estimating lipid peroxidation in plant tissues containing anthocyanin and other interfering compounds. Planta 207(4):604–611

Hossain MA, Piyatida P, da Silva JAT, Fujita M (2012) Molecular mechanism of heavy metal toxicity and tolerance in plants: central role of glutathione in detoxification of reactive oxygen species and methylglyoxal and in heavy metal chelation. J Bot 2012:872875

Ismael MA, Elyamine AM, Moussa MG, Cai M, Zhao X, Hu C (2019) Cadmium in plants: uptake, toxicity, and its interactions with selenium fertilizers. Metallomics 11(2):255–277

Jaskulak M, Rorat A, Grobelak A, Kacprzak M (2018) Antioxidative enzymes and expression of rbcL gene as tools to monitor heavy metal-related stress in plants. J Environ Manag 218:71–78

Jawad Hassan M, Ali Raza M, Ur Rehman S, Ansar M, Gitari H, Khan I, Wajid M, Ahmed M, Abbas Shah G, Peng Y (2020) Effect of cadmium toxicity on growth, oxidative damage, antioxidant defense system and cadmium accumulation in two sorghum cultivars. Plan Theory 9(11):1575

Jiang N, Li Z, Yang J, Zu Y (2022) Responses of antioxidant enzymes and key resistant substances in perennial ryegrass (*Lolium perenne* L.) to cadmium and arsenic stresses. BMC Plant Biol 22(1):145

Jouili H, Bouazizi H, El Ferjani E (2011) Plant peroxidases: biomarkers of metallic stress. Acta Physiol Plant 33:2075–2082

Jozefczak M, Bohler S, Schat H, Horemans N, Guisez Y, Remans T, Vangronsveld J, Cuypers A (2015) Both the concentration and redox state of glutathione and ascorbate influence the sensitivity of arabidopsis to cadmium. Ann Bot 116(4):601–612

Kabata-Pendias A, Pendias H (2001) Trace elements in soils and plants. CRC Press, Boca Raton, FL

Karalija E, Selović A (2018) The effect of hydro and proline seed priming on growth, proline and sugar content, and antioxidant activity of maize under cadmium stress. Environ Sci Pollut Res 25(33):33370–33380

Karalija E, Selović A, Bešta-Gajević R, Šamec D (2022) Thinking for the future: phytoextraction of cadmium using primed plants for sustainable soil clean-up. Physiol Plant 174(4):13739

Karimpour M, Ashrafi SD, Taghavi K, Mojtahedi A, Roohbakhsh E, Naghipour D (2018) Adsorption of cadmium and lead onto live and dead cell mass of *Pseudomonas aeruginosa*: a dataset. Data Brief 18:1185–1192

Khan MH (2007) Induction of oxidative stress and antioxidant metabolism in *Calamus Tenuis* leaves under chromium and zinc toxicity. Indian J Plant Physiol 12(4):353–359

Labidi O, Vives-Peris V, GómezCadenas A, Pérez-Clemente RM, Sleimi N (2021) Assessing of growth, antioxidant enzymes, and phytohormone regulation in *Cucurbita pepo* under cadmium stress. Food Sci Nutr 9:2021–2031

Loix C, Huybrechts M, Vangronsveld J, Gielen M, Keunen E, Cuypers A (2017) Frontiers reciprocal interactions between cadmium-induced cell wall responses and oxidative stress in plants. Front Plant Sci 8:1867

Luo H, Li H, Zhang X, Fu J (2011) Antioxidant responses and gene expression in perennial ryegrass (*Lolium perenne* L.) under cadmium stress. Ecotoxicology 20:770–778

Mäder M (1992) Compartmentation of peroxidase isoenzymes in plant cells. In: Penel C, Gaspar T, Greppin M (eds) Topic and detailed literature on molecular, biochemical and physiological aspects. University of Geneva, Geneva

Mahmoud A, AbdElgawad H, Hamed BA, Beemster GT, El-Shafey NM (2021) Differences in cadmium accumulation, detoxification and antioxidant defenses between contrasting maize cultivars implicate a role of superoxide dismutase in Cd tolerance. Antioxidants 10(11):1812

Mohanpuria P, Rana NK, Yadav SK (2007) Cadmium induced oxidative stress influence on glutathione metabolic genes of Camellia sinensis (L.) O. Kuntze. Environ Toxicol 22(4): 368–374

Noriega GO, Balestrasse KB, Batlle A, Tomaro ML (2007) Cadmium induced oxidative stress in soybean plants also by the accumulation of delta-aminolevulinic acid. Biometals 20:841

Otter T, Polle A (1994) The influence of apoplastic ascorbate on the activities of cell-wall associated peroxidase and NADH oxidase in needles of Norway spruce (*Picea abies* L.). Plant Cell Physiol 35(8):1231–1238

Pandey VP, Awasthi M, Singh S, Tiwari S, Dwivedi UN (2017) A comprehensive review on function and application of plant peroxidases. Biochem Anal Biochem 6:308

Pathak N, Khandelwal S (2006) Oxidative stress and apoptotic changes in murine splenocytes exposed to cadmium. Toxicology 220(1):26–36

Peijnenburg W, Baerselman R, de Groot A, Jager T, Leenders D, Posthuma L, Van Veen R (2000) Quantification of metal bioavailability for lettuce (*Lactuca sativa* L.) in field soils. Arch Environ Contam Toxicol 39(4):420–430

Ramel F, Birtic S, Cuin S, Triantaphylides C, Ravanat JL, Havaux M (2012) Chemical quenching of singlet oxygen by carotenoids in plants. Plant Physiol 158(3):1267–1278

Rodríguez-Serrano M, Romero-Puertas MC, Zabalza A, Corpas FJ, Gómez M, Del Río LA, Sandalio LM (2006) Cadmium effect on oxidative metabolism of pea (*Pisum sativum* L.) roots. Imaging of reactive oxygen species and nitric oxide accumulation *in vivo*. Plant Cell Environ 29:1532–1544

Shanying HE, Xiaoe YANG, Zhenli HE, Baligar VC (2017) Morphological and physiological responses of plants to cadmium toxicity: a review. Pedosphere 27(3):421–438

Shapiguzov A, Vainonen JP, Wrzaczek M, Kangasjärvi J (2012) ROS-talk-how the apoplast, the chloroplast, and the nucleus get the message through. Front Plant Sci 3:292

Sharkey T (2005) Effects of moderate heat stress on photosynthesis: importance of thylakoid reactions, rubisco deactivation, reactive oxygen species, and thermotolerance provided by isoprene. Plant Cell Environ 28:269–277

Smeets K, Cuypers A, Lambrechts A, Semane B, Hoet P, Van L, Vangrosveld J (2005) Induction of oxidative stress and antioxidative mechanisms in *Phaseolus vulgaris* after Cd application. Plant Physiol Biochem 43:437–444

Smeets K, Ruytinx J, Semane B, Van Belleghem F, Remans T, Van Sanden S, Vangronsveld J, Cuypers A (2008) Cadmium-induced transcriptional and enzymatic alterations related to oxidative stress. Environ Exp Bot 63:1–8

Sparks DL (2012) Advances in agronomy. Academic Press

Srivastava RK, Pandey P, Rajpoot R, Rani A, Dubey RS (2014) Cadmium and lead interactive effects on oxidative stress and antioxidative responses in rice seedlings. Protoplasma 251:1047–1065

Subašić M, Šamec D, Selović A, Karalija E (2022) Phytoremediation of cadmium polluted soils: current status and approaches for enhancing. Soil Syst 6(1):3

Szarka A, Tomasskovics B, Bánhegyi G (2012) The ascorbate-glutathione-α-tocopherol triad in abiotic stress response. Int J Mol Sci 13(4):4458–4483

Takahashi A, Masuda A, Sun M, Centonze VE, Herman B (2004) Oxidative stress-induced apoptosis is associated with alterations in mitochondrial caspase activity and Bcl-2-dependent alterations in mitochondrial pH (pH_m). Brain Res Bull 62:497–504

Tchounwou PB, Yedjou CG, Patlolla AK, Sutton DJ (2012) Heavy metals toxicity and the environment. In: Luch A (ed) Molecular, clinical and environmental toxicology. Springer, Basel, pp 133–164

Volpe MG, Di Nazzaro M, Stasio FS, Coppola R, De Marco A (2015) Content of micronutrients, mineral and trace elements in some Mediterranean spontaneous edible herbs. Chem Cent J 9:57

Xu SS, Lin SZ, Lai ZX (2015) Cadmium impairs iron homeostasis in *Arabidopsis thaliana* by increasing the polysaccharide contents and the iron-binding capacity of root cell walls. Plant Soil 392:71–85

Zhang H, Lian C, Shen Z (2009) Proteomic identification of small, copper-responsive proteins in germinating embryos of *Oryza sativa*. Ann Bot 103:923–930

Zhou YJ, Zhang SP, Liu CW, Cai YQ (2009) The protection of selenium on ROS-mediated apoptosis by mitochondria dysfunction in cadmium-induced LLC-PK1 cells. Toxicol In Vitro 23(2):288–294

Phytoremediation: A Promising Approach for Re-vegetation of Cadmium-Polluted Land

Arwa Abdulkreem AL-Huqail, Mahmoud F. Seleiman, Maha Aljabri, Awais Ahmad, Majed Alotaibi, and Martin L. Battaglia

Abstract Cadmium (Cd), a non-essential trace element for plants, is commonly found in the environment due to human activities such as waste disposal, mining, metal manufacturing, and the use of synthetic phosphate fertilizers. These activities contribute to the accumulation of Cd in the soil, posing a carcinogenic risk to human health. In plants, Cd toxicity has detrimental effects on nutrient and water uptake, leading to deficiencies in essential elements. It also induces oxidative damage in plants, disrupts plant metabolism, and inhibits overall plant growth and physiological processes. In this chapter, we described the various sources of Cd contamination, the mechanisms of Cd uptake and translocation in plants, and its toxic effects on plant growth and physiology. Among different remediation approaches for Cd toxicity in soil, phytoremediation is one of the most promising and likely the one that has spurred the greatest interest over the last 10 years. This approach involves using plants to extract, stabilize, and degrade contaminants accumulated in soil, thereby reducing their concentration. Over the past several years, various plants have been studied for their effectiveness in Cd phytoremediation. Some plants, known as

A. A. AL-Huqail
Department of Biology, College of Science, Princess Nourah bint Abdulrahman University, Riyadh, Saudi Arabia

M. F. Seleiman (✉)
Department of Plant Production, College of Food and Agriculture Sciences, King Saud University, Riyadh, Saudi Arabia

Department of Crop Sciences, Faculty of Agriculture, Menoufia University, Shibin El-kom, Egypt
e-mail: mseleiman@ksu.edu.sa

M. Aljabri
Department of Biology, Faculty of Applied Science, Umm Al-Qura University, Makkah, Saudi Arabia

A. Ahmad · M. Alotaibi
Department of Plant Production, College of Food and Agriculture Sciences, King Saud University, Riyadh, Saudi Arabia

M. L. Battaglia
Corporate Engagement, The Nature Conservancy, Arlington, VA, USA

A. K. Jha, N. Kumar (eds.), *Cadmium Toxicity Mitigation*,
https://doi.org/10.1007/978-3-031-47390-6_9

hyperaccumulators, have a natural ability to take up high concentrations of heavy metals in soils, including Cd, without significant damage to their growth and development. These hyperaccumulator plants show potential for Cd phytoextraction, where the metal is absorbed by the plant's roots and translocated to the above-ground parts. Besides hyperaccumulator plants, other species have been investigated for their ability to enhance Cd phytoremediation through various mechanisms. For example, certain plants can enhance the bioavailability of Cd in the soil by releasing organic acids or enzymes that solubilize or chelate the metal, making it more accessible for plant uptake. This chapter summarizes the use of different plant species in Cd phytoremediation, highlighting their bioaccumulation capacities, biomass production, and Cd tolerance. It also discusses various approaches that have recently been investigated to enhance the effectiveness of phytoremediation. These findings contribute to the development of sustainable and eco-friendly strategies for Cd remediation in contaminated soils.

Keywords Cadmium · Remediation · Contaminated soils · Human health

1 Introduction

Cadmium (Cd) is a non-essential toxic heavy metal which means that plants do not require it for their growth and development (Ali et al. 2020a, b; Imran et al. 2021). Cd can pose significant toxic effects on human health, animals, and plants (Wang et al. 2023). Cd exists naturally in the atmosphere, sedimentary rocks, and soils. Human activities have also contributed to various forms of Cd exposure, leading to its presence in the environment over the past century (AL-Huqail 2022; Khatun et al. 2022; Bouida et al. 2022). Industrial applications utilize Cd as a corrosive reagent, as well as a stabilizer in color pigments PVC products and Ni-Cd batteries, etc. In areas where soils are contaminated, house dust can be a potential pathway source for Cd exposure (Puthran and Patil 2023). Alike, anthropogenic activities such as copper and nickel smelting, refining processes, combustion of fossil fuels, and the use of phosphate fertilizers are the main sources of Cd contamination in the environment (Priyadarshanee et al. 2022; Ramesh et al. 2023). Cd is also introduced as a pollutant in non-ferrous metal smelters and through electronic waste recycling (Wei et al. 2022; Luo et al. 2023). Various other factors such as forest fires also contribute to Cd exposure in the living environment like soil, water, and atmosphere (El-Naggar et al. 2021; Luo et al. 2021; Sarvestani and Aghasi 2022).

In terms of natural sources, Cd contamination can occur as a result of the natural breakdown of rocks through geological weathering, gradual erosion, and abrasion of rocks and volcanic activity where Cd releases into the environment, and it can accumulate in soils and potentially affect plants and animals (Liu et al. 2021, 2022). Direct and indirect sources of Cd contribute to the contamination of air, water, and soil, leading to various economic, public, and environmental implications (Kishor et al. 2021). Cd exhibits resistance to degradation by microbial activity, thus persisting in the environment and causing detrimental effects (Priyadarshanee et al.

2022; Salama et al. 2022). Since 1972, Cd has been recognized as a food contaminant, and its high levels of exposure can lead to renal failure, bone demineralization, and an increased risk of cancer (Subašić et al. 2022). Cd does not exist in the human body at birth, but accumulates over time, primarily in the kidneys and liver (Suciu et al. 2022). The general population is mainly exposed to Cd through contaminated food, Cd industry workers, or individuals with high industrial or people living in contaminated areas (Lamkarkach et al. 2021; Zhao et al. 2023). In humans, short-term exposure to low levels of Cd can pose adverse health risks (El Kholy et al. 2021), and long-term exposure to Cd can cause lung cancer (Cui et al. 2021; Nordberg et al. 2022). Much evidence supports the carcinogenic properties of Cd metal and its compounds (i.e., cadmium chloride, oxide, sulfate, and sulfide) in living organisms, particularly in animals (Lamkarkach et al. 2021). Incidences of prostate and lung cancer have been observed at elevated rates in animals exposed to Cd (Lee et al. 2022; Sevimli et al. 2022).

Cd pollution in soils and groundwater is a widespread issue globally. In uncontaminated soils, Cd is naturally found at concentrations ranging from 0.01 to 1 mg kg^{-1}, with an average global concentration of 0.36 mg kg^{-1} (Subašić et al. 2022). Among heavy metals, Cd is relatively highly mobile in the environment, depending on various factors, and may be released into groundwater more rapidly (Hussein et al. 2021). It can easily migrate from the soil into vegetation, thus entering the food chain (Bouida et al. 2022). Several strategies have been suggested for preventing Cd contamination and remediating affected areas. Actions such as wastewater treatment, controlling Cd levels in landfills and mines, and reducing the use of phosphate fertilizers contaminated with Cd can help to mitigate soil and water pollution (Haider et al. 2021). Various methods such as soil washing, chemical stabilization, electrokinetic remediation, bioremediation, phytoremediation, and thermal treatment can be employed to eliminate Cd from the soil and prevent contamination of the food chain (Abdel-Salam et al. 2015; Salem et al. 2018; Kumar et al. 2021; Rajendran et al. 2022; Subašić et al. 2022). It is important to note that the selection of the appropriate remediation method depends on various factors such as the level of contamination, soil characteristics, site conditions, and regulatory requirements. Each method has its advantages and limitations, and a combination of techniques may be necessary for effective cadmium removal from contaminated soil. However, among these methods, phytoremediation is an economically feasible and environmentally friendly approach that is garnering considerable attention in recent years. By consolidating the current knowledge on Cd phytoremediation, this chapter provides valuable insights into the potential of the phytoremediation approach and the exploration of different strategies to enhance Cd phytoremediation.

2 Cd Sources in the Environment

Cd contamination has gained global attention due to its potential for bioaccumulation and harm to living organisms (Moiseenko and Gashkina 2020). The presence of Cd in the environment has profound effects on biological systems in both terrestrial and marine ecosystems (Trombulak and Frissell 2000). Contamination of soil with Cd can occur through natural as well as human activities. In contaminated areas, Cd often originates from the parent soil, which is known as a lithogenic source. Natural occurrences like wind-blown soil particles, ocean salt sprays, volcanic emissions, rock weathering, and forest fires contribute to the introduction of Cd into the soil (Ali et al. 2021). These natural sources establish the background levels of Cd in the environment. However, human activities have also significantly intensified Cd levels beyond natural concentrations. Activities such as mining, smelting, and manufacturing processes release substantial amounts of Cd into the environment (Duruibe et al. 2007; Bhat et al. 2022). These anthropogenic actions are responsible for the contamination of soil, water, and air with Cd, surpassing the levels that would occur naturally. The contamination of soil and water with Cd from human sources poses severe risks to both human health and the environment. Cd is a toxic heavy metal that can accumulate in living organisms, including plants, animals, and humans. Long-term exposure to elevated Cd levels can lead to detrimental health effects, including kidney damage, respiratory issues, and even cancer (Subašić et al. 2022). Moreover, the presence of Cd in the environment disrupts the delicate ecological balance and threatens biodiversity in affected ecosystems. Therefore, it is essential to implement remediation strategies like phytoremediation to address Cd contamination and mitigate its adverse impacts on human and environmental well-being. By comprehending the sources and consequences of Cd contamination, appropriate measures can be taken to prevent further contamination and safeguard ecosystems and human populations from the harmful effects of Cd exposure. Industrial processes, electronic products, paints, sewage alloys, and wastewater from mining operations and agricultural practices are among the key anthropogenic activities resulting in soil contamination over time (Seleiman and Kheir 2018; Borah et al. 2020; Madhav et al. 2020; Seleiman et al. 2021). Accidental spills or leaks from industrial sites or during transportation can also contribute to soil contamination with Cd (Gautam et al. 2023).

3 Cd in the Soil and Uptake by Plants

Cd can persist in soil for a long period of time, and its persistence in soil is influenced by various factors, including soil properties, environmental conditions, and the presence of other substances (Zhang et al. 2020). In general, Cd has a strong affinity for soil particles and tends to bind tightly to soil minerals and organic matter (Subašić et al. 2022; Deng et al. 2023). This binding can reduce its mobility and bioavailability, making it less accessible to plants and organisms. However, Cd can still remain in the soil for extended periods due to its slow degradation rate and limited natural processes for its removal (Aparicio et al. 2022). The persistence of Cd in the soil can be influenced by factors such as pH, temperature, moisture content, and organic matter content (Subašić et al. 2022; Lasota et al. 2020). For example, acidic soil conditions can increase the mobility of Cd and facilitate its leaching into groundwater, while alkaline conditions can reduce its solubility and limit its movement (Wang et al. 2020). The adsorptive capacity of soils for Cd increases threefold for each unit increase in pH within the range of 4–7 (Yu and Tang 2022). This suggests that higher pH levels in this range enhance the ability of soils to effectively adsorb Cd. Organic matter has a significant role in Cd binding and its conversion into an organically bound fraction, thereby decreasing the bioavailability of Cd (Xu et al. 2023). The study by Xu et al. (2023) supports the notion that organic matter in the soil acts as a valuable agent for immobilizing Cd and limiting its movement, making it less accessible for uptake by plants and organisms. Among metals, Cd demonstrates a higher affinity for low molecular weight organic matter, with the following affinity order: $Cu^{2+} > Cd^{2+} > Fe^{2+} > Pb^{2+} > Ni^{2+} > Co^{2+} > Mn^{2+} > Zn^{2+}$ (Ware et al. 2003; Bolan et al. 2022). When present in high concentrations, organic acids containing carboxyl groups facilitate the complexation of Cd. The degree of Cd solubilization follows the order: of fumaric > citric > oxalic > acetic ≈ succinic acid (Krishnamurti et al. 1997). The presence of certain anions chloride (Cl^-) and nitrate (NO^{3-}) restricts Cd sorption by forming soluble inorganic complexes, while phosphate (H_2PO^{4-}) and sulfate ($HSO4^-$) enhance Cd sorption through surface precipitation (Sipos et al. 2019; Subašić et al. 2022).

The absorption and distribution of Cd in plants involve several regulatory processes and can vary depending on plant species, environmental conditions, and the concentration of Cd in the soil (Subašić et al. 2022). In the plasma membrane of root cells, as well as processes related to leaf/shoot sequestration, xylem, and phloem loading/unloading and detoxification, metal transporters play important roles in Cd uptake and redistribution (Subašić et al. 2022; Asare et al. 2023). Cd is absorbed by plant roots, and its availability can be influenced by factors such as pH, the rhizosphere, and organic acids (Krishnamurti et al. 1997; Yu and Tang 2022). Plants can take up Cd in different forms, such as $CdCl^-$, $CdHCO3^+$, $CdCO3^+$, and CdCln, particularly at pH levels between 6 and 7 (Smolders and McLaughlin 1996; Ismael et al. 2019). The uptake of Cd involves two stages: apoplastic adsorption and symplastic uptake. During apoplastic adsorption, metal ions, including Cd, accumulate in the root apoplast through electrostatic interactions. This process occurs due to the attraction between positively charged metal cations and negatively charged carboxyl groups present in the apoplastic space. (Ismael et al. 2019). While in symplastic uptake, Cd must cross the cell membrane, facilitated by various channels and metal transporters (Ismael et al. 2019). The transport of metals, including Cd (cadmium), from the extracellular space into the plant cytoplasm involves specific transporter families. The ZIP (ZRT and IRT-like proteins) transporter family plays a crucial role in facilitating the uptake and transport of metals. ZIP transporters are integral membrane proteins that are responsible for the movement of metal ions across cell membranes. They are involved in the transport of a wide range of metal ions, including Cd, from the extracellular space into the cytoplasm of plant cells. These transporters have been identified in various plant species and are highly conserved. In addition to the ZIP transporter family, the OPT (oligopeptide transporter) family is also involved in metal transport across plant cell membranes. This family includes transporters known as YSL (yellow stripe-like) transporters. YSL transporters have been found to play a role in the transport of metal-nicotinamide complexes, including Cd-nicotinamide complexes, across the plasma membrane (Ishimaru et al. 2012; Subašić et al. 2022). The NRAMP (natural resistance-associated macrophage protein) family comprises proton-coupled metal ion transporters that participate in Cd transport (Ishimaru et al. 2012). There is evidence to suggest that Cd (cadmium) can enter plant cells through calcium (Ca^{2+}) channels. Reports indicate that Cd ions may utilize Ca^{2+} channels as a pathway for their entry into cells. This suggests that Cd ions exploit the same channels that are responsible for the transport of Ca^{2+} ions across cell membranes (White 2000). Within the root cells' plasma membrane, the dissociation of H_2CO_3 during respiration leads to the release of H^+ and HCO^{3-}. Subsequent to the electrostatic interactions between Cd ions and carboxyl groups, there is a rapid exchange of hydrogen ions (H^+) with Cd^{2+} ions. This exchange process leads to the adsorption of Cd^{2+} ions on the cell surface, preparing it for the apoplastic absorption pathway. The exchange of H^+ ions with Cd^{2+} ions occurs due to the higher affinity of Cd^{2+} for binding sites on the cell surface. As a result, Cd^{2+} ions replace the H^+ ions present on the surface of the root cells. This priming step facilitates the subsequent movement of Cd^{2+} ions toward the apoplast (Song et al. 2017; Subašić et al. 2022). Cd ions can enter plant cells through specific ion channels that are primarily intended for the transport of other metal ions such as Fe^{2+}, Zn^{2+}, and Ca^{2+}. These channels may have some level of selectivity, allowing Cd to utilize them as a pathway for entry into the cell. Additionally, plants have the ability to secrete low molecular weight compounds that can aid in Cd uptake and transport. For example, mugineic acid and malic acid are known to be secreted by plant roots that enhance ion availability and form metal chelates, which are then subsequently absorbed by the plants (Curie et al. 2009; Ismael et al. 2019; Subašić et al. 2022). Once Cd ions cross the root membrane and enter the root symplast, they can be further transported toward the xylem vessels, and this process is regulated by various factors such as ion transporters, concentration gradient, and plant physiology (Curie et al. 2009; Subašić et al. 2022). Once inside the plant cell, Cd can accumulate up to a specific level based on plant tolerance, influenced by chelating molecules such as proline, glutathione, and nicotinamide as well as the presence and selectivity of transporters (Curie et al. 2009; Subašić et al. 2022). To maintain plant functions, surplus Cd is drained out from the cytosol through chelation and compartmentalization in the vacuole or plant cell walls (Wei et al. 2021). Chelating agents, along with vacuolar sequestration and apoplastic barriers, play a role in Cd detoxification (Riaz et al. 2021). Additionally, the loading activity to the xylem and its higher CEC (cation exchange capacity) are significant reasons that influence Cd xylem loading and subsequent transport (Mendoza-Cózatl et al. 2011; Viehweger 2014). Hyperaccumulating plants employ various detoxification mechanisms to mitigate Cd toxicity. These mechanisms include Cd chelation and Cd vacuolar sequestration, where Cd is bound to sulfur-containing ligands such as glutathione, phytochelatins, and metallothioneins (DalCorso et al. 2008; Sigel et al. 2015).

The transfer of Cd from the xylem to the phloem is a critical process for its transport within plants. This movement allows for Cd to be distributed to various plant tissues and organs. The transfer of Cd from roots to shoots through the xylem is tightly regulated and involves the coordination of sulfur and acetate ligands (Ueno et al. 2008; Song et al. 2017). Sulfur and acetate ligands play important roles in Cd transport within the plant. Sulfur ligands, such as phytochelatins and glutathione, form complexes with Cd ions, facilitating their movement through the xylem. These complexes enhance the solubility and stability of Cd, preventing its precipitation or immobilization within the plant (Wong and Cobbett 2009; Vogel-Mikuš et al. 2010). When metals are transported through the phloem, they must be bound to ligands such as nicotianamine, glutathione, or phytochelatins (Mendoza-Cózatl et al. 2008). Among these ligands, phytochelatins have a high affinity for binding with Cd. Cadmium is thought to be loaded into the phloem as Cd-thiolate complexes, where the stability of the Cd-S bond minimizes its toxicity (Mendoza-Cózatl et al. 2008; Ismael et al. 2019). Despite significant progress, several aspects of Cd transport and accumulation in plants, particularly in those with varying levels of Cd accumulation capacity, resistance, and tolerance, remain unclear. A deeper understanding of these mechanisms is crucial for optimizing the use of hyperaccumulating plants in phytoremediation processes.

4 Toxic Effects of Cd on Plants

Cadmium exhibits detrimental effects on various metabolic and physiological processes in plants. It is important to note that the severity of Cd toxicity depends on factors such as Cd concentration, exposure duration, plant species, and environmental conditions (Sarkar et al. 2013). Even at low doses, Cd can result in leaf chlorosis, which is characterized by the yellowing or whitening of plant leaves. Cadmium toxicity disrupts chlorophyll synthesis or breakdown, impairing the plant's ability to carry out photosynthesis effectively. High levels of Cd can also lead to the formation of necrotic lesions on plant tissues. Such lesions impair the affected plant parts' functionality and overall plant health (Tiryakioglu et al. 2006; Dias et al. 2013; Dubey et al. 2020). Cadmium can damage chloroplasts, and disrupts the structure and function of chloroplasts, hindering their ability to capture light energy and convert it into chemical energy. Cd toxicity induces water stress in plants, resulting in reduced water uptake by roots or increased water loss through transpiration, leading to dehydration and wilting (Rucińska-Sobkowiak 2016).

Upon entering plant cells, Cd triggers the production of free radicals, which in turn initiates apoptosis-like processes via oxidative stress (Perrone et al. 2008; Ranjan et al. 2021). To counteract the toxic effects of Cd, plants employ various mechanisms that begin at the root membrane, which serves as the first point of Cd entry (Gallego et al. 2012). Plants have the ability to transport Cd in the form of metal-organic complexes. These complexes involve the binding of Cd ions with organic ligands, such as phytochelatins, organic acids, and amino acids. These metal-organic complexes enable the transport of Cd within the plant, facilitating its movement from roots to shoots and various plant tissues (Uchimiya et al. 2020; Rai et al. 2021). In the rhizosphere, Cd competes with essential metal ions for uptake by plant roots. This competition occurs because Cd shares similar transport pathways with essential metals such as iron (Fe), zinc (Zn), and manganese (Mn). The presence of high concentrations of Cd can interfere with the uptake and utilization of these essential metals by plants (Uchimiya et al. 2020; Rai et al. 2021). Upon reaching plant leaves, Cd is accumulated in vacuoles as a protective measure to mitigate its toxic effects on cellular processes and photosynthesis of plants. Alternatively, Cd can be detoxified through the action of chelating compounds, such as metallothioneins, phytochelatins, and glutathione and other cysteine-rich membrane proteins (Irfan et al. 2013). The level of Cd toxicity in plants varies significantly among different plant species. Some species exhibit sensitivity to even low concentrations of Cd. While others such as *Chromolaena odorata, Arabis gramminifera, Nitella opaca, Chara aculeata*, and *Silene sendtneri* demonstrate high tolerance and can store substantial amounts of Cd in shoots, exceeding 100–2000 mg Cd kg^{-1}, thus being classified as Cd hyperaccumulators (McGrath and Zhao 2003; Tanhan et al. 2007; Krämer 2010; Sooksawat et al. 2013). Plants that are not tolerant to Cd are highly susceptible to the toxic effects of Cd. Cadmium can have detrimental impacts on various aspects of plant development, including mineral uptake and translocation, carbon assimilation, biomass production, and photosynthesis as well as the development of reproductive tissues. Cd toxicity in shoots manifests primarily through noticeable changes in leaves, including desiccation, necrosis, chlorosis, and stunting (Solís-Domínguez et al. 2007). Cd adversely affects the photosystems (PSI and PSII) of plants through multiple mechanisms. It inhibits the photosystem II (PSII) directly by interfering with the activity of PSII, which is responsible for capturing light energy and driving the electron transport chain. Cd can bind to the reaction center of PSII, displacing essential cofactors such as magnesium (Mg), which disrupts the flow of electrons and impairs the light-capturing process (Sigfridsson et al. 2004; Faller et al. 2005). Furthermore, Cd alters the ultrastructure of chloroplasts and disrupts enzymes involved in the Calvin cycle, resulting in a decrease in the photosynthetic rate. Cd has the ability to displace Mg^{2+} ions in chlorophyll pigments and Ca^{2+} ions in oxygen-evolving complexes leading to a reduction in chlorophyll and carotenoid content (Küpper et al. 2007). Cadmium negatively impacts the absorption of various minerals, including potassium, iron, zinc, manganese, calcium, copper, magnesium, and silicon (Jinadasa et al. 2016). This interference with mineral absorption is attributed to molecular competition between cations and Cd in channels responsible for the uptake of essential metals from the soil to the roots, ultimately resulting in a deficiency of essential elements (Clemens et al. 2001).

5 Phytoremediation of Cd from Polluted Soils

Phytoremediation is an eco-friendly approach for soil remediation that utilizes fast-growing plants to eliminate toxic contaminants. This green remediation method relies on the ability of plants to uptake, sequester, and transform contaminants present in the soil or water. Phytoremediation offers a sustainable and versatile approach for addressing soil and water contamination, providing tailored strategies for different types of contaminants and environmental conditions. The list of the plants which have shown potential for application in phytoremediation and their rate of cadmium accumulation is described in Table 1. Depending on the specific mechanisms involved in eliminating heavy metals, phytoremediation can be classified into the following types.

5.1 Phytostabilization or Phytosequestration

Phytostabilization involves the plant's ability to immobilize metals through various mechanisms, such as chemical or physical processes, performed primarily by the plant roots (Khalid et al. 2017). This approach offers a sustainable and environmentally friendly solution for managing heavy metal contamination, providing long-term stability and minimizing the potential risks associated with metal exposure. In the phytostabilization method, plants play a crucial role in reducing water percolation, thereby limiting the contact between heavy metals and decreasing the movement of contaminants (Shackira and Puthur 2019). This method can be employed for the remediation of soil contaminated with Cd, as well as other pollutants. Various plant species, such as *Virola surinamensis*, *Thlaspi arvense*, Miscanthus x giganteus, *Avena sativa*, *Commelina communis*, and *Sinapis alba*, have shown potential for phytostabilization of Cd-contaminated soil and are classified as heavy metal excluders or hypertolerant plants (Andrade Júnior et al. 2019; Boros-Lajszner et al. 2020; Bilandžija et al. 2022). These plants demonstrate potential in reducing the metals' availability to plants, but it is important to note that phytostabilization does not actually eliminate metals from soil which is a significant drawback of phytostabilization technique (Vangronsveld et al. 2009).

5.2 Phytostimulation or Phytotransformation

Phytostimulation describes the degradation process of organic pollutants in the rhizosphere (Abdel-Shafy and Mansour 2018). This degradation occurs due to the increased microbial activity facilitated by root exudates (Sangeeta and Maiti 2010). Plant roots release a variety of compounds such as sugars, carbohydrates, amino acids, acetates, and enzymes into the rhizosphere. These exudates enrich the

Table 1 The list of the plants which have shown potential for application in phytoremediation and their rate of cadmium accumulation

Cd conc. in the plant parts (mg kg^{-1})	Plant species	Mechanism	References
186 in roots	*Aerva sanguinolenta*	Phytostabilization	Phaenark et al. (2009)
242 in shoots	*Amaranthus hybridus*	Phytoextraction	Zhang et al. (2010)
217 in leaf	*Amaranthus hypochondriacus*	Phytoextraction	Yu et al. (2020)
102–604 in shoots, roots, leaf	*Amaranthus mangostanus*	Phytoextraction Phytostabilization	Fan and Zhou 2009
228–5722 in shoots, roots	*Arabidopsis halleri*	Phytoextraction Phytostabilization	Zhao et al. (2006)
1810 in leaf	*Arabis gemmifera*	Phytoextraction	Kubota and Takenaka (2003)
1662–8670 in leaves, roots	*Arabis paniculate*	Phytoextraction Phytostabilization	Qiu et al. (2008)
685 in leaves	*Arabis yokoscense*	Phytoextraction	Kubota and Takenaka (2003)
217–606 in shoot, roots	*Atriplex halimus*	Phytoextraction Phytostabilization	Nedjimi and Daoud (2009)
740 in whole plant	*Azolla pinnata*	Phytoextraction	Rai (2008)
314.17–4547 in shoots, roots	*Beta vulgaris*	Phytoextraction Phytostabilization	Li et al. (2015)
400 in leaf	*Bidens pilosa*	Phytoextraction	Dai et al. (2017)
186 in shoots	*Brachiaria mutica*	Phytoextraction	Dai et al. (2017)
137.3–647 in shoot, roots	*Brachiaria sp.*	Phytoextraction Phytostabilization	Ullah et al. (2019)
112.62 in shoots	*Cosmos bipinnata*	Phytoextraction	Huang et al. (2017)
>100 in shoots	*Chromolaena odorata*	Phytoextraction	Wei et al. (2018)
165 in roots	*Calendula calypso*	Phytostabilization	Farooq et al. (2020)
>101 in shoots	*Callisia fragrans*	Phytoextraction	Simek et al. (2018)
148–236 in leaves	*Carthamus tinctorius*	Phytoextraction	Angelova et al. (2016)
159 in roots	*Cassia alata*	Phytostabilization	Bhargava and Singh (2017)
121–236 in leaves, roots	*Celosia argentea*	Phytoextraction Phytostabilization	Yu et al. (2019)
>100 in shoots	*Chlorophytum comosum*	Phytoextraction	Simek et al. (2018)
102–1440 in leaves, roots	*Chromolaena odorata*	Phytoextraction Phytostabilization	Tanhan et al. (2007)
312 in shoots	*Desmostachya bipinnata*	Phytoextraction	Ullah et al. (2019)
10.5 in roots	*Eucalyptus camaldulensis*	Phytostabilization	Madejon et al. (2017)

(continued)

Table 1 (continued)

Cd conc. in the plant parts (mg kg^{-1})	Plant species	Mechanism	References
150 in roots	*Eleusine indica*	Phytostabilization	Phaenark et al. (2009)
120 in leaves	*Eucalyptus camaldulensis*	Phytostabilization	Madejon et al. (2017)
5.11 in roots	*Eucalyptus globulus*	Phytostabilization	Luo et al. (2016)
74.8–290 in leaves	*Glycine max*	Phytoextraction	Guo et al. (2020)
65.7 in roots and shoot	*Helianthus annuus*	Phytoextraction Phytostabilization	Alaboudi et al. (2018)
328–2167 in leaves, roots	*Helianthus tuberosus*	Phytoextraction Phytostabilization	Liang et al. (2011)
128 in shoots	*Hydrocotyle sibthorpioides*	Phytoextraction	Pan et al. (2019)
212.3 in shoots	*Impatiens violaeflora*	Phytoextraction	Phaenark et al. (2009)
133.2 in roots	*Imperata cylindrica*	Phytostabilization	Phaenark et al. (2009)
548 in shoots	*Justicia procumbens*	Phytoextraction	Phaenark et al. (2009)
>100 in shoots	*Lantana camara*	Phytoextraction	Murakami et al. (2007)
245 in shoots	*Leptochloa fusca*	Phytoextraction	Ullah et al. (2019)
106 in leaves	*Lolium multiflorum*	Phytoextraction	Guo et al. (2020)
402 in shoots	*Lonicera japonica*	Phytoextraction	Liu et al. (2009)

microbial populations already present in the soil, serving as a source of nutrients and energy for their growth and activity (Anderson et al. 1993). Plant root systems supply oxygen to the rhizosphere through root respiration. This oxygen availability creates aerobic conditions that support the metabolic transformations carried out by aerobic microorganisms (Kumar et al. 2018). Furthermore, fine-root biomass contributes to an increase in organic carbon content in the rhizosphere. As these fine roots and associated organic matter decompose, they provide a nutrient-rich environment that fosters microbial growth and activity (Anderson et al. 1993; Kumar et al. 2018). The presence of plants creates a favorable habitat for increased microbial populations and activity in the rhizosphere. The combination of root exudates, organic carbon, and conducive conditions creates a niche that supports diverse microbial communities and their activities, including nutrient cycling and organic matter decomposition (Abdel-Shafy and Mansour 2018; Kleber et al. 2021).

Phytotransformation involves the breakdown of organic compounds either through plant metabolism or by plant enzymes, independent of microbial communities (McGuinness and Dowling 2009). In this process, plants have the ability to degrade heavy metals and certain volatile compounds, leading to their release into the atmosphere, a phenomenon known as phytovolatilization (Dhanwal et al. 2017). Phytotransformation occurs through internal plant metabolic processes or external plant body processes. Plant's enzymes such as peroxidase, dehydrogenase, dehalogenase, and oxidoreductase facilitate the transformation or degradation of toxic metals. For example, *Liriodendron tulipifera* has been observed to grow in environment with higher concentration of metals (Jadia and Fulekar 2009).

5.3 Phytofiltration

Phytofiltration utilizes plant roots to address lower levels of heavy metal contamination in soil surfaces, groundwater, and wastewater. The process involves the absorption or precipitation of contaminants through the release of root exudates and changes in pH (Pinto et al. 2015). By employing phytofiltration, plants can absorb or adsorb contaminants present in polluted surface waters or wastewater, effectively preventing their movement into underground water sources (Muthusaravanan et al. 2018). This remediation approach can be implemented in situ, meaning that the plants are directly grown in the contaminated water body, resulting in reduced costs and logistical challenges. Phytofiltration has shown promising results in the removal of heavy metals such as Cd, Cr, Cu, and Zn. Several plant species, including *Limnocharis flava* and *Arundo donax*, have demonstrated potential for Cd phytofiltration (Subašić et al. 2022). Gomes et al. (2016) have identified three main types of phytofiltration techniques: rhizofiltration, caulofiltration, and blastofiltration. Rhizofiltration involves the use of plant roots to absorb pollutants from contaminated soils and water, effectively cleansing the environment through processes like accumulation, adsorption, absorption, and precipitation into the plant biomass (Mahajan and Kaushal 2018). Both terrestrial and fast-growing aquatic plants can be utilized in rhizofiltration to extract heavy metals such as Cd, Cr, Cu, Ni, Pb, and Zn. In the context of biogas production, a phytofiltration lagoon incorporating the invasive species *Pistia stratiotes* proves particularly suitable as it provides year-round biomass and successfully treats polluted water (Olguín et al. 2017). *Eichhornia crassipes*, along with other invasive aquatic plants like *Pistia stratiotes* and *Myriophyllum spicatum*, has been identified as a potential and cost-effective option for the phytoremediation of pesticide-contaminated agro-industrial wastewater (Xia and Ma 2006; Olguín et al. 2017).

5.4 Phytoextraction/Phytoaccumulation

Phytoextraction or phytoaccumulation utilizes fast-growing plants to remove heavy metals from contaminated soil or water. This process involves the absorption and accumulation of contaminants within the plant and then transported to the upper parts of the plant (Karalija and Selović 2018). The harvested plant material can subsequently be used for biomining or phytomining, which involves the recovery of metals from the plants (Jadia and Fulekar 2008). Phytoextraction is effective in eliminating heavy metals through processes such as absorption, translocation, and accumulation, particularly in plant families known as hyperaccumulators such as *Scrophulariaceae*, *Lamiaceae*, *Asteraceae*, *Euphorbiaceae*, and *Brassicaceae* (Van der Ent et al. 2013; Bhargava and Singh 2017). This is the permanent solution for HMs detoxification through the harvesting biomass. The hyperaccumulating plants have specific thresholds level for metals as follows: Se (selenium) and Cd > 100 mg kg^{-1}, Cr (chromium), Co (cobalt), and Cu (cupper) > 300 mg kg^{-1}, As (Arsenic), Pb (lead), and Ni (nickel) > 1000 mg kg^{-1}, Mn (manganese) > 10,000 mg kg^{-1}, and Zn (zinc) > 3000 mg kg^{-1} (Ghosh and Singh 2005; Van der Ent et al. 2013). It is important for a plant to maintain its growth while achieving metal hyperaccumulation in order to be suitable for phytoextraction purposes (Van der Ent et al. 2013). In cases where there are no suitable plants available for phytoextraction, chelating agents such as EDTA, citric acid, or proline can be added to the soil. These agents increase the solubility and availability of pollutants, facilitating the phytoextraction process (Ghosh and Singh 2005). In highly polluted soils, even hyperaccumulating plants may experience severely impaired growth, which reduces the effectiveness of phytoextraction (Khalid et al. 2017). Several plant species such as *Celosia argentea* (Yu et al. 2019), *Cassia alata* (Silva et al. 2018), *Vigna unguiculata*, *Solanum melongena*, *Momordica charantia* (Ali et al. 2016), *Nicotiana tabacum*, *Kummerowia striata* (Liu et al. 2011), *Swietenia macrophylla* (Fan et al. 2011), and *Silene sendtneri* (Karalija et al. 2021) have been identified as hyperaccumulators of Cd. These plants have shown the capacity to accumulate elevated concentrations of Cd, making them potentially useful for phytoextraction or phytoremediation purposes in contaminated environments.

5.5 *Phytovolatilization*

Phytovolatilization is a technique in phytoremediation that utilizes plants to uptake pollutants from the soil, convert them into less toxic volatile forms, and release them into the atmosphere through the process of plant transpiration via leaves or foliage (Dhankher et al. 2012). This approach is particularly useful for detoxifying organic pollutants and heavy metals (Kamusoko and Jingura 2017). *Brassica juncea*, a member of the Brassicaceae family, has been identified as an effective volatilizer of heavy metal (Chaturvedi et al. 2016). In this process, heavy metal is assimilated into organic acids and subsequently bio-methylated to form a volatile compound with reduced toxicity compared to inorganic heavy metal (Igiri et al. 2018). Phytovolatilization offers the advantage of removing heavy metal and metalloid contaminants from the site and dispersing them as gaseous compounds, eliminating the need for plant harvesting and disposal (Awa and Hadibarata 2020). However, it is important to note that phytovolatilization does not completely eliminate pollutants; instead, it transfers them from the soil to the atmosphere, potentially leading to contamination of the ambient air. There is also a risk of redeposition to the soil through precipitation, necessitating a thorough risk assessment before its implementation in the field (Kamusoko and Jingura 2017; Awa and Hadibarata 2020).

5.6 Rhizofiltration

Rhizofiltration is an effective and sustainable technique widely used for the removal of contaminants from water and liquid waste. This innovative approach relies on harnessing the natural filtration capabilities of specific plant species. Plants with fibrous root systems and large surface areas are particularly well suited for rhizofiltration due to their enhanced capacity to absorb and accumulate contaminants (Mwegoha 2008). One notable example of a plant species that has demonstrated remarkable effectiveness in rhizofiltration is *Typha latifolia*. This aquatic perennial, commonly known as cattail, has been found to efficiently remove methyl parathion, an organophosphorus insecticide, from hydromorphic soils (Nedjimi 2021). By taking up and sequestering the harmful pesticide within its root system, *Typha latifolia* contributes to the remediation of contaminated water bodies and promotes the restoration of ecological balance. Another plant species that has shown promise in the context of rhizofiltration is *Phaseolus vulgaris*, commonly known as beans. This versatile legume has demonstrated efficient extraction of uranium and cesium from groundwater (Priya et al. 2023). Its ability to selectively uptake and accumulate these radioactive elements within its roots makes it a valuable candidate for remediation efforts in areas affected by nuclear accidents or radioactive contamination. *Arundo donax*, commonly referred to as giant reed, has also exhibited impressive capabilities in rhizofiltration, specifically in the removal of copper from constructed wetlands (Ali et al. 2020a, b). This tall perennial grass effectively takes up copper ions present in the water, mitigating the adverse effects of metal pollution and enhancing water quality in affected areas. In the realm of industrial sludge treatment, several plant species have emerged as promising options for heavy metal removal through rhizofiltration. *Eichhornia crassipes*, also known as water hyacinth, *Salvinia molesta*, and *Pistia stratiotes*, commonly referred to as water lettuce and water cabbage, respectively, have demonstrated their efficiency in extracting heavy metals from contaminated sludge (Ali et al. 2020a, b). These fast-growing aquatic plants effectively absorb and accumulate heavy metal contaminants, offering a cost-effective and environmentally friendly solution for sludge remediation.

Rhizofiltration provides numerous advantages over traditional remediation methods. It is a relatively cost-effective and sustainable approach that utilizes natural processes and avoids the use of harsh chemicals. By employing plants as living filters, rhizofiltration not only removes contaminants but also contributes to the overall ecological health and restoration of affected ecosystems. However, it is important to acknowledge that rhizofiltration may have some limitations. Compared to alternative remediation techniques such as chemical treatments or excavation, the process of significantly reducing contaminant levels through rhizofiltration may take a longer period of time. Additionally, the effectiveness of rhizofiltration can vary depending on factors such as the specific contaminant, soil conditions, and plant species used. Therefore, a thorough understanding of site-specific factors and careful selection of suitable plant species are crucial for achieving optimal results in rhizofiltration projects.

5.7 Rhizodegradation

Rhizodegradation harnesses the collaborative efforts of plants and their associated root-zone microorganisms to degrade contaminants within the soil (Etim 2012). This approach has proven to be effective and cost-efficient in remediating contaminated soils. The process involves intricate interactions between plant roots, microorganisms, and the pollutants present (Vishnoi and Srivastava 2007). In the rhizospheric region, microorganisms have the ability to interact with both the plant and the contaminants, playing a vital role in the degradation process. The selection of appropriate plant species is a critical factor in facilitating successful rhizodegradation. Different plants release distinct types and quantities of root exudates, which influence the microbial community and their capacity to degrade contaminants (Kumar et al. 2020). Therefore, careful consideration must be given to choosing plant species that can foster the desired microbial activity and enhance the degradation process. Rhizodegradation offers several advantages over traditional remediation methods. It is a cost-effective approach that minimizes environmental impact and has the potential for long-term effectiveness (Gabriele et al. 2021). By utilizing the natural processes and interactions between plants and microorganisms, rhizodegradation avoids the need for expensive and potentially harmful chemical interventions. However, the efficiency of rhizodegradation is influenced by various factors. The specific contaminants present in the soil, the plant species selected, and the environmental conditions all play crucial roles in determining the success of the process. It is essential to carefully evaluate the soil composition and contaminant profile to identify the most suitable plant species and associated microorganisms for effective remediation (Cheng et al. 2017). Different plant species and their unique root-zone microorganisms possess varying abilities to degrade specific types of pollutants and adapt to diverse environmental conditions.

6 Approaches for Enhancing Cadmium Phytoremediation

Phytoremediation can be improved by employing various methods and techniques, which aimed at improving phytoremediation through soil amelioration, plant performance, tolerance, and accumulation properties.

6.1 Soil Amelioration

The mobility of Cd in soil can be influenced or increased by different methods and treatments such as the use of specific chemicals, surfactants, soil amendments, and addition of biological enhancers like bacteria, fungi, or intercropping. To address Cd mobility, certain chemicals and surfactants like ethylene diamine tetraacetic acid (EDTA), urea, ethylenediamine disuccinic acid (EDDS), and citric acid are employed (Miller 1995; Elkhatib et al. 2001). Among these, EDTA is widely recognized as a chemical chelator that enhances the solubility of Cd, facilitating its uptake and translocation to the shoots (Usman and Mohamed 2009). However, the interaction between EDTA and pH is intricate, as the effectiveness of EDTA in extracting cations depends on the soil pH. Additionally, the efficiency of EDTA treatment may vary depending on soil characteristics, with calcium-rich soil rapidly depleting EDTA due to calcite dissolution (Malaviya and Singh 2012). Citric acid has shown promise as an effective mobilizer of Cd when used in smaller dosages, facilitating the process of phytoextraction (Gomes et al. 2016). The beneficial effects of citric acid on Cd uptake by plants are likely attributed to its ameliorative effects. Studies have observed changes in root structure and shape, as well as the activation of ATPases in the root plasma membrane, leading to alterations in ion transport and an increased uptake of Cd, in both the symplast and apoplast of plants grown in citric acid-amended soils (Clemens et al. 2002; Gomes et al. 2016). Similarly, humic acids have shown potential for enhancing phytoextraction when supplemented to the soil but at higher pH levels. Their carboxyl and hydroxyl functional groups play a crucial role in the transportation, bioavailability, and solubility of heavy metals (Li and Gong 2021). In the context of sustainable bioavailability improvement, the application of acidified manure can be utilized to enhance phytoextraction efficiency (Ashraf et al. 2022). Numerous studies have demonstrated the potential of microorganisms in phytoremediation by influencing the bioavailability of Cd (Rajkumar et al. 2010). Some soil bacteria are capable of transforming metals into soluble and bioavailable forms through the production of siderophores (Rajkumar et al. 2010). Various bacterial groups, such as *Pseudomonas* sp., *Microbacterium* sp., *Bacillus* sp., *Rahnella* sp., *Burkholderia* sp., and *Enterobacter* sp., have been identified to exhibit such activity (Mishra et al. 2021). Certain strains of these bacteria have shown resistance to Cd and can be utilized to enhance the tolerance and accumulation capacities of hyperaccumulating plants, thereby improving phytoremediation efficiency.

6.2 Improvement in Plant Capacities

To enhance phytoremediation efficiency specifically for cadmium (Cd) remediation, increasing plant biomass production is a straightforward approach that has shown positive results (Subašić et al. 2022). This can be accomplished through various means, including soil amendments or direct modifications to the plants themselves. One effective method to promote plant biomass growth is seed pre-treatment, commonly referred to as priming, prior to sowing. Priming entails subjecting seeds to controlled rehydration (imbibition) that stimulates metabolic activity without initiating radicle emergence. By priming seeds, their physiological state can be altered in a way that promotes better germination and subsequent plant growth (Farooq et al. 2019). This technique has been demonstrated to enhance the overall phytoremediation capacity of plants for Cd-contaminated sites. Through priming, seeds are pre-conditioned to optimize their potential for nutrient uptake and tolerance to Cd stress. The controlled rehydration process activates various metabolic pathways and enzyme systems within the seeds, facilitating better utilization of resources and promoting early seedling vigor (Farooq et al. 2019). Seed priming techniques can vary depending on the plant species and the desired outcomes. Common approaches include soaking seeds in water, nutrient solutions, or specific chemical solutions for a specific duration. The priming duration and conditions are carefully determined to strike a balance between promoting seed viability and avoiding excessive germination, as the objective is to induce metabolic activity without triggering radical emergence (Hussian et al. 2013; Dutta 2018). Seed priming also has the potential to improve plant establishment in Cd-contaminated areas, leading to higher biomass production and subsequently enhancing the remediation rates. Indeed, while seed priming has shown promise in enhancing phytoremediation efficiency, there is still a limited understanding of how it specifically affects tolerance levels and the underlying mechanisms by which seed priming induces changes in metabolism that can be transferred to growing plants under heavy metal stress. Further research is needed to delve deeper into these aspects and gain a comprehensive understanding of the processes involved.

6.3 Genetic Engineering

Genetic engineering of plants holds great potential for the development of advanced, efficient, and robust hyperaccumulators that can effectively remediate heavy metal contamination. When selecting candidate plants for genetic engineering, those with high biomass production and existing capabilities for heavy metal accumulation are typically preferred (McIntyre 2003). Through genetic engineering techniques, genes can be manipulated to induce overexpression and enhance heavy metal accumulation in plants. One such gene is glutamylcysteine synthetase (gshl), which has been shown to increase heavy metal accumulation (Fulekar et al. 2009; Subašić et al. 2022). For instance, in *Brassica juncea*, the introduction of the gshl gene from *Escherichia coli* resulted in higher concentrations of phytochelatins, glutathione, and nonprotein thiols, leading to increased heavy metal tolerance (Saxena and Misra 2010; de Mello-Farias et al. 2011). Another genetic engineering approach involves introducing the gene for nicotianamine synthase (HcNAS1) from *Hordeum vulgare* into Arabidopsis, which stimulates heavy metal accumulation (Subašić et al. 2022).

Similarly, the incorporation of the metallothionein gene (IlMt2a) from Iris lactea into the Arabidopsis genome enhances Cd tolerance (Subašić et al. 2022). Manipulation of genes involved in phytochelatin synthesis, such as phytochelatin synthetase and γ-glutamylcysteine synthetase, can also result in enhanced heavy metal tolerance (Yadav 2010). For example, transgenic tobacco plants (*Nicotiana tabacum* and *Nicotiana glauca*) overexpressing these genes have shown higher Cd accumulation (Das et al. 2016). Furthermore, recent studies have demonstrated the improvement of heavy metal tolerance in plants through the overexpression of metallothionein transgenes (Sunitha et al. 2013). Despite the potential benefits, genetic manipulation and genetic transformation face significant challenges, particularly in public acceptance. There are concerns and fear surrounding genetically modified organisms, and the process of introducing such plants into open fields is lengthy and complex.

7 Challenges and Difficulties

7.1 Lack of Rapid Application of Phytoremediation Techniques

Phytoremediation, although a promising approach, can indeed be a slow process, especially for certain metals that may require years or even centuries to be effectively remediated. The presence and adaptability of foreign substances in the soil, as well as the ability of plants to retain and accumulate them, can limit the uptake of heavy metals (Tangahu et al. 2011). Another factor that impacts phytoremediation is the extent and depth of root growth of selected plants. If the root systems do not cover the entire contaminated area, it can hinder the overall effectiveness of the remediation process (Simek et al. 2018; Subašić et al. 2022). These challenges in terms of time and spatial coverage can pose significant obstacles in achieving satisfactory results in a reasonable timeframe. To address these limitations and expedite the remediation process, a combination of phytoremediation with other remediation techniques may be necessary. Integrating phytoremediation with methods such as soil amendments, chemical treatments, or physical removal can enhance the overall efficiency and speed up the remediation process (Subašić et al. 2022). However, it is important to carefully consider the cost and time implications of such combined approaches. The rate of biomass production by plants and the accumulation rate of toxins are crucial factors that determine the efficacy and duration of phytoremediation (Cheng et al. 2017). The amount of biomass produced by plants directly influences their capacity to remove or stabilize pollutants in the soil. Additionally, the rate at which toxins accumulate in plant tissues determines how quickly they can be eliminated from the contaminated site. However, it is worth noting that the slow growth rate of many hyperaccumulating plants can limit their effectiveness in achieving rapid phytoremediation (Qiu et al. 2008).

7.2 Removal of Contaminated Biomass

Phytoremediation is an environmentally friendly and cost-effective approach that can be used alone or in combination with other remediation techniques. However, one challenge of phytoextraction is the production of contaminated biomass, which must be properly disposed of to prevent additional environmental pollution (Subašić et al. 2022). It is essential to consider the environmental impacts not only of the phytoremediation process itself but also of the disposal of the contaminated biomass (Rai 2008). There are six standard methods for disposing of contaminated biomass: composting, landfilling, pyrolysis, leaching, cremation, and direct disposal. Each method has its advantages and disadvantages, and the most suitable disposal method should be carefully assessed based on the specific circumstances of each site (Tangahu et al. 2011). Soil treatment involves the use of microorganisms to degrade contaminants in the biomass, while compaction aims to compress the biomass into a dense and stable material. Pyrolysis, on the other hand, is a thermal degradation process that converts biomass into solid, liquid, and gas fractions (Li et al. 2015). Leaching is a disposal method that relies on the movement of soluble heavy metals through a medium, allowing them to migrate away from the contaminated biomass. On the other hand, cremation involves subjecting the biomass to high temperatures, resulting in the thermal degradation of the organic matter and the formation of treatable-metal containing ash. However, it is important to note that direct disposal of contaminated biomass is not recommended due to the potential environmental risks associated with releasing the contaminants into the environment without proper treatment or containment measures. Considering the complexities and potential environmental impacts associated with the disposal of contaminated biomass, further research is necessary to develop effective and environmentally friendly methods specifically tailored for the proper disposal of biomass generated from phytoremediation (Dai et al. 2017).

8 Conclusion

Cadmium excess in soils is a significant concern due to its potential threat to human health through the contamination of the food chain. The increased mobility of Cd in soil, exacerbated by global climate change, poses a risk of Cd leakage into underground freshwater reservoirs. Therefore, effective remediation strategies for Cd-contaminated soils are necessary, and phytoremediation, particularly phytoextraction, is an eco-friendly and cost-effective method for Cd removal. Phytoremediation relies on plant species with the ability to hyperaccumulate Cd, but the number of known hyperaccumulators is relatively small. Ongoing research aims to identify new hyperaccumulating plant species and develop methods to enhance the efficiency of phytoremediation. One such method is seed priming, which involves treating plant seeds to enhance their tolerance and accumulation potential without negatively affecting soil properties. Seed priming offers a safe and eco-friendly approach to improve plant performance in Cd-contaminated soil.

References

Abdel-Salam AA, Salem HM, Abdel-Salam MA, Seleiman MF (2015) Phytochemical removal of heavy metal-contaminated soils. In: Sherameti I, Varma A (eds) Heavy metal contamination of soils. Soil biology, vol 44. Springer, Cham. https://doi.org/10.1007/978-3-319-14526-6_16

Abdel-Shafy HI, Mansour MS (2018) Phytoremediation for the elimination of metals, pesticides, PAHs, and other pollutants from wastewater and soil. In: Phytobiont and ecosystem restitution. Springer, pp 101–136

Alaboudi KA, Ahmed B, Brodie G (2018) Phytoremediation of Pb and Cd contaminated soils by using sunflower (Helianthus annuus) plant. Ann Agric Sci 63(1):123–127

AL-Huqail AA (2022) Biochar derived from cow bones and corn stalks reduced the release of Cd and Pb and the human health risk index of quinoa grown in contaminated soils. J Soil Sci Plant Nutr 22:4024–4034. https://doi.org/10.1007/s42729-022-01005-5

Ali SY, Banerjee SN, Chaudhury S (2016) Phytoextraction of cadmium and lead by three vegetable-crop plants. Plant Sci Today 3:298–303

Ali S, Abbas Z, Rizwan M, Zaheer IE, Yavaş İ, Ünay A et al (2020a) Application of floating aquatic plants in phytoremediation of heavy metals polluted water: a review. Sustainability 12(5):1927

Ali S, Abbas Z, Seleiman MF, Rizwan M, Yavaş İ, Alhammad BA, Shami A, Hasanuzzaman M, Kalderis D (2020b) Glycine betaine accumulation, significance and interests for heavy metal tolerance in plants. Plants 9:896. https://doi.org/10.3390/plants9070896

Ali MM, Hossain D, Khan MS, Begum M, Osman MH, Nazal MK, Zhao H (2021) Environmental pollution with heavy metals: a public health concern. IntechOpen, pp 1–20

Anderson TA, Guthrie EA, Walton BT (1993) Bioremediation in the rhizosphere. Environ Sci Technol 27(13):2630–2636

Andrade Júnior WV, de Oliveira Neto CF, Santos Filho BGD, do Amarante CB, Cruz ED, Okumura RS et al (2019) Effect of cadmium on young plants of Virola surinamensis. AoB Plants 11(3):plz022

Angelova VR, Perifanova-Nemska MN, Uzunova GP, Kolentsova EN (2016) Accumulation of heavy metals in safflower (Carthamus tinctorius L.). Int J Agric Biosyst Eng 10(7):410–415

Aparicio JD, Raimondo EE, Saez JM, Costa-Gutierrez SB, Alvarez A, Benimeli CS, Polti MA (2022) The current approach to soil remediation: a review of physicochemical and biological technologies, and the potential of their strategic combination. J Environ Chem Eng 10(2): 107141

Asare MO, Száková J, Tlustoš P (2023) The fate of secondary metabolites in plants growing on Cd-, As-, and Pb-contaminated soils—a comprehensive review. Environ Sci Pollut Res 30(5): 11378–11398

Ashraf S, Ahmad SR, Ali Q, Ashraf S, Majid M, Zahir ZA (2022) Acidified Cow dung-assisted phytoextraction of heavy metals by ryegrass from contaminated soil as an eco-efficient technique. Sustainability 14(23):15879

Awa SH, Hadibarata T (2020) Removal of heavy metals in contaminated soil by phytoremediation mechanism: a review. Water Air Soil Pollut 231(2):47

Bhargava M, Singh V (2017) Phytomining: principles and applications. In: Biotechnology. CRC Press, pp 141–159

Bhat SA, Bashir O, Haq SAU, Amin T, Rafiq A, Ali M et al (2022) Phytoremediation of heavy metals in soil and water: an eco-friendly, sustainable and multidisciplinary approach. Chemosphere 303:134788

Bilandžija N, Zgorelec Ž, Pezo L, Grubor M, Velaga AG, Krička T (2022) Solid biofuels properties of Miscanthus X giganteus cultivated on contaminated soil after phytoremediation process. J Energy Inst 101:131–139

Bolan N, Srivastava P, Kirkham MB (2022) Microbial responses to immobilization of potentially toxic elements in soils. In: Appraisal of Metal(loids) in the Ecosystem. Elsevier, pp 315–330

Borah P, Kumar M, Devi P (2020) Types of inorganic pollutants: metals/metalloids, acids, and organic forms. In: Inorganic pollutants in water. Elsevier, pp 17–31

Boros-Lajszner E, Wyszkowska J, Kucharski J (2020) Application of white mustard and oats in the phytostabilisation of soil contaminated with cadmium with the addition of cellulose and urea. J Soils Sediments 20:931–942

Bouida L, Rafatullah M, Kerrouche A, Qutob M, Alosaimi AM, Alorfi HS, Hussein MA (2022) A review on cadmium and lead contamination: sources, fate, mechanism, health effects and remediation methods. Water 14(21):3432

Chaturvedi R, Varun M, Paul MS (2016) Phytoremediation: uptake and role of metal transporters in some members of Brassicaceae. In: Phytoremediation: management of environmental contaminants, vol 3. Springer, pp 453–468

Cheng M, Zeng G, Huang D, Yang C, Lai C, Zhang C, Liu Y (2017) Advantages and challenges of Tween 80 surfactant-enhanced technologies for the remediation of soils contaminated with hydrophobic organic compounds. Chem Eng J 314:98–113

Clemens S, Schroeder JI, Degenkolb T (2001) Caenorhabditis elegans expresses a functional phytochelatin synthase. Eur J Biochem 268(13):3640–3643

Clemens S, Palmgren MG, Krämer U (2002) A long way ahead: understanding and engineering plant metal accumulation. Trends Plant Sci 7(7):309–315

Cui ZG, Ahmed K, Zaidi SF, Muhammad JS (2021) Ins and outs of cadmium-induced carcinogenesis: mechanism and prevention. Cancer Treat Res Commun 27:100372

Curie C, Cassin G, Couch D, Divol F, Higuchi K, Le Jean M et al (2009) Metal movement within the plant: contribution of nicotianamine and yellow stripe 1-like transporters. Ann Bot 103(1): 1–11

Dai H, Wei S, Twardowska I, Han R, Xu L (2017) Hyperaccumulating potential of Bidens pilosa L. for Cd and elucidation of its translocation behavior based on cell membrane permeability. Environ Sci Pollut Res 24:23161–23167

DalCorso G, Farinati S, Maistri S, Furini A (2008) How plants cope with cadmium: staking all on metabolism and gene expression. J Integr Plant Biol 50(10):1268–1280

Das N, Bhattacharya S, Maiti MK (2016) Enhanced cadmium accumulation and tolerance in transgenic tobacco overexpressing rice metal tolerance protein gene OsMTP1 is promising for phytoremediation. Plant Physiol Biochemy 105:297–309

de Mello-Farias PC, Chaves ALS, Lencina CL (2011) Transgenic plants for enhanced phytoremediation–physiological studies. In: Genetic transformation. Intech, Rijeca, pp 305–328

Deng Y, Ren C, Chen N, Huang Y, Zhu G, Zhang X et al (2023) Effects of pH and phosphate on cadmium adsorption onto goethite and a paddy soil: experiments and NOM-CD model. J Soils Sediments 23(5):2072–2082

Dhankher OP, Pilon-Smits EA, Meagher RB, Doty S (2012) Biotechnological approaches for phytoremediation. In: Plant biotechnology and agriculture. Academic Press, pp 309–328

Dhanwal P, Kumar A, Dudeja S, Chhokar V, Beniwal V (2017) Recent advances in phytoremediation technology. Adv Environ Biotechnol 2017:227–241

Dias MC, Monteiro C, Moutinho-Pereira J, Correia C, Gonçalves B, Santos C (2013) Cadmium toxicity affects photosynthesis and plant growth at different levels. Acta Physiol Plant 35:1281–1289

Dubey R, Gupta DK, Sharma GK (2020) Chemical stress on plants. In: New frontiers in stress management for durable agriculture. Springer, pp 101–128

Duruibe JO, Ogwuegbu MOC, Egwurugwu JN (2007) Heavy metal pollution and human biotoxic effects. Int J Phys Sci 2(5):112–118

Dutta P (2018) Seed priming: new vistas and contemporary perspectives. In: Advances in seed priming. Springer, pp 3–22
El Kholy S, Giesy JP, Al Naggar Y (2021) Consequences of a short-term exposure to a sub lethal concentration of CdO nanoparticles on key life history traits in the fruit fly (Drosophila melanogaster). J Hazard Mater 410:124671
Elkhatib EA, Thabet AG, Moharem KL (2001) Sorption of cadmium in surfactant-amended soils. Arid Land Res Manag 15(4):385–394
El-Naggar A, Chang SX, Cai Y, Lee YH, Wang J, Wang SL et al (2021) Mechanistic insights into the (im)mobilization of arsenic, cadmium, lead, and zinc in a multi-contaminated soil treated with different biochars. Environ Int 156:106638
Etim EE (2012) Phytoremediation and its mechanisms: a review. Int J Environ Bioenergy 2(3): 120–136
Faller P, Kienzler K, Krieger-Liszkay A (2005) Mechanism of Cd2+ toxicity: Cd2+ inhibits photoactivation of Photosystem II by competitive binding to the essential Ca2+ site. Biochim Biophys Acta 1706(1–2):158–164
Fan HL, Zhou W (2009) Screening of amaranth cultivars (Amaranthus mangostanus L.) for cadmium hyperaccumulation. Agric Sci China 8(3):342–351
Fan KC, His HC, Chen CW, Lee HL, Hseu ZY (2011) Cadmium accumulation and tolerance of mahogany (Swietenia macrophylla) seedlings for phytoextraction applications. J Environ Manag 92:2818–2822
Farooq M, Usman M, Nadeem F, ur Rehman H, Wahid A, Basra SM, Siddique KH (2019) Seed priming in field crops: potential benefits, adoption and challenges. Crop Pasture Sci 70(9): 731–771
Farooq A, Nadeem M, Abbas G, Shabbir A, Khalid MS, Javeed HMR et al (2020) Cadmium partitioning, physiological and oxidative stress responses in marigold (Calendula calypso) grown on contaminated soil: Implications for phytoremediation. Bull Environ Contam Toxicol 105:270–276
Fulekar MH, Singh A, Bhaduri AM (2009) Genetic engineering strategies for enhancing phytoremediation of heavy metals. Afr J Biotechnol 8(4)
Gabriele I, Race M, Papirio S, Esposito G (2021) Phytoremediation of pyrene-contaminated soils: a critical review of the key factors affecting the fate of pyrene. J Environ Manag 293:112805
Gallego SM, Pena LB, Barcia RA, Azpilicueta CE, Iannone MF, Rosales EP et al (2012) Unravelling cadmium toxicity and tolerance in plants: insight into regulatory mechanisms. Environ Exp Bot 83:33–46
Gautam K, Sharma P, Dwivedi S, Singh A, Gaur VK, Varjani S et al (2023) A review on control and abatement of soil pollution by heavy metals: emphasis on artificial intelligence in recovery of contaminated soil. Environ Res 115592
Ghosh M, Singh SP (2005) A review on phytoremediation of heavy metals and utilization of it's by products. Asian J Energy Environ 6(4):18
Gomes MA, Hauser-Davis RA, de Souza AN, Vitória AP (2016) Metal phytoremediation: general strategies, genetically modified plants and applications in metal nanoparticle contamination. Ecotoxicol Environ Saf 134:133–147
Guo JK, Zhao J, Ren XH, Jia HL, Muhammad H, Lv X et al (2020) Effects of Burkholderia sp. D54 on growth and cadmium uptake of tomato, ryegrass and soybean plants. Int J Environ Sci Technol 17:1149–1158
Haider FU, Liqun C, Coulter JA, Cheema SA, Wu J, Zhang R et al (2021) Cadmium toxicity in plants: Impacts and remediation strategies. Ecotoxicol Environ Saf 211:111887
Huang J, Yang Z, Li J, Liao MA, Lin L, Wang J et al (2017) Cadmium accumulation characteristics of floricultural plant Cosmos bipinnata. Chem Ecol 33(9):807–816
Hussein M, Yoneda K, Mohd-Zaki Z, Amir A, Othman N (2021) Heavy metals in leachate, impacted soils and natural soils of different landfills in Malaysia: an alarming threat. Chemosphere 267:128874

Hussian I, Ahmad R, Farooq M, Wahid A (2013) Seed priming improves the performance of poor quality wheat seed. Int J Agric Biol 15(6)

Igiri BE, Okoduwa SI, Idoko GO, Akabuogu EP, Adeyi AO, Ejiogu IK (2018) Toxicity and bioremediation of heavy metals contaminated ecosystem from tannery wastewater: a review. J Toxicol 2018:2568038

Imran K, Seleiman MF, Chattha MU et al (2021) Enhancing antioxidant defense system of mung bean with a salicylic acid exogenous application to mitigate cadmium toxicity. Not Botan Hort Agrob Cluj-Napoca 49(2):12303–12303. https://doi.org/10.15835/nbha49212303

Irfan M, Hayat S, Ahmad A, Alyemeni MN (2013) Soil cadmium enrichment: allocation and plant physiological manifestations. Saudi J Biol Sci 20(1):1–10

Ishimaru Y, Takahashi R, Bashir K, Shimo H, Senoura T, Sugimoto K et al (2012) Characterizing the role of rice NRAMP5 in manganese, iron and cadmium transport. Sci Rep 2(1):286

Ismael MA, Elyamine AM, Moussa MG, Cai M, Zhao X, Hu C (2019) Cadmium in plants: uptake, toxicity, and its interactions with selenium fertilizers. Metallomics 11(2):255–277

Jadia CD, Fulekar MH (2008) Phytoremediation: the application of vermicompost to remove zinc, cadmium, copper, nickel and lead by sunflower plant. Environ Eng Manag J 7(5)

Jadia CD, Fulekar MH (2009) Phytoremediation of heavy metals: recent techniques. Afr J Biotechnol 8(6)

Jinadasa N, Collins D, Holford P, Milham PJ, Conroy JP (2016) Reactions to cadmium stress in a cadmium-tolerant variety of cabbage (Brassica oleracea L.): is cadmium tolerance necessarily desirable in food crops? Environ Sci Pollut Res 23:5296–5306

Kamusoko R, Jingura RM (2017) Utility of Jatropha for phytoremediation of heavy metals and emerging contaminants of water resources: a review. Clean Soil Air Water 45(11):1700444

Karalija E, Selović A (2018) The effect of hydro and proline seed priming on growth, proline and sugar content, and antioxidant activity of maize under cadmium stress. Environ Sci Pollut Res 25(33):33370–33380

Karalija E, Selovic A, Dahija S, Demir A, Samardžic J, Vrobel O, Zeljkovic SC, Paric A (2021) Use of seed priming to improve Cd accumulation and tolerance in Silene sendtneri, novel Cd hyper-accumulator. Ecotoxicol Environ Saf 210:111882

Khalid S, Shahid M, Niazi NK, Murtaza B, Bibi I, Dumat C (2017) A comparison of technologies for remediation of heavy metal contaminated soils. J Geochem Explor 182:247–268

Khatun J, Intekhab A, Dhak D (2022) Effect of uncontrolled fertilization and heavy metal toxicity associated with arsenic (As), lead (Pb) and cadmium (Cd), and possible remediation. Toxicology 477:153274

Kishor R, Purchase D, Saratale GD, Saratale RG, Ferreira LFR, Bilal M et al (2021) Ecotoxicological and health concerns of persistent coloring pollutants of textile industry wastewater and treatment approaches for environmental safety. J Environ Chem Eng 9(2):105012

Kleber M, Bourg IC, Coward EK, Hansel CM, Myneni SC, Nunan N (2021) Dynamic interactions at the mineral–organic matter interface. Nat Rev Earth Environ 2(6):402–421

Krämer U (2010) Metal hyperaccumulation in plants. Annu Rev Plant Biol 61:517–534

Krishnamurti GSR, Cieslinski G, Huang PM, Van Rees KCJ (1997) Kinetics of cadmium release from soils as influenced by organic acids: implication in cadmium availability. J Environ Qual 26(1):271–277

Kubota H, Takenaka C (2003) Field note: Arabis gemmifera is a hyperaccumulator of Cd and Zn. Int J Phytoremediation 5(3):197–201

Kumar V, Shahi SK, Singh S (2018) Bioremediation: an eco-sustainable approach for restoration of contaminated sites. In: Microbial bioprospecting for sustainable development. Springer, pp 115–136

Kumar A, Devi S, Agrawal H, Singh S, Singh J (2020) Rhizoremediation: a unique plant microbiome association of biodegradation. Plant Microbe Symbiosis 2020:203–220

Kumar A, Subrahmanyam G, Mondal R, Cabral-Pinto MMS, Shabnam AA, Jigyasu DK et al (2021) Bio-remediation approaches for alleviation of cadmium contamination in natural resources. Chemosphere 268:128855

Küpper H, Parameswaran A, Leitenmaier B, Trtílek M, Šetlík I (2007) Cadmium-induced inhibition of photosynthesis and long-term acclimation to cadmium stress in the hyperaccumulator Thlaspi caerulescens. New Phytol 175(4):655–674

Lamkarkach F, Ougier E, Garnier R, Viau C, Kolossa-Gehring M, Lange R, Apel P (2021) Human biomonitoring initiative (HBM4EU): human biomonitoring guidance values (HBM-GVs) derived for cadmium and its compounds. Environ Int 147:106337

Lasota J, Błońska E, Łyszczarz S, Tibbett M (2020) Forest humus type governs heavy metal accumulation in specific organic matter fractions. Water Air Soil Pollut 231:1–13

Lee NW, Wang HY, Du CL, Yuan TH, Chen CY, Yu CJ, Chan CC (2022) Air-polluted environmental heavy metal exposure increase lung cancer incidence and mortality: a population-based longitudinal cohort study. Sci Total Environ 810:152186

Li Y, Gong X (2021) Effects of dissolved organic matter on the bioavailability of heavy metals during microbial dissimilatory iron reduction: a review. Rev Environ Contam Toxicol 257:69–92

Li Y, Hu X, Song X, Hou Y, Sun L (2015) Phytoextraction potential of Solanum nigrum L. and Beta vulgaris L. Var. cicla L. in Cd-contaminated water. Pol J Environ Stud 24(4)

Liang C, Xiao-Hua L, Zhang ZH, Zheng XT, Rengel Z, Zhao-Pu L (2011) Cadmium accumulation and translocation in two Jerusalem artichoke (Helianthus tuberosus L.) cultivars. Pedosphere 21(5):573–580

Liu Z, He X, Chen W, Yuan F, Yan K, Tao D (2009) Accumulation and tolerance characteristics of cadmium in a potential hyperaccumulator—Lonicera japonica Thunb. J Hazard Mater 169(1–3): 170–175

Liu L, Li Y, Tang J, Hu L, Chen X (2011) Plant coexistence can enhance phytoextraction of cadmium by tobacco (Nicotiana tabacum L.) in contaminated soil. J Environ Sci 23:453–460

Liu Y, Xiao T, Zhu Z, Ma L, Li H, Ning Z (2021) Geogenic pollution, fractionation and potential risks of Cd and Zn in soils from a mountainous region underlain by black shale. Sci Total Environ 760:143426

Liu Y, Chen Z, Xiao T, Zhu Z, Jia S, Sun J et al (2022) Enrichment and environmental availability of cadmium in agricultural soils developed on Cd-rich black shale in southwestern China. Environ Sci Pollut Res 29(24):36243–36254

Luo J, Qi S, Peng L, Wang J (2016) Phytoremediation efficiency of Cd by Eucalyptus globulus transplanted from polluted and unpolluted sites. Int J Phytoremediation 18(4):308–314

Luo H, Gu R, Ouyang H, Wang L, Shi S, Ji Y et al (2021) Cadmium exposure induces osteoporosis through cellular senescence, associated with activation of NF-κB pathway and mitochondrial dysfunction. Environ Pollut 290:118043

Luo Y, Wang Z, Zhang ZL, Huang FY, Jia WJ, Zhang JQ, Feng XY (2023) Characteristics and source analysis of potentially toxic elements pollution in atmospheric fallout around non-ferrous metal smelting slag sites—taking southwest China as an example. Environ Sci Pollut Res 30(3): 7813–7824

Madejon P, Maranon T, Navarro-Fernandez CM, Dominguez MT, Alegre JM, Robinson B, Murillo JM (2017) Potential of Eucalyptus camaldulensis for phytostabilization and biomonitoring of trace-element contaminated soils. PLoS One 12(6):e0180240

Madhav S, Ahamad A, Singh AK, Kushawaha J, Chauhan JS, Sharma S, Singh P (2020) Water pollutants: sources and impact on the environment and human health. In: Sensors in water pollutants monitoring: role of material. Springer, pp 43–62

Mahajan P, Kaushal J (2018) Role of phytoremediation in reducing cadmium toxicity in soil and water. J Toxicol 2018

Malaviya P, Singh A (2012) Phytoremediation strategies for remediation of uranium-contaminated environments: a review. Crit Rev Environ Sci Technol 42(24):2575–2647

McGrath SP, Zhao FJ (2003) Phytoextraction of metals and metalloids from contaminated soils. Curr Opin Biotechnol 14(3):277–282

McGuinness M, Dowling D (2009) Plant-associated bacterial degradation of toxic organic compounds in soil. Int J Environ Res Public Health 6(8):2226–2247

McIntyre T (2003) Phytoremediation of heavy metals from soils. Adv Biochem Eng Biotechnol 2003:97–123

Mendoza-Cózatl DG, Butko E, Springer F, Torpey JW, Komives EA, Kehr J, Schroeder JI (2008) Identification of high levels of phytochelatins, glutathione and cadmium in the phloem sap of Brassica napus. A role for thiol-peptides in the long-distance transport of cadmium and the effect of cadmium on iron translocation. Plant J 54(2):249–259

Mendoza-Cózatl DG, Jobe TO, Hauser F, Schroeder JI (2011) Long-distance transport, vacuolar sequestration, tolerance, and transcriptional responses induced by cadmium and arsenic. Curr Opin Plant Biol 14(5):554–562

Miller RM (1995) Biosurfactant-facilitated remediation of metal-contaminated soils. Environ Health Perspect 103(suppl 1):59–62

Mishra S, Lin Z, Pang S, Zhang Y, Bhatt P, Chen S (2021) Biosurfactant is a powerful tool for the bioremediation of heavy metals from contaminated soils. J Hazard Mater 418:126253

Moiseenko TI, Gashkina NA (2020) Distribution and bioaccumulation of heavy metals (Hg, Cd and Pb) in fish: Influence of the aquatic environment and climate. Environ Res Lett 15(11):115013

Murakami M, Ae N, Ishikawa S (2007) Phytoextraction of cadmium by rice (Oryza sativa L.), soybean (Glycine max (L.) Merr.), and maize (Zea mays L.). Environ Pollut 145(1):96–103

Muthusaravanan S, Sivarajasekar N, Vivek JS, Paramasivan T, Naushad M, Prakashmaran J et al (2018) Phytoremediation of heavy metals: mechanisms, methods and enhancements. Environ Chem Lett 16:1339–1359

Mwegoha WJ (2008) The use of phytoremediation technology for abatement soil and groundwater pollution in Tanzania: opportunities and challenges. J Sustain Dev Africa 10(1):140–156

Nedjimi B (2021) Phytoremediation: a sustainable environmental technology for heavy metals decontamination. SN Appl Sci 3(3):286

Nedjimi B, Daoud Y (2009) Cadmium accumulation in Atriplex halimus subsp. schweinfurthii and its influence on growth, proline, root hydraulic conductivity and nutrient uptake. Flora 204(4): 316–324

Nordberg GF, Åkesson A, Nogawa K, Nordberg M (2022) Cadmium. In: Handbook on the toxicology of metals. Academic Press, pp 141–196

Olguín EJ, García-López DA, González-Portela RE, Sánchez-Galván G (2017) Year-round phytofiltration lagoon assessment using Pistia stratiotes within a pilot-plant scale biorefinery. Sci Total Environ 592:326–333

Pan P, Lei M, Qiao P, Zhou G, Wan X, Chen T (2019) Potential of indigenous plant species for phytoremediation of metal (loid)-contaminated soil in the Baoshan mining area, China. Environ Sci Pollut Res 26:23583–23592

Perrone GG, Tan SX, Dawes IW (2008) Reactive oxygen species and yeast apoptosis. Biochim Biophys Acta 1783(7):1354–1368

Phaenark C, Pokethitiyook P, Kruatrachue M, Ngernsansaruay C (2009) Cd and Zn accumulation in plants from the Padaeng zinc mine area. Int J Phytoremediation 11(5):479–495

Pinto AP, De Varennes A, Fonseca R, Teixeira DM (2015) Phytoremediation of soils contaminated with heavy metals: techniques and strategies. In: Phytoremediation: management of environmental contaminants, vol 1. Springer, pp 133–155

Priya AK, Muruganandam M, Ali SS, Kornaros M (2023) Clean-up of heavy metals from contaminated soil by phytoremediation: a multidisciplinary and eco-friendly approach. Toxics 11(5):422

Priyadarshanee M, Mahto U, Das S (2022) Mechanism of toxicity and adverse health effects of environmental pollutants. In: Microbial biodegradation and bioremediation. Elsevier, pp 33–53

Puthran D, Patil D (2023) Usage of heavy metal-free compounds in surface coatings. J Coat Technol Res 20(1):87–112

Qiu RL, Zhao X, Tang YT, Yu FM, Hu PJ (2008) Antioxidative response to Cd in a newly discovered cadmium hyperaccumulator, Arabis paniculata F. Chemosphere 74(1):6–12

Rai PK (2008) Phytoremediation of Hg and Cd from industrial effluents using an aquatic free floating macrophyte Azolla pinnata. Int J Phytoremediation 10(5):430–439

Rai GK, Bhat BA, Mushtaq M, Tariq L, Rai PK, Basu U et al (2021) Insights into decontamination of soils by phytoremediation: a detailed account on heavy metal toxicity and mitigation strategies. Physiol Plant 173(1):287–304

Rajendran S, Priya TAK, Khoo KS, Hoang TK, Ng HS, Munawaroh HSH et al (2022) A critical review on various remediation approaches for heavy metal contaminants removal from contaminated soils. Chemosphere 287:132369

Rajkumar M, Ae N, Prasad MNV, Freitas H (2010) Potential of siderophore-producing bacteria for improving heavy metal phytoextraction. Trends Biotechnol 28(3):142–149

Ramesh B, Saravanan A, Kumar PS, Yaashikaa PR, Thamarai P, Shaji A, Rangasamy G (2023) A review on algae biosorption for the removal of hazardous pollutants from wastewater: limiting factors, prospects and recommendations. Environ Pollut 121572

Ranjan A, Sinha R, Sharma TR, Pattanayak A, Singh AK (2021) Alleviating aluminum toxicity in plants: implications of reactive oxygen species signaling and crosstalk with other signaling pathways. Physiol Plant 173(4):1765–1784

Riaz M, Kamran M, Rizwan M, Ali S, Parveen A, Malik Z, Wang X (2021) Cadmium uptake and translocation: selenium and silicon roles in Cd detoxification for the production of low Cd crops: a critical review. Chemosphere 273:129690

Rucińska-Sobkowiak R (2016) Water relations in plants subjected to heavy metal stresses. Acta Physiol Plant 38:1–13

Salama FM, AL-Huqail AA, Ali M, Abeed AHA (2022) Cd Phytoextraction potential in halophyte *Salicornia fruticosa*: Salinity impact. Plants 11:2556. https://doi.org/10.3390/plants11192556

Salem HM, Abdel-Salam A, Abdel-Salam MA, Seleiman MF (2018) Phytoremediation of metal and metalloids from contaminated soil. In: Hasanuzzaman M, Nahar K, Fujita M (eds) Plants under metal and metalloid stress. Springer, Singapore. https://doi.org/10.1007/978-981-13-2242-6_9

Sangeeta M, Maiti SK (2010) Phytoremediation of metal enriched mine waste: a review. American-Eurasian J Agric Environ Sci 9(5):560–575

Sarkar A, Ravindran G, Krishnamurthy V (2013) A brief review on the effect of cadmium toxicity: from cellular to organ level. Int J Biotechnol Res 3(1):17–36

Sarvestani RA, Aghasi M (2022) Exposure and health risk assessment of PHEs (Pb, Cd, Cu, Fe) in agricultural soil in Kerman City, Southeast of Iran

Saxena P, Misra N (2010) Remediation of heavy metal contaminated tropical land. In: Soil heavy metals. Springer, pp 431–477

Seleiman MF, Kheir AM (2018) Maize productivity, heavy metals uptake and their availability in contaminated clay and sandy alkaline soils as affected by inorganic and organic amendments. Chemosphere 204:514–522

Seleiman MF, Al-Suhaibani N, El-Hendawy S, Abdella K, Alotaibi M, Alderfasi A (2021) Impacts of long- and short-term of irrigation with treated wastewater and synthetic fertilizers on the growth, biomass, heavy metal content, and energy traits of three potential bioenergy crops in arid regions. Energies 14:3037. https://doi.org/10.3390/en14113037

Sevimli M, Bayram D, Özgöçmen M, Armağan I, Sevimli TS (2022) Boric acid suppresses cell proliferation by TNF signaling pathway mediated apoptosis in SW-480 human colon cancer line. J Trace Elem Med Biol 71:126958

Shackira AM, Puthur JT (2019) Phytostabilization of heavy metals: understanding of principles and practices. In: Plant-metal interactions. Springer, pp 263–282

Sigel A, Sigel H, Sigel RK (eds) (2015) Metallothioneins and related chelators, vol 5. Walter de Gruyter

Sigfridsson KG, Bernát G, Mamedov F, Styring S (2004) Molecular interference of Cd2+ with Photosystem II. Biochim Biophys Acta 1659(1):19–31

Silva J, Fernandes A, Junior MS, Santos C, Lobato A (2018) Tolerance mechanisms in Cassia alata exposed to cadmium toxicity– potential use for phytoremediation. Photosynthetica 56:495–504

Simek J, Kovalikova Z, Dohnal V, Tuma J (2018) Accumulation of cadmium in potential hyperaccumulators Chlorophytum comosum and Callisia fragrans and role of organic acids under stress conditions. Environ Sci Pollut Res 25:28129–28139

Sipos P, Tóth A, Kis VK, Balázs R, Kovács I, Németh T (2019) Partition of Cd, Cu, Pb and Zn among mineral particles during their sorption in soils. J Soils Sediments 19:1775–1787

Smolders E, McLaughlin MJ (1996) Chloride increases cadmium uptake in Swiss chard in a resin-buffered nutrient solution. Soil Sci Soc Am J 60(5):1443–1447

Solís-Domínguez FA, González-Chávez MC, Carrillo-González R, Rodríguez-Vázquez R (2007) Accumulation and localization of cadmium in Echinochloa polystachya grown within a hydroponic system. J Hazard Mater 141(3):630–636

Song Y, Jin L, Wang X (2017) Cadmium absorption and transportation pathways in plants. Int J Phytoremediation 19(2):133–141

Sooksawat N, Meetam M, Kruatrachue M, Pokethitiyook P, Nathalang K (2013) Phytoremediation potential of charophytes: bioaccumulation and toxicity studies of cadmium, lead and zinc. J Environ Sci 25(3):596–604

Subašić M, Šamec D, Selović A, Karalija E (2022) Phytoremediation of cadmium polluted soils: current status and approaches for enhancing. Soil Syst 6(1):3

Suciu NA, Devivo R, Rizzati N, Capri E (2022) Cd content in phosphate fertilizer: which potential risk for the environment and human health? Curr Opin Environ Sci Health 30:100392

Sunitha MS, Prashant S, Kumar SA, Rao SK, Narasu ML, Kishor PK (2013) Cellular and molecular mechanisms of heavy metal tolerance in plants: a brief overview of transgenic plants overexpressing phytochelatin synthase and metallothionein genes. Plant Cell Biotechnol Mol Biol 14(2):33–48

Tangahu BV, Sheikh Abdullah SR, Basri H, Idris M, Anuar N, Mukhlisin M (2011) A review on heavy metals (As, Pb, and Hg) uptake by plants through phytoremediation. Int J Chem Eng 2011:939161

Tanhan P, Kruatrachue M, Pokethitiyook P, Chaiyarat R (2007) Uptake and accumulation of cadmium, lead and zinc by Siam weed [Chromolaena odorata (L.) King & Robinson]. Chemosphere 68(2):323–329

Tiryakioglu M, Eker S, Ozkutlu F, Husted S, Cakmak I (2006) Antioxidant defense system and cadmium uptake in barley genotypes differing in cadmium tolerance. J Trace Elem Med Biol 20(3):181–189

Trombulak SC, Frissell CA (2000) Review of ecological effects of roads on terrestrial and aquatic communities. Conserv Biol 14(1):18–30

Uchimiya M, Bannon D, Nakanishi H, McBride MB, Williams MA, Yoshihara T (2020) Chemical speciation, plant uptake, and toxicity of heavy metals in agricultural soils. J Agric Food Chem 68(46):12856–12869

Ueno D, Iwashita T, Zhao FJ, Ma JF (2008) Characterization of Cd translocation and identification of the Cd form in xylem sap of the Cd-hyperaccumulator Arabidopsis halleri. Plant Cell Physiol 49(4):540–548

Ullah S, Mahmood T, Iqbal Z, Naeem A, Ali R, Mahmood S (2019) Phytoremediative potential of salt-tolerant grass species for cadmium and lead under contaminated nutrient solution. Int J Phytoremediation 21(10):1012–1018

Usman ARA, Mohamed HM (2009) Effect of microbial inoculation and EDTA on the uptake and translocation of heavy metal by corn and sunflower. Chemosphere 76(7):893–899

Van der Ent A, Baker AJ, Reeves RD, Pollard AJ, Schat H (2013) Hyperaccumulators of metal and metalloid trace elements: facts and fiction. Plant Soil 362:319–334

Vangronsveld J, Herzig R, Weyens N, Boulet J, Adriaensen K, Ruttens A et al (2009) Phytoremediation of contaminated soils and groundwater: lessons from the field. Environ Sci Pollut Res 16:765–794

Viehweger K (2014) How plants cope with heavy metals. Bot Stud 55(1):1–12

Vishnoi SR, Srivastava PN (2007, October) Phytoremediation–green for environmental clean. In Proceedings of Taal2007: the 12th World lake conference, vol 1016, p 1021

Vogel-Mikuš K, Arčon I, Kodre A (2010) Complexation of cadmium in seeds and vegetative tissues of the cadmium hyperaccumulator Thlaspi praecox as studied by X-ray absorption spectroscopy. Plant Soil 331:439–451
Wang Y, Xu Y, Liang X, Sun Y, Huang Q, Peng Y (2020) Leaching behavior and efficiency of cadmium in alkaline soil by adding two novel immobilization materials. Sci Total Environ 710: 135964
Wang R, Zhu Z, Cheng W, Chang C, Song X, Huang F (2023) Cadmium accumulation and isotope fractionation in typical protozoa tetrahymena: a new perspective on remediation of Cd pollution in wastewater. J Hazard Mater 454:131517
Ware GW, Albert LA, Bro-Rasmussen F, Crosby DG, de Voogt P, Frehse H et al (2003) Role of phosphorus in (im)mobilization and bioavailability of heavy metals in the soil-plant system. Springer, New York, pp 1–44
Wei T, Lv X, Jia H, Hua L, Xu H, Zhou R et al (2018) Effects of salicylic acid, Fe(II) and plant growth-promoting bacteria on Cd accumulation and toxicity alleviation of Cd tolerant and sensitive tomato genotypes. J Environ Manag 214:164–171
Wei W, Peng H, Xie Y, Wang X, Huang R, Chen H, Ji X (2021) The role of silicon in cadmium alleviation by rice root cell wall retention and vacuole compartmentalization under different durations of Cd exposure. Ecotoxicol Environ Saf 226:112810
Wei C, Lei M, Chen T, Zhou C, Gu R (2022) Method on site-specific source apportionment of domestic soil pollution across China through public data mining: a case study on cadmium from non-ferrous industries. Environ Pollut 295:118605
White PJ (2000) Calcium channels in higher plants. Biochim Biophys Acta 1465(1–2):171–189
Wong CKE, Cobbett CS (2009) HMA P-type ATPases are the major mechanism for root-to-shoot Cd translocation in Arabidopsis thaliana. New Phytol 181(1):71–78
Xia H, Ma X (2006) Phytoremediation of ethion by water hyacinth (Eichhornia crassipes) from water. Bioresour Technol 97(8):1050–1054
Xu Z, Bai X, Li Y, Weng Y, Li F (2023) New insights into the decrease in Cd2+ bioavailability in sediments by microplastics: role of geochemical properties. J Hazard Mater 442:130103
Yadav SK (2010) Heavy metals toxicity in plants: an overview on the role of glutathione and phytochelatins in heavy metal stress tolerance of plants. South Afr J Bot 76(2):167–179
Yu Z, Tang J (2022) Quality improvement in vegetable greenhouse by cadmium pollution remediation. J Food Qual 2022(4):1–10
Yu G, Liu J, Long Y, Chen Z, Sunahara GI, Jiang P, You S, Lin H, Xiao H (2019) Phytoextraction of cadmium-contaminated soils: comparison of plant species and low molecular weight organic acids. Int J Phytoremediat 22:383–391
Yu G, Liu J, Long Y, Chen Z, Sunahara GI, Jiang P et al (2020) Phytoextraction of cadmium-contaminated soils: comparison of plant species and low molecular weight organic acids. Int J Phytoremediation 22(4):383–391
Zhang X, Zhang S, Xu X, Li T, Gong G, Jia Y et al (2010) Tolerance and accumulation characteristics of cadmium in Amaranthus hybridus L. J Hazard Mater 180(1–3):303–308
Zhang H, Yuan X, Xiong T, Wang H, Jiang L (2020) Bioremediation of co-contaminated soil with heavy metals and pesticides: influence factors, mechanisms and evaluation methods. Chem Eng J 398:125657
Zhao FJ, Jiang RF, Dunham SJ, McGrath SP (2006) Cadmium uptake, translocation and tolerance in the hyperaccumulator Arabidopsis halleri. New Phytol 172(4):646–654
Zhao D, Wang P, Zhao FJ (2023) Dietary cadmium exposure, risks to human health and mitigation strategies. Crit Rev Environ Sci Technol 53(8):939–963

Cadmium Toxicity in Plants and Its Amelioration

Jesús Rubio-Santiago, Gisela Adelina Rolón-Cárdenas, Alejandro Hernández-Morales, Jackeline Lizzeta Arvizu-Gómez, and Ruth Elena Soria-Guerra

Abstract Cadmium (Cd) is a toxic heavy metal primarily released from human industrial activities and agriculture. It contaminates soil and water and then is absorbed by the plant roots to enter the aerial tissues. Although Cd is not an essential nutrient for plants, it can enter plant cells, affecting their physiology and compromising their growth and development. However, some plant species have developed strategies to withstand Cd toxic effects, so they are classified as indicators, excluders, and hyperaccumulators plants according to their coping mechanisms. There is an increased interest in developing and applying strategies to reduce Cd concentration in the environment and avoid crop accumulation. Therefore, this chapter presents an overview of Cd effects in plants, the biochemical responses of different plants to cope with Cd stress, and advances using different compounds to reduce Cd toxicity in plants, improve phytoremediation, and avoid the accumulation of Cd in crops.

Keywords Cadmium · Toxic effect · Hyperaccumulator

J. Rubio-Santiago · R. E. Soria-Guerra
Facultad de Ciencias Químicas, Universidad Autónoma de San Luis Potosí, San Luis Potosí, San Luis Potosí, Mexico

G. A. Rolón-Cárdenas
Facultad de Estudios Profesionales Zona Huasteca, Universidad Autónoma de San Luis Potosí, Ciudad Valles, San Luis Potosí, Mexico

A. Hernández-Morales (✉)
Facultad de Ciencias Químicas, Universidad Autónoma de San Luis Potosí, San Luis Potosí, San Luis Potosí, Mexico

Facultad de Estudios Profesionales Zona Huasteca, Universidad Autónoma de San Luis Potosí, Ciudad Valles, San Luis Potosí, Mexico
e-mail: alejandro.hernandez@uaslp.mx

J. L. Arvizu-Gómez
Secretaría de Investigación y Posgrado, Centro Nayarita de Innovación y Transferencia de Tecnología (CENITT), Universidad Autónoma de Nayarit, Tepic, Nayarit, Mexico

A. K. Jha, N. Kumar (eds.), *Cadmium Toxicity Mitigation*,
https://doi.org/10.1007/978-3-031-47390-6_10

1 Introduction

Cadmium (Cd) is a heavy metal of relatively poor abundance in the earth's crust (0.1–0.5 ppm) (Sarkar et al. 2013). It is present in Zn-, Cu-, and Pb-bearing ores or can form hawleyite otavite ($CdCO_3$), monteponite (CdO), greenockite (CdS), xanthochroite ($CdS(H_2O)_x$), and cadmoselite (CdSe) minerals (Kubier et al. 2019; Moiseenko and Gashkina 2018; Sarkar et al. 2013). Although Cd minerals do not form deposits, they are released into the environment by extracting and processing zinc, copper, and lead (Moiseenko and Gashkina 2018; Sarkar et al. 2013).

Cd is used in electroplating, welding, electrical, and nuclear fission applications, producing colorants, plastic stabilizers, solders, alloys, rods, and batteries (Kubier et al. 2019; Prasad 1995). Therefore, industrial processes are the primary source of anthropogenic Cd emissions (Yuan et al. 2019). Besides, synthetic phosphate fertilizers containing Cd impurities are another cause of Cd pollution (Haider et al. 2021; Kubier et al. 2019; Prasad 1995).

Cd is not essential heavy metal for organisms, being one of the most dangerous even at low concentrations (Das et al. 1997). In humans, Cd oral ingestion causes salivation, vomiting, abdominal pain, vertigo, loss of consciousness, and painful spasm of the anal sphincter. Cd inhalation causes dry throat, cough, headache, flu-like symptoms, fever, vomiting, chest pains, acute pulmonary edema, asthma-like bronchospasm pneumonitis, muscle weakness, and leg pains (Johri et al. 2010).

Long-term Cd exposure (Rafati Rahimzadeh et al. 2017) causes hepatocyte necrosis and liver apoptosis (Godt et al. 2006) and renal damage (Johri et al. 2010). Occupational exposure to 0.006–0.015 mg of Cd per m^3 causes proteinuria and renal tubular damage in occupationally exposed personnel (Choi et al. 2020). Since kidneys are the main organs for long-term Cd accumulation, with approximately ten years half-life, it is eventually causing tubule cell necrosis (Godt et al. 2006).

Cd has been classified as a human carcinogen (Group 1) by the International Agency for Research on Cancer based on the high incidence of lung cancer in occupationally exposed workers (Hartwig 2013; Johri et al. 2010; Stayner et al. 1992). Cd induces oncogenesis in various cellular model systems (Waalkes 2003) while in the human causes prostate (Vinceti et al. 2007), bladder (Kellen et al. 2007), endometrium (Åkesson et al. 2008), breast (Gallagher et al. 2010), and stomach cancer (Kim et al. 2019), and in mice, it affects sperm quality by reducing motility and viability (Wang et al. 2017; Zhao et al. 2015).

Because of its toxicity, physicochemical and biological remediation strategies have been developed to remove Cd from contaminated sites (Changjia et al. 2021). Physical methods include barriers made with impermeable materials such as steel, cement, bentonite, and grout to avoid Cd dispersion from the polluted site to the land, water, and crops (Li et al. 2019). Soil replacement dilutes the pollutant concentration by replacing polluted soil with other non-contaminated soil (Yao et al. 2012).

Chemical approaches include soil washing by using freshwater, chelation agents, and surfactants that remove the pollutant from the soil. Besides, ion exchange, precipitation, adsorption, and chelation remove heavy metals from polluted soil (Yao et al. 2012). Chitosan, calcium oxide, phosphate salts, and bentonite have been applied in polluted soils to immobilize heavy metals by forming insoluble compounds. So, it decreases the migration of pollutants to water, plants, and other environmental matrices (Guo et al. 2006; Yao et al. 2012).

Biological techniques use plants and microorganisms to remove or immobilize hazardous contaminants from a polluted environment (Li et al. 2019). They include phytoremediation and bioremediation, which are considered eco-friendly and cost-effective strategies for heavy metal removal (Li et al. 2019; Yao et al. 2012).

Phytoremediation includes many plant strategies to remove contaminants from the soil, water, and sediments (Shen et al. 2022; Yao et al. 2012). Phytoextraction consists of adsorption, transference, and storage of heavy metals in aerial tissues (Yao et al. 2012). In rhizofiltration, heavy metals are absorbed and accumulated mainly in the plant roots and, to a lesser extent, in aerial parts (Mahajan and Kaushal 2018). Plants immobilize heavy metals in phytostabilization through adsorption, precipitation, and reduction reactions. So, plants restrict the contaminants into the rhizosphere and reduce their migration to the groundwater and food chain (Bolan et al. 2011; Yao et al. 2012). On the other hand, bioremediation uses the metabolic potential of microorganisms to clean up contaminated environments (Watanabe 2001). Microorganisms change heavy metals′ physical and chemical properties, affecting their environmental mobility (Yao et al. 2012).

Biological methods to remove Cd from polluted environments are an eco-friendly, low-cost, and effective technique that overcome the disadvantages of chemical methods. Therefore, this chapter discusses the information on the toxic effect of Cd in plants, strategies to increase phytoextraction, and new strategies to immobilize Cd to avoid entering the food chain.

2 Mineral Nutrients and Cadmium Uptake by Plants

Plant roots assimilate nitrogen (N) (nitrates or ammonium), phosphorus (P) (phosphate), potassium (K), calcium (Ca), sulfur (sulfate), magnesium (Mg) (macronutrients), iron (Fe), zinc (Zn), copper (Cu), boron (B), manganese (Mn), molybdenum (Mo), chloride (Cl), nickel (Ni), and sodium (Na) (micronutrients) from the soil (Hopkins and Hüner 2009; Kumar et al. 2021; Jacoby and Moran 2001). Nutrients are transferred from the root surface to the xylem (Sondergaard et al. 2004) by diffusing between extracellular spaces and cell walls via the apoplastic pathway (Ricachenevsky et al. 2018), which allows mineral nutrients to reach the xylem in young roots (Lux et al. 2011), while in mature root tissue, the Casparian band makes the passage of ions and small hydrophilic molecules to the cortex and stele difficult (Sterckeman and Thomine 2020). Therefore, nutrients must cross the plasma membrane of the cortical or endodermal cells via a symplastic pathway (Hopkins and Hüner 2009).

Specific proteins transport solutes into the cell against a mineral concentration gradient using the electrochemical gradient established by H^+-ATPases (Reid and Hayes 2003; Sondergaard et al. 2004). The transporter proteins that allow the entrance of specific ions or molecules are classified into channels and carriers (Mitra 2015; Reid and Hayes 2003). Channels act as selective pores that let molecules or ions diffuse across the membrane (Taiz and Zeiger 2002). In contrast, carrier proteins bind a solute and deliver it into the cytoplasm (White 2017). Carriers are classified as uniporters, symporters (cotransporters), and antiporters (exchangers) (Mitra 2015). Uniporters transport only one type of ion by diffusion across the membrane. Symporters or cotransporters move two substances in the same direction, and the antiporters exchange ions across the membrane (Mitra 2015; White 2017).

Ca and K uptake occurs via channels, while Cl^-, NO_3^-, SO_4^{2-}, $H_2PO_4^-$, and HPO_4^{2-} are actively taken up by anion/proton symport systems operating at the plasma membrane (Barbier-Brygoo et al. 2000; Reid and Hayes 2003; Yang and Jie 2005). It is hypothesized that the non-selective Ca channels also allow the intake of other divalent cations and traces of heavy metals (Reid and Hayes 2003). In solution, Cd occurs as Cd^{2+}, so it has been suggested that Cd competes with divalent cations for the same transmembrane carriers and crosses via non-selective cation channels, such as depolarization-activated calcium channels (DACC), hyperpolarization-activated calcium channels (HACC), and voltage-insensitive cation channels (VICC) (Andresen and Küpper 2013; Chen et al. 2018; Qin et al. 2020).

In mineral deficiency, plants express other transporter systems such as zinc-regulated transporters (ZRT) and iron-regulated transporter proteins (IRT) (Eide et al. 1996; Grotz et al. 1998; Reid and Hayes 2003). IRT1 mediates the uptake of Fe^{2+} and other divalent cations such as Mn^{2+}, Zn^{2+}, and Co^{2+} (Korshunova et al. 1999), while ZRT is involved in transporting Zn^{2+} and other transition metal cations such as Mn^{2+}, Fe^{2+}, Co^{2+}, Cu^{2+}, and Ni^{2+} (Ajeesh Krishna et al. 2020; Pedas et al. 2009). In addition, Cd uptake has been related to these two metal transporter systems (Benavides et al. 2005).

Natural resistance-associated macrophage proteins (NRAMP) are transporters involved in the cation uptake and transport of metal into the vacuoles and the xylem. NRAMP were identified in the phagosomes of infected mouse macrophages, bacteria, fungi, plants, and animals (Hall and Williams 2003). NRAMP functional studies showed that it could be involved in the transport of Mn^{2+}, Zn^{2+}, Cu^{2+}, Fe^{2+}, Cd^{2+}, Ni^{2+}, Co^{2+}, and Al^{3+} (Nevo and Nelson 2006; Qin et al. 2020; Xia et al. 2010). Once Cd enters root cells, it is transported to the vacuole by NRAMP (Lux et al. 2011). Since the loss of function of vacuolar transporters causes an increase in the translocation of ions to the shoot, it is suggested that root vacuoles control the trace element concentrations in the root symplast and the metal movement into the xylem (Ricachenevsky et al. 2018).

In addition, Cd can be loaded into the xylem by heavy metal P_{1B}-ATPases and possibly also by YSL (Yellow-Stripe 1-Like) proteins, a family of transporter proposed to participate in metal uptake and long-range transport in model plants

(Feng et al. 2017; Lux et al. 2011; Song et al. 2022). The P-type ATPases form a large family of more than 50 membrane proteins responsible for the active transport of various cations across cell membranes (Hall and Williams 2003; Lutsenko and Kaplan 1995). P_{1B}-ATPases, or heavy metal ATPases (HMAs), are a subgroup of this family that transport heavy metals (Cu^{+}, Cu^{2+}, Zn^{2+}, Co^{2+}) across biological membranes. Because of the chemical similarities among transition metals, these pumps can transport non-physiological substrates; for instance, Cu^{+}-ATPases transport Ag^{+} (Cu/Ag ATPases), while Zn^{2+-}ATPases can transport Cd^{2+} and Pb^{2+} (Zn/Co/Cd/Pb ATPases) (Argüello et al. 2007; Qin et al. 2020).

The cation diffusion facilitator (CDF)/metal tolerance protein (MTP) family is another specialized family of transporters implicated in metal homeostasis in plant cells which compartmentalize or efflux metal ions, maintaining cytosolic concentrations within a narrow range (Menguer et al. 2013).

3 Toxic Effects of Cadmium in Plants

Reactive oxygen species (ROS) such as hydroxyl radical ($HO^{\bullet}$), superoxide radical ($O_2^{\bullet-}$), or hydrogen peroxide (H_2O_2) are produced in organelles such as plastids, mitochondria, chloroplast, and peroxisomes during cell respiration and photosynthesis. ROS are involved in defense against infectious agents and cellular signaling pathways. However, high levels of ROS cause oxidative stress, damaging DNA, membrane lipids, and proteins (Bailey-Serres and Mittler 2006; Benavides et al. 2005; Ercal et al. 2001; Jomova et al. 2012; Lin et al. 2007; Martins et al. 2011; Pinto et al. 2017).

Micronutrients (metallic ions) are essential in plant nutrition. However, they produce hydroxyl radicals by Fenton or Haber Weiss reactions generating oxidative stress (Mendoza-Cózatl et al. 2011; Thomine and Vert 2013). Redox-active micronutrients such as iron (Fe) and copper (Cu) and heavy metals including chromium (Cr), vanadium (V), and cobalt (Co) induce hydroxyl radical formation by H_2O_2 decomposition (Jomova et al. 2012; Valko et al. 2005). Meanwhile, mercury (Hg), cadmium (Cd), and nickel (Ni), considered redox-inactive metals, deplete antioxidant systems such as glutathione and antioxidant enzymes, causing oxidative stress (Cuypers et al. 2010; Nevo and Nelson 2006; Valko et al. 2005).

Cd has been shown to cause DNA damage, leaf lipidic peroxidation, and root H_2O_2 accumulation in *Vicia faba* (Lin et al. 2007). In plant cells, Cd binds to sulfhydryl groups of proteins, replaces the essential ions in metalloenzymes, and depletes the reduced glutathione (GSH) pool resulting in a disturbance of the redox balance (Ahmad et al. 2019; Cuypers et al. 2010). The accumulation of high levels of ROS causes membrane lipid peroxidation, mtDNA cleavage, and mitochondrial damage, inducing apoptosis and aging of cells (Cuypers et al. 2010).

Furthermore, Cd binds to thiol and carboxylic groups, causing misfolding of cytosolic proteins such as enzymes, transporters, and regulatory proteins (Van Assche and Clijsters 1990). Cd decreases enzymatic activities of antioxidant enzymes, including catalase (CAT), ascorbate peroxidase (APOX), and superoxide dismutase (SOD) (Haider et al. 2021; Lin et al. 2007; Palma et al. 2013). Cd displaces metal cofactors like Zn, Fe, and Ca of these metalloproteins causing the loss of catalytic function (Das et al. 1997; Wan and Zhang 2012).

Cd decreases nutrient availability by inhibiting soil enzyme activity (Das et al. 1997) or reducing the soil microbes population (Benavides et al. 2005). Soil enzymes play a prominent role in the decomposition of organic matter and nutrient recycling by oxidation (dehydrogenase, hydrolase, glucosidase) and mineralization (protease, amidase, urease, phosphatase, and sulfatase) (Inamdar et al. 2022). Moreover, Cd interferes with the uptake, transport, and use of several elements (Ca, Mg, P, and K) and water by plants (Das et al. 1997), probably by competing for transporters or by downregulating transporter gene expression. Therefore, Cd reduces plant growth by interfering with the mechanisms involved in soil nutrient uptake (Andresen and Küpper 2013).

Cd is accumulated in the plant roots, and only a small amount is translocated to the shoots, whereby it causes chlorosis, leaf rolls, and stunting (Andresen and Küpper 2013; Benavides et al. 2005; Das et al. 1997). Leaf chlorosis could be caused by a deficiency of nitrogen, potassium, magnesium, or iron (Li et al. 2021b). Iron is part of the catalytic group for many redox enzymes involved in photosynthesis, respiration, and nitrogen fixation. In addition, iron is required to synthesize chlorophyll, and its deficiency leads to a simultaneous loss of chlorophyll and degeneration of chloroplast structure (Hopkins and Hüner 2009). Another proposed mechanism of Cd-induced damage to the photosynthetic system is the substitution of the central Mg^{2+} in chlorophylls with Cd^{2+}. This replacement forms a chlorophyll unsuitable for photosynthesis since Cd can reduce electron flow through PS II and enhance non-photochemical exciton quenching. Substituting Ca^{2+} by Cd^{2+} in the water-splitting complex also inhibits the PS II reaction (Andresen and Küpper 2013).

4 Vegetal Strategies to Cope with Cd Toxic Effects

Plant root exudates are the initial strategy to counteract Cd toxicity (Song et al. 2017). Exudates can mobilize or immobilize Cd from the soil according to chemical composition. Pinto et al. (2008) demonstrated that malate and citrate exuded from the roots of sorghum and maize reduce the Cd bioavailability by inducing Cd–organic acid complexation.

Rhizospheric acidification increases Cd bioavailability and uptake by the plant (Shahid et al. 2017). Oxalic acid, malic acid, and tartaric acid are the predominant organic acids in the rhizosphere of *Sedum alfredii* exposed to Cd. Among them, tartaric acid significantly increased the solubility of Cd (Tao et al. 2020). Moreover, other organic compounds such as mucilage, proteins, sugars, amino acids, and mono and di-saccharides are found in root exudates of Cd-exposed plants (Lapie et al. 2019). In sunflowers (*Helianthus annuus* L.), root exudates include proteins, polysaccharides, and phenolic compounds with strong Cd adsorption ability (Yang and Pan 2013). In addition, Yao et al. (2022) found that amino acids reduce the bioavailability of Cd.

Cd immobilization in the cell wall is another vegetal strategy to avoid its toxicity. The cell wall is a good biosorbent for heavy metals that limits their entrance into the intracellular space (Parrotta et al. 2015). Subcellular fractionation of Cd-contained *Bechmeria nivea* tissues indicated that 48.2–61.9% of Cd is adsorbed in cell walls, 30.2–38.1% in the soluble fraction, and trace amounts in cellular organelles (Wang et al. 2008). Similarly in *Brassica parachinensis* L. Cd was associated with the cell wall fraction, 55% in the root cell walls and 47% in the shoot cell walls, followed by 34% in the soluble fraction and 12–20% in organelles (Qiu et al. 2011). These results indicate that the cell wall prevents the uptake and transport of Cd (Qin et al. 2020). The primary cell wall comprises cellulose, hemicellulose, pectin, some functional proteins, and several aromatic compounds (Chebli and Geitmann 2017; Xiao et al. 2020). The presence of different functional groups (e.g., carboxyl and hydroxyl) favors ion-exchange mechanisms with the cell wall counter-ions (Benavides et al. 2005; Parrotta et al. 2015; Sanità di Toppi and Gabbrielli 1999). Meyer et al. (2015) studied the cadmium-induced cell wall modifications in the metal hyperaccumulator *Arabidopsis halleri* and concluded that Cd increases pectin polysaccharide levels in plant cell walls.

Once within plant cells, Cd shows a high affinity for thiols and binds to S-containing ligands such as glutathione (GSH), metallothioneins (MTs), and phytochelatins (PCs) (Cuypers et al. 2010; Lux et al. 2011). Glutathione (-γ-glutamyl-cysteinyl-glycine) has a dual role; it can help deal with oxidative stress and form complexes with Cd (Cuypers et al. 2010). Delalande et al. (2010) studied the structure of GSH complexes. They found that the formation of $Cd(GS)_2$ complexes is spontaneous and rapid at ambient temperature, pH from 5.6 to 7.2, and GSH concentration close to physiological conditions. In addition, computational studies have proven glutathione´s coordination properties with Cd^{2+} (Belcastro et al. 2009; Liu et al. 2014).

Metallothioneins (MTs) are small cysteine-rich proteins involved in the homeostasis of Zn^{2+} and Cu^{+} and the detoxification of heavy metals (Freisinger and Vašák 2013; Zimeri et al. 2005). Initially, MTs were identified in the horse kidney cortex, and their elemental composition showed a content of 2.9% cadmium, 0.6% zinc, and 4.1% sulfur per gram of protein dried weight (Kägi and Vallee 1960; Margoshes and Vallee 1957). It has been proposed that MTs form 7 metal-thiolate complexes for each of 20 cysteine residues (Bernhard et al. 1983; Kägi et al. 1984). In addition to protection against toxic metals and their homeostasis, MTs are involved in cell growth and proliferation regulation, free radical scavenging, protection against oxidative stress, DNA repair, and antiapoptotic defense (Joshi et al. 2016).

Phytochelatins (PCs) are other cysteine-rich peptides involved in the chelation and sequestration of heavy metals (Benavides et al. 2005; Merlos et al. 2014). PCs are enzymatically synthesized from the tripeptide glutathione (GSH), producing a general structure of (γ-Glu-Cys)n-Gly, where n varies from 2 to 11 (PC2, PC3…PC11) (Merlos et al. 2014). The complexes Cd-PCs (PC2–PC6) structure has been studied regarding their complex stoichiometries. It has been found that Cd^{2+} interacts with PC2 and PC3 to form bis complexes of a CdL_2 fashion, where L is the ligand molecule. PC4–PC6 form mono complexes (CdL) at sub-equimolar metal concentration and form more complicated clustered species when the metal ion concentration exceeds the PC concentration. In the case of PC4, the complex formed is $Cd_3(PC4)_2$ and dimeric complexes for PC5 and PC6 (Wątły et al. 2021).

Hossain et al. (2012) conducted a comparative proteomic study to unravel the protein networks involved in Cd stress response in two soybean cultivars (a high and a low Cd-accumulator cultivar). The leaf proteome analysis revealed that Cd enhances the expression of chaperones in the high Cd accumulating cultivar, suggesting that refolding is a defense mechanism to maintain cellular homeostasis and fix misfolded proteins induced by Cd. Likewise, overexpression of metallochaperone OsHIPP56 reduces Cd toxicity and Cd accumulation in rice (Zhao et al. 2022).

Heavy metals chelation by cysteine-rich MTs and PCs forms non-toxic complexes compartmentalized in vacuoles to avoid free circulation in the cytosol and prevent their potential damage (Ahmad et al. 2019; Benavides et al. 2005; Clemens 2006; Cobbett 2000; Luo and Zhang 2021; Wan and Zhang 2012). Park et al. (2012) showed that ABCC-type phytochelatin transporters, AtABCC1 and AtABCC2, are essential for vacuolar Cd sequestration. They found that single *atabcc1* or double knockout mutants (*atabcc1 atabcc2*) exhibit a hypersensitive phenotype in the presence of Cd^{2+} and Hg^{2+}, while AtABCC1 overexpression in *Arabidopsis thaliana* enhanced Cd^{2+} tolerance and accumulation.

On the other hand, plants have antioxidant defense systems to avoid oxidative stress and restore redox balance (Loix et al. 2017). For instance, superoxide dismutase (SOD), catalase (CAT), ascorbate peroxidase (APOX), dehydroascorbate reductase (DHAR), monodehydroascorbate reductase (MDHAR), glutathione reductase (GR), and proteins such as thioredoxin and peroxiredoxin are critical antioxidant systems that decrease oxidative stress by Cd (Ahmad et al. 2019). Therefore, enzymatic and non-enzymatic antioxidant defense systems reduce oxygen-reactive species (Fryzova et al. 2018; Hasanuzzaman et al. 2020). The enzymatic antioxidant defense initiates with the superoxide dismutase (SOD) that converts $O_2^{\bullet-}$ into H_2O_2. Then, peroxidase (POD), catalase (CAT), ascorbate peroxidase (APX), and glutathione peroxidase (GPX) convert the H_2O_2 generated by the SOD into H_2O (Hasanuzzaman et al. 2020). SOD, POD, and CAT are inhibited by Cd stress (Hassan et al. 2020), while ascorbate peroxidase (APX), monodehydroascorbate reductase (MDHAR), dehydroascorbate reductase (DHAR), and glutathione reductase (GR) are enhanced during Cd stress (Li et al. 2022b).

In plant cells, the ascorbate and glutathione (AsA, and GSH) are non-enzymatic antioxidants that are implied in redox-coupled reactions whose primary function is the scavenging of H_2O_2 (Hasanuzzaman et al. 2020; Hernández et al. 2017). Ascorbate peroxidase (APX) and glutathione peroxidase (GPX) use the AsA and GSH to reduce H_2O_2 to water with the formation of oxidized glutathione (GSSG) and monodehydroascorbate (MDHA). To restore the reductive power of oxidized forms, monodehydroascorbate reductase (MDHAR), dehydroascorbate reductase

(DHAR), and glutathione reductase (GR) regenerate ASA and GSH by using nicotinamide adenine dinucleotide phosphate (NADPH). Dehydroascorbic acid (DHA) produced from MDHA can be reduced to AsA by the dehydroascorbate reductase (DHAR) and the electrons provided by GSH. This cycle is known as the AsA-GSH cycle, and it is the primary antioxidant pathway to detoxify H_2O_2 (Hasanuzzaman et al. 2020; Hernández et al. 2017; Li et al. 2010b). Qin et al. (2018) found that Cd increases the concentrations of MDA, GSH, and AsA and the activities of antioxidant enzymes (SOD, POD, and CAT) in the leaves and roots of wheat seedlings. They suggested that wheat tolerates Cd by enhancing the antioxidant enzyme activities and increasing the concentration of ascorbate and glutathione. Similar results were observed by Li et al. (2022b) in honeysuckle leaves (*Lonicera japonica* Thunb.) exposed to Cd.

Non-enzymatic scavengers like tocopherols, carotenoids, and phenolic acids help to restore the redox balance (Benavides et al. 2005; Haider et al. 2021; Hasanuzzaman et al. 2020). Manquián-Cerda et al. (2016) evaluated the effect of Cd^{2+} on the production of *Vaccinium corymbosum* L. phenolic compounds. They found that applying different Cd concentrations to the blueberry plantlet culture increases phenolic compound contents, which may be involved in Cd detoxification.

5 Response of Hyperaccumulator and No Accumulator Plants to Cd

Cd distribution in the vegetal tissue depends on plant species, genotype, and cultivar (Rizwan et al. 2017; Shamsi et al. 2014; Sterckeman and Thomine 2020; Wang et al. 2015). Plants can be divided into indicators, excluders, and hyperaccumulators according to their capacity to uptake, transport, and accumulate heavy metals in their tissues (Baker 1981).

Indicator plants are sensitive to heavy metals. The shoots´ internal heavy metal content is proportional to the contaminant concentration in the soil, water, or sediments. Because of this, they are used as indicators of heavy metal pollution in the soil (Baker 1981).

Excluder plants contain low heavy metal concentrations in their shoot regarding the concentration found in the soil. However, the mechanism breaks down, and the accumulation in the shoot increases when the plant is exposed to high heavy metal concentrations; even in this condition, the roots exceed the heavy metal concentration regarding the content of the above tissues (Baker 1981; Seregin and Kozhevnikova 2004). Excluder plants avoid metal accumulation by blocking the uptake in the roots or by active (energy-dependent) efflux pumps (Leitenmaier and Küpper 2013).

Hyperaccumulator plants tolerate high heavy metal concentrations and accumulate them in their above-ground parts (Baker 1981; Leitenmaier and Küpper 2013). Despite the heavy metal accumulation in their tissues, hyperaccumulator plants carry out normal physiological functions without showing visible stress symptoms when grown under excess metal concentrations (Sytar et al. 2021). Brooks et al. (1977) used the concept "Nickel hyperaccumulator" to refer to plants that accumulate more than 1000 mg Ni/kg dry weight (DW) in their shoots. Concerning Cd hyperaccumulator plants, they must accumulate more than 100 mg Cd/kg DW in the aerial tissue (Chaney et al. 1997; Farooqi et al. 2022; Reeves et al. 2018; Wei et al. 2005). The detoxification efficiency of hyperaccumulator plants depends on how quickly they immobilize metal ions from the cytoplasm into vacuoles or cell walls in aerial tissues without compromising their metabolic function (Sytar et al. 2021). Cd accumulation has been proposed as a vegetal strategy to avoid herbivores (Jiang et al. 2005). *Arabidopsis halleri*, *Sedum alfredii*, *Sedum plumbizincicola*, *Thlaspi caerulescens*, and *Solanum nigrum* are the most studied Cd hyperaccumulators. However, other vegetal species have been characterized as hyperaccumulators (Table 1).

Basic heavy metal tolerance mechanisms are observed in many plant species, but most are non-tolerant to extreme metal concentrations in soils and waters (Sytar et al. 2021). Comparative studies have been carried out to evaluate the response to Cd in hyperaccumulators and non-accumulator plants. Ebbs et al. (2002) evaluated the role of phytochelatins (PCs) in the Cd tolerance in *T. caerulescens* and non-accumulator *T. arvense*. They demonstrated that PCs are not involved in the *T. caerulescens* response to metal stress. Although both species have a positive correlation between PCs levels and Cd concentration in their tissue, the total PC levels in the *T. caerulescens* (hyperaccumulator) were lower than in *T. arvense* (non-accumulator).

High Cd translocation in hyperaccumulator plants from the roots to the aerial tissue is correlated with low Cd sequestration into the root vacuoles and efficient transport to the shoot via the xylem (Rascio and Navari-Izzo 2011). Some studies have evaluated the chemical nature of the metal ligands implied in the Cd complexation. Huguet et al. (2012) studied the Cd speciation in *Arabidopsis halleri* exposed to Cd. The results showed that this heavy metal was predominantly bound to COOH/OH groups belonging to organic acids and/or cell wall components in the leaves, and less than 25% was bound to thiol groups. Similarly, a comparative study between *A. halleri* and *A. lyrata*, a Cd hyperaccumulator and non-hyperaccumulator plant, respectively, showed that in *A. lyrata*, the Cd was coordinated mainly to sulfur compounds. While in *A. halleri*, the amount of Cd-oxygen ligands was higher than Cd-sulfur ligands (Isaure et al. 2015). These results suggest that Cd strongly binds to ligands, such as sulfur compounds, in non-hyperaccumulator plants. In contrast, the Cd in hyperaccumulators binds to weak ligands, such as oxygen compounds, instead of strong sulfur ligands (Sytar et al. 2021). The mechanisms implied in the speciation of Cd could be different in other species. Studies in *Solanum nigrum* and *Solanum melongena* showed that Cd in roots and fresh leaves is mainly bound to thiol ligands. However, it is also suggested that Cd bound to O ligands is a transient form in younger leaves that could be associated with the translocation and transport of Cd from the roots to the leaves, where Cd would eventually be stored bound to S ligands (Pons et al. 2021).

Table 1 Vegetal species cataloged as Cd hyperaccumulators and their Cd-accumulation characteristics

Number	Cd-hyperaccumulator plant	Medium	Cd treatment	Cd concentration in plant tissues (mg Cd kg^{-1} D.W.)			References
				Leaves/ shoots	Stem	Root	
1	*Thlaspi caerulescens*	Nutrient solution	56 mg L^{-1}	14187	NR	28051	Lombi et al. (2000)
2	*Sedum alfredii*	Nutrient solution	22 mg L^{-1}	7800	5600	2300	Yang et al. (2004)
3	*Solanum nigrum* L.	Soil	50 mg kg^{-1}	356	387	337	Sun et al. (2008)
4	*Sedum plumbizincicola*	Nutrient solution	11 mg L^{-1}	7010	NR	840	Cao et al. (2014)
5	*Arabidopsis halleri*	Nutrient solution	11 mg L^{-1}	~4000	NR	12800	Zhao et al. (2006)
6	*Solanum photeinocarpum*	Soil	100 mg kg^{-1}	292	372	845	Zhang et al. (2011)
7	*Sphagneticola calendulacea*	Soil	100 mg kg^{-1}	294	297	194	Lu et al. (2020b)
8	*Phytolacca americana*	Nutrient solution	11 mg L^{-1}	637	NR	1185	Liu et al. (2010)
9	*Thlaspi praecox*	Nutrient solution	5.6 mg L^{-1}	~3000	NR	~2500	Tolrà et al. (2006)
10	*Rorippa globosa (Turcz.) Thell*	Soil	100 mg kg^{-1}	433	411	142	Sun et al. (2011)
11	*Viola Baoshanensis*	Nutrient solution	50 mg L^{-1}	4825	NR	~2500	Liu et al. (2004)
12	*Amaranthus hypochondriacus L.*	Soil	100 mg kg^{-1}	300	NR	NR	Xie et al. (2019)
13	*Picris divaricata*	Nutrient solution	11 mg L^{-1}	1482	NR	17419	Hu et al. (2012)
14	*Arabidopsis arenosa*	Nutrient solution	22.5 mg L^{-1}	7140	NR	NR	Szopiński et al. (2020)
15	*Pfaffia glomerata*	Nutrient solution	10.1 mg L^{-1}	~200	NR	~1200	Pedrosa-Gomes et al. (2013)
16	*Microsorum pteropus*	Nutrient solution	56 mg L^{-1}	890	NR	1117	Lan et al. (2018)
17	*Lantana camara L.*	Soil	200 mg kg^{-1}	423	392	293	Liu et al. (2019)
18	*Abelmoschus manihot*	Soil	100 mg kg^{-1}	185	126	210	Wu et al. (2018)
19	*Brassica juncea*	Nutrient solution	0.6 mg L^{-1}	~300	NR	~1500	Salt et al. (1995)

D.W., dry weight; *NR*, data is not reported

Under Cd stress, the behavior of hyperaccumulator and non-hyperaccumulator plants differs in morphological and physiological responses (Andresen and Küpper 2013; He et al. 2017). However, it has been proposed that particular genes are not involved in this feature; instead, hyperaccumulation depends on the regulation and expression of genes present in both hyperaccumulator and non-hyperaccumulator plants (Rascio and Navari-Izzo 2011). The constitutive high expression of genes related to metal uptake, transport, synthesis of metal ligands, and oxidative stress responses has been suggested as the molecular mechanisms of metal tolerance and hyperaccumulation (Sytar et al. 2021).

6 Alleviation of Cd Stress and Strategies to Diminish or Increase Cd Accumulation in Plants

Cd in soil and water disturbs the growth of plants and reduces crop yield (Huang et al. 2022). Plant growth and development require a proper balance of inorganic nutrients under optimal and stressful environmental conditions. It is known that the optimal supply of mineral nutrients alleviates Cd effects in plants and avoids its entry into the food chain (Nazar et al. 2012; Qin et al. 2020; Sarwar et al. 2010). Since mineral nutrient deficiency increases Cd uptake (Mubeen et al. 2023; Nakanishi et al. 2006; Qin et al. 2020), the supply of Fe (Jian et al. 2019), B (Riaz et al. 2021), Ca (Rahman et al. 2016), Mo (Imran et al. 2020), Se (Saidi et al. 2014), Si (Li et al. 2018; Thind et al. 2020; Vaculík et al. 2009), and Zn (Hassan et al. 2005) inhibit Cd accumulation in several plants probably by competing with membrane transporters (Sarwar et al. 2010).

Bacteria also contribute to plant nutrient acquisition through phosphate solubilization, nitrogen fixation, and siderophore production for Fe acquisition (Ahemad and Kibret 2014). Plant growth-promoting rhizobacteria (PGPR) are soil bacteria colonizing plant rhizosphere, whereby they secrete regulatory molecules that promote plant growth and development (Vocciante et al. 2022). PGPR contribute to withstand salinity, drought, and diseases by producing extracellular polymeric substances (EPS) (Saha et al. 2020). EPS are biopolymers composed of polysaccharides, proteins, and DNA that adsorb Cd, limiting its availability for the plant (Costa et al. 2018; Wang et al. 2020; Wei et al. 2011). Gu et al. (2023) applied EPS isolated from *Bacillus* sp. on rice seedlings exposed to Cd. They demonstrated that 100 mg EPS/L reduced oxidative damage and Cd accumulation and increased rice biomass.

Exogenous application of plant growth regulators (PGRs), such as auxins, gibberellins, abscisic acid, jasmonic acid, brassinosteroids, and melatonin, is another strategy to reduce Cd toxicity, since phytohormones play critical roles in helping plants to adapt to adverse environmental conditions and mediate defense responses (Asgher et al. 2015; Han et al. 2016; Verma et al. 2016).

Abscisic acid (ABA) is involved in the plant response to abiotic stress and is vital in mitigating Cd^{2+} toxicity in herbaceous species (Han et al. 2016). ABA application on lettuce and *Arabidopsis thaliana* decreases Cd accumulation in their tissues (Fan et al. 2014; Tang et al. 2020). Likewise, ABA application in *Arabidopsis thaliana* inhibits IRT1 transporters, reducing Cd uptake in plants exposed to Cd (Fan et al. 2014). Besides, ABA diminished oxidative stress in *P. euphratica* cells exposed to Cd by increasing antioxidant enzyme activities such as catalase, superoxide dismutase, and ascorbate peroxidase (Han et al. 2016), while in *Brassica campestris* seedlings exposed to Cd, ABA alleviates Cd toxicity by reducing the reactive oxygen species (Shen et al. 2017).

Salicylic and jasmonic acid are associated with plant responses to biotic stress because their content in vegetal tissue increases when pathogens attack plants (Verma et al. 2016). Besides, salicylic and jasmonic acid effects have been demonstrated on Cd stress alleviation. Agami and Mohamed (2013) demonstrated that seed treatment with salicylic acid increases the superoxide dismutase (SOD), catalase (CAT), and peroxidase (POX) activity, promotes the growth of wheat plants exposed to Cd, and decreases Cd content in roots and leaves. In addition, seed pre-treatments (priming) with salicylic acid before sowing increase seedling vigor and biomass production (Subašić et al. 2022). Similar results in Cd accumulation were found in rice seedlings. These results indicated that salicylic acid promotes pectin and lignin synthesis, strengthening the cell wall and increasing Cd deposition, preventing its entrance into the vegetal tissue (Pan et al. 2021). In contrast with these findings, in *Medicago sativa* the treatment with salicylic acid increased the Cd content in plants exposed to Cd (Drazic et al. 2006).

Exogenous application of jasmonic acid or methyl jasmonate on rice (Li et al. 2022c), faba bean (Ahmad et al. 2017), soybean (Keramat et al. 2010), and *Arabidopsis thaliana* (Lei et al. 2020) alleviates Cd stress by regulating antioxidant response. In faba bean, external supplementation of jasmonic acid minimized the accumulation of Cd in roots, shoots, and leaves (Ahmad et al. 2017). A similar response was observed in *Arabidopsis thaliana*. Moreover, exogenous methyl jasmonate (MeJA) application downregulates *AtIRT1, AtHMA2,* and *AtHMA4* genes involved in Cd uptake and long-distance translocation, respectively. These results could explain the decrement of the Cd concentration in the root cell sap and shoot (Lei et al. 2020).

Brassinosteroids (BRs) are steroid hormones essential in plant growth, metabolism, and stress alleviation (Manghwar et al. 2022). Li et al. (2022a) demonstrated that 2,4-epibrassinolide application on Cd-exposed grape plants reduces oxidative stress by decreasing the superoxide anions ($O2^{\cdot-}$), hydrogen peroxide (H_2O_2), and malondialdehyde (product of lipid peroxidation), and increases antioxidant compounds and guaiacol peroxidase (POD) and superoxide dismutase (SOD) activities. Similarly, 2,4-epibrassinolide decreases oxidative stress in *Solanum nigrum* plants exposed to Cd and reduces the Cd content in roots and leaves (Peng et al. 2020).

Exogenous application of gibberellic acid (GA) on Cd-exposed rice plants decreases Cd content in roots and leaves. This effect is due to the downregulation of transporter genes involved in Cd absorption (*OsNRAMP5* and *OsCd1*) and the upregulation of transporters responsible for vacuolar Cd sequestering and Cd efflux outside the cell (*OsHMA3* and *OsCAL1*, respectively) (Liu et al. 2022). In *Vigna radiata*, foliar application of gibberellic acid alleviates oxidative stress and decreases the Cd concentration in the root and shoot (Sadiq et al. 2021). Gibberellic acid treatment decreases Cd content in the roots and downregulates *IRT1* in *Arabidopsis thaliana* exposed to Cd. In addition, gibberellic acid decreases the nitric oxide production induced by Cd^{2+} in roots and leaves of *A. thaliana* exposed to Cd (Zhu et al. 2012).

Melatonin (MT) phytohormone regulates plant growth and development, circadian rhythms, photoperiodic responses, antioxidation, and stress resistance (Zhang and Zhang 2021). MT application decreases Cd content in Cd-exposed *Oryza sativa* (Huang et al. 2023), *Fragaria x ananassa* (Saqib et al. 2023), *Camelia sinensis* (Tan et al. 2022), *Triticum aestivum* (Kaya et al. 2019), *Solanum lycopersicum* (Altaf et al. 2022), *Nicotiana tabacum* (Song et al. 2022), *Brassica chinensis* (Wang et al. 2021), *Brassica juncea* (Zargar et al. 2022), *Carthamus tinctorius L.* (Amjadi et al. 2021), *Malus baccata,* and *Malus micromalus* (He et al. 2020). In *Oryza sativa*, low Cd content is associated with the downregulation of metal transporter genes (*OsNRAMP1, OsNRAMP5, OsIRT1*, and *OsCd1*) implicated in Cd intake and upregulation of *OsCAL1* and *OsHMA3* involved in the expulsion of Cd from the cell and vacuolar chelation of Cd. Moreover, MT treatment increases the cell´s wall hemicellulose level and then increases the cell wall´s Cd binding capacity (Huang et al. 2023). Similarly, in *Malus baccata* and *Malus micromalus* MT application downregulates *HA7, NRAMP1, NRAMP3, HMA4, PCR2, NAS1, MT2, ABCC1*, and *MHX* genes involved in uptake, transport, and detoxification of Cd (He et al. 2020). In Chinese cabbage *(Brassica chinensis)*, the expression levels of *IRT1* decrease with the MT application (Wang et al. 2021). On the other hand, MT application decreases the levels of reactive oxygen species and malondialdehyde and enhances the enzymatic activity and the content of antioxidant compounds such as phenols, anthocyanin, and flavonoids (He et al. 2020; Kaya et al. 2019; Saqib et al. 2023; Tan et al. 2022; Zargar et al. 2022).

Strigolactones (SLs) are classified as plant hormones that regulate plant growth in response to several abiotic stresses (Alvi et al. 2022). The exogenous application of SL in switchgrass, soybean, and barley plants exposed to Cd modulates the antioxidant system for redox homeostasis (Qiu et al. 2021; Shah et al. 2023; Tai et al. 2017). While in switchgrass and barley plants, it improves mineral nutrition, reducing Cd uptake (Qiu et al. 2021; Tai et al. 2017).

Exogenous application of antioxidants (glutathione and ascorbic acid), chelating agents (polyaspartic and malic acid), osmolytes (glycinebetaine and betaine), carbon sources (glucose), or amino acids (glutamate) alleviates Cd stress. Exogenous GSH application modulates metal transporter genes, increases antioxidant enzymatic activity, and diminishes reactive oxygen species and malondialdehyde, contributing to Cd stress alleviating (Chen et al. 2010; Jung et al. 2021; Li et al. 2017, 2021a). Applying ascorbic acid (ASA) alleviates Cd stress in maize and wheat plants. This effect has been associated with the upregulation of antioxidant enzymes and non-enzymatic antioxidants, thus diminishing oxidative damage (Zhang et al. 2019; Zhou et al. 2022). In rice, the application of glutamate alleviates Cd toxicity by modulating the expression of root metal transporter genes (*OsNramp1,*

OsNramp5, OsIRT1, OsIRT2, OsHMA2, and *OsHMA3*), the oxidative stress, and the antioxidant enzyme activity (Jiang et al. 2020). Adding glucose into the nutrient solution alleviates Cd stress in *A. thaliana* by increasing Cd fixation in the root cell wall and upregulating tonoplast-localized metal transporters genes, increasing vacuolar compartmentalization (Shi et al. 2015). The addition of chelating agents such as polyaspartic acid and malic acid increases the enzymatic activity to scavenge ROS in *Brassica napus* and *Miscanthus sacchariflorus* exposed to Cd (Guo et al. 2017; Wu et al. 2021). The osmolytes glycinebetaine (GB) and betaine modulate oxidative stress in maize and spinach plants exposed to Cd, diminishing the reactive oxygen species and increasing the antioxidant activity of some enzymes (Aamer et al. 2018; Li et al. 2016; Zhang et al. 2020a).

Intercropping hyperaccumulator plants with non-hyperaccumulator plants is another strategy to reduce Cd content in crops and vegetables. Intercropping with *Solanum nigrum*, a hyperaccumulator plant, decreases Cd accumulation in grape plants (Hu et al. 2019).

Genetic modification has also been used to reduce Cd content in grains. Overexpression of the *OsNRAMP5* gene encoding a Cd influx transporter in rice promotes Cd accumulation mainly in the roots and, to a lesser extent, in shoots, getting rice grains with low Cd content (Chang et al. 2020). Similarly, overexpression of the *OsLCT2* gene involved in Cd transport promotes Cd accumulation in the roots and decreases its translocation to the rice grains (Tang et al. 2021). On the other hand, rice *OsHMA3* overexpression in wheat plants increases Cd sequestration in wheat roots while decreasing root-to-shoot Cd translocation and Cd accumulation in wheat grains (Zhang et al. 2020b). In soybean, transgenic plants overexpressing GmWRKY172, a transcription factor associated with abiotic stress response, showed high Cd tolerance and low Cd content in shoots and seeds (Xian et al. 2023).

Since few plants have been characterized as Cd hyperaccumulators, there is a constant search for new hyperaccumulating species and methods for improving the phytoremediation process. Therefore, strategies to enhance phytoremediation can be based on the mobilization of Cd in the soil by adding compounds to solubilize it or by increasing plant biomass production to subsequently enhance the Cd phytoremediation rates (Subašić et al. 2022). The exogenous application of plant growth promoters, chelating agents, mineral nutrients, and PGPR is also a strategy to improve phytoremediation in Cd hyperaccumulators.

Cd-hyperaccumulator *Sedum alfredii* has been extensively studied to increase its Cd removal capacity and understand biochemical and genetic mechanisms involved in phytoremediation. Exogenous citric and tartaric acid application improves Cd accumulation in *Sedum alfredii* (Lu et al. 2013). Citric acid contributes to Cd uptake, translocation, and tolerance; meanwhile, tartaric acid mainly increases Cd root uptake. Exogenous silicon application also improves Cd phytoextraction in *S. alfredii* by enhancing plant growth and resistance to Cd (Hu et al. 2023). Under Cd stress conditions, adding exogenous abscisic acid affects growth and Cd accumulation in hyperaccumulators and common plants (Tang et al. 2020). In *S. alfredii* the abscisic acid application enhances Cd accumulation and hormetic plant growth (Lu et al. 2020a).

Bacteria associated with heavy metal hyperaccumulators contribute significantly to mobilizing heavy metals and enhancing phytoextraction. Several bacterial strains isolated from *S. alfredii* rhizosphere increase the soluble Cd concentration. This effect was correlated with the bacterial production of short-chain organic acids, such as oxalic acid, tartaric acid, formic acid, and acetic acid, which decreased the pH from 7 to 2.6 (Li et al. 2010a). In addition, bacteria inoculation in *S. alfredii* improves the root absorption of Cd by increasing root-soil contact area and root organic acid secretion (Chen et al. 2014). Bacterial inoculation also regulates the expression of metal transporter genes, such as *SaZIP, SaIRT1, SaHMA2, SaHMA3,* and *SaHMA4,* which enhance Cd uptake and Cd root-to-shoot translocation (Chen et al. 2017).

Acknowledgments This work was funded by grants from CONAHCYT, Programa Presupuestario F003 (Formerly Fondo Sectorial de Investigación para la Educación) CB2017-2018, A1-S-40454, to Alejandro Hernández-Morales. Jesús Rubio-Santiago (CVU 787127) thanks CONAHCYT-Mexico for the financial support given to carry out his Ph.D. studies.

References

Aamer M, Muhammad UH, Li Z, Abid A, Su Q, Liu Y, Adnan R, Muhammad AUK, Tahir AK, Huang G (2018) Foliar application of glycinebetaine (GB) alleviates the Cadmium (Cd) toxicity in spinach through reducing Cd uptake and improving the activity of anti-oxidant system. Appl Ecol Environ Res 16(6):7575–7583. https://doi.org/10.15666/aeer/1606_75757583

Agami RA, Mohamed GF (2013) Exogenous treatment with indole-3-acetic acid and salicylic acid alleviates cadmium toxicity in wheat seedlings. Ecotoxicol Environ Saf 94:164–171. https://doi.org/10.1016/j.ecoenv.2013.04.013

Ahemad M, Kibret M (2014) Mechanisms and applications of plant growth promoting rhizobacteria: Current perspective. J King Saudi Univ Sci 26(1):1–20. https://doi.org/10.1016/j.jksus.2013.05.001

Ahmad P, Alyemeni MN, Wijaya L, Alam P, Ahanger MA, Alamri SA (2017) Jasmonic acid alleviates negative impacts of cadmium stress by modifying osmolytes and antioxidants in faba bean (Vicia faba L.). Arch Agron Soil Sci 63(13):1889–1899. https://doi.org/10.1080/03650340.2017.1313406

Ahmad J, Ali AA, Baig MA, Iqbal M, Haq I, Irfan Qureshi M (2019) Chapter 8—Role of phytochelatins in cadmium stress tolerance in plants. In: Hasanuzzaman M, Prasad MNV, Fujita M (eds) Cadmium toxicity and tolerance in plants. Academic Press, pp 185–212. https://doi.org/10.1016/B978-0-12-814864-8.00008-5

Ajeesh Krishna TP, Maharajan T, Victor Roch G, Ignacimuthu S, Antony Ceasar S (2020) Structure, function, regulation and phylogenetic relationship of ZIP family transporters of plants. Front Plant Sci 11. https://www.frontiersin.org/articles/10.3389/fpls.2020.00662

Åkesson A, Julin B, Wolk A (2008) Long-term dietary cadmium intake and postmenopausal endometrial cancer incidence: a population-based prospective cohort study. Cancer Res 68(15):6435–6441. https://doi.org/10.1158/0008-5472.CAN-08-0329

Altaf MA, Shahid R, Ren M-X, Naz S, Altaf MM, Khan LU, Lal MK, Tiwari RK, Shakoor A (2022) Melatonin mitigates cadmium toxicity by promoting root architecture and mineral homeostasis of tomato genotypes. J Soil Sci Plant Nutr 22(1):1112–1128. https://doi.org/10.1007/s42729-021-00720-9

Alvi AF, Sehar Z, Fatma M, Masood A, Khan NA (2022) Strigolactone: an emerging growth regulator for developing resilience in plants. Plants 11(19):19. https://doi.org/10.3390/plants11192604

Amjadi Z, Namdjoyan S, Abolhasani Soorki A (2021) Exogenous melatonin and salicylic acid alleviates cadmium toxicity in safflower (Carthamus tinctorius L.) seedlings. Ecotoxicology 30(3):387–401. https://doi.org/10.1007/s10646-021-02364-y

Andresen E, Küpper H (2013) Cadmium toxicity in plants. In: Sigel A, Sigel H, Sigel RK (eds) Cadmium: from toxicity to essentiality. Springer, Dordrecht, The Netherlands, pp 395–413. https://doi.org/10.1007/978-94-007-5179-8_13

Argüello JM, Eren E, González-Guerrero M (2007) The structure and function of heavy metal transport P1B-ATPases. BioMetals 20(3):233–248. https://doi.org/10.1007/s10534-006-9055-6

Asgher M, Khan MIR, Anjum NA, Khan NA (2015) Minimising toxicity of cadmium in plants—role of plant growth regulators. Protoplasma 252(2):399–413. https://doi.org/10.1007/s00709-014-0710-4

Bailey-Serres J, Mittler R (2006) The roles of reactive oxygen species in plant cells. Plant Physiol 141(2):311. https://doi.org/10.1104/pp.104.900191

Baker AJM (1981) Accumulators and excluders -strategies in the response of plants to heavy metals. J Plant Nutr 3(1–4):643–654. https://doi.org/10.1080/01904168109362867

Barbier-Brygoo H, Vinauger M, Colcombet J, Ephritikhine G, Frachisse J-M, Maurel C (2000) Anion channels in higher plants: Functional characterization, molecular structure and physiological role. Biochim Biophys Acta 1465(1):199–218. https://doi.org/10.1016/S0005-2736(00)00139-5

Belcastro M, Marino T, Russo N, Toscano M (2009) The role of glutathione in cadmium ion detoxification: Coordination modes and binding properties – a density functional study. J Inorg Biochem 103(1):50–57. https://doi.org/10.1016/j.jinorgbio.2008.09.002

Benavides MP, Gallego SM, Tomaro ML (2005) Cadmium toxicity in plants. Braz J Plant Physiol 17:21–34. https://doi.org/10.1590/S1677-04202005000100003

Bernhard W, Good M, Vašák M, Kägi JHR (1983) Spectroscopic studies and characterization of metallothioneins containing mercury, lead and bismuth. Inorg Chim Acta 79:154–155. https://doi.org/10.1016/S0020-1693(00)95191-0

Bolan NS, Park JH, Robinson B, Naidu R, Huh KY (2011) Chapter four - Phytostabilization: a green approach to contaminant containment. In: Sparks DL (ed) Advances in agronomy, vol 112. Academic Press, pp 145–204. https://doi.org/10.1016/B978-0-12-385538-1.00004-4

Brooks RR, Lee J, Reeves RD, Jaffre T (1977) Detection of nickeliferous rocks by analysis of herbarium specimens of indicator plants. J Geochem Explor 7:49–57. https://doi.org/10.1016/0375-6742(77)90074-7

Cao D, Zhang H, Wang Y, Zheng L (2014) Accumulation and distribution characteristics of zinc and cadmium in the hyperaccumulator plant sedum plumbizincicola. Bull Environ Contam Toxicol 93(2):171–176. https://doi.org/10.1007/s00128-014-1284-8

Chaney RL, Malik M, Li YM, Brown SL, Brewer EP, Angle JS, Baker AJ (1997) Phytoremediation of soil metals. Curr Opin Biotechnol 8(3):279–284. https://doi.org/10.1016/S0958-1669(97)80004-3

Chang J-D, Huang S, Konishi N, Wang P, Chen J, Huang X-Y, Ma JF, Zhao F-J (2020) Overexpression of the manganese/cadmium transporter OsNRAMP5 reduces cadmium accumulation in rice grain. J Exp Bot 71(18):5705–5715. https://doi.org/10.1093/jxb/eraa287

Changjia J, Shuang C, Qi F, Xiuli S (2021) Advances in Cd remediation techniques for contaminated soils. IOP Conf Ser Earth Environ Sci 621(1):012129. https://doi.org/10.1088/1755-1315/621/1/012129

Chebli Y, Geitmann A (2017) Cellular growth in plants requires regulation of cell wall biochemistry. Curr Opin Cell Biol 44:28–35. https://doi.org/10.1016/j.ceb.2017.01.002

Chen F, Fang W, Wu F, Mao W, Zhang G, Zhou M (2010) Modulation of exogenous glutathione in antioxidant defense system against Cd stress in the two barley genotypes differing in Cd tolerance. Plant Physiol Biochem 48:663–672. https://doi.org/10.1016/j.plaphy.2010.05.001

Chen B, Zhang Y, Rafiq MT, Khan KY, Pan F, Yang X, Feng Y (2014) Improvement of cadmium uptake and accumulation in Sedum alfredii by endophytic bacteria Sphingomonas SaMR12: effects on plant growth and root exudates. Chemosphere 117:367–373. https://doi.org/10.1016/j.chemosphere.2014.07.078

Chen B, Luo S, Wu Y, Ye J, Wang Q, Xu X, Pan F, Khan KY, Feng Y, Yang X (2017) The effects of the endophytic bacterium Pseudomonas fluorescens Sasm05 and IAA on the plant growth and cadmium uptake of Sedum alfredii Hance. Front Microbiol 8. https://www.frontiersin.org/articles/10.3389/fmicb.2017.02538

Chen X, Ouyang Y, Fan Y, Qiu B, Zhang G, Zeng F (2018) The pathway of transmembrane cadmium influx via calcium-permeable channels and its spatial characteristics along rice root. J Exp Bot 69(21):5279–5291. https://doi.org/10.1093/jxb/ery293

Choi W-J, Kang S-K, Ham S, Chung W, Kim AJ, Kang M (2020) Chronic cadmium intoxication and renal injury among workers of a small-scale silver soldering company. Saf Health Work 11(2):235–240. https://doi.org/10.1016/j.shaw.2020.03.005

Clemens S (2006) Toxic metal accumulation, responses to exposure and mechanisms of tolerance in plants. Biochimie 88(11). https://doi.org/10.1016/j.biochi.2006.07.003

Cobbett CS (2000) Phytochelatins and their roles in heavy metal detoxification. Plant Physiol 123(3):825–832. https://doi.org/10.1104/pp.123.3.825

Costa OYA, Raaijmakers JM, Kuramae EE (2018) Microbial extracellular polymeric substances: ecological function and impact on soil aggregation. Front Microbiol 9. https://www.frontiersin.org/articles/10.3389/fmicb.2018.01636

Cuypers A, Plusquin M, Remans T, Jozefczak M, Keunen E, Gielen H, Opdenakker K, Nair AR, Munters E, Artois TJ, Nawrot T, Vangronsveld J, Smeets K (2010) Cadmium stress: an oxidative challenge. BioMetals 23(5):927–940. https://doi.org/10.1007/s10534-010-9329-x

Das P, Samantaray S, Rout GR (1997) Studies on cadmium toxicity in plants: A review. Environ Pollut 98(1):29–36. https://doi.org/10.1016/S0269-7491(97)00110-3

Delalande O, Desvaux H, Godat E, Valleix A, Junot C, Labarre J, Boulard Y (2010) Cadmium – glutathione solution structures provide new insights into heavy metal detoxification. FEBS J 277(24):5086–5096. https://doi.org/10.1111/j.1742-4658.2010.07913.x

Drazic G, Mihailovic N, Lojic M (2006) Cadmium accumulation in Medicago sativa seedlings treated with salicylic acid. Biol Plant 50(2):239–244. https://doi.org/10.1007/s10535-006-0013-5

Ebbs S, Lau I, Ahner B, Kochian L (2002) Phytochelatin synthesis is not responsible for Cd tolerance in the Zn/Cd hyperaccumulator Thlaspi caerulescens (J. & C. Presl). Planta 214(4): 635–640. https://doi.org/10.1007/s004250100650

Eide D, Broderius M, Fett J, Guerinot M (1996) A novel iron-regulated metal transporter from plants identified by functional expression in yeast. Proc Natl Acad Sci 93(11):5624–5628. https://doi.org/10.1073/pnas.93.11.5624

Ercal N, Gurer-Orhan H, Aykin-Burns N (2001) Toxic metals and oxidative stress part I: mechanisms involved in metal-induced oxidative damage. Curr Top Med Chem 1(6):529–539. https://doi.org/10.2174/1568026013394831

Fan SK, Fang XZ, Guan MY, Ye YQ, Lin XY, Du ST, Jin CW (2014) Exogenous abscisic acid application decreases cadmium accumulation in Arabidopsis plants, which is associated with the inhibition of IRT1-mediated cadmium uptake. Front Plant Sci 5. https://www.frontiersin.org/articles/10.3389/fpls.2014.00721

Farooqi ZUR, Hussain MM, Ayub MA, Qadir AA, Ilic P (2022) Chapter 2 - Potentially toxic elements and phytoremediation: opportunities and challenges. In: Bhat RA, Tonelli FMP, Dar GH, Hakeem K (eds) Phytoremediation. Academic Press, pp 19–36. https://doi.org/10.1016/B978-0-323-89874-4.00020-0

Feng S, Tan J, Zhang Y, Liang S, Xiang S, Wang H, Chai T (2017) Isolation and characterization of a novel cadmium-regulated Yellow Stripe-Like transporter (SnYSL3) in Solanum nigrum. Plant Cell Rep 36(2):281–296. https://doi.org/10.1007/s00299-016-2079-7

Freisinger E, Vašák M (2013) Cadmium in metallothioneins. In: Sigel A, Sigel H, Sigel RK (eds) Cadmium: from toxicity to essentiality. Springer, Dordrecht, The Netherlands, pp 339–371. https://doi.org/10.1007/978-94-007-5179-8_11

Fryzova R, Pohanka M, Martinkova P, Cihlarova H, Brtnicky M, Hladky J, Kynicky J (2018) Oxidative stress and heavy metals in plants. Rev Environ Contam Toxicol 245:129–156. https://doi.org/10.1007/398_2017_7

Gallagher CM, Chen JJ, Kovach JS (2010) Environmental cadmium and breast cancer risk. Aging (Albany NY) 2(11):804–814

Godt J, Scheidig F, Grosse-Siestrup C, Esche V, Brandenburg P, Reich A, Groneberg DA (2006) The toxicity of cadmium and resulting hazards for human health. J Occup Med Toxicol 1(1):22. https://doi.org/10.1186/1745-6673-1-22

Grotz N, Fox T, Connolly E, Park W, Guerinot ML, Eide D (1998) Identification of a family of zinc transporter genes from Arabidopsis that respond to zinc deficiency. Proc Natl Acad Sci 95(12): 7220–7224. https://doi.org/10.1073/pnas.95.12.7220

Gu T, Lu Y, Li F, Zeng W, Shen L, Yu R, Li J (2023) Microbial extracellular polymeric substances alleviate cadmium toxicity in rice (Oryza sativa L.) by regulating cadmium uptake, subcellular distribution and triggering the expression of stress-related genes. Ecotoxicol Environ Saf 257: 114958. https://doi.org/10.1016/j.ecoenv.2023.114958

Guo G, Zhou Q, Ma LQ (2006) Availability and assessment of fixing additives for the in situ remediation of heavy metal contaminated soils: a review. Environ Monit Assess 116(1): 513–528. https://doi.org/10.1007/s10661-006-7668-4

Guo H, Chen H, Hong C, Jiang D, Zheng B (2017) Exogenous malic acid alleviates cadmium toxicity in Miscanthus sacchariflorus through enhancing photosynthetic capacity and restraining ROS accumulation. Ecotoxicol Environ Saf 141:119–128. https://doi.org/10.1016/j.ecoenv.2017.03.018

Haider FU, Liqun C, Coulter JA, Cheema SA, Wu J, Zhang R, Wenjun M, Farooq M (2021) Cadmium toxicity in plants: impacts and remediation strategies. Ecotoxicol Environ Saf 211: 111887. https://doi.org/10.1016/j.ecoenv.2020.111887

Hall JL, Williams LE (2003) Transition metal transporters in plants. J Exp Bot 54(393):2601–2613. https://doi.org/10.1093/jxb/erg303

Han Y, Wang S, Zhao N, Deng S, Zhao C, Li N, Sun J, Zhao R, Yi H, Shen X, Chen S, liang. (2016) Exogenous abscisic acid alleviates cadmium toxicity by restricting Cd2+ influx in Populus euphratica cells. J Plant Growth Regul 35:827–837. https://doi.org/10.1007/s00344-016-9585-2

Hartwig A (2013) Cadmium and cancer. In: Sigel A, Sigel H, Sigel RK (eds) Cadmium: from toxicity to essentiality. Springer, Dordrecht, The Netherlands, pp 491–507. https://doi.org/10.1007/978-94-007-5179-8_15

Hasanuzzaman M, Bhuyan MHMB, Zulfiqar F, Raza A, Mohsin SM, Mahmud JA, Fujita M, Fotopoulos V (2020) Reactive oxygen species and antioxidant defense in plants under abiotic stress: revisiting the crucial role of a universal defense regulator. Antioxidants 9(8):8. https://doi.org/10.3390/antiox9080681

Hassan MJ, Zhang G, Wu F, Wei K, Chen Z (2005) Zinc alleviates growth inhibition and oxidative stress caused by cadmium in rice. J Plant Nutr Soil Sci 168(2):255–261. https://doi.org/10.1002/jpln.200420403

Hassan MJ, Raza MA, Rehman SU, Ansar M, Gitari H, Khan I, Wajid M, Ahmed M, Abbas Shah G, Peng Y, Li Z (2020) Effect of cadmium toxicity on growth, oxidative damage, antioxidant defense system and cadmium accumulation in two sorghum cultivars. Plants 9(11):1575. https://doi.org/10.3390/plants9111575

He S, Yang X, He Z, Baligar VC (2017) Morphological and physiological responses of plants to cadmium toxicity: a review. Pedosphere 27(3):421–438. https://doi.org/10.1016/S1002-0160(17)60339-4

He J, Zhuang X, Zhou J, Sun L, Wan H, Li H, Lyu D (2020) Exogenous melatonin alleviates cadmium uptake and toxicity in apple rootstocks. Tree Physiol 40(6):746–761. https://doi.org/10.1093/treephys/tpaa024

Hernández JA, Barba-Espín G, Diaz-Vivancos P (2017) Glutathione-mediated biotic stress tolerance in plants. In: Hossain MA, Mostofa MG, Diaz-Vivancos P, Burritt DJ, Fujita M, Tran L-SP (eds) Glutathione in plant growth, development, and stress tolerance. Springer, pp 309–329. https://doi.org/10.1007/978-3-319-66682-2_14

Hopkins WG, Hüner NPA (2009) Introduction to plant physiology, 4th edn. Wiley

Hossain Z, Hajika M, Komatsu S (2012) Comparative proteome analysis of high and low cadmium accumulating soybeans under cadmium stress. Amino Acids 43(6):2393–2416. https://doi.org/10.1007/s00726-012-1319-6

Hu P-J, Gan Y-Y, Tang Y-T, Zhang Q-F, Jiang D, Yao N, Qiu R-L (2012) Cellular tolerance, accumulation and distribution of cadmium in leaves of hyperaccumulator Picris divaricata. Pedosphere 22(4):497–507. https://doi.org/10.1016/S1002-0160(12)60034-4

Hu R, Zhang Z, Lin L, Liao M, Tang Y, Liang D, Xia H, Wang J, Wang X, Lv X, Ren W (2019) Intercropping with hyperaccumulator plants decreases the cadmium accumulation in grape seedlings. Acta Agric Scand B Soil Plant Sci 69(4):304–310. https://doi.org/10.1080/09064710.2018.1564786

Hu Y, Zhou X, Shi A, Yu Y, Rensing C, Zhang T, Xing S, Yang W (2023) Exogenous silicon promotes cadmium (Cd) accumulation in Sedum alfredii Hance by enhancing Cd uptake and alleviating Cd toxicity. Front Plant Sci 14. https://www.frontiersin.org/articles/10.3389/fpls.2023.1134370

Huang Y, Mubeen S, Yang Z, Wang J (2022) Cadmium contamination in agricultural soils and crops. In: Yang Z, He C, Xin J (eds) Theories and methods for minimizing cadmium pollution in crops: case studies on water spinach. Springer Nature, pp 1–30. https://doi.org/10.1007/978-981-16-7751-9_1

Huang J, Jing HK, Zhang Y, Chen SY, Wang HY, Cao Y, Zhang Z, Lu YH, Zheng QS, Shen RF, Zhu XF (2023) Melatonin reduces cadmium accumulation via mediating the nitric oxide accumulation and increasing the cell wall fixation capacity of cadmium in rice. J Hazard Mater 445:130529. https://doi.org/10.1016/j.jhazmat.2022.130529

Huguet S, Bert V, Laboudigue A, Barthès V, Isaure M-P, Llorens I, Schat H, Sarret G (2012) Cd speciation and localization in the hyperaccumulator Arabidopsis halleri. Environ Exp Bot 82:54–65. https://doi.org/10.1016/j.envexpbot.2012.03.011

Imran M, Hussain S, El-Esawi MA, Rana MS, Saleem MH, Riaz M, Ashraf U, Potcho MP, Duan M, Rajput IA, Tang X (2020) Molybdenum supply alleviates the cadmium toxicity in fragrant rice by modulating oxidative stress and antioxidant gene expression. Biomolecules 10(11):11. https://doi.org/10.3390/biom10111582

Inamdar A, Sangawe V, Adhapure N (2022) Chapter 14—Enzymes in rhizosphere engineering. In: Dubey RC, Kumar P (eds) Rhizosphere engineering. Academic Press, pp 259–272. https://doi.org/10.1016/B978-0-323-89973-4.00009-0

Isaure M-P, Huguet S, Meyer C-L, Castillo-Michel H, Testemale D, Vantelon D, Saumitou-Laprade P, Verbruggen N, Sarret G (2015) Evidence of various mechanisms of Cd sequestration in the hyperaccumulator Arabidopsis halleri, the non-accumulator Arabidopsis lyrata, and their progenies by combined synchrotron-based techniques. J Exp Bot 66(11):3201–3214. https://doi.org/10.1093/jxb/erv131

Jian L, Jingchun L, Chongling Y, Daolin D, Haoliang L (2019) The alleviation effect of iron on cadmium phytotoxicity in mangrove A. marina. Alleviation effect of iron on cadmium phytotoxicity in mangrove Avicennia marina (Forsk.) Vierh. Chemosphere 226:413–420. https://doi.org/10.1016/j.chemosphere.2019.03.172

Jiang RF, Ma DY, Zhao FJ, McGrath SP (2005) Cadmium hyperaccumulation protects Thlaspi caerulescens from leaf feeding damage by thrips (Frankliniella occidentalis). New Phytol 167(3):805–814. https://doi.org/10.1111/j.1469-8137.2005.01452.x

Jiang M, Jiang J, Li S, Li M, Tan Y, Song S, Shu Q, Huang J (2020) Glutamate alleviates cadmium toxicity in rice via suppressing cadmium uptake and translocation. J Hazard Mater 384:121319. https://doi.org/10.1016/j.jhazmat.2019.121319

Johri N, Jacquillet G, Unwin R (2010) Heavy metal poisoning: the effects of cadmium on the kidney. BioMetals 23(5):783–792. https://doi.org/10.1007/s10534-010-9328-y

Jomova K, Baros S, Valko M (2012) Redox active metal-induced oxidative stress in biological systems. Transit Met Chem 37(2):127–134. https://doi.org/10.1007/s11243-012-9583-6

Joshi R, Pareek A, Singla-Pareek SL (2016) Chapter 9 - Plant metallothioneins: classification, distribution, function, and regulation. In: Ahmad P (ed) Plant metal interaction. Elsevier, pp 239–261. https://doi.org/10.1016/B978-0-12-803158-2.00009-6

Jung H, Lee T-G, Lee J, Chae M-J, Lee E-J, Kim M-S, Jung G-B, Emmanuel A, Jeon S, Lee B-R (2021) Foliar-applied glutathione mitigates cadmium-induced oxidative stress by modulating antioxidant-scavenging, redox-regulating, and hormone-balancing systems in Brassica napus. Front Plant Sci 12:700413. https://doi.org/10.3389/fpls.2021.700413

Kägi JHR, Vallee BL (1960) Metallothionein: a cadmium- and zinc-containing protein from equine renal cortex. J Biol Chem 235(12):3460–3465. https://doi.org/10.1016/S0021-9258(18)64490-4

Kägi JH, Vasák M, Lerch K, Gilg DE, Hunziker P, Bernhard WR, Good M (1984) Structure of mammalian metallothionein. Environ Health Perspect 54:93–103

Kaya C, Okant M, Ugurlar F, Alyemeni MN, Ashraf M, Ahmad P (2019) Melatonin-mediated nitric oxide improves tolerance to cadmium toxicity by reducing oxidative stress in wheat plants. Chemosphere 225:627–638. https://doi.org/10.1016/j.chemosphere.2019.03.026

Kellen E, Zeegers MP, Hond ED, Buntinx F (2007) Blood cadmium may be associated with bladder carcinogenesis: the Belgian case-control study on bladder cancer. Cancer Detect Prev 31(1): 77–82. https://doi.org/10.1016/j.cdp.2006.12.001

Keramat B, Kalantari KM, Arvin MJ (2010) Effects of methyl jasmonate treatment on alleviation of cadmium damages in soybean. J Plant Nutr 33(7):1016–1025. https://doi.org/10.1080/01904161003728685

Kim H, Lee J, Woo HD, Kim DW, Choi IJ, Kim Y-I, Kim J (2019) Association between dietary cadmium intake and early gastric cancer risk in a Korean population: a case–control study. Eur J Nutr 58(8):3255–3266. https://doi.org/10.1007/s00394-018-1868-x

Korshunova YO, Eide D, Gregg Clark W, Lou Guerinot M, Pakrasi HB (1999) The IRT1 protein from Arabidopsis thaliana is a metal transporter with a broad substrate range. Plant Mol Biol 40(1):37–44. https://doi.org/10.1023/A:1026438615520

Kubier A, Wilkin RT, Pichler T (2019) Cadmium in soils and groundwater: A review. Appl Geochem 108:1–16. https://doi.org/10.1016/j.apgeochem.2019.104388

Kumar S, Kumar S, Mohapatra T (2021) Interaction between macro- and micro-nutrients in plants. Front Plant Sci 12. https://www.frontiersin.org/articles/10.3389/fpls.2021.665583

Lan X-Y, Yang B, Yan Y-Y, Li X-Y, Xu F-L (2018) Resistance mechanisms and their difference between the root and leaf of Microsorum pteropus – a novel potential aquatic cadmium hyperaccumulator. Sci Total Environ 616–617:480–490. https://doi.org/10.1016/j.scitotenv.2017.10.271

Lapie C, Leglize P, Paris C, Buisson T, Sterckeman T (2019) Profiling of main metabolites in root exudates and mucilage collected from maize submitted to cadmium stress. Environ Sci Pollut Res 26(17):17520–17534. https://doi.org/10.1007/s11356-019-05168-0

Lei GJ, Sun L, Sun Y, Zhu XF, Li GX, Zheng SJ (2020) Jasmonic acid alleviates cadmium toxicity in Arabidopsis via suppression of cadmium uptake and translocation. J Integr Plant Biol 62(2): 218–227. https://doi.org/10.1111/jipb.12801

Leitenmaier B, Küpper H (2013) Compartmentation and complexation of metals in hyperaccumulator plants. Front Plant Sci 4. https://www.frontiersin.org/articles/10.3389/fpls.2013.00374

Li WC, Ye ZH, Wong MH (2010a) Metal mobilization and production of short-chain organic acids by rhizosphere bacteria associated with a Cd/Zn hyperaccumulating plant, Sedum alfredi. Plant Soil 326(1):453–467. https://doi.org/10.1007/s11104-009-0025-y

Li Y, Liu Y, Zhang J (2010b) Advances in the research on the AsA-GSH cycle in horticultural crops. Front Agric China 4(1):84–90. https://doi.org/10.1007/s11703-009-0089-8
Li M, Wang G, Li J, Cao F (2016) Foliar application of betaine alleviates cadmium toxicity in maize seedlings. Acta Physiol Plant 38(4):95. https://doi.org/10.1007/s11738-016-2116-8
Li M, Hao P, Cao F (2017) Glutathione-induced alleviation of cadmium toxicity in Zea mays. Plant Physiol Biochem 119:240–249. https://doi.org/10.1016/j.plaphy.2017.09.005
Li L, Ai S, Li Y, Wang Y, Tang M (2018) Exogenous silicon mediates alleviation of cadmium stress by promoting photosynthetic activity and activities of antioxidative enzymes in Rice. J Plant Growth Regul 37(2):602–611. https://doi.org/10.1007/s00344-017-9758-7
Li C, Zhou K, Qin W, Tian C, Qi M, Yan X, Han W (2019) A review on heavy metals contamination in soil: effects, sources, and remediation techniques. Soil Sediment Contam Int J 28(4):380–394. https://doi.org/10.1080/15320383.2019.1592108
Li G-Z, Chen S-J, Li N-Y, Wang Y-Y, Kang G-Z (2021a) Exogenous glutathione alleviates cadmium toxicity in wheat by influencing the absorption and translocation of cadmium. Bull Environ Contam Toxicol 107(2):320–326. https://doi.org/10.1007/s00128-021-03283-8
Li J, Cao X, Jia X, Liu L, Cao H, Qin W, Li M (2021b) Iron deficiency leads to chlorosis through impacting chlorophyll synthesis and nitrogen metabolism in Areca catechu L. Front Plant Sci 12:710093. https://doi.org/10.3389/fpls.2021.710093
Li B-B, Fu Y-S, Li X-X, Yin H-N, Xi Z-M (2022a) Brassinosteroids alleviate cadmium phytotoxicity by minimizing oxidative stress in grape seedlings: toward regulating the ascorbate-glutathione cycle. Sci Hortic 299:111002. https://doi.org/10.1016/j.scienta.2022.111002
Li C, Tang Y, Gu F, Wang X, Yang W, Han Y, Ruan Y (2022b) Phytochemical analysis reveals an antioxidant defense response in Lonicera japonica to cadmium-induced oxidative stress. Sci Rep 12(1):1. https://doi.org/10.1038/s41598-022-10912-7
Li Y, Zhang S, Bao Q, Chu Y, Sun H, Huang Y (2022c) Jasmonic acid alleviates cadmium toxicity through regulating the antioxidant response and enhancing the chelation of cadmium in rice (Oryza sativa L.). Environ Pollut 304:119178. https://doi.org/10.1016/j.envpol.2022.119178
Lin A, Zhang X, Chen M, Cao Q (2007) Oxidative stress and DNA damages induced by cadmium accumulation. J Environ Sci 19(5):596–602. https://doi.org/10.1016/S1001-0742(07)60099-0
Liu W, Shu W, Lan C (2004) Viola baoshanensis, a plant that hyperaccumulates cadmium. Chin Sci Bull 49(1):29–32. https://doi.org/10.1007/BF02901739
Liu X, Peng K, Wang A, Lian C, Shen Z (2010) Cadmium accumulation and distribution in populations of Phytolacca americana L. and the role of transpiration. Chemosphere 78(9): 1136–1141. https://doi.org/10.1016/j.chemosphere.2009.12.030
Liu J, Liu H, Li Y, Wang H (2014) Probing the coordination properties of glutathione with transition metal ions (Cr^{2+},Mn^{2+},Fe^{2+},Co^{2+}, Ni^{2+},Cu^{2+},Zn^{2+},Cd^{2+},Hg^{2+}) by density functional theory. J Biol Phys 40(4):313–323. https://doi.org/10.1007/s10867-014-9350-3
Liu S, Ali S, Yang R, Tao J, Ren B (2019) A newly discovered Cd-hyperaccumulator Lantana camara L. J Hazard Mater 371:233–242. https://doi.org/10.1016/j.jhazmat.2019.03.016
Liu YS, Tao Y, Yang XZ, Liu YN, Shen RF, Zhu XF (2022) Gibberellic acid alleviates cadmium toxicity in rice by regulating NO accumulation and cell wall fixation capacity of cadmium. J Hazard Mater 439:129597. https://doi.org/10.1016/j.jhazmat.2022.129597
Loix C, Huybrechts M, Vangronsveld J, Gielen M, Keunen E, Cuypers A (2017) Reciprocal interactions between cadmium-induced cell wall responses and oxidative stress in plants. Front Plant Sci 8. https://www.frontiersin.org/articles/10.3389/fpls.2017.01867
Lombi E, Zhao FJ, Dunham SJ, McGRATH SP (2000) Cadmium accumulation in populations of Thlaspi caerulescens and Thlaspi goesingense. New Phytol 145(1):11–20. https://doi.org/10.1046/j.1469-8137.2000.00560.x
Lu L, Tian S, Yang X, Peng H, Li T (2013) Improved cadmium uptake and accumulation in the hyperaccumulator Sedum alfredii: the impact of citric acid and tartaric acid. J Zhejiang Univ Sci B 14(2):106–114. https://doi.org/10.1631/jzus.B1200211
Lu Q, Chen S, Li Y, Zheng F, He B, Gu M (2020a) Exogenous abscisic acid (ABA) promotes cadmium (Cd) accumulation in Sedum alfredii Hance by regulating the expression of Cd stress

response genes. Environ Sci Pollut Res 27(8):8719–8731. https://doi.org/10.1007/s11356-019-07512-w

Lu R-R, Hu Z-H, Zhang Q-L, Li Y-Q, Lin M, Wang X-L, Wu X-N, Yang J-T, Zhang L-Q, Jing Y-X, Peng C-L (2020b) The effect of Funneliformis mosseae on the plant growth, Cd translocation and accumulation in the new Cd-hyperaccumulator Sphagneticola calendulacea. Ecotoxicol Environ Saf 203:110988. https://doi.org/10.1016/j.ecoenv.2020.110988

Luo J-S, Zhang Z (2021) Mechanisms of cadmium phytoremediation and detoxification in plants. Crop J 9(3):521–529. https://doi.org/10.1016/j.cj.2021.02.001

Lutsenko S, Kaplan JH (1995) Organization of P-type ATPases: significance of structural diversity. Biochemistry 34(48):15607–15613. https://doi.org/10.1021/bi00048a001

Lux A, Martinka M, Vaculík M, White PJ (2011) Root responses to cadmium in the rhizosphere: a review. J Exp Bot 62(1):21–37. https://doi.org/10.1093/jxb/erq281

Mahajan P, Kaushal J (2018) Role of phytoremediation in reducing cadmium toxicity in soil and water. J Toxicol 2018:e4864365. https://doi.org/10.1155/2018/4864365

Manghwar H, Hussain A, Ali Q, Liu F (2022) Brassinosteroids (BRs) role in plant development and coping with different stresses. Int J Mol Sci 23(3):3. https://doi.org/10.3390/ijms23031012

Manquián-Cerda K, Escudey M, Zúñiga G, Arancibia-Miranda N, Molina M, Cruces E (2016) Effect of cadmium on phenolic compounds, antioxidant enzyme activity and oxidative stress in blueberry (Vaccinium corymbosum L.) plantlets grown in vitro. Ecotoxicol Environ Saf 133: 316–326. https://doi.org/10.1016/j.ecoenv.2016.07.029

Margoshes M, Vallee BL (1957) A cadmium protein from equine kidney cortex. J Am Chem Soc 79(17):4813–4814. https://doi.org/10.1021/ja01574a064

Martins LL, Mourato MP, Cardoso AI, Pinto AP, Mota AM, de Lurdes S, Gonçalves M, de Varennes A (2011) Oxidative stress induced by cadmium in Nicotiana tabacum L.: effects on growth parameters, oxidative damage and antioxidant responses in different plant parts. Acta Physiol Plant 33(4):1375–1383. https://doi.org/10.1007/s11738-010-0671-y

Mendoza-Cózatl DG, Jobe TO, Hauser F, Schroeder JI (2011) Long-distance transport, vacuolar sequestration and transcriptional responses induced by cadmium and arsenic. Curr Opin Plant Biol 14(5):554–562. https://doi.org/10.1016/j.pbi.2011.07.004

Menguer PK, Farthing E, Peaston KA, Ricachenevsky FK, Fett JP, Williams LE (2013) Functional analysis of the rice vacuolar zinc transporter OsMTP1. J Exp Bot 64(10):2871–2883. https://doi.org/10.1093/jxb/ert136

Merlos MA, Michálek P, Kryštofová O, Zítka O, Adam V, Kizek R (2014) The role of phytochelatins in plant and animals: a review. J Met Nano 1(3)

Meyer C-L, Juraniec M, Huguet S, Chaves-Rodriguez E, Salis P, Isaure M-P, Goormaghtigh E, Verbruggen N (2015) Intraspecific variability of cadmium tolerance and accumulation, and cadmium-induced cell wall modifications in the metal hyperaccumulator Arabidopsis halleri. J Exp Bot 66(11):3215–3227. https://doi.org/10.1093/jxb/erv144

Mitra GN (2015) Introduction and uptake of water and nutrient ions. In: Mitra GN (ed) Regulation of nutrient uptake by plants: a biochemical and molecular approach. Springer India, pp 1–7. https://doi.org/10.1007/978-81-322-2334-4_1

Moiseenko TI, Gashkina NA (2018) Biogeochemistry of cadmium: anthropogenic dispersion, bioaccumulation, and ecotoxicity. Geochem Int 56(8):798–811. https://doi.org/10.1134/S0016702918080062

Jacoby B, Moran N (2001) Mineral nutrient transport in plants. In: Handbook of plant and crop physiology, 2nd edn. CRC Press

Mubeen S, Ni W, He C, Yang Z (2023) Agricultural strategies to reduce cadmium accumulation in crops for food safety. Agriculture 13(2):2. https://doi.org/10.3390/agriculture13020471

Nakanishi H, Ogawa I, Ishimaru Y, Mori S, Nishizawa NK (2006) Iron deficiency enhances cadmium uptake and translocation mediated by the Fe2+ transporters OsIRT1 and OsIRT2 in rice. Soil Sci Plant Nutr 52(4):464–469. https://doi.org/10.1111/j.1747-0765.2006.00055.x

Nazar R, Iqbal N, Masood A, Khan MIR, Syeed S, Khan NA (2012) Cadmium toxicity in plants and role of mineral nutrients in its alleviation. Am J Plant Sci 03(10):1476–1489. https://doi.org/10.4236/ajps.2012.310178

Nevo Y, Nelson N (2006) The NRAMP family of metal-ion transporters. Biochim Biophys Acts 1763(7):609–620. https://doi.org/10.1016/j.bbamcr.2006.05.007

Palma JM, Gupta DK, Corpas FJ (2013) Metalloenzymes involved in the metabolism of reactive oxygen species and heavy metal stress. In: Gupta DK, Corpas FJ, Palma JM (eds) Heavy metal stress in plants. Springer, pp 1–17. https://doi.org/10.1007/978-3-642-38469-1_1

Pan J, Guan M, Xu P, Chen M, Cao Z (2021) Salicylic acid reduces cadmium (Cd) accumulation in rice (Oryza sativa L.) by regulating root cell wall composition via nitric oxide signaling. Sci Total Environ 797:149202. https://doi.org/10.1016/j.scitotenv.2021.149202

Park J, Song W-Y, Ko D, Eom Y, Hansen TH, Schiller M, Lee TG, Martinoia E, Lee Y (2012) The phytochelatin transporters AtABCC1 and AtABCC2 mediate tolerance to cadmium and mercury. Plant J 69(2):278–288. https://doi.org/10.1111/j.1365-313X.2011.04789.x

Parrotta L, Guerriero G, Sergeant K, Cai G, Hausman J-F (2015) Target or barrier? The cell wall of early- and later-diverging plants vs cadmium toxicity: differences in the response mechanisms. Front Plant Sci 6. https://www.frontiersin.org/articles/10.3389/fpls.2015.00133

Pedas P, Schjoerring JK, Husted S (2009) Identification and characterization of zinc-starvation-induced ZIP transporters from barley roots. Plant Physiol Biochem 47(5):377–383. https://doi.org/10.1016/j.plaphy.2009.01.006

Pedrosa-Gomes M, Marques TCLLS, Soares AM (2013) Cadmium effects on mineral nutrition of the Cd-hyperaccumulator Pfaffia glomerata. Biologia 68(2):223–230. https://doi.org/10.2478/s11756-013-0005-9

Peng R, Sun W, Jin X, Yu L, Chen C, Yue Z, Dong Y (2020) Analysis of 2,4-epibrassinolide created an enhancement tolerance on Cd toxicity in Solanum nigrum L. Environ Sci Pollut Res 27(14):16784–16797. https://doi.org/10.1007/s11356-020-08228-y

Pinto AP, Simotildees I, Mota AM (2008) Cadmium impact on root exudates of sorghum and maize plants: a speciation study. J Plant Nutr 31(10):1746–1755. https://doi.org/10.1080/01904160802324829

Pinto FR, Mourato MP, Sales JR, Moreira IN, Martins LL (2017) Oxidative stress response in spinach plants induced by cadmium. J Plant Nutr 40(2):268–276. https://doi.org/10.1080/01904167.2016.1240186

Pons M-L, Collin B, Doelsch E, Chaurand P, Fehlauer T, Levard C, Keller C, Rose J (2021) X-ray absorption spectroscopy evidence of sulfur-bound cadmium in the Cd-hyperaccumulator Solanum nigrum and the non-accumulator Solanum melongena. Environ Pollut 279:116897. https://doi.org/10.1016/j.envpol.2021.116897

Prasad MNV (1995) Cadmium toxicity and tolerance in vascular plants. Environ Exp Bot 35(4): 525–545. https://doi.org/10.1016/0098-8472(95)00024-0

Qin S, Liu H, Nie Z, Gao W, Li C, Lin Y, Zhao P (2018) AsA–GSH cycle and antioxidant enzymes play important roles in Cd tolerance of wheat. Bull Environ Contam Toxicol 101(5):684–690. https://doi.org/10.1007/s00128-018-2471-9

Qin S, Liu H, Nie Z, Rengel Z, Gao W, Li C, Zhao P (2020) Toxicity of cadmium and its competition with mineral nutrients for uptake by plants: a review. Pedosphere 30(2):168–180. https://doi.org/10.1016/S1002-0160(20)60002-9

Qiu Q, Wang Y, Yang Z, Yuan J (2011) Effects of phosphorus supplied in soil on subcellular distribution and chemical forms of cadmium in two Chinese flowering cabbage (Brassica parachinensis L.) cultivars differing in cadmium accumulation. Food Chem Toxicol 49(9): 2260–2267. https://doi.org/10.1016/j.fct.2011.06.024

Qiu C-W, Zhang C, Wang N-H, Mao W, Wu F (2021) Strigolactone GR24 improves cadmium tolerance by regulating cadmium uptake, nitric oxide signaling and antioxidant metabolism in barley (Hordeum vulgare L.). Environ Pollut 273:116486. https://doi.org/10.1016/j.envpol.2021.116486

Rafati Rahimzadeh M, Rafati Rahimzadeh M, Kazemi S, Moghadamnia A (2017) Cadmium toxicity and treatment: an update. Caspian J Intern Med 8(3):135–145. https://doi.org/10.22088/cjim.8.3.135

Rahman A, Mostofa MG, Nahar K, Hasanuzzaman M, Fujita M (2016) Exogenous calcium alleviates cadmium-induced oxidative stress in rice (Oryza sativa L.) seedlings by regulating the antioxidant defense and glyoxalase systems. Braz J Bot 39(2):393–407. https://doi.org/10.1007/s40415-015-0240-0

Rascio N, Navari-Izzo F (2011) Heavy metal hyperaccumulating plants: How and why do they do it? And what makes them so interesting? Plant Sci 180(2):169–181. https://doi.org/10.1016/j.plantsci.2010.08.016

Reeves RD, Baker AJM, Jaffré T, Erskine PD, Echevarria G, van der Ent A (2018) A global database for plants that hyperaccumulate metal and metalloid trace elements. New Phytol 218(2):407–411. https://doi.org/10.1111/nph.14907

Reid R, Hayes J (2003) Mechanisms and control of nutrient uptake in plants. Int Rev Cytol 229:73–114. https://doi.org/10.1016/s0074-7696(03)29003-3

Riaz M, Kamran M, Fang Y, Yang G, Rizwan M, Ali S, Zhou Y, Wang Q, Deng L, Wang Y, Wang X (2021) Boron supply alleviates cadmium toxicity in rice (Oryza sativa L.) by enhancing cadmium adsorption on cell wall and triggering antioxidant defense system in roots. Chemosphere 266:128938. https://doi.org/10.1016/j.chemosphere.2020.128938

Ricachenevsky FK, de Araújo Junior AT, Fett JP, Sperotto RA (2018) You shall not pass: root vacuoles as a symplastic checkpoint for metal translocation to shoots and possible application to grain nutritional quality. Front Plant Sci 9:412. https://doi.org/10.3389/fpls.2018.00412

Rizwan M, Ali S, Adrees M, Ibrahim M, Tsang DCW, Zia-ur-Rehman M, Zahir ZA, Rinklebe J, Tack FMG, Ok YS (2017) A critical review on effects, tolerance mechanisms and management of cadmium in vegetables. Chemosphere 182:90–105. https://doi.org/10.1016/j.chemosphere.2017.05.013

Sadiq R, Maqbool N, Hussain M, Tehseen S, Naseer M, Rafique T, Zikrea A, Naqve M, Mahmood A, Javaid A, Wali M (2021) Boosting antioxidant defense mechanism of mungbean with foliar application of gibberellic acid to alleviate cadmium toxicity. Plant Physiol Rep 26(4):741–748. https://doi.org/10.1007/s40502-021-00630-z

Saha I, Datta S, Biswas D (2020) Exploring the role of bacterial extracellular polymeric substances for sustainable development in agriculture. Curr Microbiol 77(11):3224–3239. https://doi.org/10.1007/s00284-020-02169-y

Saidi I, Chtourou Y, Djebali W (2014) Selenium alleviates cadmium toxicity by preventing oxidative stress in sunflower (Helianthus annuus) seedlings. J Plant Physiol 171(5):85–91. https://doi.org/10.1016/j.jplph.2013.09.024

Salt DE, Prince RC, Pickering IJ, Raskin I (1995) Mechanisms of cadmium mobility and accumulation in Indian mustard. Plant Physiol 109(4):1427–1433. https://doi.org/10.1104/pp.109.4.1427

Sanità di Toppi L, Gabbrielli R (1999) Response to cadmium in higher plants. Environ Exp Bot 41(2):105–130. https://doi.org/10.1016/S0098-8472(98)00058-6

Saqib M, Shahzad U, Zulfiqar F, Tiwari RK, Lal MK, Naz S, Jahan MS, Awan ZA, El-Sheikh MA, Altaf MA (2023) Exogenous melatonin alleviates cadmium-induced inhibition of growth and photosynthesis through upregulating antioxidant defense system in strawberry. South Afr J Bot 157:10–18. https://doi.org/10.1016/j.sajb.2023.03.039

Sarkar A, Ravindran G, Krishnamurthy V (2013) A brief review on the effect of cadmium toxicity: from cellular to organ level. Int J Biotechnol Res 3:17–36

Sarwar N, Saifullah, Malhi SS, Zia MH, Naeem A, Bibi S, Farid G (2010) Role of mineral nutrition in minimizing cadmium accumulation by plants. J Sci Food Agric 90(6):925–937. https://doi.org/10.1002/jsfa.3916

Seregin IV, Kozhevnikova AD (2004) Strontium transport, distribution, and toxic effects on maize seedling growth. Russ J Plant Physiol 51(2):215–221. https://doi.org/10.1023/B:RUPP.0000019217.89936.e7

Shah T, Asad M, Khan Z, Amjad K, Alsahli AA, D'amato R (2023) Strigolactone decreases cadmium concentrations by regulating cadmium localization and glyoxalase defense system: Effects on nodules organic acids and soybean yield. Chemosphere 335:139028. https://doi.org/10.1016/j.chemosphere.2023.139028

Shahid M, Dumat C, Khalid S, Niazi NK, Antunes PMC (2017) Cadmium bioavailability, uptake, toxicity and detoxification in soil-plant system. In: de Voogt P, Gunther FA (eds) Reviews of environmental contamination and toxicology, vol 241. Springer, pp 73–137. https://doi.org/10.1007/398_2016_8

Shamsi I, Zhang G, Hu H, Xue Q, Hussain N, Ali E, Shen Q, Zheng W, Zhang Q, Liu X, Lin X (2014) Assessment of the hazardous effects of Cd on physiological and biochemical characteristics of soybean genotypes. Int J Agric Biol 16:1560–8530

Shen G, Niu J, Deng Z (2017) Abscisic acid treatment alleviates cadmium toxicity in purple flowering stalk (Brassica campestris L. ssp. Chinensis var. Purpurea Hort.) seedlings. Plant Physiol Biochem 118:471–478. https://doi.org/10.1016/j.plaphy.2017.07.018

Shen X, Dai M, Yang J, Sun L, Tan X, Peng C, Ali I, Naz I (2022) A critical review on the phytoremediation of heavy metals from environment: performance and challenges. Chemosphere 291:132979. https://doi.org/10.1016/j.chemosphere.2021.132979

Shi Y-Z, Zhu X-F, Wan J-X, Li G-X, Zheng S-J (2015) Glucose alleviates cadmium toxicity by increasing cadmium fixation in root cell wall and sequestration into vacuole in Arabidopsis. J Integr Plant Biol 57(10):830–837. https://doi.org/10.1111/jipb.12312

Sondergaard TE, Schulz A, Palmgren MG (2004) Energization of transport processes in plants. roles of the plasma membrane H+-ATPase. Plant Physiol 136(1):2475–2482. https://doi.org/10.1104/pp.104.048231

Song Y, Jin L, Wang X (2017) Cadmium absorption and transportation pathways in plants. Int J Phytoremediation 19(2):133–141. https://doi.org/10.1080/15226514.2016.1207598

Song Z, Wang P, Chen X, Peng Y, Cai B, Song J, Yin G, Jia S, Zhang H (2022) Melatonin alleviates cadmium toxicity and abiotic stress by promoting glandular trichome development and antioxidant capacity in Nicotiana tabacum. Ecotoxicol Environ Saf 236:113437. https://doi.org/10.1016/j.ecoenv.2022.113437

Stayner L, Smith R, Thun M, Schnorr T, Lemen R (1992) A dose-response analysis and quantitative assessment of lung cancer risk and occupational cadmium exposure. Ann Epidemiol 2(3):177–194. https://doi.org/10.1016/1047-2797(92)90052-R

Sterckeman T, Thomine S (2020) Mechanisms of cadmium accumulation in plants. Crit Rev Plant Sci 39(4):322–359. https://doi.org/10.1080/07352689.2020.1792179

Subašić M, Šamec D, Selović A, Karalija E (2022) Phytoremediation of cadmium polluted soils: current status and approaches for enhancing. Soil Syst 6(1):1. https://doi.org/10.3390/soilsystems6010003

Sun Y, Zhou Q, Diao C (2008) Effects of cadmium and arsenic on growth and metal accumulation of Cd-hyperaccumulator Solanum nigrum L. Bioresour Technol 99(5):1103–1110. https://doi.org/10.1016/j.biortech.2007.02.035

Sun R, Zhou Q, Wei S (2011) Cadmium accumulation in relation to organic acids and nonprotein thiols in leaves of the recently found Cd hyperaccumulator Rorippa globosa and the Cd-accumulating plant Rorippa islandica. J Plant Growth Regul 30(1):83–91. https://doi.org/10.1007/s00344-010-9176-6

Sytar O, Ghosh S, Malinska H, Zivcak M, Brestic M (2021) Physiological and molecular mechanisms of metal accumulation in hyperaccumulator plants. Physiol Plant 173(1):148–166. https://doi.org/10.1111/ppl.13285

Szopiński M, Sitko K, Rusinowski S, Zieleźnik-Rusinowska P, Corso M, Rostański A, Rojek-Jelonek M, Verbruggen N, Małkowski E (2020) Different strategies of Cd tolerance and accumulation in Arabidopsis halleri and Arabidopsis arenosa. Plant Cell Environ 43(12):3002–3019. https://doi.org/10.1111/pce.13883

Tai Z, Yin X, Fang Z, Shi G, Lou L, Cai Q (2017) Exogenous GR24 alleviates cadmium toxicity by reducing cadmium uptake in switchgrass (Panicum virgatum) seedlings. Int J Environ Res Public Health 14(8):8. https://doi.org/10.3390/ijerph14080852

Taiz L, Zeiger E (2002) Plant physiology, 3rd edn. Sinauer Associates

Tan X, Huang J, Lin L, Tang Q (2022) Exogenous melatonin attenuates Cd toxicity in tea (Camellia sinensis). Agronomy 12(10):10. https://doi.org/10.3390/agronomy12102485

Tang Y, Wang L, Xie Y, Yu X, Li H, Lin L, Liao M, Wang Z, Sun G, Wang X, Liang D, Xia H, Tu L (2020) Effects of exogenous abscisic acid on the growth and cadmium accumulation of lettuce under cadmium-stress conditions. Int J Environ Anal Chem 100(6):720–731. https://doi.org/10.1080/03067319.2019.1639686

Tang L, Dong J, Tan L, Ji Z, Li Y, Sun Y, Chen C, Lv Q, Mao B, Hu Y, Zhao B (2021) Overexpression of OsLCT2, a low-affinity cation transporter gene, reduces cadmium accumulation in shoots and grains of rice. Rice 14(1):89. https://doi.org/10.1186/s12284-021-00530-8

Tao Q, Zhao J, Li J, Liu Y, Luo J, Yuan S, Li B, Li Q, Xu Q, Yu X, Huang H, Li T, Wang C (2020) Unique root exudate tartaric acid enhanced cadmium mobilization and uptake in Cd-hyperaccumulator Sedum alfredii. J Hazard Mater 383:121177. https://doi.org/10.1016/j.jhazmat.2019.121177

Thind S, Hussain I, Ali S, Hussain S, Rasheed R, Ali B, Hussain HA (2020) Physiological and biochemical bases of foliar silicon-induced alleviation of cadmium toxicity in wheat. J Soil Sci Plant Nutr 20(4):2714–2730. https://doi.org/10.1007/s42729-020-00337-4

Thomine S, Vert G (2013) Iron transport in plants: better be safe than sorry. Curr Opin Plant Biol 16(3):322–327. https://doi.org/10.1016/j.pbi.2013.01.003

Tolrà R, Pongrac P, Poschenrieder C, Vogel-Mikuš K, Regvar M, Barceló J (2006) Distinctive effects of cadmium on glucosinolate profiles in Cd hyperaccumulator Thlaspi praecox and non-hyperaccumulator Thlaspi arvense. Plant Soil 288(1):333–341. https://doi.org/10.1007/s11104-006-9124-1

Vaculík M, Lux A, Luxová M, Tanimoto E, Lichtscheidl I (2009) Silicon mitigates cadmium inhibitory effects in young maize plants. Environ Exp Bot 67(1):52–58. https://doi.org/10.1016/j.envexpbot.2009.06.012

Valko M, Morris H, Cronin MTD (2005) Metals, toxicity and oxidative stress. Curr Med Chem 12(10):1161–1208. https://doi.org/10.2174/0929867053764635

Van Assche F, Clijsters H (1990) Effects of metals on enzyme activity in plants. Plant Cell Environ 13(3):195–206. https://doi.org/10.1111/j.1365-3040.1990.tb01304.x

Verma V, Ravindran P, Kumar PP (2016) Plant hormone-mediated regulation of stress responses. BMC Plant Biol 16:86. https://doi.org/10.1186/s12870-016-0771-y

Vinceti M, Venturelli M, Sighinolfi C, Trerotoli P, Bonvicini F, Ferrari A, Bianchi G, Serio G, Bergomi M, Vivoli G (2007) Case-control study of toenail cadmium and prostate cancer risk in Italy. Sci Total Environ 373(1):77–81. https://doi.org/10.1016/j.scitotenv.2006.11.005

Vocciante M, Grifoni M, Fusini D, Petruzzelli G, Franchi E (2022) The role of plant growth-promoting rhizobacteria (PGPR) in mitigating plant's environmental stresses. Appl Sci 12(3):3. https://doi.org/10.3390/app12031231

Waalkes MP (2003) Cadmium carcinogenesis. Mutat Res 533(1):107–120. https://doi.org/10.1016/j.mrfmmm.2003.07.011

Wan L, Zhang H (2012) Cadmium toxicity. Plant Signal Behav 7(3):345–348. https://doi.org/10.4161/psb.18992

Wang X, Liu Y, Zeng G, Chai L, Song X, Min Z, Xiao X (2008) Subcellular distribution and chemical forms of cadmium in Bechmeria nivea (L.) Gaud. Environ Exp Bot 62(3):389–395. https://doi.org/10.1016/j.envexpbot.2007.10.014

Wang F, Wang M, Liu Z, Shi Y, Han T, Ye Y, Gong N, Sun J, Zhu C (2015) Different responses of low grain-Cd-accumulating and high grain-Cd-accumulating rice cultivars to Cd stress. Plant Physiol Biochem 96:261–269. https://doi.org/10.1016/j.plaphy.2015.08.001

Wang H-F, Chang M, Peng T-T, Yang Y, Li N, Luo T, Cheng Y-M, Zhou M-Z, Zeng X-H, Zheng L-P (2017) Exposure to cadmium impairs sperm functions by reducing CatSper in mice. Cell Physiol Biochem 42(1):44–54. https://doi.org/10.1159/000477113

Wang X, Dong G, Liu X, Zhang S, Li C, Lu X, Xia T (2020) Poly-γ-glutamic acid-producing bacteria reduced Cd uptake and effected the rhizosphere microbial communities of lettuce. J Hazard Mater 398:123146. https://doi.org/10.1016/j.jhazmat.2020.123146

Wang T, Song J, Liu Z, Liu Z, Cui J (2021) Melatonin alleviates cadmium toxicity by reducing nitric oxide accumulation and IRT1 expression in Chinese cabbage seedlings. Environ Sci Pollut Res 28(12):15394–15405. https://doi.org/10.1007/s11356-020-11689-w

Watanabe K (2001) Microorganisms relevant to bioremediation. Curr Opin Biotechnol 12(3): 237–241. https://doi.org/10.1016/S0958-1669(00)00205-6

Wątły J, Łuczkowski M, Padjasek M, Krężel A (2021) Phytochelatins as a dynamic system for Cd (II) buffering from the micro- to femtomolar range. Inorg Chem. https://doi.org/10.1021/acs.inorgchem.0c03639

Wei S, Zhou Q, Wang X, Zhang K, Guo G, Ma LQ (2005) A newly-discovered Cd-hyperaccumulator Solatium nigrum L. Chin Sci Bull 50(1):33–38. https://doi.org/10.1360/982004-292

Wei X, Fang L, Cai P, Huang Q, Chen H, Liang W, Rong X (2011) Influence of extracellular polymeric substances (EPS) on Cd adsorption by bacteria. Environ Pollut 159(5):1369–1374. https://doi.org/10.1016/j.envpol.2011.01.006

White PJ (2017) Ion transport. In: Thomas B, Murray BG, Murphy DJ (eds) Encyclopedia of applied plant sciences, 2nd edn. Academic Press, pp 238–245. https://doi.org/10.1016/B978-0-12-394807-6.00216-1

Wu M, Luo Q, Zhao Y, Long Y, Liu S, Pan Y (2018) Physiological and biochemical mechanisms preventing Cd toxicity in the new hyperaccumulator Abelmoschus manihot. J Plant Growth Regul 37(3):709–718. https://doi.org/10.1007/s00344-017-9765-8

Wu X, Tian H, Li L, Wang X (2021) Polyaspartic acid alleviates cadmium toxicity in rapeseed leaves by affecting cadmium translocation and cell wall fixation of cadmium. Ecotoxicol Environ Saf 224:112685. https://doi.org/10.1016/j.ecoenv.2021.112685

Xia J, Yamaji N, Kasai T, Ma JF (2010) Plasma membrane-localized transporter for aluminum in rice. Proc Natl Acad Sci 107(43):18381–18385. https://doi.org/10.1073/pnas.1004949107

Xian P, Yang Y, Xiong C, Guo Z, Alam I, He Z, Zhang Y, Cai Z, Nian H (2023) Overexpression of GmWRKY172 enhances cadmium tolerance in plants and reduces cadmium accumulation in soybean seeds. Front Plant Sci 14. https://www.frontiersin.org/articles/10.3389/fpls.2023.1133892

Xiao Y, Wu X, Liu D, Yao J, Liang G, Song H, Ismail AM, Luo J-S, Zhang Z (2020) Cell wall polysaccharide-mediated cadmium tolerance between two Arabidopsis thaliana ecotypes. Front Plant Sci 11. https://www.frontiersin.org/articles/10.3389/fpls.2020.00473

Xie M, Chen W, Lai X, Dai H, Sun H, Zhou X, Chen T (2019) Metabolic responses and their correlations with phytochelatins in Amaranthus hypochondriacus under cadmium stress. Environ Pollut 252:1791–1800. https://doi.org/10.1016/j.envpol.2019.06.103

Yang H, Jie Y-L (2005) Uptake and transport of calcium in plants. Zhi Wu Sheng Li Yu Fen Zi Sheng Wu Xue Xue Bao 31:227–234

Yang J, Pan X (2013) Root exudates from sunflower (Helianthus annuus L.) show a strong adsorption ability toward Cd(II). J Plant Interact 8(3):263–270. https://doi.org/10.1080/17429145.2012.737030

Yang XE, Long XX, Ye HB, He ZL, Calvert DV, Stoffella PJ (2004) Cadmium tolerance and hyperaccumulation in a new Zn-hyperaccumulating plant species (Sedum alfredii Hance). Plant Soil 259(1):181–189. https://doi.org/10.1023/B:PLSO.0000020956.24027.f2

Yao Z, Li J, Xie H, Yu C (2012) Review on remediation technologies of soil contaminated by heavy metals. Procedia Environ Sci 16:722–729. https://doi.org/10.1016/j.proenv.2012.10.099

Yao W, Yang Z, Huang L, Su C (2022) Complexation of amino acids with cadmium and their application for cadmium-contaminated soil remediation. Appl Sci 12(3):3. https://doi.org/10.3390/app12031114

Yuan Z, Luo T, Liu X, Hua H, Zhuang Y, Zhang X, Zhang L, Zhang Y, Xu W, Ren J (2019) Tracing anthropogenic cadmium emissions: from sources to pollution. Sci Total Environ 676: 87–96. https://doi.org/10.1016/j.scitotenv.2019.04.250

Zargar T, Mir A, Alam P, Hayat S (2022) Melatonin alleviates cadmium-induced toxicity by modulating antioxidant defence mechanisms, growth and photosynthesis in Brassica juncea. Russ J Plant Physiol 69. https://doi.org/10.1134/S1021443722060279

Zhang Z, Zhang Y (2021) Melatonin in plants: what we know and what we don't. Food Qual Saf 5: fyab009. https://doi.org/10.1093/fqsafe/fyab009

Zhang X, Xia H, Li Z, Zhuang P, Gao B (2011) Identification of a new potential Cd-hyperaccumulator Solanum photeinocarpum by soil seed bank-metal concentration gradient method. J Hazard Mater 189(1):414–419. https://doi.org/10.1016/j.jhazmat.2011.02.053

Zhang K, Wang G, Bao M, Wang L, Xie X (2019) Exogenous application of ascorbic acid mitigates cadmium toxicity and uptake in Maize (Zea mays L.). Environ Sci Pollut Res Int 26(19): 19261–19271. https://doi.org/10.1007/s11356-019-05265-0

Zhang G, Ba Q, Chen S, Liu F, Li G (2020a) Exogenous application of glycine betaine alleviates cadmium toxicity in super black waxy maize by improving photosynthesis, the antioxidant system and glutathione-ascorbic acid cycle metabolites. Cereal Res Commun 48(4):449–458. https://doi.org/10.1007/s42976-020-00062-9

Zhang L, Gao C, Chen C, Zhang W, Huang X-Y, Zhao F-J (2020b) Overexpression of rice OsHMA3 in wheat greatly decreases cadmium accumulation in wheat grains. Environ Sci Technol 54(16):10100–10108. https://doi.org/10.1021/acs.est.0c02877

Zhao FJ, Jiang RF, Dunham SJ, McGrath SP (2006) Cadmium uptake, translocation and tolerance in the hyperaccumulator Arabidopsis halleri. New Phytol 172(4):646–654. https://doi.org/10.1111/j.1469-8137.2006.01867.x

Zhao X, Cheng Z, Zhu YI, Li S, Zhang L, Luo Y (2015) Effects of paternal cadmium exposure on the sperm quality of male rats and the neurobehavioral system of their offspring. Exp Ther Med 10(6):2356–2360. https://doi.org/10.3892/etm.2015.2777

Zhao YN, Wang MQ, Li C, Cao HW, Rono JK, Yang ZM (2022) The metallochaperone OsHIPP56 gene is required for cadmium detoxification in rice crops. Environ Exp Bot 193:104680. https://doi.org/10.1016/j.envexpbot.2021.104680

Zhou Z, Wei C, Liu H, Jiao Q, Li G, Zhang J, Zhang B, Jin W, Lin D, Chen G, Yang S (2022) Exogenous ascorbic acid application alleviates cadmium toxicity in seedlings of two wheat (Triticum aestivum L.) varieties by reducing cadmium uptake and enhancing antioxidative capacity. Environ Sci Pollut Res 29(15):21739–21750. https://doi.org/10.1007/s11356-021-17371-z

Zhu XF, Jiang T, Wang ZW, Lei GJ, Shi YZ, Li GX, Zheng SJ (2012) Gibberellic acid alleviates cadmium toxicity by reducing nitric oxide accumulation and expression of IRT1 in Arabidopsis thaliana. J Hazard Mater 239–240:302–307. https://doi.org/10.1016/j.jhazmat.2012.08.077

Zimeri AM, Dhankher OP, McCaig B, Meagher RB (2005) The plant MT1 metallothioneins are stabilized by binding cadmiums and are required for cadmium tolerance and accumulation. Plant Mol Biol 58(6):839–855. https://doi.org/10.1007/s11103-005-8268-3

Priming, Cd Tolerance, and Phytoremediation

Erna Karalija, Mirel Subašić, and Alisa Selović

Abstract Cadmium is a non-essential toxic soil pollutant with toxic effects on plants, especially crop plants, and can affect human health through accumulation in crop plants, thus entering the food chain. Under Cd stress, plants experience oxidative stress, lipid peroxidation, reduction in chlorophyll content, and reduced growth as well as biomass production and yield. Seed priming is a simple, cost-effective technique used for improvement of seed and seedling performance. In case of amelioration of Cd stress through seed priming, there is significant research that confirms positive effects of seed priming using salicylic acid as well as other priming agents to enhance plants tolerance towards Cd stress.

Keywords Phytoremediation · Cd toxicity · Seed priming

1 Introduction

Depletion and degradation of natural resources due to climate change events, soil pollution, and human activities as well as rapid human population growth is increasing the risk of human hunger as well as political and economic crisis (Beddington et al. 2012). Additionally, decrease in germination and poor seedling emergence due to increased stress constrain crop productivity while disruptions in precipitation lead to unfavourable soil moisture combined with drought adding on to poor crop yield (Farooq et al. 2019).

E. Karalija
Laboratory for Plant Physiology, Department of Biology, Faculty of Science, University of Sarajevo, Sarajevo, Bosnia and Herzegovina

M. Subašić
Faculty of Forestry, University of Sarajevo, Sarajevo, Bosnia and Herzegovina

A. Selović (✉)
Laboratory for Analytical Chemistry, Department of Chemistry, Faculty of Science, University of Sarajevo, Sarajevo, Bosnia and Herzegovina
e-mail: alisa.selovic@pmf.unsa.ba

A. K. Jha, N. Kumar (eds.), *Cadmium Toxicity Mitigation*,
https://doi.org/10.1007/978-3-031-47390-6_11

Seed priming is a low-cost and effective method that can be applied to improve seed and seedling performance and involves seed rehydration followed by dehydration before radicle protrusion. For seed priming, solutions with low water potential are used to ensure controlled rehydration, while the duration of priming varied depending on the seed used as well as the priming agent (Subašić et al. 2022). In the process of priming, seeds are rehydrated and metabolic processes, DNA repair, and mRNA synthesis are activated initiating different processes that contribute to increased seedling vigour as well as increasing plants' adaptability and stress tolerance (Karalija and Selović 2018).

Cadmium (Cd) belongs to the group of non-essential elements that are toxic to living organisms, are persistent in the environment, and can bioaccumulate; it is included in the first group of carcinogens. Elevated levels of Cd in the environment cause serious hazard to humans, animals, and plants (Lacave et al. 2020). Background amounts of cadmium in soil vary depending on the parent rock materials, but usually the content of this heavy metal is below 1 mg/kg. In Europe, for example, median value of Cd in topsoil is 0.15 mg/kg with maximum level of 14.1 mg/kg (UNEP 2013). Cadmium is used in many industries and products such as electroplating, pigments, telluride solar cells, plastics, and rechargeable batteries, and it is found in cigarette smoke. Cadmium isotope 113Cd is used in the nuclear power plant to capture thermal neutrons. Further, activities like mining, fossil fuel combustion, and non-proper use of pesticides and fertilizers can also lead to an increased Cd content in the environment. Soil polluted with cadmium has been identified in agricultural soils in Europe (about 21 percent) as well as in many areas in South America and sub-Saharan Africa (FAO and UNEP 2021).

Crop plants are usually sensitive to Cd even in low concentrations and can show signs of Cd toxicity such as leaf chlorosis, growth retardation, and root damage resulting in significant decreases of crop yield. Seed priming can contribute to increase of plants to Cd (Karalija and Selović 2018), and it can elevate the lethal threshold in plants tolerant to Cd and increase levels of accumulation (Karalija et al. 2021).

2 Mechanism and Techniques of Seed Priming

Seed germination is one of the most critical processes in plants growth and development; it is the most vulnerable plant state considering that complete first steps of seedling growth are at the expense of seed storage reserves. Emerging plumule and radicle are fragile and highly susceptible to stress factors. The process itself includes different events, from metabolic processes (activation of food reserves) to growth processes driven by the molecular patterns embedded in the seed. Beginning of the seed germination is marked by the process of rehydration of the seed by water uptake (Phase I) that subsequently initiates metabolic changes through activation of translation processes (Phase II) and finally by initiation of growth of plumule and radicle (phase III). Priming is a technique used for improvement of seeds, seedling, and plant performance and involves imbibition of seeds under controlled conditions resulting in initiation of early germination events followed by desiccation to original moisture content. When using priming, it is important to stop the process of priming before the entrance to phase III while completing phase I and II. During phase I, seeds are prone to uptake water rapidly which can quickly lead to germination; in priming the goal is to slow down lag phase by slowing down and controlling the rate of water uptake, usually using osmotic solutions. Changes in protein synthesis under priming have been recorded as well as changes related to starch metabolism, especially related to a-amylase activity (Pawar and Laware 2018). Additionally, seed priming can improve the activity of malate synthase and isocitrate lyase, enzymes involved in conversion of lipids into carbohydrates affecting the scavenging activity of antioxidant enzymes. In this way, seed priming can increase seed longevity by diminishing lipid peroxidation and oxidative stress through increased antioxidant capacity.

Seed priming can be performed using different priming agents, but the procedure needs to be optimized often for each species and can include manipulation of various factors such as the duration of priming treatment, temperature, and light conditions, but also storage conditions of primed seed need to be monitored and optimized (Selvarani and Umarani 2011). Depending on the agent used, priming can be grouped into hydro-priming, halo-priming, osmo-priming, hormonal priming, matrix priming, nutri-priming, bio-priming, and nano-priming (Pawar and Laware 2018).

Hydro-priming process uses water to prime seeds by soaking the seeds for several hours, redrying the seeds, and then sowing in the next few days. It is recorded that such technique can improve seed germination and seedling vigour and restore the quality lost during long storage (Musa et al. 1999). The reparative effect of priming is due to events in phase II such as DNA, RNA, protein, enzyme, and membrane repair (Soleimanzadeh 2013).

Halo-priming uses inorganic salts (e.g. sodium chloride, potassium chloride, potassium nitrate, and calcium chloride). Some of the salts used for halo-priming can also provide nutritional benefits, and using halo-priming improves seed germination and yield (Bajehbaj 2010). Osmo-priming has been successfully used for improvement of seed germination and plant performance under salt stress and/or drought conditions (Abraha and Yohannes 2013). The possible mechanism for improvement of seed performance under halo-priming hypothesizes that water absorption of halo-primed seeds is more effective in synchronizing seed germination, further stimulating germination through prepared seeds for cell division facilitating germination (Singh et al. 2015). Additionally, halo-primed seed can tolerate higher concentrations of salt in medium due to accumulation of K^+ and Ca^{2+} thus reducing Na^{2+} accumulation and through accumulation of proline enhancing osmosis regulation (Pawar and Laware 2018).

Osmo-priming uses osmotic solutes, with lower water potential, for seed soaking facilitating restricted water uptake, followed by dehydration before sowing. Osmotic solutes often used in seed priming include simple molecules like sugars, polyethylene glycol (PEG), glycerol, mannitol, sorbitol, and specialized vermiculite compounds (Rehman et al. 2015). A large number of research studies have been testing PEG 6000, with reports confirming that this type of osmo-priming can enhance germination and stimulate root growth under drought or salt stress. The mechanism behind enhanced seed performance due to PEG priming lies in its inert nature and ability to maintain stable osmotic potential during imbibition, reviving seed metabolism, facilitating germination synchronicity, and reducing physiological heterogeny in germination (Pawar and Laware 2018).

Priming using plant growth regulators (*hormonal priming*) utilizes naturally occurring plant hormones such as salicylic acid, jasmonic acid, gibberellic acid (GA_3), kinetin, and others that in general are involved in plant growth and development and/or stress response (Singh et al. 2015). Seed priming using gibberellins can enhance enzymatic activities such as carbonic anhydrase (CA) as well as many biochemical processes such as protein content antioxidant enzymes, sugar, and starch metabolism, as well as growth processes (plant yield, leaf number and area, shoot and root elongation). Priming with hormones can have large-scale effects on plant processes and tolerance towards different stressors. Priming using brassinolide in lucerne seeds induced an increase in germination percentage, vigour index, and antioxidant enzyme activity (CAT, POD, and SOD) inducing greater tolerance of salt stress (Zhang et al. 2007).

Matrix priming is a process of priming that refers to incubation of seeds in a matrix (solid and non-soluble) restricting the water uptake; often used matrixes are sand and vermiculite. Using sand as matrix can improve seed germination up to 50% inducing higher seedling vigour which can be very useful for priming of crops (Selvarani and Umarani 2011).

Nutri-priming uses solutes containing plant nutrients important for normal plant growth as priming agents to improve nutrient content in seeds. Examples are micronutrients that can increase plants photosynthesis improving plant growth and yield (Farooq et al. 2012). Research using iron and boron in seed priming can induce increase of germination and essential oil yield in dill. Contrarily if boron is used in higher concentrations, it can impair seedling establishment (Mirshekari 2012). Enhancement of plant growth and increasing leaf numbers, dry mater, nodule number, and leghaemoglobin content in pigeon pea and peanuts was recorded when seeds were primed with copper sulphate or manganese sulphate, respectively (Khalid and Malik 1982; Raj 1987).

Bio-priming normally refers to seed priming using beneficial microorganisms. In *Catharanthus roseus* varieties, seed priming increased production of ajmalicine and improved number of leaves, root length, and shoot height (Karthikeyan et al. 2009). In barley, *Azospirillum brasilense* and *Pseudomonas fluorescens* were used to prime seeds resulting in improvement of yield, grain weight, and an increase in the number of grains per spike (Shirinzadeh et al. 2013). Seed priming using bacteria such as Rhizobium especially in combination with nutri-priming can activate/inhibit mRNA, membrane permeability, and pathogen defence mechanisms (Pattanayak et al. 2000).

Nano-priming is a relatively new method that utilizes nanoparticles (NP) like NP of zinc oxide, iron oxide, silver, titanium dioxide, etc. for seed priming. Besides seed priming nanoparticles have found application in mineral supplementation of plants facilitating easier utilization of minerals from the soil that would be otherwise more easily washed out if they were not delivered in the form of nanoparticles (Alam et al. 2015). Depending on the mineral used (in the form of nanoparticles) priming effects are diverse; iron NP enhance germination while Cu NP can have inhibitory effect emphasizing the importance of priming optimization procedures adjusted to species and priming agent (Yasmeen et al. 2015). A large number of studies have been analysing the effect of silver NP on wheat, barley (Cantliffe et al. 1984), corn, watermelon, and zucchini (Hojjat 2015), and all studies indicated that AG dose is crucial in obtaining optimal results; larger doses can result in the invasion of AG into the seeds damaging chromosomes resulting in different abnormalities (Pawar and Laware 2018).

Titanium NP increased germination of fenugreek seeds, increased proteins, carbohydrates, and chlorophyll, and reduced sugar content (Durairaj et al. 2015). Similarly, seed priming with TiO_2 nanoparticles in pepper improved germination and enhanced biochemical parameters through increased translation and electron transport boosting photosynthetic efficiency, increasing carbohydrate synthesis. Improved photosynthetic activity combined with enhanced nitrogen metabolism led to improved seedling growth (Dehkourdi et al. 2014).

Other nanoparticles also had positive effect on seed and seedling performance, such as zinc oxide (ZnO), with reports confirming positive effects on seed germination, adaptability, growth, and chlorophyll content. The concentration of nanoparticles is an important factor in the optimization of seed priming, and higher concentrations can give negative effects such as Zn nanoparticles in concentration of 2000 ppm that induced negative effects on plant growth (Prasad et al. 2012), reducing flowering and initiating other phytotoxic effects (Laware and Raskar 2014).

3 Effects of Seed Priming on Plants Growth

Seed priming has many beneficial effects on plant development, most prominent being faster germination rate, increase in seedling vigour, and faster and more successful crop establishment due to increased seedling adaptability. Additionally seed priming can help in germination of seeds under thermodormancy, where high temperatures prevent germination due to inhibition of amino acid accumulation and esterase activity; priming with kinetin can reverse this and lead to seed germination (Cantliffe et al. 1984). Similarly in lettuce where β-mannanase is necessary for weakening, the endosperm high temperatures can inhibit ethylene production which regulates the activity of β-mannanase and thus inhibit germination. In this case, osmo-priming with PEG can overcome thermodormancy and induce seed germination (Nascimento et al. 2004).

One beneficial effect of priming can also ameliorate DNA damage caused by ageing and increase germination of seeds after long storage, and it can reduce the soil-borne plant diseases. There are reports suggesting that seed priming using plant extracts (such as *Acacia nilotica* and *Sapindus mukorossi* extracts) can reduce infection rate and frequency with root fungi (such as *Macrophomina phaseolina*, *Fusarium* spp., and *Rhizoctonia solani*) in different crops like okra, sunflower, peanut, and chickpea seeds (Rafi et al. 2015).

The effect of seed priming on seedling establishment has been well studied, especially under stress conditions (Farooq et al. 2019; Karalija et al. 2021). Once sown, primed seeds emerge quickly with synchronized seed germination decreasing the differences in growth in later stages of plant. Quick germination contributes to plants' faster growth and fewer crop failures due to slow germination (Karalija et al. 2022a). Increased seedling vigour results in better mobilization of nutrients and enables seed to germinate faster as well as completion of other phenological phases earlier compared to control (Farooq et al. 2012; Rehman et al. 2015).

Hydro-primed rice seeds produced seedlings with stronger roots and longer roots, as well as higher biomass (Farooq et al. 2006), while seed priming using selenium and salicylic acid improved chilling tolerance in rice through an increase in soluble sugar content thanks to the improved activity of α-amylase (Wang et al. 2016). Seed priming benefits reflected in seedlings are a result of priming-induced processes being "encapsulated" as memory in seeds through desiccation process. Primed memory in seeds includes triggering of different signalling pathways and genome-level changes that are integrated into seedlings' metabolome and genome. An increase in gene expression related to ribosome has been recorded in silicone acid-primed seeds, contributing to seed-primed memory through adjustments in ribosome activity (Karalija et al. 2022b) affecting seedling growth and adaptability.

Yield can be largely affected by seed priming. Primed seeds have earlier seedling emergence and uniformity of crop stands. Plants perform better under stress resulting in increased growth rates, dry matter, and yield (Farooq et al. 2019). The real effect of priming is more evident under suboptimal and stressful conditions with primed seeds also having competitive advantage over weeds (Sarkar 2012). Competitive advantage of primed seeds includes higher drought tolerance, higher heavy metal stress tolerance, salt stress tolerance, and higher resistance to pathogens, depending on type and concentration of priming agent (Farooq et al. 2019).

Additionally seed priming can improve use of resources in terms of water as well as improve the quality of harvested produce (Bakhtavar et al. 2015). Using micronutrients as priming agents (nutri-priming) can improve micronutrient content in seeds and affect its germination; e.g. using Zn priming can improve Zn and P content in seeds which improves water use efficiency in barley improving germination and seedling growth (Ajouri et al. 2004). It was also recorded that seed response to P priming can be related to genotype with different effects. Seed priming with P of rice containing different purine permease 1 (*PUP1*) genes (low and high P content genotypes) resulted in increased P content in all seeds. Even though seeds contained higher P levels the effect was different in different genotypes. Low P genotypes demonstrated improved stand establishment with higher germination and seedling growth, while in high P genotypes it was noticed that early seedling growth was enhanced (Pame et al. 2015). The combination of seed priming with different genotypes in crops can be effectively utilized to improve crop tolerance, seedling emergence and establishment, and yield and crop quality with the possibility of also improving the shelf life of crops grown from primed seeds.

4 Effect of Cd on Plant Growth and Plant Tolerance

Heavy metals as soil pollutants affect plant growth globally, with the most prominent effect on crops considering their ability to bioaccumulate and possibility of entry into the food chain. Cadmium is a non-essential and toxic heavy metal which can influence morphological and physiological traits of plants through generation of oxidative stress (Karalija and Selović 2018; Yang et al. 2018). Research has shown that Cd can modulate different processes such as stomatal opening, photosynthesis, and transpiration (Chandra and Kang 2016).

Morphological symptoms of Cd toxicity in plants are chlorosis, leaf roll, and plant growth retardation, while physiological symptoms include effects on photosynthetic rate and decrease in nitrogen and primary ammonia assimilation (Kadioglu et al. 2012; Hussain et al. 2020). Transport of different minerals can be affected by Cd such as Ca, K, Mg, and P (Mourato et al. 2019). Under Cd toxicity nitrate reductase activity is decreased affecting absorption of nitrate and its transport from roots to shoots.

The photosynthetic rate is reduced due to Cd effects on chlorophyll quantity (Cheng et al. 2002). Chronic exposure even to low concentrations of Cd can induce a significant decrease of chlorophyll content and affect biomass production (Kumar et al. 2021). Cd affects the ultrastructure of chloroplast and decreases photosynthesis; at the same time, it decreases stomatal conductance resulting in decreased transpiration rate (Chen et al. 2020). The effect on photosynthesis is also reflected in Cd^{2+} ion effect on the activity of Rubisco by substituting Mg^{2+} ions, cofactors of carboxylation reactions, shifting the Rubisco activity towards oxygenation reactions (Figlioli et al. 2019). Furthermore, accumulated Cd in plant cells can induce retrogression of mitochondria, accelerating necrosis, chlorosis, and retardation of root and shoot growth (Rizwan et al. 2019).

Main toxic effect of Cd is exerted through oxidative stress through indirect creation of ROS which can be damaging to cell membranes and lipid peroxidation (Jawad Hassan et al. 2020). The synthesis of free amino acids and soluble sugars in roots and shoots is also affected by Cd (Zhu et al. 2018).

Adding to the changes in photosynthesis and increase of ROS production, accumulation of Cd also affects nutrient uptake in plants, especially by modulation of N metabolism (Khan et al. 2016). Reduction of Ca, Mg, K, and micronutrients in tomato under Cd stress (Street et al. 2010) as well as decrease in Fe and Mn content in sunflower leaves has been recorded (Nada et al. 2007). Modulation of N metabolisms is crucial ion plant tolerance towards Cd through synthesis of N-containing metabolites like proline, phytochelatins, and glutathione, which play crucial role in plants tolerance to Cd (Sharma and Dietz 2006).

5 Phytoremediation of Cd

5.1 Bioavailability of Soil Cd

Two important atomic properties, ionic potential (IP) and electronegativity, influence cadmium biogeochemical behaviour in soil system. The Pauling radius of 97 pm and high polarizability define the tendency of Cd to react with ligands containing nitrogen or sulphur and form strong complexes through more covalent bonding. Acting as a soft Lewis acid, cadmium creates stable solution complexes with inorganic and organic ligands such as halide ions, acetate, citrate, succinate, organic sulphides, and thiols. Different chemical species of Cd like free cation, hydrated cation, complexed with ligands, bound to carbonates, Fe/Mn oxides, organic matter, or silicate minerals (residual form) can be found in soil. Considering that the soil solution is in constant contact with the solid phase, the concentration of metal species can change depending on the conditions in the soil. This partition is particularly important since the soil species of Cd directly affects bioavailability, mobility, and uptake of this heavy metal by plants. Primary modes of Cd occurrence in silicate minerals are sulphide inclusions and isomorphic substitutions for Zn, Cu, Pb, and Hg in sulphides. As a trace element in soil, Cd can be coprecipitated with secondary soil minerals such as iron and manganese oxides and carbonates as well as in association with soil humus (Sposito 2008). Similarity of ionic radii between Cd and Ca (109 pm) facilitates the cation substitutions and results in relatively high content of Cd in carbonate and phosphate minerals. Weathering of these soil solids will determine the Cd potential hazard to living organisms (Xia et al. 2020).

Plants are exposed to Cd via soil solution, and cadmium species that are available to plants for root uptake are only Cd^{2+} ions (Chavez et al. 2016). Soil cadmium bioavailability is influenced by physicochemical soil properties such as soil pH, clay content, amount of organic matter, ionic strength, soil texture, and redox potential. Among biotic factors, plant roots and soil microbes can affect metal bioavailability through secretion of root exudates, siderophores, amino acids, and enzymes (Chandra et al. 2018). In soil, key processes governing the transfer of metal from the solid to the solution phase include sorption, complexation, and precipitation.

Sorption can be considered an effective way to control the migration and transformation of heavy metals in the soil. The major sorbents in soils are silicate clays, iron/manganese oxides, and organic matter. Layer silicates such as kaolinite, smectites, and vermiculites play a significant role in Cd adsorption due to permanent negative charge on their mineral surface. Based on this, clay minerals can reduce the mobility of cadmium in contaminated soils as confirmed for montmorillonite (Suzuki et al. 2020), illite (Qi et al. 2022), and kaolinite (Li et al. 2021a). However, the presence of organic ligands in soil can influence the immobilization capacity of clay minerals for heavy metals. For example, dissolved organic matter from animal manure was found to reduce the capacity of bentonite and zeolite for cadmium adsorption (Zhou et al. 2017).

Iron and manganese oxides are the most abundant oxides in soils. Further, large surface area and sorption capacity of iron and manganese oxides can influence the mobility of cadmium through sorption. The heavy metal retention mechanism by metal oxides is well known (Shi et al. 2021). Sorption of Cd onto iron/manganese oxides can be reduced in the presence of other cations, and it decreases in the following order Ca2+ > K+ > Na + due to competition. Furthermore, co-composition of iron minerals and organic matter can affect the surface charge and surface pore size of iron oxides resulting in different adsorption properties of heavy metals from pure minerals (Yang et al. 2022). Cadmium sorption can increase in the presence of sulphate and phosphate because of the influences of cation exchange capacity and surface precipitation (Caporale and Violante 2016; Liang et al. 2014; Zaman et al. 2009).

Inorganic and organic ligands in the soil solution may decrease soil adsorption by formation of dissolved Cd complexes. Cd easily creates complexes with chloride ligands, and some soils may have high chloride concentrations to result in significant complexation. In saline soils, dominant chloro-complexes of Cd are $CdCl^+$, $CdCl_2^0$, $CdCl_3^-$, and $CdCl_4^-$. Chloride's ability to increase the bioavailability of cadmium has been proven (Guo et al. 2022; Li et al. 2019). Natural low molecular weight organic acids (LMWOAs) such as acetic, oxalic, succinic, tartaric, and citric acids may also complex cadmium and increase its mobility (Lu et al. 2021). On the other hand, high molecular weight organic acid (HMWOAs) principally composed of fulvic and humic acid can immobilize Cd in soil (Yao et al. 2022).

The formation of Cd precipitates is one of the processes governing the partitioning of cadmium between the soil solid phase and the soil solution. The most important solid species whose formation results in the immobilization of cadmium in the soil are phosphates, carbonates, sulphides, and hydroxides. Cadmium immobilization with phosphates is covered by several processes in the soil, as stated in the work of Ruangcharus et al. (2020). On the other hand, mechanisms of Cd immobilization with carbonate are not completely understood but may involve nucleation, chemisorption, and coprecipitation (Li et al. 2022b). Sun et al. (2019) demonstrated that cadmium hydroxide phases were formed at the first phase of Cd sorption on the surfaces of γ-Al2O3 and lately transformed to Cd carbonates.

The key factor controlling the solubility and chemical forms of cadmium is soil pH. With increasing acidity, Cd is transferred from immobile to a readily mobile and phytoavailable form. However, alkaline medium promotes a negative effect on plant uptake as sorption and precipitation of cadmium decrease free cation concentration in soil solution. Lu et al. (2022) demonstrated that the incorporation of crop residue biochars led to an increase in pH and a decrease in available Cd content in soil. It seems that pH 6 acts as a threshold point for the solubility of Cd in soil due to complex formation and its adsorption on mineral surfaces (Sullivan et al. 2013). Higher pH values cause a decrease in cadmium availability (Zhang et al. 2023) in both lightly and moderately polluted soils. According to Kang (2018), there is no simple linear relationship between pH and soil cadmium content. The behaviour of cadmium in soil solution in terms of dominant chemical species can also be affected by the addition of chelating agents.

Organic chelators usually form stable compounds with cadmium and increase its mobility (Guo et al. 2018). The use of degradable chelating agents to improve phytoextraction efficiency is a promising tool for remediation of cadmium-contaminated soil. As a novel green organic chelator, tetrasodium glutamate diacetate significantly increases the Cd content in French marigold (Li et al. 2022a). The application of different types of organic chelators in a growth medium contaminated with Cd to increase phytoextraction capacity is summarized in the results of the review by Zulfiqar et al. (2022). In the soil-plant system, root exudates can affect speciation and bioavailability of heavy metals by reducing the pH in the rhizosphere. Among other, this influence depends on the properties of the soil. He et al. (2022) showed that the translocation factors of Cd in Sorghum sudanense grown in soils of different textural grades were as follows: sandy soil > clay soil > clay soil.

It is worrying that some of the persistent pollutants, such as microplastics, can cause cadmium mobilization in soil and increase its bioavailability due to their large specific surface area as well as strong hydrophobicity (Huang et al. 2023).

5.2 *Phytoremediation Strategies*

Many methods have been available for soil remediation, but phytoremediation has attracted much attention due to its efficient, low-cost, environmentally friendly, and sustainable features. Phytoremediation is based on the use of plants for the extraction, degradation, or sequestration of contaminants from soil, sediment, sludge, or water. There are several phytoremediation strategies: phytostabilization, rhizofiltration, phytovolatilization, phytodegradation based on used plants, type of contaminants, and site characteristics (Nedjimi 2021). Phytoextraction and phytostabilization are especially useful for the remediation of soil contaminated with inorganic contaminants such as heavy metals, and the other strategies are more applicable for the remediation of organic contaminants.

Phytostabilization refers to the ability of used plants to immobilize heavy metals in soil through root exudates, thereby preventing their leaching into groundwater and reducing the likelihood of metals entering the food chain. Phytostabilization can be enhanced by chemical stabilization of metals using various soil amendments and plant species that are tolerant to elevated levels of heavy metals. For example, root to shoot translocation of Cd was reduced in perennial ryegrass grown in moderately contaminated soil after addition of synthetic or organic amendment (Epelde et al. 2009). Lan et al. (2020) showed that the simultaneous use of coconut shell biochar, organic fertilizer, and Fe-Si-Ca material on highly polluted soils can significantly increase soil pH, reduce Cd concentrations, and improve the growth of *Boehmeria nivea* L. The addition of mineral amendments significantly increased Festuca rubra biomass, decreased Cd content in the soil, and increased Cd concentrations in the roots (Radziemska et al. 2020).

Phytoextraction of heavy metals pertains to the removal of these contaminants from the soil through uptake by a plant root and their translocation to the aboveground plant parts. Based on the concentrations of heavy metals in aboveground plant parts, plants can be differentiated as metal excluders, indicators, and (hyper)-accumulators. Metal concentrations in excluders are very low with the ratio of metal level in the plant to that in soil far below 1. In indicators plants, the uptake of heavy metals and their translocation to the shoots are regulated in a such way that the ratio of metal level in plant to that in the soil is close to 1. Hyperaccumulators are plants which in quantities well above those contained in the soil or in non-accumulating plants can concentrate heavy metals in their aboveground parts (Farooqi et al. 2022). Regarding cadmium uptake and translocation, the term hyperaccumulator defines plants that accumulate more than 100 mg/kg of Cd in their harvestable parts. About 400 plant species are known to hyperaccumulate cadmium and some other heavy metals (Imperiale et al. 2022). Phytoextraction of heavy metals occurs in the root zone of plant, so numerous factors including soil pH, mobility and bioavailability of metals, and plant species and their physiology can affect the remediation process and phytoextraction efficiency. To increase heavy metal mobility and improve the metal accumulation capacities and uptake speed of plants, chelate-assisted phytoextraction can be used (Chengatt et al. 2022). Cadmium phytoavailability can be increased by using different natural and synthetic chelating agents, e.g. nitrilotriacetic acid (NTA), ethylene diamine disuccinate (EDDS), sodium glutamate tetraacetate (GLDA), citric acid (CA), oxalic acid (OA), diethylene triamine pentaacetic acid (DTPA), and ethylene diamine tetraacetic acid (EDTA) (Gul et al. 2021). For example, the addition of 30 mg/L of citric acid enhanced the translocation of Cd to aboveground parts of sunflower and elevated the growth of the plant (Niu et al. 2023). Different chelating agents can exhibit varying extraction efficiencies for Cd in the soil. In the work of Dong et al. (2023), EDTA, GLDA, and DTPA exhibited the extraction efficiencies for Cd of 67.67% while the CA had a medium extraction efficiency (15.25%), and the extraction efficiency of OA for Cd was less than 5.0%. Although chelating agents can improve the efficiency of heavy metal extraction, their addition to contaminated soil can lead to heavy metal leaching and groundwater contamination (Huang et al. 2019). Further, plant exposure to elevated levels of heavy metals can cause a broad range of physiological and biochemical changes.

Cd toxicity is known to reduce the uptake and translocation of water and nutrients, induce oxidative stress, and disrupt plant metabolism, resulting in stunted growth (Waheed et al. 2022; Haider et al. 2021). To survive and sustain proper growth under heavy metal stress, plants must develop/adopt stress defence mechanisms which can be triggered through exogenously applied treatments in a process called priming. Seed priming is a pre-sowing rehydration approach that allows the metabolic processes of germination to take place without actual germination (Marthandan et al. 2020).

6 Seed Priming and Cd Remediation/Tolerance

Enhanced ability of plants to tolerate Cd due to seed priming is a favourable trait in crop plants. Accelerated emergence of seedling under Cd stress is desirable considering that Cd^{2+} can affect and inhibit seed germination processes (Kumar et al. 2021). Research has demonstrated that several priming agents and techniques can alleviate Cd toxicity and some of them are summarized in Table 1. Using plant-associated bacteria as a priming agent, Jan et al. (2019) concluded that inoculation of *Bacillus cereus* under Cd treatments reduced Cd^{2+} uptake and increased antioxidant enzyme activities in rice cultivars. Nanoparticles such as multiwall carbon nanotubes can be used as a nutrient source for the protection of plants against metal stress and to improve crop plant growth. Chen et al. (2021a, b) conducted a pot experiment with multiwall carbon nanotubes and suggested that this seed priming agent could alleviate cadmium toxicity by promoting root and shoot fresh weight. Biologically active compounds such as S-methylmethionine (SMM), nonproteinogenic amino acid, are known to have the potential to minimize damage in plants exposed to abiotic stress. Pretreatment of one-month-old plants with 0.02 and 0.05 mM SMM lowered shoot Cd concentrations by 45 and 46%, respectively (Rana et al. 2022). Karrikinolide is a plant growth regulator that has been found in plant-derived smoke, and it can be used as a seed priming agent. In the work of Sardar et al. (2021), seed priming with karrikinolide showed improved germination rates and enhancements in growth, photosynthesis, and biomass production of coriander under cadmium stress.

Priming using salicylic acid, KNO_3, or KCl showed a positive effect on germination percentage and rate and plumule and radicle emergence and elongation, as well as an increase in fresh and dry weight, soluble protein, electrical conductivity, malondialdehyde (MDA), proline content, soluble sugars, etc. (Espanany et al. 2016).

Cadmium affects plant growth, and plants sensitive to cadmium toxicity can experience severe growth retardation due to disruption of photosynthesis, mineral metabolism, and their uptake (Haider et al. 2021). Seed priming using salicylic acid can decrease uptake and transport of Cd alleviating Cd toxicity by lowering accumulation rates. Additionally salicylic acid priming can reduce electrolyte leakage, as well as lipid peroxidation (reduction of MDA accumulation), and it can improve membrane repair and stability (Belkadhi et al. 2015; Gul et al. 2021).

Table 1 Some examples of alleviation of Cd stress by seed priming

Plant species	Priming agent	Priming method	Priming effect on plants under Cd stress	References
Oryza sativa cultivars cv. Basmati 385 (B-385) and S. Basmati (S-Basmati)	*Bacillus cereus*	Pot experiment, bacteria added in the form of bacterial suspension (10^8 CFU) directly to pots to 21 days old seedlings	Reduced Cd^{2+} uptake and increased antioxidant enzymes activities Enhanced plant growth, biomass production, photosynthetic pigments, micronutrients, and lowered electrolytes leakage	Jan et al. (2019)
Maize varieties, Yuebaitiannuo7 and Yuecainuo2	Multiwall carbon nanotubes (MWCNTs)	Seed priming with three levels of MWCNT (0, 100, and 200) Plants cultivated in pots	MWCNT application promoted the seed germination rate MWCNTs enhanced root and shoot fresh weight and antioxidant enzyme activities, including peroxidase (POD), superoxide dismutase (SOD), and catalase (CAT) activities, and reduced the malonaldehyde (MDA) content under Cd across MWCNT treatments	Chen et al. (2021a, b)
Elymus elongatus subsp. *ponticus* cv. Szarvasi-1	S-methylmethionine (SMM)	Four weeks old seedlings treated with SMM for 24 h Plants grown in Hoagland	SMM priming decreased the translocation and accumulation of Cd	Rana et al. (2022)
Coriandrum sativum L.	Karrikinolide (KAR1)	Seed primed with KAR1 solution Seeds cultivated in petri dishes	Improved relative water content, leaf osmotic potential, and membranous stability index Improved gas exchange attributes and metal tolerance index	Sardar et al. (2021)
Oryza sativa cultivars Basmati 385 and Shaheen Basmati	Smoke water of *Moringa oleifera*		Increase of biochemical, metabolic activities, and antioxidant	Shah et al. (2022)

(continued)

Table 1 (continued)

Plant species	Priming agent	Priming method	Priming effect on plants under Cd stress	References
		28 days old seedlings exposed to smoke water (1:1000) for one week in hydroponic culture	metabolites Improved plant growth	
Coriandrum sativum	Putrescine (Put)	Seeds primed with Put	Improved seed germination, gas exchange, root growth, and shoot growth Enhanced activity of superoxide dismutase, catalase, and peroxidase Increase of proline content	Sardar et al. (2022a)
Vigna radiata (NM 2006, NM 19–19, NM 2011, NM 20–21, NM 121–123, AZRI-2006, NM 13–1 and NM-51)	Gibberellic acid (GA3), salicylic acid (SA), and proline	Seeds primed with GA_3, SA, and proline solutions for 12 h Seed cultivated in petri dishes	Improved morphological parameters Changes in antioxidant enzymes activities Decrease of MDA and proline contents	Hassan et al. (2021)
Wheat (Lasani-2008)	Silicon nanoparticles (Si NPs)	Seed primed with Si NPs for 24 h under continuous aeration	Increase of growth, chlorophyll content Increase of antioxidant enzyme activity Decreased Cd concentrations in wheat	Hussain et al. (2019)
Mentha arvensis L.	Salicylic acid (SA)	Seed primed with SA in the dark for 30 min seeds cultivated in pots	Better regulation of stomatal conductance	Zaid et al. (2022)
Oryza sativa cultivars, e.g. Xiangyaxiangzhan (XYXZ) and Yuxiangyouzhan (YXYZ)	ZnO nanoparticle (ZnO NPs with 99.9% purity, and particle size of 30 ± 10 nm)	Seeds soaked in ZnO NPs solutions for 20 h at 25 °C in the dark with continuous aeration. Seeds cultivated in petri dishes	Improvement of early seedling growth and related physio-biochemical attributes	Li et al. (2021b)

Zea mays L.	Salep gum (SG), *Spirulina platensis* (SP), and combination of SG and SP	Seeds soaked in SG, SP, and SG + SP and dried for 2 to 3 h at room temperature (23 ± 2 °C). The experiment was set in coco-peat	Increase in plant growth, photosynthetic capacity, and Cd accumulation and translocation	Seifikalhor et al. (2020)
Cajanus cajan L. *(AL-201, AL-882)*	Methyl jasmonate (Me-JA)	Seeds soaked in Me-JA for 8 h.	Reduction of heavy metal-induced oxidative stress Increased proline production, glutathione, and ascorbic acid content	Kaushik et al. (2022)
Soybean cultivar *Giza-35*	*Serratia marcescens* BM1 suspension	The 6-day-old plants inoculated with *Serratia marcescens* BM1 suspension for 25 min and transferred into Hoagland solution	Increase of plant growth, biomass, gas exchange, nutrients uptake, antioxidant capacity, and the contents of chlorophyll, total phenolics, flavonoids, soluble sugars, and proteins Enhanced activity of antioxidant enzymes and the expression of stress-related genes	El-Esawi et al. (2020)
Hordeum vulgare L. cv. Slaven	Indole-3-butyric acid (IBA)	Seedlings placed between two sheets of filter paper soaked with IBA for 3 h	Activation of Cd tolerance mechanisms in roots	Demecsová et al. (2020)
Sedum alfredii	*Pseudomonas fluorescens*	The roots soaked in bacterial suspension	Enhanced photosynthetic and carbon-fixation processes	Wu et al. (2020)
Oryza sativa cultivars, i.e. Xiangyaxiangzhan and Yuxiangyouzhan	Lanthanum oxide nanoparticles (La_2O_3 NPs)	Seed primed with La_2O_3 NPs	Enhanced early seedlings growth	Sun et al. (2023)
Brassica napus L. (cultivars, namely Puriga and MS-007)	Calcium chloride ($CaCl_2$), hydrogen peroxide (H_2O_2), selenium (Se), and salicylic acid (SA)	Seeds soaked in appropriate solution for 8 h Seeds sown in pots	Improved growth and chlorophyll content	Ashraf et al. (2022)

(continued)

Table 1 (continued)

Plant species	Priming agent	Priming method	Priming effect on plants under Cd stress	References
Sesbania sesban L.	*Bacillus anthracis* PM21	Seeds immersed in bacterial inoculum for 3–4 h. Seeds sown in pots	Increased shoot and root length, and biomass (fresh and dry weight) Increased chlorophyll contents	Ali et al. (2021)
Azolla filiculoides	TiO_2 nanoparticles (TiO_2 NPs)	Seeds sonicated in solution containing TiO_2 NPs for 3 days	Higher translocation of cadmium to the aerial plant part Higher efficiency of the antioxidants (proline, GPX, and CAT) Induced decrease in H_2O_2 content	Spanò et al. (2019)
Nicotiana tabacum, cv. K326	Sodium chloride (NaCl)	60-days old seedlings exposed to gradually increasing salt stress in culture medium	Reduction of Cd absorption	Yang et al. (2020)
Coriandrum sativum L.	Selenium nanoparticles (SeNPs)	Seeds soaked in SeNPs solution for 24 h under dim light at 25 ± 1 °C Seeds cultivated in pots	Improved antioxidative systems Reduction of oxidative injury Increased plant growth	Sardar et al. (2022b)
Cucumis sativus	2-Hydroxymelatonin (2-OHMT)	Seeds soaked in 2-OHMT (50 mM, 100 mM, and 150 mM) for 4 h Seeds sown in pots	Enhanced accumulation of non-enzymatic antioxidants' upregulation of the expression of stress-responsive genes	Shah et al. (2020)
Wheat (Punjab-2011; Sammar) and rice (Kisan Basmati; Chenab)	Zn-lysine (Zn-lys)	Foliar spray with Zn-lys (0, 12.5, and 25 mM) solution	Reduced toxicity of Cd Improved plants' health Increase of photosynthesis and antioxidants	Ali et al. (2022)
Coriandrum sativum L.	Triacontanol (Tria) solutions	Seeds primed with triacontanol (Tria) solutions (5, 10, and	Alleviation of phytotoxic effects of Cd exposure	Sardar et al. (2022c)

		20 μmol L^{-1}) for 15 h Seeds were cultivated in pots		
Brassica juncea	Castasterone and citric acid	Seeds soaked for 8 h in castasterone solution (0, 0.01, 1, or 100 nM) Seeds cultivated in petri plate	Decreased ROS overproduction Improved photosynthetic pigment content Restored stomatal opening in stressed plants	Kaur et al. (2018)
Hordeum vulgare L.	*Bacillus mycoides* PM35 in combination with titanium dioxide (TiO_2) –nanoparticles	Seeds primed with titanium dioxide nanoparticles (TiO_2–NPs)	Decreased Cd concentration in the plant tissues Reduction of ROS Increased organic acids' exudation Increased antioxidant activities Increased content of essential nutrients in the plants	Ma et al. (2023)
Linum usitatissimum L.	Salicylic acid	Seeds soaked for 8 h in SA solution Grown in hydroponics	Preservation of lipids in chloroplast Improved photosynthetic capacity	Belkadhi et al. (2015)
Coriandrum sativum	Titanium dioxide nanoparticles (TiO_2-NPs)	Seeds primed for 24 h with titanium dioxide nanoparticles (TiO_2-NPs)	Improved function of osmotic regulators Decreased damages induced by oxidative stress	Sardar et al. (2022d)

References

Abraha B, Yohannes G (2013) The role of seed priming in improving seedling growth of maize (Zea mays L.) under salt stress at field conditions. Agric Sci 4(12):667–672

Ajouri A, Asgedom H, Becker M (2004) Seed priming enhances germination and seedling growth of barley under conditions of P and Zn deficiency. J Plant Nutr Soil Sci 167(5):630–636

Alam MJ, Sultana F, Iqbal MT (2015) Potential of iron nanoparticles to increase germination and growth of wheat seedling. J Nanosci Adv Technol 1:14–20

Ali J, Wang X, Rafique M, Ahmad I, Fiaz S, Munis MFH, Chaudhary HJ (2021) Phytoremediation of cadmium contaminated soil using Sesbania sesban L. in association with bacillus anthracis PM21: a biochemical analysis. Sustainability 13(24):13529

Ali S, Mfarrej MFB, Hussain A, Akram NA, Rizwan M, Wang X, Maqbool A, Nafees M, Ali B (2022) Zinc fortification and alleviation of cadmium stress by application of lysine chelated zinc on different varieties of wheat and rice in cadmium stressed soil. Chemosphere 295:133829

Ashraf MA, Rasheed R, Hussain I, Hafeez A, Adrees M, Ur Rehman MZ, Rizwan M, Ali S (2022) Effect of different seed priming agents on chromium accumulation, oxidative defense, glyoxalase system and mineral nutrition in canola (Brassica napus L.) cultivars. Environ Pollut 309:119769

Bajehbaj AA (2010) The effects of NaCl priming on salt tolerance in sunflower germination and seedling grown under salinity conditions. Afr J Biotechnol 9(12):1764–1770

Bakhtavar MA, Afzal I, Basra SMA, Ahmad AUH, Noor MA (2015) Physiological strategies to improve the performance of spring maize (Zea mays L.) planted under early and optimum sowing conditions. PLoS One 10(4):e0124441

Beddington JR, Asaduzzaman M, Clark ME, Bremauntz AF, Guillou MD, Jahn MM, Lin E, Mamo T, Negra C, Nobre CA, Scholes RJ (2012) The role for scientists in tackling food insecurity and climate change. Agric Food Secur 1(1):1–9

Belkadhi A, De Haro A, Obregon S, Chaïbi W, Djebali W (2015) Positive effects of salicylic acid pretreatment on the composition of flax plastidial membrane lipids under cadmium stress. Environ Sci Pollut Res 22:1457–1467

Cantliffe DJ, Fischer JM, Nell TA (1984) Mechanism of seed priming in circumventing thermodormancy in lettuce. Plant Physiol 75(2):290–294

Caporale AG, Violante A (2016) Chemical processes affecting the mobility of heavy metals and metalloids in soil environments. Curr Pollut Rep 2:15–27

Chandra R, Kang H (2016) Mixed heavy metal stress on photosynthesis, transpiration rate, and chlorophyll content in poplar hybrids. For Sci Technol 12:55–61

Chandra R, Dubey NK, Kumar V (2018) Phytoremediation of environmental pollutants. CRC Press

Chavez E, He ZL, Stoffella PJ, Mylavarapu RS, Li YC, Baligar VC (2016) Chemical speciation of cadmium: an approach to evaluate plant-available cadmium in Ecuadorian soils under cacao production. Chemosphere 150:57–62

Chen HC, Zhang SL, Wu KJ, Li R, He XR, He DN, Huang C, Wei H (2020) The effects of exogenous organic acids on the growth, photosynthesis and cellular ultrastructure of Salix variegata Franch under Cd stress. Ecotoxicol Environ Saf 187:109790

Chen J, Zeng X, Yang W, Xie H, Ashraf U, Mo Z, Liu J, Li G, Li W (2021a) Seed priming with multiwall carbon nanotubes (MWCNTs) modulates seed germination and early growth of maize under cadmium (Cd) toxicity. J Soil Sci Plant Nutr 21:1793–1805

Chen X, Wang J, Hayat K, Zhang D, Zhou P (2021b) Small structures with big impact: multi-walled carbon nanotubes enhanced remediation efficiency in hyperaccumulator Solanum nigrum L. under cadmium and arsenic stress. Chemosphere 276:130130

Cheng S, Ren F, Grosse W, Wu Z (2002) Effects of cadmium on chlorophyll content, photochemical efficiency, and photosynthetic intensity of Canna indica Linn. Int J Phytoremediation 4: 239–246

Chengatt AP, Sarath NG, Sebastian DP, Mohanan NS, Sindhu ES, George S, Puthur JT (2022) Chelate assisted phytoextraction for effective rehabilitation of heavy metal(loid)s contaminated lands. Int J Phytoremediation 25:1–16
Dehkourdi EH, Chehrazi M, Hosseini H, Hosseini M (2014) The effect of anatase nanoparticles (TiO2) on pepper seed germination (Capsicum annum L.). Int J Biosci 4(5):141–145
Demecsová L, Zelinová V, Liptáková Ľ, Valentovičová K, Tamás L (2020) Indole-3-butyric acid priming reduced cadmium toxicity in barley root tip via NO generation and enhanced glutathione peroxidase activity. Planta 252:1–16
Dong W, Wang R, Li H, Yang X, Li J, Wang H, Jiang C, Wang Z (2023) Effects of chelating agents addition on ryegrass extraction of cadmium and Lead in artificially contaminated soil. Water 15(10):1929
Durairaj B, Muthu S, Xavier T (2015) Effect of fungal generated N-TiO2 on fenugreek (Trigonella foenumgraecum) seed germination and seedling growth. J Global Biosci 4(5):2319–2325
El-Esawi MA, Elkelish A, Soliman M, Elansary HO, Zaid A, Wani SH (2020) Serratia marcescens BM1 enhances cadmium stress tolerance and phytoremediation potential of soybean through modulation of osmolytes, leaf gas exchange, antioxidant machinery, and stress-responsive genes expression. Antioxidants 9(1):43
Epelde L, Becerril JM, Mijangos I, Garbisu C (2009) Evaluation of the efficiency of a phytostabilization process with biological indicators of soil health. J Environ Qual 38(5): 2041–2049
Espanany A, Fallah S, Tadayyon A (2016) Seed priming improves seed germination and reduces oxidative stress in black cumin (Nigella sativa) in presence of cadmium. Ind Crop Prod 79:195–204
FAO, UNEP (2021) Global assessment of soil pollution: report. Rome. https://doi.org/10.4060/cb4894en
Farooq M, Basra SMA, Afzal I, Khaliq A (2006) Optimization of hydropriming techniques for rice seed invigoration. Seed Sci Technol 34(2):507–512
Farooq M, Wahid A, Siddique KH (2012) Micronutrient application through seed treatments: a review. J Soil Sci Plant Nutr 12(1):125–142
Farooq M, Usman M, Nadeem F, Ur Rehman H, Wahid A, Basra SM, Siddique KH (2019) Seed priming in field crops: potential benefits, adoption and challenges. Crop Pasture Sci 70(9): 731–771
Farooqi ZUR, Hussain MM, Ayub MA, Qadir AA, Ilic P (2022) Potentially toxic elements and phytoremediation: opportunities and challenges. Phytoremediation 2022:19–36
Figlioli F, Sorrentino MC, Memoli V, Arena C, Maisto G, Giordano S, Capozzi F, Spagnuolo V (2019) Overall plant responses to Cd and Pb metal stress in maize: growth pattern, ultrastructure, and photosynthetic activity. Environ Sci Pollut Res 26:1781–1790
Gul I, Manzoor M, Hashim N, Shah GM, Waani SPT, Shahid M, Antoniadis V, Rinklebe J, Arshad M (2021) Challenges in microbially and chelate-assisted phytoextraction of cadmium and lead – a review. Environ Pollut 287:117667
Guo SH, Hu N, Li QS, Yang P, Wang LL, Xu ZM, Chen HJ, He BY, Zeng EY (2018) Response of edible Amaranth cultivar to salt stress led to Cd mobilization in rhizosphere soil: a metabolomic analysis. Environ Pollut 241:422–431
Guo J, Chen M, Huang Y, Xie S, Zhang X, Zuo T, Hu C, Wang G (2022) Chloride application weakens cadmium immobilization by lime in paddy rice soil. Ecotoxicol Environ Saf 241: 113761
Haider FU, Liqun C, Coulter JA, Cheema SA, Wu J, Zhang R, Wenjun M, Farooq M (2021) Cadmium toxicity in plants: impacts and remediation strategies. Ecotoxicol Environ Saf 211: 111887
Hassan M, Israr M, Mansoor S, Hussain SA, Basheer F, Azizullah A, Ur Rehman S (2021) Acclimation of cadmium-induced genotoxicity and oxidative stress in mung bean seedlings by priming effect of phytohormones and proline. PLoS One 16(9):e0257924

He M, Li Z, Chen C, Mei P (2022) Impact of soil types and root exudates on cadmium and petroleum hydrocarbon phytoremediation by sorghum Sudanense, Festuca Arundinace, and Lolium Perenne. Front Ecol Evol 10:1036765

Hojjat SS (2015) Impact of silver nanoparticles on germinated fenugreek seed. Int J Agric Crop Sci 8(4):627–630

Huang X, Luo D, Chen X, Wei L, Liu Y, Wu Q, Xiao T, Mai X, Liu G, Liu L (2019) Insights into heavy metals leakage in chelator-induced phytoextraction of Pb-and Tl-contaminated soil. Int J Environ Res Public Health 16(8):1328

Huang F, Hu J, Chen L, Wang Z, Sun S, Zhang W, Jiang H, Luo Y, Wang L, Zeng Y, Fang L (2023) Microplastics may increase the environmental risks of Cd via promoting Cd uptake by plants: a meta-analysis. J Hazard Mater 448:130887

Hussain A, Rizwan M, Ali Q, Ali S (2019) Seed priming with silicon nanoparticles improved the biomass and yield while reduced the oxidative stress and cadmium concentration in wheat grains. Environ Sci Pollut Res 26:7579–7588

Hussain S, Khaliq A, Noor MA, Tanveer M, Hussain HA, Hussain S, Shah T, Mehmood T (2020) Metal toxicity and nitrogen metabolism in plants: an overview. In: Carbon and nitrogen cycling in soil. Springer, pp 221–248

Imperiale D, Lencioni G, Marmiroli M, Zappettini A, White JC, Marmiroli N (2022) Interaction of hyperaccumulating plants with Zn and Cd nanoparticles. Sci Total Environ 817:152741

Jan M, Shah G, Masood S, Iqbal Shinwari K, Hameed R, Rha ES, Jamil M (2019) Bacillus cereus enhanced phytoremediation ability of rice seedlings under cadmium toxicity. Biomed Res Int 2019:8134651

Jawad Hassan M, Ali Raza M, Ur Rehman S, Ansar M, Gitari H, Khan I, Wajid M, Ahmed M, Abbas Shah G, Peng Y, Li Z (2020) Effect of cadmium toxicity on growth, oxidative damage, antioxidant defense system and cadmium accumulation in two sorghum cultivars. Plan Theory 9:1575

Kadioglu A, Terzi R, Saruhan N, Saglam A (2012) Current advances in the investigation of leaf rolling caused by biotic and abiotic stress factors. Plant Sci 182:42–48

Kang LS (2018) Cadmium reduction effects of soil pH, flood control, and low cadmium accumulation variety in Rice field production. J Anhui Agric Sci 46:110–112

Karalija E, Selović A (2018) The effect of hydro and proline seed priming on growth, proline and sugar content, and antioxidant activity of maize under cadmium stress. Environ Sci Pollut Res 25(33):33370–33380

Karalija E, Selović A, Dahija S, Demir A, Samardžić J, Vrobel O, Zeljković SĆ, Parić A (2021) Use of seed priming to improve Cd accumulation and tolerance in Silene sendtneri, novel Cd hyperaccumulator. Ecotoxicol Environ Saf 210:111882

Karalija E, Vergata C, Basso MF, Negussu M, Zaccai M, Grossi-de-Sa MF, Martinelli F (2022a) Chickpeas' tolerance of drought and heat: current knowledge and next steps. Agronomy 12(10): 2248

Karalija E, Demir A, Samardžić J, Parić A, Dahija S, Contaldi F, Martinelli F (2022b) RNA sequencing indicates gene expression changes in Silene sendtneri seeds after seed priming with silicic acid. Genet Appl 6(2):79–88

Karthikeyan B, Jaleel CA, Azooz MM (2009) Individual and combined effects of Azospirillum brasilense and Pseudomonas fluorescens on biomass yield and Ajmalicine production in Catharanthus roseus. Acad J Plant Sci 2(2):69–73

Kaur R, Yadav P, Thukral AK, Sharma A, Bhardwaj R, Alyemeni MN, Wijaya L, Ahmad P (2018) Castasterone and citric acid supplementation alleviates cadmium toxicity by modifying antioxidants and organic acids in Brassica juncea. J Plant Growth Regul 37:286–299

Kaushik S, Sharma P, Kaur G, Singh AK, Al-Misned FA, Shafik HM, Sirhindi G (2022) Seed priming with methyl jasmonate mitigates copper and cadmium toxicity by modifying biochemical attributes and antioxidants in Cajanus cajan. Saudi J Biol Sci 29(2):721–729

Khalid BY, Malik NS (1982) Presowing soaking of wheat seeds in copper and manganese solutions. Commun Soil Sci Plant Anal 13(11):981–986

Khan A, Khan S, Alam M, Khan MA, Aamir M, Qamar Z, Rehman ZU, Perveen S (2016) Toxic metal interactions affect the bioaccumulation and dietary intake of macro-and micro-nutrients. Chemosphere 146:121–128

Kumar N, Kumar V, Bose B, Singhal RK (2021) Cadmium toxicity in plants and alleviation through seed priming approach. Plant Physiol Rep 26:1–14

Lacave JM, Bilbao E, Gilliland D, Mura F, Dini L, Cajaraville MP, Orbea A (2020) Bioaccumulation, cellular and molecular effects in adult zebrafish after exposure to cadmium sulphide nanoparticles and to ionic cadmium. Chemosphere 238:124588

Lan MM, Liu C, Liu SJ, Qiu RL, Tang YT (2020) Phytostabilization of Cd and Pb in highly polluted farmland soils using ramie and amendments. Int J Environ Res Public Health 17(5): 1661

Laware SL, Raskar S (2014) Influence of zinc oxide nanoparticles on growth, flowering and seed productivity in onion. Int J Curr Microbiol Appl Sci 3(7):874–881

Li H, Li Z, Khaliq MA, Xie T, Chen Y, Wang G (2019) Chlorine weaken the immobilization of Cd in soil-Rice systems by biochar. Chemosphere 235:1172–1179

Li J, Zhong Z, Du H, Li Q, Wang N, Zhao H, Huang J (2021a) Theoretical study on the adsorption mechanism of PbCl2/CdCl2 by kaolinite during municipal solid waste pyrolysis. Chemosphere 267:129184

Li Y, Liang L, Li W, Ashraf U, Ma L, Tang X, Pan S, Tian H, Mo Z (2021b) ZnO nanoparticle-based seed priming modulates early growth and enhances physio-biochemical and metabolic profiles of fragrant rice against cadmium toxicity. J Nanobiotechnol 19(1):1–19

Li H, Kong D, Zhang B, Kalkhajeh YK, Zhao Y, Huang J, Hu H (2022a) Chemical forms of cadmium in soil and its distribution in French Marigold sub-cells in response to chelator GLDA. Sci Rep 12(1):17577

Li W, Zhang Y, Achal V (2022b) Mechanisms of cadmium retention on enzyme-induced carbonate precipitation (EICP) of Ca/Mg: nucleation, chemisorption, and co-precipitation. J Environ Chem Eng 10(3):107507

Liang Y, Cao X, Zhao L, Arellano E (2014) Biochar- and phosphate-induced immobilization of heavy metals in contaminated soil and water: implication on simultaneous remediation of contaminated soil and groundwater. Environ Sci Pollut Res 21:4665–4674

Lu H, Qiao D, Han Y, Zhao Y, Bai F, Wang Y (2021) Low molecular weight organic acids increase Cd accumulation in sunflowers through increasing Cd bioavailability and reducing Cd toxicity to plants. Fortschr Mineral 11(3):243

Lu HL, Li KW, Nkoh JN, He X, Hong ZN, Xu RK (2022) Effects of the increases in soil pH and pH buffering capacity induced by crop residue biochars on available Cd contents in acidic Paddy soils. Chemosphere 301:134674

Ma J, Li Y, Chen F, Sun Y, Zhu Y, Wang L (2023) Bacillus mycoides PM35 in combination with titanium dioxide (TiO2) nanoparticles enhanced morpho-physio-biochemical attributes in barley (Hordeum vulgare L.) under cadmium stress. Chemosphere 323:138224

Marthandan V, Geetha R, Kumutha K, Renganathan VG, Karthikeyan A, Ramalingam J (2020) Seed priming: a feasible strategy to enhance drought tolerance in crop plants. Int J Mol Sci 21(21):8258

Mirshekari B (2012) Seed priming with iron and boron enhances germination and yield of dill (Anethum graveolens). Turk J Agric For 36(1):27–33

Mourato M, Pinto F, Moreira I, Sales J, Leitão I, Martins LL (2019) The effect of Cd stress in mineral nutrient uptake in plants. In: Cadmium toxicity and tolerance in plants. Academic Press, pp 327–348

Musa AM, Johansen C, Kumar J, Haris D (1999) Response of chickpea to seed priming in the high Barind tract of Bangladesh. Int Chickpea Pigeonpea Newsl 6:20–22

Nada E, Ferjani BA, Ali R, Bechir BR, Imed M, Makki B (2007) Cadmium-induced growth inhibition and alteration of biochemical parameters in almond seedlings grown in solution culture. Acta Physiol Plant 29:57–62

Nascimento WM, Cantliffe DJ, Huber DJ (2004) Ethylene evolution and endo-β-mannanase activity during lettuce seed germination at high temperature. Sci Agric 61(2):156–163
Nedjimi B (2021) Phytoremediation: a sustainable environmental technology for heavy metals decontamination. SN Appl Sci 3(3):286
Niu Z, Li X, Mahamood M (2023) Accumulation potential cadmium and Lead by sunflower (Helianthus annuus L.) under citric and glutaric acid-assisted phytoextraction. Int J Environ Res Public Health 20(5):4107
Pame AR, Kreye C, Johnson D, Heuer S, Becker M (2015) Effects of genotype, seed P concentration and seed priming on seedling vigor of rice. Exp Agric 51(3):370–381
Pattanayak SK, Dash D, Jena MK, Nayak RK (2000) Seed treatment of green gram with molybdenum and cobalt: effect on nodulation, biomass production and N uptake in an acid soil. J Indian Soc Soil Sci 48(4):769–773
Pawar VA, Laware SL (2018) Seed priming a critical review. Int J Sci Res Biol Sci 5(5):94–101
Prasad TNVKV, Sudhakar P, Sreenivasulu Y, Latha P, Munaswamy V, Reddy KR, Sreeprasad TS, Sajanlal PR, Pradeep T (2012) Effect of nanoscale zinc oxide particles on the germination, growth and yield of peanut. J Plant Nutr 35(6):905–927
Qi C, Xu X, Chen Q, Liu H, Min X, Fourie A, Chai L (2022) Ab initio calculation of the adsorption of As, Cd, Cr, and Hg heavy metal atoms onto the Illite(001) surface: implications for soil pollution and reclamation. Environ Pollut 312:120072
Radziemska M, Bęś A, Gusiatin ZM, Cerdà A, Jeznach J, Mazur Z, Brtnický M (2020) Assisted phytostabilization of soil from a former military area with mineral amendments. Ecotoxicol Environ Saf 188:109934
Rafi H, Dawar S, Zaki MJ (2015) Seed priming with extracts of Acacia nilotica (L.) Willd. ex Delile and Sapindus mukorossi (L.) plant parts in the control of root rot fungi and growth of plants. Pak J Bot 47:1129–1113
Raj AS (1987) Cobalt nutrition of pigeonpea and peanut in relation to growth and yield. J Plant Nutr 10:2137–2145
Rana D, Arcoverde Cerveira Sterner V, Potluri AK, May Z, Müller B, Solti Á, Rudnóy S, Sipos G, Gyuricza C, Fodor F (2022) S-Methylmethionine effectively alleviates stress in Szarvasi-1 energy grass by reducing root-to-shoot cadmium translocation. Plan Theory 11(21):2979
Rehman HU, Kamran M, Basra SMA, Afzal I, Farooq M (2015) Influence of seed priming on performance and water productivity of direct seeded rice in alternating wetting and drying. Rice Sci 22(4):189–196
Rizwan M, Ali S, ur Rehman MZ, Maqbool A (2019) A critical review on the effects of zinc at toxic levels of cadmium in plants. Environ Sci Pollut Res 26:6279–6289
Ruangcharus C, Kim SU, Hong CO (2020) Mechanism of cadmium immobilization in phosphate-amended arable soils. Appl Biol Chem 63:36. https://doi.org/10.1186/s13765-020-00522-0
Sardar R, Ahmed S, Yasin NA (2021) Seed priming with karrikinolide improves growth and physiochemical features of Coriandrum sativum under cadmium stress. Environ Adv 5:100082
Sardar R, Ahmed S, Yasin NA (2022a) Role of exogenously applied putrescine in amelioration of cadmium stress in Coriandrum sativum by modulating antioxidant system. Int J Phytoremediation 24(9):955–962
Sardar R, Ahmed S, Shah AA, Yasin NA (2022b) Selenium nanoparticles reduced cadmium uptake, regulated nutritional homeostasis and antioxidative system in Coriandrum sativum grown in cadmium toxic conditions. Chemosphere 287:132332
Sardar R, Ahmed S, Akbar M, Yasin NA, Li G (2022c) Alleviation of cadmium phytotoxicity in triacontanol treated Coriandrum sativum L. by modulation of physiochemical attributes, oxidative stress biomarkers and antioxidative system. Chemosphere 295(133924):133924
Sardar R, Ahmed S, Yasin NA (2022d) Titanium dioxide nanoparticles mitigate cadmium toxicity in Coriandrum sativum L. through modulating antioxidant system, stress markers and reducing cadmium uptake. Environ Pollut 292:118373
Sarkar RK (2012) Seed priming improves agronomic trait performance under flooding and non-flooding conditions in rice with QTL SUB1. Rice Sci 19(4):286–294

Seifikalhor M, Hassani SB, Aliniaeifard S (2020) Seed priming by cyanobacteria (Spirulina platensis) and salep gum enhances tolerance of maize plant against cadmium toxicity. J Plant Growth Regul 39:1009–1021

Selvarani K, Umarani R (2011) Evaluation of seed priming methods to improve seed vigour of onion (Allium cepa cv. aggreratum) and carrot (Daucus carota). J Agric Sci Technol 7(3): 857–867

Shah AA, Ahmed S, Yasin NA (2020) 2-Hydroxymelatonin induced nutritional orchestration in Cucumis sativus under cadmium toxicity: modulation of non-enzymatic antioxidants and gene expression. Int J Phytoremediation 22(5):497–507

Shah G, Tu J, Fayyaz M, Masood S, Ullah H, Jamil M (2022) Moringa oleifera smoke induced positive changes in biochemical, metabolic, and antioxidant profile of rice seedling under cadmium stress. Int J Phytoremediation 25:1–11

Sharma SS, Dietz KJ (2006) The significance of amino acids and amino acid-derived molecules in plant responses and adaptation to heavy metal stress. J Exp Bot 57:711–726

Shi M, Min X, Ke Y, Lin Z, Yang Z, Wang S, Peng N, Yan X, Luo S, Wu J, Wei Y (2021) Recent Progress in understanding the mechanism of heavy metals retention by iron (oxyhydr)oxides. Sci Total Environ 752:141930

Shirinzadeh A, Soleimanzadeh H, Shirinzadeh Z (2013) Effect of seed priming with plant growth promoting rhizobacteria (PGPR) on agronomic traits and yield of barley cultivars. World Appl Sci J 21(5):727–731

Singh H, Jassal RK, Kang JS, Sandhu SS, Kang H, Grewal K (2015) Seed priming techniques in field crops - a review. Agric Rev 36(4):251–264

Soleimanzadeh H (2013) Effect of seed priming on germination and yield of corn. Int J Agric Crop Sci 5(4):366–369

Spanò C, Bottega S, Sorce C, Bartoli G, Ruffini Castiglione M (2019) TiO2 nanoparticles may alleviate cadmium toxicity in co-treatment experiments on the model hydrophyte Azolla filiculoides. Environ Sci Pollut Res 26:29872–29882

Sposito G (2008) The chemistry of soils. Oxford University Press

Street RA, Kulkarni MG, Stirk WA, Southway C, Van Staden J (2010) Effect of cadmium on growth and micronutrient distribution in wild garlic (Tulbaghia violacea). South Afr J Bot 76: 332–336

Subašić M, Šamec D, Selović A, Karalija E (2022) Phytoremediation of cadmium polluted soils: current status and approaches for enhancing. Soil Syst 6(1):3

Sullivan TS, McBride MB, Thies JE (2013) Soil bacterial and archaeal community composition reflects high spatial heterogeneity of pH, bioavailable Zn, and Cu in a metalliferous peat soil. Soil Biol Biochem 66:102–109. https://doi.org/10.1016/j.soilbio.2013.06.021

Sun Q, Liu C, Cui P, Fan T, Zhu M, Alves ME, Siebecker MG, Sparks DL, Wu T, Li W, Zhou D (2019) Formation of Cd precipitates on γ-Al2O3: implications for Cd sequestration in the environment. Environ Int 126:234–241

Sun F, Chen W, Ren Y, Cheng S, Ashraf U, Zheng J, Lin L, Ma Y, Tang X, Mo Z (2023) Nano-priming with La2O3 improves early growth and regulates physio-biochemical mechanisms in fragrant Rice against cadmium toxicity. J Soil Sci Plant Nutr 23:1–17

Suzuki T, Nakase K, Tamenishi T, Niinae M (2020) Influence of pH and cations contained in rainwater on leaching of Cd(II) from artificially contaminated montmorillonite. J Environ Chem Eng 8(5):104080

UNEP, United Nations Environment Programme (2013) Environmental risks and challenges of anthropogenic metals flows and cycles. International Resource Panel. https://wedocs.unep.org/20.500.11822/8451

Waheed A, Haxim Y, Islam W, Ahmad M, Ali S, Wen X, Khan KA, Ghramh HA, Zhang Z, Zhang D (2022) Impact of cadmium stress on growth and physio-biochemical attributes of Eruca sativa Mill. Plan Theory 11(21):2981

Wang W, Peng S, Chen Q, Mei J, Dong H, Nie L (2016) Effects of pre-sowing seed treatments on establishment of dry direct-seeded early rice under chilling stress. AoB Plants 8:plw074

Wu Y, Ma L, Liu Q, Sikder MM, Vestergård M, Zhou K, Wang Q, Yang X, Feng Y (2020) Pseudomonas fluorescens promote photosynthesis, carbon fixation and cadmium phytoremediation of hyperaccumulator Sedum alfredii. Sci Total Environ 726:138554

Xia X, Ji J, Yang Z, Han H, Huang C, Li Y, Zhang W (2020) Cadmium risk in the soil-plant system caused by weathering of carbonate bedrock. Chemosphere 254:126799

Yang LP, Zhu J, Wang P, Zeng J, Tan R, Yang YZ, Liu ZM (2018) Effect of Cd on growth, physiological response, Cd subcellular distribution and chemical forms of Koelreuteria paniculata. Ecotoxicol Environ Saf 160:10–18

Yang JS, Dai Y, Liu Y, Duan S, Li YY, Hu R, Zhou Z, Shi Y, Liu H, Wang S (2020) Reduced cadmium accumulation in tobacco by sodium chloride priming. Environ Sci Pollut Res 27: 37410–37418

Yang W, Huang C, Wan X, Zhao Y, Bao Z, Xiang W (2022) Enhanced adsorption of Cd on iron–organic associations formed by laccase-mediated modification: implications for the immobilization of cadmium in paddy soil. Int J Environ Res Public Health 19(23):15650

Yao WB, Huang L, Yang ZH, Zhao FP (2022) Effects of organic acids on heavy metal release or immobilization in contaminated soil. Trans Nonferr Metal Soc 32(4):1277–1289

Yasmeen F, Razzaq A, Iqbal MN, Jhanzab HM (2015) Effect of silver, copper and iron nanoparticles on wheat germination. Int J Biosci 6(4):112–117

Zaid A, Mohammad F, Siddique KH (2022) Salicylic acid priming regulates stomatal conductance, trichome density and improves cadmium stress tolerance in Mentha arvensis L. Front Plant Sci 13:895427

Zaman MI, Mustafa S, Khan S, Xing B (2009) Effect of phosphate complexation on Cd2+ sorption by manganese dioxide (β-MnO2). J Colloid Interface Sci 330(1):9–19

Zhang S, Hu J, Zhang Y, Xie XY, Knap A (2007) Seed priming with brassinolide improves lucerne (Medicago sativa L.) seed germination and seedling growth in relation to physiological changes under salinity stress. Aust J Agric Res 58(8):811–815

Zhang Z, Chen X, Qin X, Xu C, Yan X (2023) Effects of soil pH on the growth and cadmium accumulation in Polygonum hydropiper (L.) in low and moderately cadmium-contaminated paddy soil. Land 12:652. https://doi.org/10.3390/land12030652

Zhou W, Ren L, Zhu L (2017) Reducement of cadmium adsorption on clay minerals by the presence of dissolved organic matter from animal manure. Environ Pollut 223:247–254

Zhu G, Xiao H, Guo Q, Zhang Z, Zhao J, Yang D (2018) Effects of cadmium stress on growth and amino acid metabolism in two Compositae plants. Ecotoxicol Environ Saf 158:300–308

Zulfiqar U, Jiang W, Xiukang W, Hussain S, Ahmad M, Maqsood MF, Ali N, Ishfaq M, Kaleem M, Haider FU, Farooq N (2022) Cadmium phytotoxicity, tolerance, and advanced remediation approaches in agricultural soils; a comprehensive review. Front Plant Sci 13:773815

Unraveling the Adsorption Process of Cd^{2+} on Bio-Adsorbents: Experimental and Theoretical Points of View

A. Forgionny, C. Jimenez-Orozco, E. Flórez, and N. Acelas

Abstract In this chapter, we concentrate on the study of Cd^{2+} adsorption from both experimental and computational sides. The adsorption mechanisms of Cd^{2+} on biochar remain an enigmatic aspect in the current research field. While numerous studies have explored the adsorption capacities of different materials, a comprehensive understanding of the underlying adsorption mechanisms is still lacking. Kinetics models such as pseudo-first-order, pseudo-second-order, and intraparticle diffusion offer valuable insights into kinetics and rate constants but may fall short in predicting the complex adsorption mechanisms, owing to certain assumptions that may not hold in real-world systems. Isotherm studies serve as an essential approach for predicting the adsorption mechanisms in various systems, providing information about the nature of adsorption through different isotherm curves. However, a single equation cannot satisfactorily explain all the mechanisms. In the context of heavy metal adsorption on carbonaceous materials, experimental techniques have proposed interactions, i.e., electrostatic, π–π interaction, precipitation, ion exchange, and complexation. Computational approaches, however, have predominantly relied on adsorption energies to define interactions, with a limited focus on the kinetics and thermodynamics aspects of the adsorption process using molecular modeling. Indeed, several studies are limited to small models, and some of them lack complete and systematic methodologies. The absence of studies exploring computationally dynamic systems further restricts a comprehensive understanding. Bridging these gaps through experimental and computational efforts is crucial to unravel the intricate adsorption mechanisms and optimize the adsorption process for environmental remediation and industrial applications.

Keywords Biochar · Adsorption · Mechanism · Kinetics · Computational

A. Forgionny · C. Jimenez-Orozco · E. Flórez · N. Acelas (✉)
Grupo de Investigación Materiales con Impacto, MAT&MPAC, Facultad de Ciencias Básicas, Universidad de Medellín, Medellín, Colombia
e-mail: nyacelas@udemedellin.edu.co

A. K. Jha, N. Kumar (eds.), *Cadmium Toxicity Mitigation*,
https://doi.org/10.1007/978-3-031-47390-6_12

1 Introduction

Cadmium (Cd) is commonly used in battery manufacturing, steel galvanization, printing inks, and the production of artificial phosphate fertilizers. The continuous discharge of Cd^{2+} and other heavy metals from various industrial sectors pollutes water, implying a remarkable environmental concern. Due to their high toxicity, this has led to the deterioration of aquatic ecosystems and impacts negatively on human health, resulting in a wide range of diseases, some of which can even be fatal. Particularly, cadmium (found as Cd^{2+} in aqueous systems) is a toxic heavy metal that can cause serious health problems in humans and animals, even at low concentrations. Also, it can cause damage to the liver, kidneys, and other organs, due to its accumulation in the food chain. Hence, great attention has been paid to the pollution caused by Cd, and significant measures have been taken by establishing international regulations with a permissible discharge limit of Cd in wastewater between 0.003 and 0.005 mg/L (Pyrzynska 2019; World Health Organization 2011). Therefore, it is necessary to develop effective methods for removing Cd from contaminated water.

One of the most commonly used techniques for contaminant removal from water is the adsorption process, which is preferred due to its cost-effectiveness and ease of operation (Kumar et al. 2023). The challenge with this technique is to find an adsorbent material that has low production cost and high efficiency in removing various contaminants. Among a wide range of materials, those obtained from the thermal transformation of agro-industrial waste (biochar) stand out (Giraldo et al. 2022; Yang et al. 2022). The current research focuses on finding effective solutions to harness the high amounts of generated agro-industrial wastes. Hence, biomass is an attractive alternative to produce adsorbent materials due to its high availability, low cost of transformation, and potential as raw material for generating carbonaceous materials with suitable physicochemical properties for adsorption applications. Different agro-industrial wastes such as fruit and vegetable peels, seeds, and wood sawdust, among others, have shown promise for the development of efficient and low-cost adsorbent materials (Forgionny et al. 2022; Gao et al. 2023; Giraldo et al. 2022; Hu et al. 2023; Kumar et al. 2023; Ramirez-Muñoz et al. 2021).

Biochar refers to a carbonaceous material synthesized by the pyrolysis of biomass under limited oxygen conditions. Recent studies have indicated that biochar has a strong affinity for the removal of heavy metals due to its porous structure and aromatic surface (Liu et al. 2022b; Zhang et al. 2023). However, pristine biochar has a limited specific surface area; nonetheless, it contains different functional groups allowing the adsorption of Cd (Wang et al. 2022). Some studies have suggested that chemical modifications of biochar could significantly improve the adsorption capacity by enhancing the pore structure and improving the chemical activity (Giraldo et al. 2020, 2022). For instance, biochar showed excellent adsorption performance after modification with NaOH, H_3PO_4, and $ZnCl_2$ (Qiu et al. 2022; Chen et al. 2019; Abolfazli Behrooz et al. 2023; Bai et al. 2023).

According to the above-mentioned issues, the physicochemical properties influencing the adsorption behavior of different adsorbent materials include surface area, porosity, pH, surface charge density, functional groups, and mineral content. These characteristics provide insights into the interaction between the adsorbent and the contaminant during the adsorption process (Ahmad et al. 2022; Baskaran and Abraham 2022; Forgionny et al. 2022).

Though biochar is an effective adsorbent for the removal of Cd^{2+}, the mechanisms underlying this process are not yet fully understood. Most of the existing research focuses on investigating the adsorption capacities of different materials, but it fails to explain the adsorption mechanisms in depth (Liu et al. 2022b; Pang et al. 2022; Wang et al. 2022). Therefore, it is crucial to understand the adsorption process in these systems from both experimental and theoretical points of view (Forgionny et al. 2022; Giraldo et al. 2020, 2022; Forgionny et al. 2021). A deeper knowledge of the interactions between the adsorbate and adsorbent contributes to optimizing the design and development of more efficient adsorbents and improving operating conditions in real-world adsorption processes. It can also provide insights into the transport and final destination of Cd^{2+} in the environment and help to develop strategies for Cd^{2+} remediation.

Based on the earlier issues presented above, the development of this chapter is structured by covering the study of the adsorption process from both point of views: (i) experimental and (ii) computational.

2 Experimental Point of View

2.1 Adsorption Mechanisms from Kinetic and Isotherm Models

Adsorption kinetics: kinetic studies can be used to evaluate the dynamic trends of adsorbate migration to adsorbents for the parameters and characteristics determination of the adsorption process. The kinetics of the adsorption process is crucial during adsorption studies because these can predict the contaminant's removal from aqueous solutions and provide valuable data for understanding the adsorption mechanism (Ramírez Muñoz et al. 2017). Among the most commonly used kinetic models are the pseudo-first-order, pseudo-second-order, and intraparticle diffusion models. To determine which of the three models better fits the experimental data, the coefficient of correlation (R^2) is used to indicate the correlation between the experimental data and the calculated data using the parameters of each model (Aboua et al. 2015). The kinetic models enable us to determine whether the step that limits the adsorption process is physisorption or chemisorption. The pseudo-first-order model was proposed by Lagergren (1898) to explain the adsorption of a liquid adsorbate on a solid adsorbent at different time intervals (Simonin 2016). It is based on the assumption that each adsorbate is assigned to an adsorption site on the adsorbent material (Tejada Tovar et al. 2020) and that the interaction occurs through a physical process (Eq. 1 in Table 1) (Aboua et al. 2015). The pseudo-second-order model was

Table 1 Kinetics and isotherms adsorption models

Model	Equation	Lineal equation	Equation
Pseudo-first-order	$q_t = q_e\left(1 - e^{-k_1 t}\right)$	$\mathrm{Log}(q_e - q_t) = \mathrm{Log}\left(q_e - \frac{k_1}{2.303}t\right)$	(1)
Pseudo-second-order	$q_t = \frac{q_e{}^2 k_2 t}{1+q_e k_2 t}$	$\frac{t}{q_t} = \frac{1}{k_2 q_e{}^2} + \frac{t}{q_e}$	(2)
Intraparticle diffusion	$q_t = kd_i\sqrt{t} + C_i$		(3)

Terms description: q_t is uptake at time (mg/g). k_1 means adsorption rate constant (1/min); k_2 means rate constant (g/mg min); k_3 (mg/g min$^{1/2}$) is rate constant. C is a constant

developed by Ho and McKay (1999) and Enniya et al. (2018). It assumes that the rate of occupation of adsorption sites is proportional to the square of the number of unoccupied sites (Cruz et al. 2004). This model represents chemisorption, which occurs due to the formation of chemical bonds between the adsorbent and adsorbate in a monolayer on the surface (Doke and Khan, 2017) (Eq. 2 in Table 1). The intraparticle diffusion model assumes that the adsorption mechanism occurs through different steps: external diffusion, pore diffusion, and adsorption on the pore surface (Pang et al. 2022).

In Table 2, the kinetic parameters for the adsorption of Cd^{2+} from aqueous solutions using pristine biochar and modified biochar are presented. It can be observed that most studies on the adsorption of Cd^{2+} from aqueous solutions using biomass-derived materials (biochars) have reported that the limiting step of the adsorption process is chemisorption. This conclusion is generally supported by the good fit of the experimental data to the linear form of the pseudo-second-order model. This indicates that the adsorption mechanism of Cd^{2+} is limited by the binding strength with the biochar through electron sharing or exchange (Zhang et al. 2022), rather than the diffusion of the adsorbate into the pore (Yin et al. 2022). However, to further clarify the diffusion into the pore, most studies employ the intraparticle diffusion model, which suggests that diffusion occurs in multiple stages (Quisperima et al. 2022).

Although pseudo-first-order, pseudo-second-order, and intraparticle diffusion kinetics models can provide valuable information about the kinetics and rate constants of adsorption, these models may not be sufficient to fully predict the adsorption mechanism since these make certain assumptions that may not always hold in real-world systems. For example, (i) the pseudo-first-order model assumes a linear relationship between the amount of adsorbate adsorbed and time, which may not always be accurate. Similarly, (ii) the pseudo-second-order model assumes a homogeneous surface and a single rate-limiting step, which may not be the case for complex adsorption systems, and (iii) adsorption is a complex phenomenon that often involves multiple steps, such as transport of the adsorbate to the surface, interaction with active sites, and potential diffusion into the adsorbent pores. Kinetic models usually consider only the rate-determining step, neglecting the contribution of other possible steps. Therefore, to gain a deeper understanding of the adsorption mechanism, it is often necessary to combine kinetic studies with other techniques such as isotherm analysis, thermodynamic analysis, and characterizations of the adsorbent surface (Pang et al. 2022; Zhang et al. 2022). These complementary

Table 2 Kinetics parameters of Cd^{2+} adsorption on biochar

Biomass	Treatment	Kinetics model			Adsorption mechanism	
		Pseudo-first-order	Pseudo-second-order	Intraparticle diffusion		
Water chestnut	Pristine **$q_{t,exp}$ (mg/g)** **7.34**	$q_{t,cal}$ (mg/g) 6.82	$q_{t,cal}$ (mg/g) 7.23	K_{id1} (mg/(g·min$^{1/2}$)) = 1.85 C_1 (mg/g) = −0.07 R_1^2 = 0.977	Mineral precipitation, ion exchange, functional group complexation, π-bond coordination	Zhang et al. (2022)
		K_1 (mg/(g·min)) 0.15 R^2 = 0.887	K_2 (mg/(g·min)) 0.03 R^2 = 0.972	K_{id2} (mg/(g·min$^{1/2}$)) = 0.41 C_2 (mg/g) = 3.30 R_2^2 = 0.967 K_{id3} (mg/(g·min$^{1/2}$)) = 0.04 C_2 (mg/g) = 6.64 R_2^2 = 0.992		
Water chestnut	Chemical activate: Na_3PO_4 800 °C/N_2 **$q_{t,exp}$ (mg/g)** **15.26**	$q_{t,cal}$ (mg/g) 14.86	$q_{t,cal}$(mg/g) 15.25	K_{id1} (mg/(g·min$^{1/2}$)) = 3.19 C_1 (mg/g) = 2.15 R_1^2 = 0.970		
		K_1 (mg/(g·min))	K_2 (mg/(g·min))	K_{id2} (mg/(g·min$^{1/2}$)) = 0.80		

(continued)

Table 2 (continued)

Biomass	Treatment	Kinetics model: Pseudo-first-order	Pseudo-second-order	Intraparticle diffusion	Adsorption mechanism	
		0.31 $R^2 = 0.949$	0.03 $R^2 = 0.985$	C_2 (mg/g) = 7.90 $R_2^2 = 0.945$ K_{id3} (mg/(g·min$^{1/2}$)) = 0.04 C_3 (mg/g) = 14.54 $R_3^2 = 0.916$		
Crofton weed	$MgCl_2 \cdot 6H2O$	$q_{t,cal}$ (mg/g) 155.98	$q_{t,cal}$ (mg/g) 187.97	K_{id1} (mg/(g·min$^{1/2}$)) = 59.80 C_1 (mg/g) = −80.79 $R_1^2 = 0.9920$	Mineral precipitation, ion exchange.	Cheng et al. (2022)
		K_1 (mg/(g·min)) 0.9863 $R^2 = 0.9665$	K_2 (mg/(g·min)) 0.000078 $R^2 = 0.9992$	K_{id2} (mg/(g·min$^{1/2}$)) = 29.57 C_2 (mg/g) = 28.51 $R_2^2 = 0.9747$		
Platanus orientalis Linn (POL) leaves	Pristine BC	$q_{t,cal}$ (mg/g) 17.064 K_1 (mg/(g·min)) 0.003 $R^2 = 0.932$	$q_{t,cal}$ (mg/g) 19.580 K_2 (mg/(g·min)) 1.750×10^{-4} $R^2 = 0.959$		Complexation with MnOx and oxygen-containing groups; cation–π interactions; precipitation (cd(OH)$_2$ and $CdCO_3$); ion exchange	Yin et al. (2022)

	MBC $q_{t,exp}$ **(mg/g)** **24.95**	$q_{t,cal}$ (mg/g) 23.937 K_1 (mg/(g·min)) 0.344 $R^2 = 0.983$	$q_{t,cal}$ (mg/g) 24.321 K_2 (mg/(g·min)) 0.036 $R^2 = 0.997$			
Rice straw	$CaCl_2$ (600 ° C) N_2 $q_{t,exp}$ **(mg/g)** **43.95**	$q_{t,cal}$ (mg/g) 42.73 K_1 (mg/(g·min)) 0.4608 $R^2 = 0.9864$	$q_{t,cal}$ (mg/g) 43.206 K_2 (mg/(g·min)) 0.0343 $R^2 = 0.993$		Electrostatic adsorption, ion exchange, and complexation of oxygen functional groups	Gao et al. (2023)

approaches can provide a more comprehensive picture of the adsorption process and help elucidate the underlying mechanisms involved.

2.2 *Equilibrium Data Description: Adsorption Isotherms*

The understanding of adsorption phenomena allows proper utilization of adsorbent since this information lets us to determine the maximum limit to which an adsorbent-adsorbate exists in equilibrium (Rajahmundry et al. 2021). The isotherm models and their corresponding equations characterize the adsorption completely and thoroughly, describing the equilibrium data in adsorption studies (Hu et al. 2023; Mozaffari Majd et al. 2022). The interaction mechanisms between the adsorbent and the adsorbate at constant temperature can be described by the interpretation of the equilibrium data and the consideration of the adsorption properties of both the adsorbent and the adsorbate (Al-Ghouti and Da'ana 2020). Moreover, adsorption isotherms can describe and predict the adsorption equilibrium mechanism, adsorption capacity, and the inherent characteristics of the adsorption process (Mozaffari Majd et al. 2022). Therefore, deeper comprehension of equilibrium data modeling is a very essential way of predicting the adsorption mechanisms of various adsorption systems (Al-Ghouti and Da'ana 2020).

The several types of isotherm curves are related to differences in the adsorption mechanisms being used to identify the nature of adsorption (Hu et al. 2023). A variety of theoretical and empirical models have been proposed to represent different types of adsorption isotherms. Any single equation cannot give a satisfactory explanation of all mechanisms (Aktar 2020). For instance, chemical isotherms account for the monolayer adsorption process, physical isotherms denote the multilayer adsorption, and ion exchange isotherms describe the ion exchange adsorption process (Wang and Guo 2020). Hence, Langmuir, Freundlich, Temkin, and Sips models are some of the equations that have common use for describing the adsorption process of heavy metal species (Aktar 2020). The Langmuir model supposes that the adsorbate and adsorbent interact in an ideal manner at a particular homogeneous adsorbent surface (Langmuir 1916). The Langmuir model is mathematically expressed by the equation reported in Eq. (4), Table 3. However, the Langmuir model does not cover the differences in molecular sizes, various screening/blocking effects, multilayer formation, or partial mobility that may affect the adsorption rate (Hu et al. 2023). The Freundlich adsorption isotherm, represented by Eq. (6) in Table 3, describes adsorption as a phenomenon that happens through heterogeneous surfaces with a mechanism of multilayer adsorption (Freundlich 1907; Rajahmundry et al. 2021). The Freundlich isotherm assumes that the adsorption energy distribution in the active sites of the adsorbent is exponential and can be associated with both physical and chemical adsorption and homogeneous and heterogeneous surfaces (Aktar 2020). An important issue with the Freundlich isotherm is that it fails to predict the value of adsorption at higher concentrations since it establishes the relationship of adsorption with the lower concentration values of the adsorbate

Table 3 Adsorption isotherm models

Model	Isotherm equation	References
Langmuir	$q_e = \frac{q_{\max} K_L C_e}{1+K_L C_e}$ (Eq. 4) $R_L = \frac{1}{1+K_L C_0}$ (Eq. 5)	Langmuir (1916) Wang and Guo (2020)
Freundlich	$q_e = K_F C_e^{1/n}$ (Eq. 6)	Freundlich (1907)
Sips	$q_e = \frac{q_{MS} K_S C_e^{n_s}}{1+K_S C_e^{n_s}}$ (Eq. 7)	Sips (1948)
Temkin	$q_e = \frac{RT}{b_{TM}} \ln K_{TM} C_e$ (Eq. 8)	Temkin (1940)

Terms description: q_e (mg g^{-1}) is the amount of solute uptake per unit mass of adsorbent at equilibrium, C_e (mg L^{-1}) is the equilibrium concentration of the solute, $q_{\max}$ (mg g^{-1}) is the maximum adsorption capacity, and K_L (L mg^{-1}) is the equilibrium constant. KF and n are constants integrating all factors affecting the adsorption capability and adsorption intensity, respectively. q_{MS} (mg g^{-1}) is Sips maximum adsorbed amount. K_s and n_s are the Sips constants. b_{TM} and K_{TM} are constants of Temkin isotherm, where b_{TM} is related to sorption heat

(Aktar 2020).On the other hand, the Sips isotherm is a hybrid model, which takes into consideration some of the basic assumptions of the Langmuir and Freundlich model isotherms (Mozaffari Majd et al. 2022; Sips 1948). Hence, this model can describe homogeneous or heterogeneous systems (Wang and Guo 2020). By using this hybrid model, the adsorption behavior in materials with heterogeneous surfaces can be predicted, avoiding the restriction that occurs in the Freundlich isotherm with increasing adsorbate concentration (Liu et al. 2022b). The Temkin model assumes that the adsorption process occurs in a multilayer and considers a factor associated with the interaction's adsorbent-adsorbate (Mozaffari Majd et al. 2022; Temkin 1940). This isotherm ignores extremely high and low concentrations and assumes that the differential adsorption heat could be linearly reduced in the layer for entire molecules because of increment coverage (Mozaffari Majd et al. 2022; Wang and Guo 2020). This model is not appropriate to predict the liquid phase adsorption but to anticipate the gas phase adsorption (Mozaffari Majd et al. 2022). The Temkin model equation is expressed by Eq. (8) in Table 3.

Biochar has diversity in elemental composition, structure, functional groups, surface charge, pore-size distribution, and surface area, among other physicochemical properties, due to differences in the raw materials, the pyrolysis temperatures, and the employed active agent (Qiu et al. 2021). All these properties are responsible for the removal efficiency of the adsorbent material. Hence, the adsorption capacity determined by the adjustment to different isotherm models is a fundamental parameter for comparing the adsorbent efficiencies. These properties are also associated with Cd^{2+} adsorption mechanisms since their variation supports the occurrence of one or more interactions during the adsorption process (Qiu et al. 2021; Wang et al. 2019). Hence, the nature's identification of the adsorption processes by different isotherm types is a significant tool to establish the involved adsorption mechanisms commonly reported.

The adsorption isotherm models, maximum adsorption capacity, and the main mechanisms that occur during the Cd^{2+} adsorption process on biochars (modified and unmodified) are summarized in Table 4, where Langmuir and Freundlich's

Table 4 Equilibrium models applied for Cd^{2+} adsorption on biochar

Material	Modification	Equilibrium applied models	Best fit equilibrium model/isotherm classification	Q_{max} (mg g^{-1})	Adsorption mechanisms	References
Pine bark tree residues	Pyrolysis (600 °C)	Langmuir Freundlich	Langmuir/chemical adsorption model	85.5	Cation exchange, precipitation, and complex formation with phosphate	Park et al. (2019)
Cedar sawdust	Calcination (400 °C)	Langmuir Freundlich	Freundlich/adsorption empirical model	36.0	Complexation, cation—aromatic π interactions, and ionic exchange	Forgionny et al. (2022)
Enteromorpha prolifera	$ZnCl_2$ + Pyrolysis (500 ° C) H_3PO_4 + Pyrolysis (500 ° C) $KMnO_4$ + Pyrolysis (500 °C)	Langmuir Freundlich	Langmuir/chemical adsorption model	Not reported 423.0 142.0	Complexation and precipitation	Li et al. (2020)
Platanus orientalis Linn leaves	(Pyrolysis 400 °C) Pyrolysis + $KMnO_4$ Pyrolysis + H_2O Pyrolysis + H_2O_2 Pyrolysis + $K_2Cr_2O_7$	Langmuir Freundlich	Freundlich/adsorption empirical model	27.0 52.5 14.6 15.8 56.0	Complexation, precipitation, cation-π interaction, and ion exchange	Yin et al. (2022)
C. oleifera shells	$Mg(NO_3)_2$ + microwave calcination at 800 °C	Langmuir Freundlich Temkin	Langmuir/chemical adsorption model	649.9	Cd and Mg complexation	Xu et al. (2022)
Peanut shells	Pyrolysis (500 °C) $[NH_4{\cdot}PO_3]_n$ + pyrolysis H_3PO_4 + pyrolysis $NH_4H_2PO_4$ + pyrolysis	Langmuir Freundlich	Langmuir/chemical adsorption model	36.0 155.2 110.1 99.3	Precipitation	Huang et al. (2022)
Water chestnut shell	Pyrolysis (800 °C) Na_3PO_4 + pyrolysis Na_2CO_3 + pyrolysis	Langmuir Freundlich Sips	Sips/adsorption empirical model	36.1 99.8 86.6	Complexation and cation-π interactions	Zhang et al. (2022)

	Na_2HPO_4 + pyrolysis CH_3COONa + pyrolysis			65.9 57.1		
Wheat and rice husks	**Wheat husk:** Pyrolysis (300–600 °C) $Ca_5(PO_4)_3Cl$ + pyrolysis (500 °C)	Langmuir Freundlich Sips	Langmuir/chemical adsorption model	1.06–4.93 183.5	Cd^{2+}-phosphate precipitation, complexation, and cation exchange	Yuan et al. (2023)
	Rice husks: Pyrolysis (300–600 °C) $Ca_5(PO_4)_3Cl$ + pyrolysis (500 °C)			1.25–3.75 106.9		
Corn stalks	Pyrolysis (300 °C) + HNO_3/H_2SO_4/ $Na_2S_2O_4$	Langmuir Freundlich Sips Temkin	Sips/adsorption empirical model Freundlich/adsorption empirical model	375.6	Complexation and electrostatic attraction	Ma et al. (2023)

Pyrolysis in oxygen-limited conditions

model corresponds to the best fit in most reported studies. The results shown in Table 4 indicate that Cd^{2+} adsorption for the evaluated materials was substantially affected by the different modification methods. For instance, in biochars obtained by physical and thermochemical treatments, the equilibrium adsorption experimental data for Cd^{2+} fitted well to the Langmuir model, assuming monolayer adsorption. The biochar materials obtained by physical treatments showed Cd^{2+} maximum adsorption capacities from 1.06 to 85.5 mg/g (see Table 2) (Forgionny et al. 2022; Huang et al. 2022; Park et al. 2019; Yin et al. 2022; Yuan et al. 2023; Zhang et al. 2022). On the other hand, biochars modified with acid, oxidizing agents, and metal salt addition increased the maximum adsorption capacities up to 649.9 mg/g (Huang et al. 2022; Xu et al. 2022; Yin et al. 2022; Yuan et al. 2023; Zhang et al. 2022). This behavior has been associated with changes in elemental composition, a rise of surface area and porous structure generation, or more diversity and number of functional groups (active sites). A recent study reported that biochar derived from rice and wheat husks and modified by nano-chlorapatite (nClAP) exhibited a monolayer chemical process in Cd^{2+} adsorption since equilibrium experimental data showed the best fit to the Langmuir model, with regression coefficients (R^2) in the range 0.851–0.998 (Yuan et al. 2023). The Langmuir fitting results let us determine the maximum Cd^{2+} adsorption capacity of the prepared biochars, obtaining values between 2.18 and 26.44 mg/g. The rise of pyrolysis temperatures and the modification with nClAP particles significantly increased the maximum adsorption capacity values of the biochars (Yuan et al. 2023).

Some modified biochars showed a better fit of the data to the Freundlich model, indicating that the adsorption of Cd^{2+} takes place on a surface with a heterogeneous energy distribution with the formation of multilayers. For instance, Yin et al. (2022) reported that the biochar modified with the oxidizing agent ($KMnO_4$) derived from *Platanus orientalis* Linn leaves showed the highest Cd^{2+} removal efficiency with a removal percentage of 98.6%, compared to biochars modified with H_2O, H_2O_2, and $K_2Cr_2O_7$. This behavior was attributed to a larger specific surface area with distributed MnO_x particles (confirmed by XPS analysis) on the surface before and after adsorption. Besides, adsorption experimental data adjusted well to the Freundlich model, which suggests that the adsorption process occurs as multilayer sorption over a heterogeneous surface. The Freundlich constant ($1/n$) obtained for the oxidant-modified biochar was smaller than 1, indicating that the adsorption of Cd^{2+} by this material was a favorable process at the evaluated temperatures (20 °C, 30 °C, and 40 °C) (Yin et al. 2022). Similarly, biochar obtained by calcination of cedar sawdust at 400 °C was evaluated in the Cd^{2+} adsorption process (Forgionny et al. 2022). The experimental data adjusted well to the Freundlich model which was supported by a higher correlation coefficient (R^2 closer to 1) compared to the Langmuir model. The parameter $1/n$ indicated that the adsorption of Cd^{2+} on biochar derived from cedar sawdust was a favorable physical process with multilayer adsorption on heterogeneous active sites. On the other hand, other studies have reported that Cd^{2+} adsorption processes are modeled for two isotherms. For instance, Ma et al. (2023) prepared biochar derived from corn stalks by nitrification and reduction in two stages using HNO_3/H_2SO_4 and $Na_2S_2O_4$, respectively.

The R^2 values obtained by fitting the experimental data to the adsorption isotherm models showed that both the Freundlich and Sips models adjust well. The Freundlich parameter $1/n$ indicated that the adsorption of Cd^{2+} on biochar is favorable, while the maximum adsorption amount determined by the Sips model was 375.6 mg/g (Ma et al. 2023).

2.3 Thermodynamic Studies

Thermodynamic parameters such as Gibbs free energy change (ΔG), enthalpy change (ΔH), and entropy change (ΔS) are crucial for understanding the adsorption process of Cd^{2+} onto biochars in aqueous solutions. ΔG provides information about the feasibility and spontaneity of the adsorption process. A negative ΔG value indicates that the adsorption process is thermodynamically feasible and spontaneous (Liu et al. 2022b). ΔH is a measure of the heat energy involved in the adsorption process. The ΔH value helps to evaluate the exothermic or endothermic nature of the adsorption process and the energy exchanges between the ions and the biochar surface. A positive ΔH value suggests an endothermic adsorption process, indicating that the adsorption is driven by chemical interactions. A negative ΔH value indicates an exothermic adsorption process, implying physical adsorption or ion exchange as dominant mechanisms (Al-Ghouti and Da'ana 2020; Liu et al. 2022b).

ΔS reflects the randomness or disorder of the system during adsorption. An increase in entropy (positive ΔS) suggests an increase in disorder, which can occur when ions are adsorbed onto biochars and released from the solution. In other words, a positive ΔS value indicates that the system becomes more disordered during adsorption (Al-Ghouti and Da'ana 2020; Liu et al. 2022b). On the other hand, thermodynamic parameters can provide insights into the adsorption mechanisms involved in adsorption processes. Moreover, thermodynamic parameters are related to the equilibrium constant (K) through thermodynamic equations. The value of K indicates the extent of adsorbent adsorption onto biochars at equilibrium (Al-Ghouti et al. 2021). Understanding the equilibrium constant and its relationship with thermodynamic parameters allows us to estimate the adsorption capacity and efficiency of biochars in removing Cd^{2+} from aqueous solutions. Additionally, thermodynamic parameters, particularly temperature-dependent parameters such as ΔH and ΔS, let us determine the effect of this variable. By analyzing these parameters at different temperatures, it is possible to optimize the temperature conditions for achieving maximum adsorption efficiency. Therefore, the study of the thermodynamic adsorption process provides valuable insights into the energetics, feasibility, and mechanisms of process adsorption. This knowledge is essential for the design and optimization of biochar-based adsorption systems for the removal of contaminants from wastewater.

The thermodynamic parameters for Cd^{+2} adsorption on different biochars are included in Table 5. In all cases, the ΔG values are negative ($\Delta G < 0$) at different temperatures, where the absolute value of ΔG increased with the rise of the

Table 5 Adsorption thermodynamics data for removal of Cd^{2+}

Material	T (K)	ΔH (kJ mol^{-1})	ΔS (kJ mol^{-1} K^{-1})	ΔG (kJ mol^{-1})	ln$K$	References
Modified corn stalks biochar by nitrification and reduction	298			−1.63	0.66	Ma et al. (2023)
	308	17.19	0.063	−2.32	0.91	
	318			−2.89	1.09	
Modified rice straw-derived biochar with $-NH_2$ groups	293	37.37		−20.58	8.45	Zhang et al. (2019)
	303		0.199	−23.39	9.29	
	313			−24.52	9.42	
MgO-modified biochar derived from *C. oleifera*	293	8.98	0.123	−27.18	11.16	Xu et al. (2022)
	303			−28.31	11.24	
	313			−29.61	11.38	
	323			−30.86	11.49	
Salts-modified biochar obtained from water chestnut:						
Pyrolysis (800 °C)	293			−0.12	–	Zhang et al. (2022)
	303	36.14	0.117	−0.43	–	
	313			−0.68	–	
Na_3PO_4 + pyrolysis	293			−1.41	–	
	303	67.83	0.238	−4.32	–	
	313			−6.13	–	
Na_2CO_3 + pyrolysis	293			−0.92	–	
	303	58.82	0.203	−2.49	–	
	313			−5.01	–	
Na_2HPO_4 + pyrolysis	293			−0.57	–	
	303	57.07	0.196	−1.57	–	
	313			−4.53	–	
CH_3COONa + pyrolysis	293			−0.39	–	
	303	60.85	0.206	−1.13	–	
	313			−3.76	–	
Titanium-modified ultrasonic biochar	293			−1.73		Luo et al. (2019)
	298	18.34	0.0686	−2.23	–	
	303			−2.42		

adsorption temperature of the system (see Table 5). These results indicated the spontaneous nature of the adsorption processes, and higher temperatures allow the improvement of spontaneity (Liu et al. 2022b; Zhang et al. 2022). It has been reported that an increase in the heat of adsorption leads to a rise in the number of adsorbed molecules. Table 5 shows ΔH values positives ($\Delta H > 0$) in all cases, indicating that the Cd^{2+} adsorption onto biochar is mainly an endothermic process. Hence, an increase in the temperature of the process will favor uptake. The enthalpy change values reported are in the range of 8.98–67.83 kJ/mol, ΔH values over 20 kJ/mol indicate that the adsorption process was an endothermic reaction dominated by chemisorption, and ΔH values lower than 20 kJ/mol suggest that physical interactions take place during Cd^{2+} adsorption processes (Liu et al. 2022b). On the other hand, the positive values of ΔS (see Table 5) suggest that the disorder of the solid–liquid interface of biochar-Cd^{2+} was increased and Cd^{2+} adsorption onto biochar is an irreversible process (Zhang et al. 2022).

2.4 Characterization of Interactions: Experimental

To conduct a more in-depth analysis of the adsorption process, it becomes essential to characterize the involved interactions. While results from kinetic and isotherm models offer insights into the type of adsorption, identifying the interactions during the adsorption process demands the application of various experimental characterization techniques to analyze the material before and after the adsorption process (Pang et al. 2022; Xu et al. 2022; Yin et al. 2022). Furthermore, several fundamental parameters play a crucial role in characterizing these interactions. These include the type of contaminant (whether it is cationic, anionic, or organic), the pH of the solution, the point of zero charge (pH_{PZC}) of the adsorbent material, the presence of functional groups, and the type of treatment and activation method applied to the agro-industrial residue, among others. These parameters significantly influence the adsorption process and need to be carefully considered in the characterization studies (Forgionny et al. 2022; Giraldo et al. 2022). Depending on the involved parameters, different interactions can be associated with the removal of a particular contaminant. For the adsorption of heavy metals on carbonaceous materials, it has been proposed that the interactions governing this process include electrostatic interaction, π–π interaction, precipitation, ion exchange, and complexation (Yang et al. 2022). The relative contribution of each mechanism depends on factors such as the nature of the adsorbent and the characteristics of the Cd-contaminated solution. For example, the pH of the solution can influence the surface charge of the adsorbent and the speciation of Cd ions, thereby affecting the dominant adsorption mechanism. Furthermore, electrostatic interaction is strongly influenced by the pH of the solution and the point of zero charge (pH_{PZC}) of the adsorbent. If the pH of the solution is higher than the pH_{PZC} value of the adsorbent, an electrostatic interaction can be established between Cd^{2+} and the adsorbent due to the attraction between the positive charge of Cd^{2+} and the negative charge of the material (Chen et al. 2019; Kumar et al. 2023). Compared to other mechanisms, electrostatic interactions are considered easier and reversible adsorption stages. Complexation by functional groups (–OOH, –OH, etc.) in biochar can complex Cd^{2+} by releasing H^+. This process leads to a noticeable difference in the pH of the solution before and after the reaction (Forgionny et al. 2022; Wang et al. 2022; Zhang et al. 2022). The amount of Cd^{2+} adsorption can be indirectly assessed by measuring the amount of H^+ released. π–π interaction occurs between the aromatic structure of the adsorbent material and the lone pairs of electrons of Cd^{2+} (Ding et al. 2016; Forgionny et al. 2022; Zhang et al. 2022). Ionic exchange is related to the replacement of ions and can be cationic in nature, involving the exchange of positively charged ions (cations) (Yuan et al. 2023). This type of adsorption is considered reversible since, in most

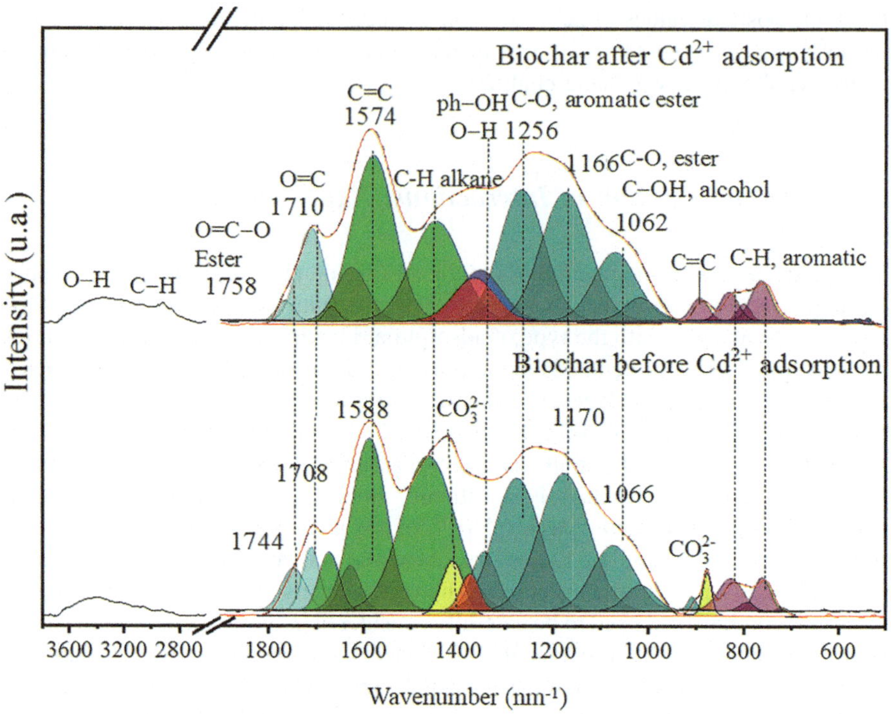

Fig. 1 FTIR spectra of biochar produced from cedar sawdust before and after adsorption of Cd^{2+}

cases, the ions can be recovered from the adsorbent material. Precipitation is another mechanism of adsorption that occurs with Cd^{2+} through the formation of insoluble compounds (Ding et al. 2016; Zhang et al. 2015). The main characterization techniques that allow the identification of adsorption mechanisms are Fourier transform infrared spectroscopy (FTIR), X-ray diffraction (XRD), and X-ray photoelectron spectroscopy (XPS). FTIR spectroscopy is used to analyze the functional groups present on the surface of the adsorbent material before and after the adsorption process. Changes in the XRD patterns before and after adsorption can provide insights into the formation of new crystalline phases or the transformation of existing phases, indicating specific adsorption mechanisms such as precipitation or complexation, and XPS enables the analysis of the elemental composition and chemical states of the adsorbent material. By analyzing the binding energy shifts or changes in the peak intensities of relevant elements (such as carbon, oxygen, or metal ions), XPS can help identify the adsorption mechanisms involved, such as ion exchange or chemical bonding (Baskaran and Abraham 2022; Ding et al. 2016; Forgionny et al. 2022; Gao et al. 2023; Liu et al. 2022a; Pang et al. 2022; Zhang et al. 2015, 2022).

Figure 1 shows the FTIR spectra of biochar derived from cedar sawdust, both before and after Cd^{2+} adsorption (Forgionny et al. 2022). The analysis revealed the involvement of various functional groups in the adsorption process of Cd^{2+}.

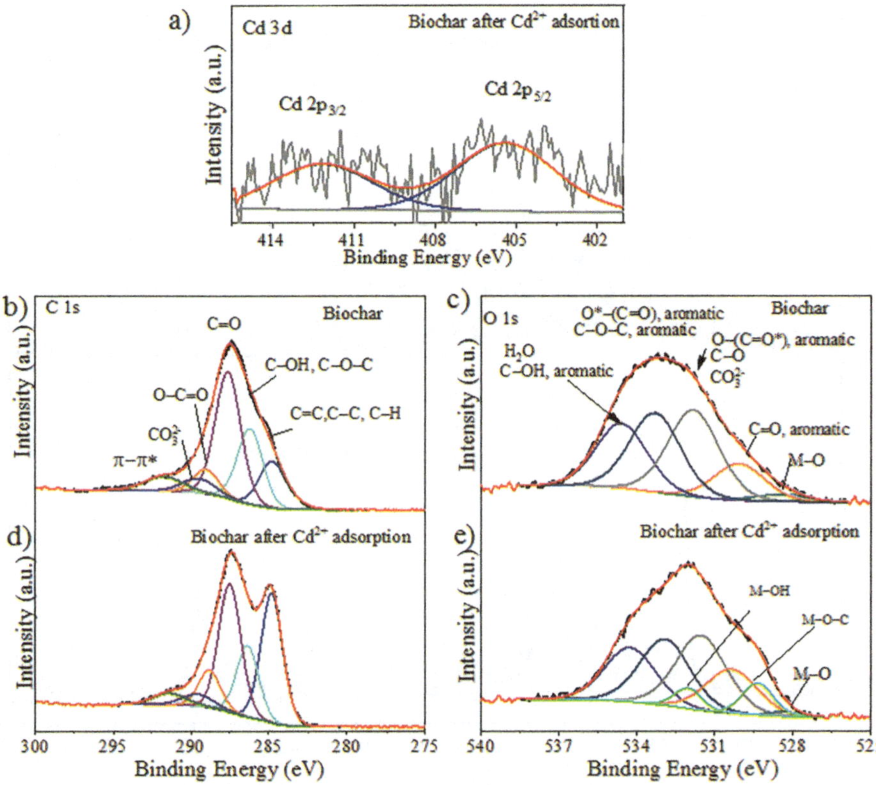

Fig. 2 XPS survey for biochar produced at 400 °C, before and after Cd^{2+} removal

Specifically, a shift in the signal at 1744 cm^{-1} was observed for the C=O group, which corresponds to esters or carboxylic acids. After adsorption, this band shifted to around 1760 cm^{-1}, indicating the occurrence of an (O–C=O)-M^{2+} interaction. The participation of these groups is further supported by the shift in the bands corresponding to the C-O bond in the ester group, ranging from 1000 to 1300 cm^{-1}. Another significant interaction was observed for the band at 1588 cm^{-1}, associated with C=C aromatic bonds, which shifted to 1576 cm^{-1} upon the introduction of Cd^{2+}, suggesting a π–M^{2+} interaction. These observations provide confirmation that the mechanism of Cd^{2+} removal on cedar biochar is complexation, involving the interaction of oxygenated functional groups such as hydroxyl, carboxyl, ester, and ether, as well as the interaction between metal ions and aromatic π bonds (Zhang et al. 2022).

Moreover, Fig. 2 illustrates the XPS survey of biochar produced at 400 °C, both before and after Cd^{2+} removal (Forgionny et al. 2022). The primary XPS signal indicating Cd^{2+} removal is the notable increase in the atomic percentage (AP) of the peak associated with carbon (C) in the C=C, C–C, and C–H bonds. This peak's AP rises from 17.9% to 33.8% following adsorption, potentially attributed to the

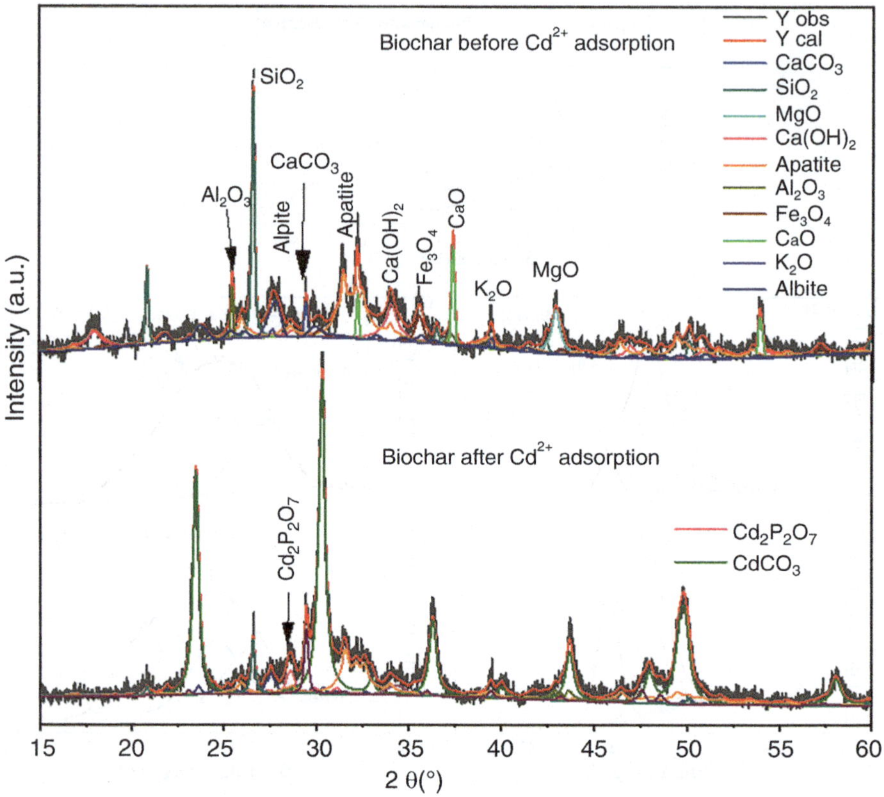

Fig. 3 XRD patterns for water hyacinth calcinated at 700 °C before and after Cd^{2+} adsorption

contribution of the C=C-M interactions signal. Conversely, the AP of oxygenated groups (C–OH/C–O–C,C = O) decreases. This reduction suggests the participation of these oxygenated groups in the adsorption process of Cd^{2+} through complexation.

In Fig. 3, the utilization of XRD analysis is demonstrated to characterize the adsorption process of Cd^{2+} on biochar produced from water hyacinth (Ramirez-Muñoz et al. 2021). The figure reveals that water hyacinth, before Cd^{2+} adsorption, contains calcium apatites, indicating that the removal of Cd^{2+} by these materials is associated with the presence of phosphate and calcium species (apatite and calcite). After adsorption, it is observed that a portion of the apatite may dissolve in the acidic aqueous solution, where the dissolved phosphates react with Cd^{2+} to form $Cd_2P_2O_7$. Moreover, the proposed mechanism involves the ion exchange of Cd^{2+} with Ca^{2+} in apatite and the precipitation of $CdCO_3$. XRD analysis confirms the involvement of ion exchange and the precipitation of $CdCO_3$ in the mechanism.

The preceding reports align with recent studies published in the literature. For instance, it was reported (Ramirez-Muñoz et al. 2021) that the mechanisms involved in Cd^{2+} adsorption on pyrochars include interactions with minerals, complexation with organic functional groups (OFGs), and π electron coordination. In contrast,

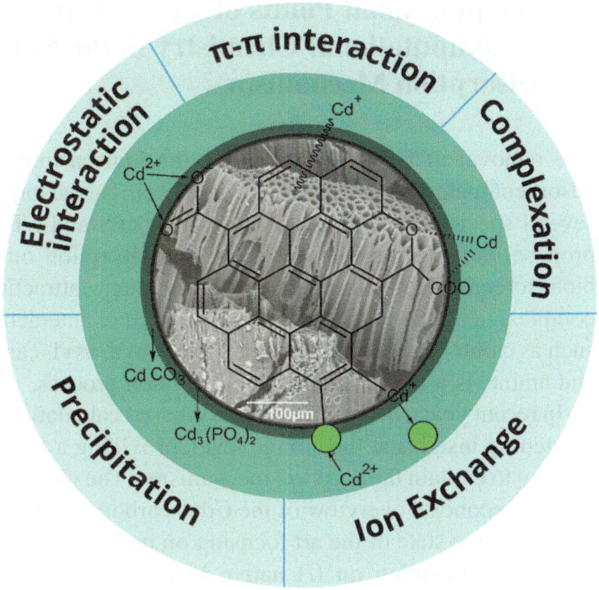

Fig. 4 Depicts a schematic representation of the interactions involved in the process of Cd^{2+} adsorption on biochar

complexation with OFGs and interactions with minerals were identified as the dominant mechanisms for Cd^{2+} adsorption on hydrochars. Besides Zongyu Gao et al. (2023), Ramirez-Muñoz et al. (2021) showed the utilization of selenium-rich straw biochar for Cd^{2+} removal. The study concluded that the primary mechanisms of adsorption involve electrostatic adsorption, ion exchange, and complexation of oxygen functional groups (OFGs). In the study, the authors investigated the adsorption mechanisms of Cd^{2+} using oxidant-modified biochar (OMB) derived from *Platanus orientalis Linn* (POL) leaves. The study revealed that complexation with MnO_x was identified as the main mechanism for Cd^{2+} adsorption. Additionally, other mechanisms such as complexation with O-containing groups, precipitation, cation–π interaction, and ion exchange were found to contribute to the adsorption process to a lesser extent. Song Cheng et al. (2022) demonstrated that mineral precipitation and ion exchange are the primary mechanisms contributing to Cd^{2+} removal on biochar derived from crofton weed. The interactions involved in Cd^{2+} adsorption on carbonaceous materials can be represented as shown in Fig. 4. However, there is a challenge in understanding the adsorption process at the molecular level, which has led to the development of techniques based on electronic structure, providing a bridge between experimental findings and fundamental understanding, based on the use of theoretical and computational methodologies.

3 Computational Points of View: Application of Computational Chemistry in the Study of Cd^{+2} Adsorption Mechanism

The removal of Cd^{2+} ions relies on the high affinity between the metal ion and adsorbent materials (Forgionny et al. 2022). The adsorption process is studied using several spectroscopic techniques, mainly XPS and FT-IR, which are conducted to propose multiple mechanisms for Cd^{2+} adsorption on bioadsorbent and modified bioadsorbent, e.g., ion exchange, electrostatic attraction, hydrogen bonding, and complexation (see Fig. 5a). The occurrence of interactions with functional groups, such as carboxyl, phenolic, lactone, ether, hydroxyl, carboxylate, sulfate, phosphate, and amine, is a key aspect of the adsorption process.

In recent years, quantum chemistry-based calculations became a fundamental tool for getting insights at the atomic level, providing a fundamental understanding, and helping to explain the sorption mechanisms. The objective of the current section is to provide a concise overview of the Cd^{2+}-carbonaceous material binding problem and the respective state of the art focusing on molecular modeling. This section contains cadmium adsorption on (*i*) native biochar, (*ii*) modified biochar, and (*iii*) other carbonaceous systems.

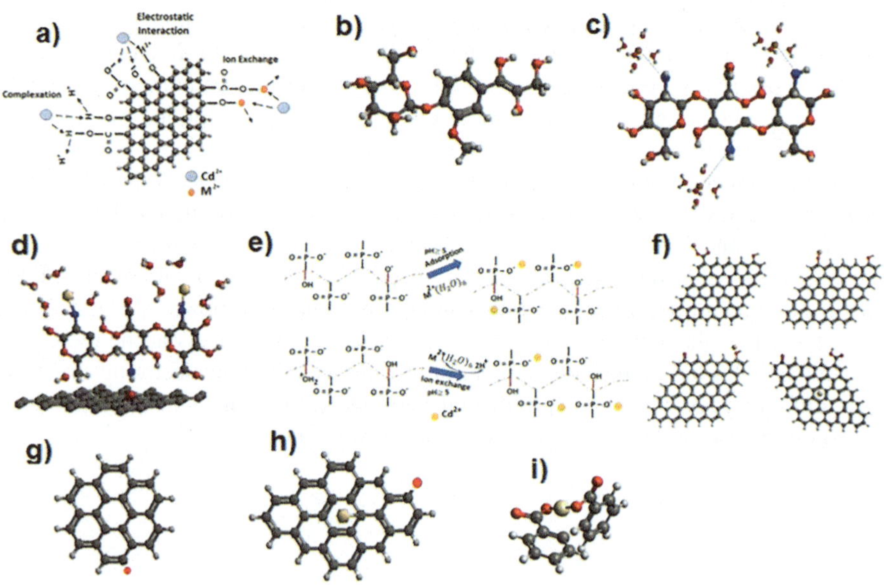

Fig. 5 Schematics for Cd^{2+} adsorption mechanisms onto several carbonaceous adsorbents

3.1 Mechanism for Adsorption on Native Biochar

The use of quantum chemistry-based methodologies has provided useful information regarding the elucidation of the Cd^{2+} adsorption mechanism on native biochar, which is explained in a detailed way below, and summarized in the schemes of Fig. 5.

C atoms are represented as black balls (Kushwaha and Singh 2023); (b) model for cellulose-lignin structure. C and O are represented as black and red balls, respectively; (c) schematic representation of structure for chitosan + Cd^{+2}. O, C, and N are presented by red, black, and blue balls, respectively; (d) schematic representation of structure for Cs/GO + Cd^{+2}. C, O, N, and Cd are represented by black, red, blue, and yellow balls, respectively; (e) schematic representation of two modes of metal ion uptake: adsorption and ion exchange by PSP at different pH values; (f) schematic representation of structures for carbonaceous surface oxygenated (red balls) + Cd^{+2} (yellow balls); (g) model for surface oxidized and aminated biochars; the white atom is hydrogen, the black atom is carbon, and the red ball represents several functional groups; (h) model for surface oxidized; the white atom is hydrogen, the black atom is carbon, and the red ball represents several functional groups. (i) Representation of Cd^{2+} chelation between two carboxylate groups of an oxygenated carbonaceous material. The black, red, light gray, and yellow balls represent C, O, H, and Cd atoms, respectively.

Ibrahim et al. (2012) studied removal of Cd^{2+} from an aqueous solution using dried water hyacinth as a biosorbent. The authors built a model using density functional theory (DFT) for the description of the plant and verified it experimentally (see Fig. 5b). They found that the water hyacinth is a mixture of cellulose and lignin, where hydrogen bonds of reactive functional groups like COOH (carboxyl) play a key role in the removal process. In another research work, Ibrahim, M, et al. (Ezzat et al. 2020) studied the Cd^{2+} removal by chitosan (Cs) using density functional theory. The proposed model (see Fig. 5c) showed that Cs is a biopolymer containing hydroxyl (OH) and amino (NH_2) groups in its structure, which can attach Cd^{2+} and serve as a chelating and reaction site. The HOMO-LUMO band gap energy was used as a descriptor of the reactivity, finding that Cs can interact with hydrated Cd, as evidenced by a decrease in the HOMO-LUMO band gap of 1.5 eV.

3.2 Mechanism for Cadmium Adsorption Over Modified Biochar Adsorption for Adsorption on Native Biochar

The use of native biochar is somewhat restricted because of its low adsorption performance and tendency to retrograde. This challenge has been overcome by the experimental modification of native biochar using chemical strategies for introducing new functionalities in starch, such as carbonyl, acetyl, hydroxypropyl, and phosphate, demonstrating that materials with better adsorption capacities can be obtained. Some theoretical studies have been addressed to study Cd^{2+} adsorption onto modified biochars as it is summarized below. Menazea et al.(2020) studied the interactions between Cd^{2+} and chitosan/graphene oxide (Cs/GO) composite. They used density functional theory (DFT) at the B3LYP level to compare the interaction Cd^{+2} – Cs/GO vs Cd^{2+} – Cs. The models constructed were (a) for Cs (three units as row materials), (b) for Cd^{+2} (the ion was hydrated with five water molecules), and (c) for GO (twelve fused rings with an oxygen atom in the basal plane) (see Fig. 5d). The authors found that GO improves the stability of Cs and its adsorption reactivity for Cd^{+2}, while the composite Cs/GO showed selectivity for hydrated metals compared to Cs alone. Bashir et al. (2020) studied Cd^{2+} adsorption on potato starch phosphate (PSP) biosorbent by combining experimental and theoretical calculations. They modeled several functional groups, such as surface hydroxyls, glycosidic hydroxyls, and phosphate hydroxyls, which are available for metal ion binding (see Fig. 5e). They found that at low pH $\leq$ 4.0, the dominant mode of metal ion binding is exchange with H^+ ions of PSP. At higher pH $>$ 5.0, the surface of PSP attains enough negative charge. Due to the negatively charged surface of PSP, cationic target metal ions get attracted and finally get adsorbed on the surface of PSP (see Fig. 5e).

Recently, Forgionny et al. (2022) reported a combined experimental and theoretical study regarding the adsorption mechanism for Cd^{+2} on an adsorbent produced through a simple thermal transformation of cedar wood sawdust. In this study, oxygenated carbonaceous structures were modeled for a 20-membered carbon ring with several functional groups (carboxylic acids, semiquinone, ester, ether, and hydroxyl); see Fig. 5f. These models were based on the experimental results obtained in this study. The authors found that the main mechanisms that govern the removal of Cd^{2+} and Cu^{2+} onto the adsorbent are (*i*) surface complexation through oxygenated functional groups (such as hydroxyl, carboxyl, ester, and ether) and (*ii*) the interaction between metal ions and aromatic π-electrons of the carbonaceous surface for both ions. Zhu et al. (2020) studied the adsorption of Cd^{+2} on the oxidized and aminated surface of biochars. The authors modeled a simple graphite structure with seven aromatic rings and various types of functional groups (-OH, -COOH, ketone and quinone types -C=O, saturated and unsaturated -COC, and -NH_2), which were modified on the surface (see Fig. 5g). The study showed that surface properties, such as the presence of functional groups, are a key factor for controlling the adsorption. From the computational results, the authors concluded that the interaction between carboxyl and Cd dominated the adsorption performance of surface oxidized biochar, while the $Cd^{2+} - \pi$ interaction was weakened by increasing the π electron electrostatic potential of aromatic rings. The lone pair electrons of the amino groups dominated the complexation of surface aminated biochar with Cd, and the π electron electrostatic potential was almost unaffected. Liu et al. (2022a) studied the removal of Cd^{2+} from water by sesame straw-derived biochar, which was prepared via alkaline hydrogen peroxide (AHP) pretreatment. For calculations, the authors selected a graphite structure (SGS) with five aromatic rings for representing the aromatic carbon structure, and various types of functional groups (-OH, -COOH, and –C=O) were modified on the surface (see Fig. 5h). From the theoretical calculations, their findings were (*i*) the hydroxyl and carboxyl groups on the surface of biochar provided preferential adsorption sites, (*ii*) the surface complexation and $Cd^{2+} - \pi$ interaction were one the dominant adsorption mechanisms, and (*iii*) the lone pair electrons during complexation and π electrons during coordination were provided by oxygen-containing functional groups and aromatic rings, respectively. Ai et al. (2020) reported the use of sugarcane bagasse with a maximum Cd^{2+} adsorption capacity of 119.3 mg/g, demonstrating that the presence of Ca^{2+} and Mg^{2+} ions in water is responsible for the poisoning of the adsorbent. The treatment with nitric acid restored the adsorption capacity of the adsorbent, where the XPS experiments indicated that carboxylate groups are responsible for the Cd^{2+} binding. The density functional theory-based calculations showed that the chelation between Cd^{2+} and carboxylate (see Fig. 5i) is feasible thermodynamically, indicating that the carboxylate groups are the active sites for the Cd^{2+} removal.

3.3 Mechanism of Cadmium Adsorption on Other Carbonaceous Materials

Shi et al. (2013) studied the Cd^{2+} binding over graphite surface by molecular dynamics simulations, where the adsorption is due to the direct interaction between Cd^{2+} and the π electron-rich system of the graphite surface. From a schematic point of view, the authors use one layer of the graphite (i.e., its surface) to perform some analysis for ion metals interactions, which strictly belongs to a graphene layer instead of graphite analysis (see Fig. 6a). The authors concluded that graphite has a high capability to adsorb Cd^{2+} ions. In a similar focus, Malhotra et al. (2022) using a model of a graphene sheet reported that there is a charge transfer from the carbonaceous model toward Cd^{2+}, creating a graphene–metal complex.

Javarani et al. (2022) analyzed carbon nanosheets functionalized with amines to analyze their adsorption capacity of Cd^{2+} via density functional theory-based

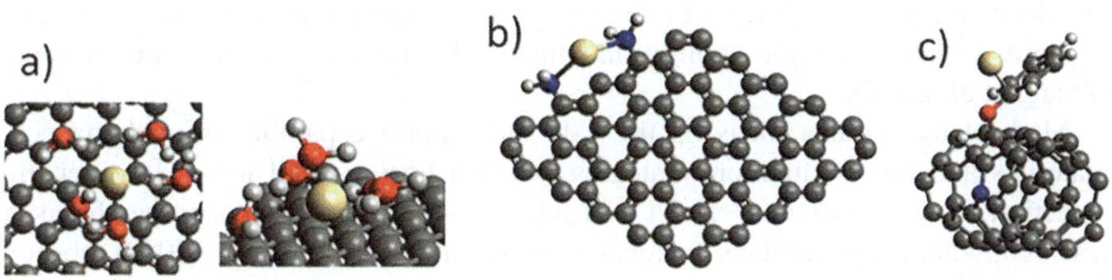

Fig. 6 (**a**) Schematics for the Cd^{2+} adsorption (yellow ball), considering explicit water molecules over a graphene sheet, representing the surface layer of graphite. C atoms as black balls. Left: top view, right: side view. (**b**) Cd^{2+} (yellow ball) adsorbed on carbon nanosheet (black balls) functionalized with amines. (**c**) Carbon nanotube N-doped (blue ball) for phenol adsorption and Cd^{2+} (yellow ball) adsorbed over phenol. C, O, and H atoms are represented with black, red, and light gray, respectively

calculations. The authors found that Cd^{2+} interacts with N atoms of amines with both mono- and bi-indentation forms (see Fig. 6b), complemented with analysis from conceptual DFT.

Diaz-Florez et al. (2009) analyzed an N-doped CNT (carbon nanotube) to adsorb simultaneously both the phenol group and Cd^{2+} ion. When phenol is pre-adsorbed, the adsorption of Cd^{2+} increases, confirmed by DFT calculations, since there is an interaction between Cd^{2+} and the carbon atom of the phenol group, where this oxygenated group is also attached to the CNT (see Fig. 6c). The authors concluded that N-doped CNTs can remove both organic and inorganic pollutants from water. On the other hand, Bastos and Camps (2014) analyzed the changes in the electronic properties of CNT after Cd adsorption; however, surprisingly, the CNT remains semiconductive with a slight change in the band gap, while adsorption of other metals like Ni and Pb modified the electronic structure and the magnetism behavior of the CNT. Similar conclusions were obtained by Aghashiri et al. (2019), which indicates a weak interaction between CNT and Cd, pointing to the need to functionalize CNT to improve the adsorption capacity. Baachaoui et al. (2021) investigated the functionalization of graphene with activated carbenes (RC(O)CH) and the capability of this material to catch Cd^{2+} ions, implying the formation of substituted cyclopropanes over graphene. Varying the R group implies a model for the pH of the system, yielding a different adsorption behavior of Cd^{2+} relative to other heavy metal ions.

Although several carbonaceous systems have shown remarkable capacity to remove Cd^{2+} and other heavy metals (particularly, those related to activated carbons), there is a problem associated with the separation of the used material from the aqueous solution. Therefore, this is a practical limitation at the end of the application and recovery of the carbonaceous material (Zhang et al. 2021). During the last few years, magnetic nanoparticles have been added during the preparation of the biochars to achieve an easy separation at the end of the application. Precursors like iron-containing salts are added during the preparation of the biochar; then during the obtaining of the char iron oxides are formed, e.g., Fe_3O_4 and γ-Fe_2O_3 (Mohan et al. 2014), producing a net magnetic effect on the carbonaceous system. Nonetheless, the formation of composites for magnetic carbonaceous materials has low stability; therefore, there are challenges in the design and preparation of magnetic carbonaceous material with high stability and high Cd^{2+} removal capacity performance (Zhang et al. 2021).

Most of the studies focus on the Cd^{2+} adsorption capability on carbonaceous materials and the modulation strategies to catch heavy metal ions. However, the detection of small amounts of Cd^{2+} in aqueous environments is another focus of research, which also involves an adsorption scheme. Banerjee et al. (2019) reported the use of rhodamine B for rapid and easy optical detection of small amounts as low as 9.6×10^{-9} M Cd^{2+}. The modeling of this system was carried out using density functional theory-based calculations, including time dependence, and the results indicated that the metal ion is located close to oxygen atoms of the carbonaceous system, rhodamine B, forming a bridge between amino and oxygenated groups O–Cd^{2+}–N.

4 Experimental and Computational Perspectives

While many research studies have investigated the adsorption mechanism in monocomponent systems, there is still a significant need to study multicomponent systems from an equilibrium, thermodynamic, and kinetic perspective. Moreover, studying the adsorption mechanism of Cd^{2+} in multicomponent systems is of significant importance when attempting to simulate a real wastewater system contaminated with multiple pollutants. It will help researchers to accurately predict the removal efficiency of cadmium under realistic conditions, accounting for the presence of other competing substances. In some cases, the presence of other pollutants can lead to synergistic effects, enhancing the adsorption of cadmium. Conversely, antagonistic interactions might hinder cadmium removal. Understanding these interactions is crucial for optimizing the treatment process and achieving effective cadmium removal.

There are very few studies employing computational chemistry for the characterization of Cd interactions with adsorbents. Existing research primarily relies on adsorption energies to define interactions, with a notable absence of articles focusing on the kinetics and thermodynamics aspects of the adsorption process using molecular modeling. Furthermore, no articles have explored dynamic systems, as the existing studies solely concentrate on static systems.

The study of Cd^{2+} adsorption on carbonaceous models lacks details from the computational side. The computational methodologies should be improved to provide a better model by including the solvent effect and van der Waals contributions to the metal–carbon interactions. The use of different carbonaceous models deserves further studies, particularly graphdiyne and fullerenes, models beyond the classical clusters of small fused rings. Moreover, a detailed study to cover pseudopotentials is necessary to achieve reasonable results. The few studies regarding cadmium adsorption focus mainly on the binding energy and changes in electronic properties; however, studies related to both thermodynamics and kinetics are still to be performed. The theoretical understanding of Cd^{2+} adsorption on carbonaceous-based systems is still in the first stages; hence, the road is clean to include several contributions from different angles from a computational point of view; i.e., the fundamental understanding is still to be built.

References

Abolfazli Behrooz B, Oustan S, Mirseyed Hosseini H, Etesami H, Padoan E, Magnacca G, Marsan FA (2023) The importance of presoaking to improve the efficiency of MgCl2-modified and non-modified biochar in the adsorption of cadmium. Ecotoxicol Environ Saf 257:114932. https://doi.org/10.1016/j.ecoenv.2023.114932

Aboua KN, Yobouet YA, Yao KB, Goné DL, Trokourey A (2015) Investigation of dye adsorption onto activated carbon from the shells of Macoré fruit. J Environ Manag 156:10–14. https://doi.org/10.1016/j.jenvman.2015.03.006

Aghashiri A, Fotooh FK, Hashemian S (2019) Density functional calculations of nickel, palladium and cadmium adsorption onto (10,0) single-walled carbon nanotube. J Mol Model 25:185. https://doi.org/10.1007/s00894-019-4062-z

Ahmad A, Priyadarshini M, Yadav S, Ghangrekar MM, Surampalli RY (2022) The potential of biochar-based catalysts in advanced treatment technologies for efficacious removal of persistent organic pollutants from wastewater: a review. Chem Eng Res Des 187:470–496. https://doi.org/10.1016/j.cherd.2022.09.024

Ai S, Huang Y, Xie T, Huang C (2020) Facile carboxylation of sugarcane bagasse and the adsorption mechanism for cadmium ions. Ind Eng Chem Res 59:8795–8804. https://doi.org/10.1021/acs.iecr.0c00037

Aktar J (2020) Batch adsorption process in water treatment. In: Intelligent environmental data monitoring for pollution management. Elsevier, pp 1–24. https://doi.org/10.1016/B978-0-12-819671-7.00001-4

Al-Ghouti MA, Da'ana DA (2020) Guidelines for the use and interpretation of adsorption isotherm models: a review. J Hazard Mater 393:122383. https://doi.org/10.1016/j.jhazmat.2020.122383

Al-Ghouti MA, Da'ana DA, Chen T, Da T, Ma Y (2021) Reasonable calculation of the thermodynamic parameters from adsorption equilibrium constant. J Mol Liq 322:114980. https://doi.org/10.1016/j.molliq.2020.114980

Baachaoui S, Aldulaijan S, Sementa L, Fortunelli A, Dhouib A, Raouafi N (2021) Density functional theory investigation of graphene functionalization with activated carbenes and its application in the sensing of heavy metallic cations. J Phys Chem C 125:26418–26428. https://doi.org/10.1021/acs.jpcc.1c07247

Bai M, Chai Y, Chen A, Yuan J, Shang C, Peng L, Peng C (2023) Enhancing cadmium removal efficiency through spinel ferrites modified biochar derived from agricultural waste straw. J Environ Chem Eng 11:109027. https://doi.org/10.1016/j.jece.2022.109027

Banerjee M, Ghosh M, Ta S, Das J, Das D (2019) A smart optical probe for detection and discrimination of Zn2+, Cd2+ and Hg2+ at nano-molar level in real samples. J Photochem Photobiol A Chem 377:286–297. https://doi.org/10.1016/j.jphotochem.2019.04.002

Bashir A, Manzoor T, Malik LA, Qureashi A, Pandith AH (2020) Enhanced and selective adsorption of Zn(II), Pb(II), Cd(II), and Hg(II) ions by a dumbbell- and flower-shaped potato starch phosphate polymer: a combined experimental and DFT calculation study. ACS Omega 5: 4853–4867. https://doi.org/10.1021/acsomega.9b03607

Baskaran P, Abraham M (2022) Adsorption of cadmium (Cd) and lead (Pb) using powdered activated carbon derived from Cocos Nucifera waste: a kinetics and equilibrium study for long-term sustainability. Sustain Energy Technol Assess 53:102709. https://doi.org/10.1016/j.seta.2022.102709

Bastos M, Camps I (2014) First-principles calculations of nickel, cadmium, and lead adsorption on a single-walled (10,0) carbon nanotube. J Mol Model 20:2094. https://doi.org/10.1007/s00894-014-2094-y

Chen H, Li W, Wang J, Xu H, Liu Y, Zhang Z, Li Y, Zhang Y (2019) Adsorption of cadmium and lead ions by phosphoric acid-modified biochar generated from chicken feather: selective adsorption and influence of dissolved organic matter. Bioresour Technol 292:121948. https://doi.org/10.1016/j.biortech.2019.121948

Cheng S, Zhao S, Guo H, Xing B, Liu Y, Zhang C, Ma M (2022) High-efficiency removal of lead/cadmium from wastewater by MgO modified biochar derived from Crofton weed. Bioresour Technol 343:126081. https://doi.org/10.1016/j.biortech.2021.126081

Cruz CCV, da Costa ACA, Henriques CA, Luna AS (2004) Kinetic modeling and equilibrium studies during cadmium biosorption by dead Sargassum sp. biomass. Bioresour Technol 91: 249–257. https://doi.org/10.1016/S0960-8524(03)00194-9

Diaz-Flores PE, López-Urı'as F, Terrones M, Rangel-Mendez JR (2009) Simultaneous adsorption of Cd2+ and phenol on modified N-doped carbon nanotubes: experimental and DFT studies. J Colloid Interface Sci 334:124–131. https://doi.org/10.1016/j.jcis.2009.02.045

Ding Y, Liu Y, Liu S, Li Z, Tan X, Huang X, Zeng G, Zhou Y, Zheng B, Cai X (2016) Competitive removal of Cd(II) and Pb(II) by biochars produced from water hyacinths: performance and mechanism. RSC Adv 6:5223–5232. https://doi.org/10.1039/c5ra26248h

Doke KM, Khan EM (2017) Equilibrium, kinetic and diffusion mechanism of Cr(VI) adsorption onto activated carbon derived from wood apple shell. Arab J Chem 10:S252–S260. https://doi.org/10.1016/j.arabjc.2012.07.031

Enniya I, Rghioui L, Jourani A (2018) Adsorption of hexavalent chromium in aqueous solution on activated carbon prepared from apple peels. Sustain Chem Pharm 7:9–16. https://doi.org/10.1016/j.scp.2017.11.003

Ezzat H, Menazea AA, Omara W, Basyouni O, Helmy S, Mohamed A, Tawfik W, Ibrahim M (2020) DFT: B3LYP/ LANL2DZ study for the removal of Fe, Ni, Cu, As, Cd and Pb with chitosan. Biointerface Res Appl Chem 10:7002–7010. https://doi.org/10.33263/BRIAC106.70027010

Forgionny A, Acelas NY, Ocampo-pérez R, Padilla-ortega E, Leyva-ramos R, Flórez E (2021) Understanding mechanisms in the adsorption of lead and copper ions on chili seed waste in single and multicomponent systems: a combined experimental and computational study. Environ Sci Pollut Res 28:23204. https://doi.org/10.1007/s11356-020-11721-z

Forgionny A, Acelas NY, Ocampo-Pérez R, Padilla-Ortega E, Pérez S, Flórez E (2022) Mechanism adsorption analysis during the removal of Cd2+ and Cu2+ onto cedar sawdust via experiment coupled with theoretical calculation: mono- and multicomponent systems. Environ Nanotechnol Monit Manag 18:100715. https://doi.org/10.1016/j.enmm.2022.100715

Freundlich H (1907) Über die adsorption in lösungen. Z Phys Chem 57:385–470

Gao Z, Shan D, He J, Huang T, Mao Y, Tan H, Shi H, Li T, Xie T (2023) Effects and mechanism on cadmium adsorption removal by CaCl2-modified biochar from selenium-rich straw. Bioresour Technol 370:128563. https://doi.org/10.1016/j.biortech.2022.128563

Giraldo S, Robles I, Ramirez A, Flórez E, Acelas N (2020) Mercury removal from wastewater using agroindustrial waste adsorbents. SN Appl Sci 2. https://doi.org/10.1007/s42452-020-2736-x

Giraldo S, Acelas NY, Ocampo-Pérez R, Padilla-Ortega E, Flórez E, Franco CA, Cortés FB, Forgionny A (2022) Application of Orange Peel waste as adsorbent for methylene blue and Cd2+ simultaneous remediation. Molecules 27. https://doi.org/10.3390/molecules27165105

Hu Q, Lan R, He L, Liu H, Pei X (2023) A critical review of adsorption isotherm models for aqueous contaminants: curve characteristics, site energy distribution and common controversies. J Environ Manag 329:117104. https://doi.org/10.1016/j.jenvman.2022.117104

Huang K, Hu C, Tan Q, Yu M, Shabala S, Yang L, Sun X (2022) Highly efficient removal of cadmium from aqueous solution by ammonium polyphosphate-modified biochar. Chemosphere 305:135471. https://doi.org/10.1016/j.chemosphere.2022.135471

Ibrahim HS, Ammar NS, Soylak M, Ibrahim M (2012) Removal of Cd(II) and Pb(II) from aqueous solution using dried water hyacinth as a biosorbent. Spectrochim Acta Part A Mol Biomol Spectrosc 96:413–420. https://doi.org/10.1016/j.saa.2012.05.039

Javarani L, Malakootian M, Hassani AH, Javid AH (2022) Removal of heavy metal ions from water using functionalized carbon nanosheet: a density-functional theory study. Jundishapur J Health Sci 14. https://doi.org/10.5812/jjhs.122840

Kumar A, Thakur A, Panesar PS (2023) A review on the industrial wastewater with the efficient treatment techniques. Chem Pap 77:4131. https://doi.org/10.1007/s11696-023-02779-3

Kushwaha J, Singh R (2023) Cellulose hydrogel and its derivatives: a review of application in heavy metal adsorption. Inorg Chem Commun 152:110721. https://doi.org/10.1016/j.inoche.2023.110721

Lagergren SK (1898) About the theory of so-called adsorption of soluble substances. Sven Vetenskapsakad Handingarl 24:1–39

Langmuir I (1916) The constitution and fundamental properties of solids and liquids. Part I. Solids. J Am Chem Soc 38:2221–2295. https://doi.org/10.1021/ja02268a002

Li X, Wang C, Tian J, Liu J, Chen G (2020) Comparison of adsorption properties for cadmium removal from aqueous solution by Enteromorpha prolifera biochar modified with different chemical reagents. Environ Res 186:109502. https://doi.org/10.1016/j.envres.2020.109502

Liu B, Chen T, Wang B, Zhou S, Zhang Z, Li Y, Pan X, Wang N (2022a) Enhanced removal of Cd2+ from water by AHP-pretreated biochar: adsorption performance and mechanism. J Hazard Mater 438:129467. https://doi.org/10.1016/j.jhazmat.2022.129467

Liu T, Lawluvy Y, Shi Y, Ighalo JO, He Y, Zhang Y, Yap PS (2022b) Adsorption of cadmium and lead from aqueous solution using modified biochar: a review. J Environ Chem Eng 10:106502. https://doi.org/10.1016/j.jece.2021.106502

Luo M, Lin H, He Y, Li B, Dong Y, Wang L (2019) Efficient simultaneous removal of cadmium and arsenic in aqueous solution by titanium-modified ultrasonic biochar. Bioresour Technol 284:333–339. https://doi.org/10.1016/j.biortech.2019.03.108

Ma F, Zhao H, Zheng X, Zhao B, Diao J, Jiang Y (2023) Enhanced adsorption of cadmium from aqueous solution by amino modification biochar and its adsorption mechanism insight. J Environ Chem Eng 11:109747. https://doi.org/10.1016/j.jece.2023.109747

Malhotra M, Puglia M, Kalluri A, Chowdhury D, Kumar CV (2022) Adsorption of metal ions on graphene sheet for applications in environmental sensing and wastewater treatment. Sensors Actuators Rep 4:100077. https://doi.org/10.1016/j.snr.2022.100077

Menazea AA, Ezzat HA, Omara W, Basyouni OH, Ibrahim SA, Mohamed AA, Tawfik W, Ibrahim MA (2020) Chitosan/graphene oxide composite as an effective removal of Ni, Cu, As, Cd and Pb from wastewater. Comput Theor Chem 1189:112980. https://doi.org/10.1016/j.comptc.2020.112980

Mohan D, Kumar H, Sarswat A, Alexandre-Franco M, Pittman CU (2014) Cadmium and lead remediation using magnetic oak wood and oak bark fast pyrolysis bio-chars. Chem Eng J 236: 513–528. https://doi.org/10.1016/j.cej.2013.09.057

Mozaffari Majd M, Kordzadeh-Kermani V, Ghalandari V, Askari A, Sillanpää M (2022) Adsorption isotherm models: a comprehensive and systematic review (2010−2020). Sci Total Environ 812:151334. https://doi.org/10.1016/j.scitotenv.2021.151334

Pang Y, Zhao C, Li Y, Li Q, Bayongzhong X, Peng D, Huang T (2022) Cadmium adsorption performance and mechanism from aqueous solution using red mud modified with amorphous MnO2. Sci Rep 12:1–18. https://doi.org/10.1038/s41598-022-08451-2

Park JH, Wang JJ, Kim SH, Kang SW, Jeong CY, Jeon JR, Park KH, Cho JS, Delaune RD, Seo DC (2019) Cadmium adsorption characteristics of biochars derived using various pine tree residues and pyrolysis temperatures. J Colloid Interface Sci 553:298–307. https://doi.org/10.1016/j.jcis.2019.06.032

Pyrzynska K (2019) Removal of cadmium from wastewaters with low-cost adsorbents. J Environ Chem Eng 7:102795. https://doi.org/10.1016/j.jece.2018.11.040

Qiu B, Tao X, Wang H, Li W, Ding X, Chu H (2021) Biochar as a low-cost adsorbent for aqueous heavy metal removal: a review. J Anal Appl Pyrolysis 155:105081. https://doi.org/10.1016/j.jaap.2021.105081

Qiu B, Shao Q, Shi J, Yang C, Chu H (2022) Application of biochar for the adsorption of organic pollutants from wastewater: modification strategies, mechanisms and challenges. Sep Purif Technol 300:121925. https://doi.org/10.1016/j.seppur.2022.121925

Quisperima A, Pérez S, Flórez E, Acelas N (2022) Valorization of potato peels and eggshells wastes: Ca-biocomposite to remove and recover phosphorus from domestic wastewater. Bioresour Technol 343:126106. https://doi.org/10.1016/j.biortech.2021.126106

Rajahmundry GK, Garlapati C, Kumar PS, Alwi RS, Vo DVN (2021) Statistical analysis of adsorption isotherm models and its appropriate selection. Chemosphere 276:130176. https://doi.org/10.1016/j.chemosphere.2021.130176

Ramírez Muñoz AP, Giraldo S, Flórez Yepes E, Acelas Soto NY (2017) Preparación de carbón activado a partir de residuos de palma de aceite y su aplicación para la remoción de colorantes. Rev Colomb Química 46(1):33–41. https://doi.org/10.15446/rev.colomb.quim.v46n1.62851

Ramirez-Muñoz A, Pérez S, Muñoz-Saldaña J, Flórez E, Acelas N (2021) Eco-friendly materials obtained through a simple thermal transformation of water hyacinth (Eichhornia Crassipes) for the removal and immobilization of Cd2+ and Cu2+ from aqueous solutions. Environ Nanotechnol Monit Manag 16:100574. https://doi.org/10.1016/j.enmm.2021.100574

Shi G, Liu J, Wang C, Song B, Tu Y, Hu J, Fang H (2013) Ion enrichment on the hydrophobic carbon-based surface in aqueous salt solutions due to cation-π interactions. Sci Rep 3:3436. https://doi.org/10.1038/srep03436

Simonin J-P (2016) On the comparison of pseudo-first order and pseudo-second order rate laws in the modeling of adsorption kinetics. Chem Eng J 300:254–263. https://doi.org/10.1016/j.cej.2016.04.079

Sips R (1948) On the structure of a catalyst surface. J Chem Phys 16:490–495. https://doi.org/10.1063/1.1746922

Tejada Tovar CN, Villabona Ortíz Á, Ramírez Vásquez PA (2020) Valorización de residuos de la obtención de almidón de ñame espino para su uso como bioadsorbente en la remoción de Cromo (VI) y Níquel (II). INGE CUC 16:22–34. https://doi.org/10.17981/ingecuc.16.1.2020.02

Temkin M (1940) Kinetics of ammonia synthesis on promoted iron catalysts. Acta Physiochim URSS 12:327–356

Wang J, Guo X (2020) Adsorption isotherm models: classification, physical meaning, application and solving method. Chemosphere 258:127279. https://doi.org/10.1016/j.chemosphere.2020.127279

Wang L, Wang Y, Ma F, Tankpa V, Bai S, Guo X, Wang X (2019) Mechanisms and reutilization of modified biochar used for removal of heavy metals from wastewater: a review. Sci Total Environ 668:1298–1309. https://doi.org/10.1016/j.scitotenv.2019.03.011

Wang J, Wang Y, Wang J, Du G, Khan KY, Song Y, Cui X, Cheng Z, Yan B, Chen G (2022) Comparison of cadmium adsorption by hydrochar and pyrochar derived from Napier grass. Chemosphere 308:136389. https://doi.org/10.1016/j.chemosphere.2022.136389

World Health Organization (2011) Guidelines for drinking-water quality, Encyclopedia of Earth Sciences Series. World Health Organization. https://doi.org/10.1007/978-1-4020-4410-6_184

Xu Y, Xia H, Zhang Q, Jiang G, Cai W, Hu W (2022) Adsorption of cadmium(II) in wastewater by magnesium oxide modified biochar. Arab J Chem 15:104059. https://doi.org/10.1016/j.arabjc.2022.104059

Yang K, Wang X, Cheng H, Tao S (2022) Enhanced immobilization of cadmium and lead adsorbed on crop straw biochars by simulated aging processes. Environ Pollut 302:119064. https://doi.org/10.1016/j.envpol.2022.119064

Yin K, Wang J, Zhai S, Xu X, Li T, Sun S, Xu S, Zhang X, Wang C, Hao Y (2022) Adsorption mechanisms for cadmium from aqueous solutions by oxidant-modified biochar derived from Platanus orientalis Linn leaves. J Hazard Mater 428:128261. https://doi.org/10.1016/j.jhazmat.2022.128261

Yuan Q, Wang P, Wang X, Hu B, Wang C, Xing X (2023) Nano-chlorapatite modification enhancing cadmium(II) adsorption capacity of crop residue biochars. Sci Total Environ 865:161097. https://doi.org/10.1016/j.scitotenv.2022.161097

Zhang F, Wang X, Yin D, Peng B, Tan C, Liu Y, Tan X, Wu S (2015) Efficiency and mechanisms of Cd removal from aqueous solution by biochar derived from water hyacinth (Eichornia crassipes). J Environ Manag 153:68–73. https://doi.org/10.1016/j.jenvman.2015.01.043

Zhang Y, Yue X, Xu W, Zhang H, Li F (2019) Amino modification of rice straw-derived biochar for enhancing its cadmium(II) ions adsorption from water. J Hazard Mater 379:120783. https://doi.org/10.1016/j.jhazmat.2019.120783

Zhang Z, Wang T, Zhang H, Liu Y, Xing B (2021) Adsorption of Pb(II) and Cd(II) by magnetic activated carbon and its mechanism. Sci Total Environ 757:143910. https://doi.org/10.1016/j.scitotenv.2020.143910

Zhang Z, Li Y, Zong Y, Yu J, Ding H, Kong Y, Ma J, Ding L (2022) Efficient removal of cadmium by salts modified-biochar: performance assessment, theoretical calculation, and quantitative mechanism analysis. Bioresour Technol 361:127717. https://doi.org/10.1016/j.biortech.2022.127717

Zhang K, Yi Y, Fang Z (2023) Remediation of cadmium or arsenic contaminated water and soil by modified biochar: a review. Chemosphere 311:136914. https://doi.org/10.1016/j.chemosphere.2022.136914

Zhu L, Tong L, Zhao N, Wang X, Yang X, Lv Y (2020) Key factors and microscopic mechanisms controlling adsorption of cadmium by surface oxidized and aminated biochars. J Hazard Mater 382:121002. https://doi.org/10.1016/j.jhazmat.2019.121002

Phytoremediation of Cadmium-Contaminated Soil

R. Sikka, Tanvi Sahni, Diksha Verma, P. Chaitra, and Annu Singh

Abstract Cadmium is a toxic metal. Its accumulation is increasing with increasing urbanization and industrialization. Thus, there is a needed for the remediation of cadmium present in soil, water, and other natural resources. To address this issue, various techniques of remediation are performed among which phytoremediation is emerging as the best method. In this chapter, we have discussed about various conventional techniques for remediating cadmium-contaminated soils and the challenges associated with these techniques. The chapter explores the application of different remediation approaches, including electro-kinetic remediation, chemical elution, stabilization and solidification, and phytoremediation, with a focus on their efficiency and practicality. Additionally, it highlights the limitations and environmental implications of each method.

Keywords Cadmium contamination · Plant · Phytoremediation · Soil · Human health

1 Introduction

Phytoremediation is a method that utilizes plants to purify contaminated environmental media, including land, waterways, and groundwater (Singh et al. 2023). Plant-assisted bioremediation, is a form of phytoremediation, in which organic contaminants present in soil are remediated with plant roots and its microbes (Shukla et al. 2010). This method offers a more environmentally friendly and efficient alternative to conventional remediation procedures, producing less secondary waste. Plants have the natural ability to extract essential nutrients, including metals, from the soil and water they grow in Roy et al. (2015). Hyperaccumulators are the plants having capacity to store excess metal ions (Chaney et al. 2007). Additionally, plants can absorb various organic molecules from the environment

R. Sikka (✉) · T. Sahni · D. Verma · P. Chaitra · A. Singh
Department of Soil Science, PAU, Ludhiana, Punjab, India
e-mail: sikkar@pau.edu

A. K. Jha, N. Kumar (eds.), *Cadmium Toxicity Mitigation*,
https://doi.org/10.1007/978-3-031-47390-6_13

and metabolize or break them down to support their physiological processes (Dusenge et al. 2019). Phytoremediation technologies are still in the early stages of development, ongoing research in laboratories and small-scale field tests aims to define and improve the processes involved. Researchers are exploring ways to enhance plants' natural ability to perform remediation tasks and investigating other plant species with potential applications in phytoremediation. This includes genetic engineering to optimize plants for specific remediation tasks (Ibañez et al. 2016). The issue of heavy metal pollution has become a significant global concern, exacerbated by increasing urbanization and disruptions to natural biogeochemical cycles (Pandey and Madhuri 2014). Heavy metals, unlike organic substances, are nonbiodegradable, leading to their accumulation in the environment. As a result, ecosystems and human health are at risk due to the buildup of heavy metals in soils and water bodies (Ayangbenro and Babalola 2017). These heavy metals can bioaccumulate in living organisms, leading to higher concentrations as organisms move up the food chain (biomagnification). In soils, heavy metals also have toxicological effects on soil microorganisms, which can result in decreased microbial activity and population decline (Abdu et al. 2017). Therefore, phytoremediation offers a promising solution to address heavy metal pollution in the environment. By harnessing the natural abilities of plants and their interaction with microbes, phytoremediation holds the potential to provide an eco-friendly and efficient method for cleaning up contaminated soils and water bodies (Mani and Kumar 2014).

1.1 Cadmium

Heavy metals are distinguished by their substantial atomic weight and density, which is approximately five times greater than that of water (Razo et al. 2004). Within this category, there are essential and non-essential heavy metals. Essential heavy metals are necessary in trace amounts and fulfill critical roles in electron transport, redox reactions, and nucleic acid metabolism. They contribute to the growth and proper function of plants, animals, and the human body (Gill 2014). However, exceeding the tolerable limits of essential heavy metals can lead to severe disruptions in the normal biological functions of organisms (Mudgal et al. 2010). Some heavy metals, such as Iron (Fe), Molybdenum (Mo), and Manganese (Mn), act as micronutrients. Their presence in trace quantities is essential for various physiological processes (He et al. 2005). Conversely, certain heavy metals like Chromium (Cr), Cobalt (Co), Copper (Cu), Nickel (Ni), Vanadium (Vn), and Zinc (Zn) are also required in small amounts but can become toxic when their concentrations surpass certain thresholds (Edelstein and Ben-Hur 2018). In contrast, non-essential heavy metals, including Antimony (Sb), Arsenic (As), Cadmium (Cd), Lead (Pb), Mercury (Hg), and Silver (Ag), have no known biological functions and are recognized as toxic to living organisms (Edelstein and Ben-Hur 2018). When these non-essential heavy metals are present in the environment, they pose significant risks to the health of ecosystems and humans alike.

Human activities are the primary sources of contamination in water and soil with heavy metal pollutants (Ye et al. 2011). These activities, driven by human actions, encompass agricultural practices involving the application of fungicides and pesticides, improper disposal of household waste, the discharge of industrial effluents, mining, smelting, electroplating operations, sewage sludge disposal, and the use of urban composts (Mrayyan and Hamdi 2006). These actions introduce heavy metals into the environment, creating a potential for bioaccumulation and biomagnification in the food chain (Ali and Khan 2019). This, in turn, magnifies the risks associated with heavy metal pollution. Among the toxic heavy metals, Cadmium stands out as an exceptionally hazardous element for plants, animals, and humans within the ecosystem (Alloway 2012). While Cd is not essential for organisms, its strong chemical reactivity in the environment, high mobility, and persistent toxicity make it particularly perilous (Khan et al. 2015). Its capacity to accumulate and become enriched in the food chain poses a significant threat to human health. Cd is renowned for its carcinogenic, teratogenic, and mutagenic effects, which can lead to severe health complications when it enters the human body through contaminated food sources.

Considering its detrimental impact on both the environment and human health, it is crucial to address the issue of Cd contamination and adopt effective remediation strategies to mitigate its adverse effects. Understanding the sources and pathways of Cd pollution is essential for developing targeted interventions that can prevent further contamination and safeguard ecosystems and public health (Gwenzi et al. 2018). Additionally, exploring innovative approaches, such as phytoremediation, may offer promising solutions to combat Cd pollution and promote sustainable environmental management. By taking concerted efforts to reduce anthropogenic contributions of Cd and other heavy metals, we can work toward preserving the integrity of our ecosystems and securing a healthier future for all living beings.

As per the 2014 National Soil Pollution Investigation Bulletin released by China's Ministry of Environmental Protection (MEP) and Ministry of Land and Resources (MLR), approximately 16.1% of Chinese arable land has been affected by pollution (Sun et al. 2019). Among the heavy metal pollutants, Cadmium (Cd) has emerged as the most serious soil contaminant, with high levels detected in 7.0% of surveyed sites. This widespread Cd contamination poses significant risks to human health (Yang et al. 2018).

The alarming increase in soil Cd pollution can be attributed to the rapid industrialization in China (Wang et al. 2019). Activities like smelting plants, mining operations, and electroplating industries contribute significantly to the release of Cd into the environment. Additionally, various anthropogenic practices, such as sewage irrigation, improper disposal of sewage sludge and unqualified phosphate fertilizer application, and atmospheric deposition, lead to elevated background levels of Cd in the soil (Yang et al. 2018).

The primary concern with soil Cd contamination lies in its ease of uptake by plant roots and subsequent transport to the edible parts of crops (Dudka and Miller 1999). The consumption of crops containing high levels of Cd can lead to severe health issues, including abnormal liver function, cancers, and diseases such as osteoporosis and renal tubular dysfunction (known as *itai-itai* disease). It is evident that excessive Cd accumulation in the body can have detrimental effects on human health.

Moreover, soil Cd contamination has detrimental effects on plant growth and development. It interferes with vital physiological processes such as photosynthesis, chlorophyll synthesis, disrupts water and nutrient uptake, and damages intracellular redox balance, leading to oxidative stress and injury (Khanna et al. 2022). Consequently, the overall plant biomass, yield, and quality are significantly reduced. Addressing the issue of soil Cd pollution is of paramount importance to safeguard human health, protect the environment, and ensure sustainable agriculture (Kollah et al. 2016). Implementing strict regulations on industrial practices and waste disposal, adopting safer agricultural practices, and conducting regular soil monitoring are essential steps toward mitigating the adverse effects of Cd contamination. By taking comprehensive and proactive measures, Countries may work toward minimizing the impact of Cd pollution and creating a safer and healthier environment for its citizens.

1.2 Properties of Cadmium

Cadmium occurs naturally within the Earth's crust. It possesses a silvery metallic appearance with a subtle bluish tint on its surface. Cadmium, due to its toxicity and its association with causing birth defects and cancer, has prompted efforts to restrict its usage. Presently, approximately 80% of the cadmium production is directed toward the production of rechargeable nickel-cadmium batteries. Nonetheless, these batteries are gradually being phased out, making way for nickel metal hydride batteries. Historically, cadmium found application in electroplating steel to provide protection against corrosion. Today, it continues to be employed in safeguarding critical components found in airplanes and on oil platforms. In the past, cadmium played a role in the creation of phosphors used in cathode ray tube color television sets, as well as in the production of pigments in shades of yellow, orange, and red. Furthermore, cadmium's neutron-absorbing properties make it suitable for use in rods within nuclear reactors to regulate atomic fission processes. Cadmium is widely recognized for its toxicity, carcinogenicity, and teratogenicity, which refers to its capacity to disrupt the development of embryos or fetuses. On average, individuals ingest as little as 0.05 milligrams of cadmium daily. However, it accumulates within the body, resulting in an average storage of approximately 50 milligrams over time.

1.3 Sources of Cadmium Pollution

Heavy metals in the environment have their origins in both natural processes and human activities. Human-induced sources, which encompass agriculture, industry, mining, transportation, fuel consumption, residual organic materials, and wastewater

Table 1 The total content ranges of major heavy metals found in different rock types and typical environments

Rock type	Heavy metals (mg/kg)
Balsitic igneous	0.006–0.6
Granite igneous	0.003–0.18
Shales and clay	0–11
Black shales	0.3–>08.4
Sandstones	–
Typic soil	0.01–0.7

discharge, are significant contributors to the pollution caused by heavy metals. Conversely, natural origins like wind-driven dust, particles resulting from volcanic activity, forest fires, plant life, and sea salt also have a role to play. The amalgamation of these origins leads to the dispersion of heavy metals across the environment, impacting soils, bodies of water, and the atmosphere. In order to protect the environment and human health, it is imperative to address and mitigate emissions of heavy metals stemming from both natural occurrences and human activities. The total content of heavy metals found in different rock types and soil is presented in Table 1.

The levels of various heavy metals in treated wastewater effluent are compared to permissible limits established by guidelines and regulations for agricultural irrigation, drinking water, and surface water. Industrial wastewater generally contains higher concentrations of heavy metals than domestic wastewater, with variations based on the specific sources of wastewater. Using this wastewater for irrigation can contribute to increased heavy metal contamination in the soil. In addition to wastewater, there are other sources of soil contamination with heavy metals, including:

1. The use of soil amendments containing relatively high levels of heavy metals is a concern. It is common practice to apply composted municipal waste and sewage sludge to agricultural lands as a cost-effective fertilizer and disposal method. However, this practice carries the risk of heavy metal contamination in cultivated soils, which can jeopardize groundwater quality and potentially harm the food chain.

   For example, a study conducted in Switzerland examined several factors, including the average concentration of heavy metals in sewage sludge applied in 1989, limit values set for sewage sludge in 1992, guide values for soils, and total concentrations in soils following various treatment applications. The study's findings indicated that the application of municipal waste and sewage sludge typically results in increased accumulation of heavy metals in the soil.
2. The rapid urbanization and industrialization witnessed in recent decades have made significant contributions to the contamination of the atmosphere with toxic heavy metals such as Arsenic (As), Chromium (Cr), Lead (Pb), Nickel (Ni), Zinc (Zn), Cadmium (Cd), and Vanadium (V). The concentrations of these heavy metals in the air exhibit substantial variations between urban and rural areas, largely dependent on the proximity to emission sources. Primary sources of atmospheric heavy metal pollution include waste incineration, domestic oil combustion, industrial factories, and vehicle emissions. These heavy metals can

be deposited from the polluted air into soils and water resources, subsequently finding their way into plants and entering the food chain.

Cadmium concentrations in rocks display a wide range of variability, 0.1–1 mg/g in soil and 100 mg/g in phosphatic rock. Cadmium exists in various forms, including oxides and carbonates, minerals like sphalerite (zinc sulfide) with concentrations ranging from 500 to 18,000 mg/g, as well as smithsonite (zinc carbonate).

Mining activities also indirectly contribute to environmental cadmium, as coal products used in thermal power plants cause Cd emissions during combustion. With the increasing demand for energy, coal burning has become more widespread, leading to elevated atmospheric emissions. Furthermore, cadmium associated with zinc ores tends to spread beyond localized mining regions, affecting nearby soils and drainage waters.

Aside from anthropogenic sources, exposures like tobacco smoke are self-induced to cadmium. One cigarette contains approximately 1–2 μg of Cd. Consequently, smoking one cigarette results in inhaling 0.1–0.2 μg of cadmium. This source of exposure to cadmium, combined with exposure from food and drinking water, can be a confounding factor in occupational studies.

1.4 Effect on Soil

Elevated levels of Cd have bad effects on soil fertility, disrupt plant physiology, and impede metabolisms, lead to stunted growth and reduced crop yields (Asati et al. 2016). Global average Cd concentration in soil stands at approximately 0.41 mg/kg. However, natural soils can exhibit varying Cd levels, ranging from less than 0.01–0.8 mg/kg, with some regions experiencing higher concentrations, reaching as much as 2.0–8.9 mg/kg (Holmgren et al. 1993). Cd in soil exists in various forms, including organic, exchangeable, oxide, and carbonate forms (Gill 2014). Therefore, In cadmium contaminated soils, It is critical to consider its bioavailability. When desorbed, Cd is released in the soil solution and subsequently taken up by plant roots through mechanisms of mass flow or diffusion (Ernstberger et al. 2005).

In the soil, rhizosphere, an environment encompassing intricate plant-associated microbial networks, plays an important role in assembly and immobilization of Cd. Within the rhizosphere, microbes have the capability to directly or indirectly influence the mobilization of Cd through various mechanisms, including chemical transformations, chelation, and the release of chelating agents. Additionally, they can induce alterations in redox conditions (Ernstberger et al. 2005). Conversely, these microbes can also contribute to the aggregation of Cd by stimulating plant growth. Nevertheless, the availability of Cd is diminished through processes such as bioprecipitation or biosorption.

Moreover, microbes assume a pivotal role in immobilizing Cd through essential processes, including bioaccumulation (the accumulation of Cd within microbial cells) and biotransformation (the conversion of the metal from a harmful valence state to a non-harmful form). Additionally, microbes bind toxic metals with intracellular proteins, such as metallothionein and metallo-chaperones, transforming them into non-bioavailable forms. This effectively diminishes the availability of Cd^{2+} (Hynninen 2010). The speciation of Cd in soils' rhizosphere is predominantly shaped by several factors. These factors encompass soil pH, fluctuations in redox potential, the presence of soil microbes, the existence of organic and inorganic ligands, specific types of metal and plant species within the ecosystem (Violante et al. 2010).

Cadmium (Cd) in soil and its accessibility to organisms is subject to influence by a range of factors, encompassing the physical, chemical, and biological characteristics of the soil. One pivotal factor in this regard is soil's pH level, which affects the adsorption of Cd by the soil. Consequently, it impacts the concentration of Cd in the soil solution, making it more or less available to organisms (Vig et al. 2003). Furthermore, the presence of organic matter and the clay content in the soil can have a notable impact on the Cd levels in soil solution. In clay-rich soils, Cd contamination of the soil has a relatively minor effect on microbial processes such as phosphatase and urease activity, likely due to its limited bioavailability. Additionally, Cd added to specific soils leads to a deceleration in the decomposition of particular organic compounds. Meanwhile, mineralization of carbon and nitrogen in different soils is closely correlated with the quantity of water-soluble Cd present in the soil (Hussain et al. 2017).

Given the intricate nature of soil systems, drawing sweeping conclusions regarding the influence of ligands on Cd sorption and availability is a complex endeavor (Payne et al. 2013). Nonetheless, numerous research studies imply that Cd fraction in soil that is bioavailable tends to diminish gradually. This transformation is influenced by various factors, including soil pH, clay content, and the presence of organic matter (Violante et al. 2010).

Soil pH exerts a substantial influence on the availability of metals in soil, particularly when it comes to cadmium (Cd) levels in rice grains. Research has shown that there exists a negative correlation between soil pH and the concentration of Cd in rice grains. Availability and mobility of Cd in soil increases as soil pH decreases. This phenomenon occurs because lower pH levels cause a shift in Cd from stable forms, such as carbonates and oxides of Fe and Mn, to more readily bioavailable forms, including exchangeable Cd.

Furthermore, reduced pH levels diminish Cd^{2+} adsorption, resulting in higher solubility of Cd^{2+} in the soil and enhanced uptake by rice plants. In alkaline solutions, Cd predominantly exists in less bioavailable forms like $CdHCO^{3+}$ or $CdCO_3$.

Flooding can result in a reduction in metal contents due to changing soil pH. High soil pH reduces the solubility of metal in submerged rice soils. For instance, $CaCl_2$-extractable Cd decreases 3 times when pH rises from 5 to 6.5.

During cycles of alternate wetting and drying, the transformation of Fe^{2+} to Fe^{3+} leads to proton release, making Cd more accessible to plants as Cd is sensitive to decreasing pH levels. Predictions from multi-surface models suggest alkaline soil in aerobic conditions, results in stronger complexation of trace metals. In anaerobic conditions, sulfide precipitation may regulate the solubility of trace metals in both soils.

2 Effect of Cadmium on Plants

Cadmium exists in various forms in soils, but not all of them are readily available for plant uptake. The plant's ability to absorb Cd depends on its speciation, the specific plant species, and the soil's physicochemical conditions. Cadmium can be easily taken up and transported to the aerial parts of plants. However, the absorption of Cd varies with morphological differences, physiological characteristics, and the plant's growth stage and age. Root characteristics, such as structure, size, and surface area, significantly influence metal uptake, with plants possessing thin, hairy roots showing higher absorption and accumulation of metals.

Typically, trace elements like Cd are taken up in their bivalent form. Cd^{2+} uses the same transporters as those of Ca^{2+}, Fe^{2+}, Mg^{2+}, Cu^{2+}, and Zn^{2+} to cross root cell walls. Within plant cells, cadmium transportation supported by both apoplastic and symplastic pathways. In hydroponic experiments, it has been observed that Cd tends to accumulate in the apoplast of barley roots after a 24-hour recovery period. In the case of rice seedlings with well-established apoplastic barriers, the presence of Cd leads to fewer root tips per unit of root surface area. This reduction ultimately results in diminished Cd uptake and translocation. Within the plant, Cd undergoes chelation processes. Specifically, it forms complexes with phosphorus in the apoplast and with sulfur in the symplast. Once absorbed, Cd traverses to the cytosol and subsequently enters pericycle cells. From there, it eventually makes its way to the xylem, facilitating its translocation to the aerial parts of the plant.

2.1 Effect on Growth of Plants and Germination

Cd accumulation in plant tissues leads to growth retardation and the onset of various toxicity symptoms. It leads to a decrease in germination, elongation of root, and shoot. For Instance, Cd significantly inhibits the cereal seedling germination on filter paper, with wheat showing a 31% reduction at concentrations of 0.03–4.8 mM, barley exhibiting a 23.0% reduction at 100 mM, and rice experiencing a complete inhibition of germination at 1.0 mM.

Recent studies have shown that providing a supplement of glutamate, but not glutamine, has a significant alleviating effect on Cd toxicity in rice plants grown hydroponically. This effect is likely achieved by suppressing Cd uptake and translocation within the plants.

Wheat (*Triticum aestivum* L.) is extensively cultivated worldwide but unfortunately accumulates higher cadmium than rice and maize due to increased translocation from roots to shoots. Application of plant growth regulators such as brassinosteroids, salicylic acid, and melatonin was found as suitable measure.

Additionally, the development of low-cd accumulating wheat varieties through breeding has been focused (Qin et al. 2021). Applying essential mineral elements like Si, Mn, Zn, Fe, Se, and B has shown promising results in limiting Cd uptake and transfer in wheat and other crops, making them potential solutions for Cd contamination (Wang et al. 2008). Among these elements, Si has gained particular attention as a beneficial element that enhances Cd tolerance and restricts Cd uptake and translocation in crops like wheat, rice, maize, tomato, and tobacco.

Apart from growth reduction, cadmium also causes adverse changes in root anatomy, ranging from parenchyma cell enlargement to the destruction of root cells in the cortical, endodermis, and pericycle regions (Shah et al. 2019). Similar results were obtained in barley, *A. bettzickiana*. In fenugreek seeds, different Cd concentrations (0.1 and 10 mM) inhibit early seedling growth and hinder root and hypocotyl elongation (Kamalvand et al. 2022).

Cicer arietinum (chickpea) also shows adverse effects of Cd contamination, like shoot stunting, chlorosis, and browning of root tips, ultimately resulting in death of plant. These findings emphasize the significant impact of cadmium exposure on plant growth and development, highlighting the need for measures to mitigate its toxic effects on various plant species (Ullah et al. 2020).

3 Effect of Cadmium on Human Health and Animal

Plant-based foods typically contain higher levels of Cd when compared to meat, eggs, milk, and dairy products, with the extent of contamination depending on soil conditions (McClements and Grossmann 2022). Examples of plant-derived foods with elevated Cd concentrations include rice, wheat, green leafy vegetables, potatoes, carrots, and celery. In Japan, a strategy to reduce Cd accumulation in rice for vegetarians and shellfish consumers involves managing water usage during specific times. However, this approach may lead to increased accumulation of another metal, arsenic (As) (Schaefer et al. 2020). Cadmium possesses unique hydrochemical characteristics, allowing it to remain in solution in environments with near-neutral pH (<6.5). This sets it apart from other heavy metals, which tend to become fixed in the soil. Consequently, areas near factories or mines may experience heavy metal contamination of drinking water and rivers with Cd. Cigarette smoking is a potential source of Cd exposure, with the amount varying depending on the type of cigarette. Inhaled cadmium from smoking has been linked to smoking-related respiratory diseases such as pulmonary disease and lung cancer (Bernhard et al. 2005). Studies have reported a significant increase in Cd levels in the serum and urine of smokers compared to non-smokers (Batáriová et al. 2006). Long-term exposure to Cd through smoking may have adverse effects on bone health, leading to bone tissue loss and an increased occurrence of bone pain in smokers compared to non-smokers (Nambunmee 2014).

Inhalation of cadmium is also done by smoke and dust, cigarette smoking, and through contaminated food. In bloodstream, it binds with proteins like albumins and metallothionein and transports to Gastrointestinal tract and liver (Olaolu et al. 2023). Cd causes apoptosis, necrosis, and disrupts autophagy flux in liver. Some of the absorbed cadmium is then excreted via biliary tract, letting cadmium turn back into the small intestine. The kidney becomes another major target organ for cadmium, and leads to tubular cell necrosis before being excreted through urine and feces. A well-known historical case demonstrating of disease called itai-itai in Japan, due to the discharge of industrial waste in rivers.

Besides its impact on the reproductive system, cadmium exposure can also affect lung function and increase the risk of lung cancer. Awareness of these various pathways of cadmium exposure and its detrimental effects on different organs is crucial in order to implement effective measures to mitigate its adverse health impacts (Inaba et al. 2005).

Low-level exposure to cadmium (Cd) can result in damage to various organs such as the kidneys, liver, skeletal system, and cardiovascular system, as well as lead to deteriorating vision and hearing (Mezynska and Brzóska 2018). Besides strong teratogenic and mutagenic effects, cadmium also exerts adverse effects on human reproductive health, affecting both males and females. Cadmium also exerts a major detrimental impact on the reproductive system due to its ability to interfere with hormonal action, earning it the label of an endocrine disruptor (Yilmaz et al. 2020). It can impact pregnancy and its outcomes by causing changes in gene expression and aberrant methylation patterns in the embryo and placenta. The binding of cadmium to thiols and depletion of the methyl donor S-adenosyl methionine lead to alterations in the methylome and DNA methyltransferase activity, potentially leading to placental and fetal development disorders (Geng and Wang 2019). Cadmium exposure can disrupt steroidogenesis, leading to menstrual cycle disorders and changes in reproductive hormones, delaying puberty and menarche, and increasing the risk of pregnancy loss, premature birth, and reduced birth weight. Postmenopausal women may experience reduced bone mineral density due to cadmium exposure. Women generally have a higher body burden of cadmium compared to men, with higher concentrations of cadmium in the blood, urine, and kidneys (Bernard 2008). The harmful effects of cadmium exposure on reproductive health and overall well-being underscore the importance of reducing environmental cadmium exposure and implementing measures to protect human health (Trentacosta and Mulligan 2020).

4 Remediation Techniques for Cadmium

Given the prevalence of farmland soil enriched with cadmium (Cd), several techniques have been employed to reduce the soil's Cd content and bring it in line with regulatory standards or decrease the labile Cd fractions. These remediation methods encompass various approaches, including dig and fill, electro-kinetic remediation, chemical elution, stabilization and solidification, phytoremediation, field management, and combined remediation. Recently, the field of remediating Cd-enriched farmland soil has seen rapid development, and in this section, we provide a concise overview of the research and application of the aforementioned remediation techniques.

4.1 Dig and Fill

Dig and fill techniques encompass borrowed soil, soil replacement, and ploughing soil methods. Borrowed soil or soil replacement involves covering the contaminated soil layer with clean soil obtained from elsewhere, while isolating the Cd contamination using a low-permeable clay layer. Ploughing, on the other hand, alters the structure of the soil layer by covering the slightly polluted soil surface with a deep and relatively clean soil layer.

These measures are suitable for remediating heavily contaminated soil (borrowed soil and soil replacement) and slightly contaminated soil (ploughing). They are known for their effectiveness in completely cleaning Cd contamination, offering a fast solution. However, dig and fill techniques require significant engineering effort and substantial investment. They may also disrupt the soil structure, pose the risk of transferring contamination to other areas, lead to soil fertility decline, and represent centralized treatments, which aim to replace contaminated soil to prevent further pollution.

As a result, dig and fill techniques are best suited for remediating small, heavily contaminated fields, while they are not commonly used for large sites with slightly to moderately contaminated soil due to their limitations and potential adverse impacts on soil and the environment (Tang et al. 2016).

4.2 Electro-Kinetic Remediation

Electro-kinetic remediation is a physical method employed to separate metals from contaminated soil. Initially designed for extracting metals and organic chemicals from polluted soils and groundwater, this technique applies a low voltage of direct current (typically in the order of mA/cm^2 of the cross-sectional area) or a low potential gradient (in the order of V/cm of the distance between electrodes) of an electric field. Through processes such as electro-migration, electro-osmotic flow, and electrophoresis, metals in the soil migrate toward the electrode chambers (Han et al. 2021).

It is favored for its non-disruptive nature, easy operation, and high efficiency in remediation. However, this method does alter the physicochemical properties of the soil and lacks selectivity for different types of pollutants, posing a potential ecological risk in terms of soil acidification (Alengebawy et al. 2021).

Under the electric field, positive Cd and cations migrate toward the cathode, while negative chloride, fluoride, nitrate, and other anions move toward the anode. The accumulated Cd on the electrode can be removed through methods like electroplating, electrode adsorption, precipitation, or co-precipitation at the electrode, as well as water pumping near the electrode, or complexation with ion exchange resins. Electro-kinetic remediation is especially well-suited for treating low-permeable clay or muddy soil, as it enables effective control over the flow direction of metal ions. However, the soil pH significantly influences the migration velocity of metal ions and the effectiveness of precipitation. Therefore, the proper control and optimization of the soil pH pose a key challenge in improving the efficiency of electro-kinetic remediation.

The majority of experiments conducted in China involved sieved artificially contaminated soil, with Cd contents typically hundreds of times higher than those found in naturally contaminated farmland soil. These bench-scale experiments were carried out in laboratory soil columns, generally less than 1 meter in length. Graphite emerged as the most commonly used electrode material for remediating Cd enrichment. The potential gradient applied during these experiments ranged from 1 to 4 V/cm, which partially falls within the optimal range of 0.2 to 2.0 V/cm for electro-kinetic remediation. The remediation efficiency varied significantly among different experimental studies, ranging from 14.7% to 95.1%. These variations were mainly attributed to differences in factors such as pH value, zeta potential, temperature, water content of the restored soil, and the choice of electrode material.

Furthermore, electro-kinetic remediation is a time-consuming and expensive process. For instance, Li (2004) successfully developed an electro-kinetic technique to remove Cd and arsenic from red soil. Due to the negative zeta potential and acidic soil pH (<6.0), Cd^{2+} exhibited high mobility in the red soil. In an indoor-based simulated experiment, the Cd removal efficiency reached approximately 79.2%, with a calculated electricity expenditure of approximately \$6.56/m^3 of soil.

4.3 Chemical Elution

Chemical elution emerges as a highly effective and permanent treatment method for eliminating metal contaminants from soils. This ex situ technique employs an eluent or co-solvent to separate contaminants, particularly metals, through desorption and dissolution processes. The key advantage of chemical elution lies in its capacity to easily address metals in the liquid phase, dissolving and separating them from the soil with the assistance of a chemical eluent or co-solvent. However, a primary challenge in chemical elution is the identification of suitable extraction agents capable of efficiently and cost-effectively extracting phytoavailable metals without harming the soil structure. This involves considering several factors, including the elution capacity, soil characteristics, extraction agents, optimal elution conditions, and overall remediation effectiveness, which encompasses the removal of both total Cd and phytoavailable Cd. For the extraction of phytoavailable Cd, commonly used extraction agents include ferric chloride chelators like ethylenediaminetetraacetic acid, natural organic acids such as citrate and tartrate, surfactants like Tween-80, saponin, and teasaponin, as well as combinations of chelating agents (Basta and McGowen 2004).

4.3.1 Farmyard Manure and Poultry Manure

The use of organic compounds like farmyard manure (FYM) and poultry manure has been shown to reduce the phytotoxicity of various heavy metals (HMs) by creating a chemical fixation sink for HMs, thereby reducing their bioavailability to plants. Adding lime, farmyard manure, and silt+clay has demonstrated a significant decrease in the Pb content of Indian mustard when grown in lead-contaminated soil. Additionally, metal toxicity can be alleviated through the application of phosphorus, as it forms complexes with metals and phosphate. This complexation process helps reduce the harmful effects of metals in the soil (Liu et al. 2018).

4.4 Solidification and Stabilization

Stabilization and solidification are physicochemical methods utilized to immobilize or render phytoavailable metals inactive in the soil by altering their chemical properties, including solubility, reducibility, and oxidizability. This effectively restricts the transfer and diffusion of metals within the soil environment. It is also known for its ease of operation, rapidity, and cost-effectiveness. Stabilization and solidification can be applied either in situ or ex situ. Based on application experience, this remediation technique can handle a daily treatment volume of metal-enriched soil ranging from 500 to 2000 m^3. For scaled-up applications on farmland, stabilization and solidification remediation have been already successfully implemented in farmland as compared to electro-kinetic remediation and chemical elution (Liu et al. 2018).

5 Phytoremediation

Certainly, here is a rephrased version of the text:

Phytoremediation offers an alternative and complementary approach to traditional physicochemical methods for cleaning up contaminated soils containing metals like Cd (Ghosh and Singh 2005). This method harnesses specific plants and their associated microorganisms in the rhizosphere to extract, transport, assimilate, modify, and/or immobilize contaminants present in the soil, thereby reducing the toxicity of these metals. Historically, efforts in phytoremediation of

Cd-contaminated soils primarily concentrated on the identification of hyperaccumulating plants or enhancing the Cd uptake capacity of such hyperaccumulators through measures like gene transfer. Nevertheless, the effectiveness of phytoremediation in stabilizing the phytoavailable Cd fractions within the soil has also been assessed (Ghosh and Singh 2005).

The primary criteria for assessing whether a plant qualifies as a proficient Cd hyperaccumulator are as outlined by Sikdar and Kundu (2018):

1. **Cd Accumulation Capacity:** The plant should demonstrate a substantial capacity to accumulate Cd, with a Cd content in its above-ground biomass exceeding 100 mg/kg.
2. **Bio-concentration Factor:** The bio-concentration factor, which signifies the concentration ratio of Cd in the plant shoot compared to the soil, should exceed 1.
3. **Translocation Factor:** The translocation factor, indicating the concentration ratio of Cd in the plant shoots versus the roots, should also surpass 1.
4. **Tolerance to Toxic Contaminants:** A hyperaccumulator plant should exhibit a notable tolerance to toxic substances.

To date, around 20 species of hyperaccumulator plants have been well-documented worldwide.

Phytoremediation of farmland soil enriched with Cd shows great promise during the experimental phase. Pot experiments are commonly employed to identify Cd hyperaccumulating plants and assess the effectiveness of phytoremediation. Typically, these small-scale pot experiments involve soils that have been artificially contaminated with chemical reagents. As a result, the experimental conditions significantly differ from those found in actual field settings. Consequently, the use of data from pot experiments to predict the potential success of phytoremediation in real-world Cd-enriched farmland soil can be questionable. There exist numerous uncertainties regarding the accuracy and reliability of the phytoremediation findings derived from such experiments (Tang et al. 2016).

In field experiments, the effectiveness of phytoremediation, measured by the rate of decline in total Cd content in the soil, exhibited significant variability. One potential explanation for this phenomenon is the notable influence of various factors such as soil water content, salinity, pH level, and organic matter content on phytoremediation efficiency. These factors can profoundly affect the availability of soil Cd to plants. Furthermore, the success of phytoremediation is intricately linked to the dynamic variability in space and time. This variability is closely intertwined with factors like the types of plants used, soil texture, levels of pollution, and prevailing climatic conditions. It is important to note that the roots of crops can facilitate the movement of mobile Cd from the upper soil layer to deeper soil layers during extended periods of phytoremediation efforts. While this may enhance Cd enrichment in the deep soil layers and underground water bodies, including groundwater, it also introduces potential risks.

Phytoremediation, in general, offers numerous technical and economic advantages and is often a suitable choice for environmental restoration. Compared to methods like excavation and replacement, electro-kinetic remediation, chemical leaching, and stabilization and solidification techniques, phytoremediation typically exhibits economic superiority. However, a notable challenge lies in the fact that most hyperaccumulator plants are generally small, slow growing, and challenging to select and cultivate. Moreover, the effectiveness of most hyperaccumulators is limited to a single metal element. As a result, phytoremediation often results in lower efficiency, leading to extended restoration periods, often exceeding 10 years. Furthermore, it is important to note that phytoremediation techniques can fully address contamination concerns only in the top 50 cm of lightly contaminated surface soil. These limitations mentioned above significantly impede the widespread commercial application of phytoremediation.

5.1 Nano-phytoremediation

Nano-phytoremediation is a burgeoning approach that synergizes nanotechnology and phytotechnology to tackle environmental pollution through pollutant removal. It has proven to be particularly effective in the degradation and removal of TNT (2,4,6-trinitrotoluene) from contaminated soil, surpassing the individual efficiencies of nanoremediation or phytoremediation techniques (Fernández-Luqueño et al. 2023). The swift rise in the utilization of nanoparticles for environmental purposes is attributed to their substantially increased surface area compared to bulk-size particles. This heightened surface reactivity significantly enhances their suitability for chemical- or biological-mediated reactions in diverse environmental applications.

Nanomaterials exhibit a promising capacity to either absorb or mitigate contaminants in the environment. Among these, iron-based nanoparticles have gained significant traction in the field of remediation. For instance, nanoscale zerovalent iron (nZVI) has demonstrated its efficacy in reducing environmental pollutants such as cis-1,2-dichloroethylene (c-DCE), tetrachloroethylene (PCE), and trichloroethylene (TCE) (Mobasser and Firoozi 2016). Recent research endeavors have been concentrated on enhancing the efficiency of nanomaterials. Additionally, there is a concerted effort to comprehend the mechanisms governing the uptake and transport of nanomaterials, in order to evaluate potential toxicological impacts on the ecosystem. Nanoparticles possess an inherent ability to adsorb pollutants, attributed to their expansive surface area. Furthermore, these nanoparticles attach to plant roots and are subsequently transported to the harvestable plant parts, thereby facilitating phytoremediation processes (Rajendran et al. 2022).

The introduction of TiO_2 nanoparticles notably heightened the *Sorghum bicolor's* capacity to accumulate antimony (Sb), with the highest accumulation of 1624.1 μg per pot observed at 250 mg/kg TiO_2 NP application, complemented by a 2.5% biochar co-application (Zand et al. 2020). The use of nanoparticles also exhibited the potential to enhance plant stress resilience, both in controlled and field conditions, thereby augmenting phytoremediation capabilities. Additionally, nanoparticles might induce alterations in plant physiology, potentially triggering mechanisms that aid in the reduction of heavy metal (HM) uptake from the soil. The incorporation of magnetic Fe_3O_4 nanoparticles at a concentration of 2000 mg/L in a solution containing Pb, Zn, Cd, and Cu (1 mM each) significantly curtailed growth inhibition in wheat seedlings by diminishing HM uptake (Ogunkunle et al. 2020). This led to the mitigation of HM-induced toxicity symptoms in the wheat seedlings.

6 Some Case Studies

Sikka and Nayyar (2012) conclude that the growth of Indian mustard is negatively affected by the application of cadmium (Cd) to soil, and the accumulation of Cd in shoots increases with increasing soil Cd concentration. The concentration and uptake of micronutrients in Indian mustard reduce significantly with the application of Cd beyond 10 mg Cd kg^{-1} soil level, suggesting that Cd not only causes toxicity to plants but also causes nutritional imbalance. The study confirms that once a soil is contaminated with Cd, it remains available to plants for a long time, and many cropping cycles are required to make it safe for cultivation. Therefore, growers should avoid using such soils for growing crops whose shoots are edible, and there is a need to find ameliorative measures for Cd-contaminated soils for safely growing crops on such soils.

Sikka et al. (2009) observed cadmium accumulation and movement in soil, plant along with soil characterization by irrigation through sewage water in three different cities. Irrigated water contains organic matter which on decomposition releases organic acids and trace metals in plants. This also decreases the pH of treated soil.

The relationship between Cd availability and the time since contamination was explored across various soil characteristics by Sikka (2011). Soil samples with different pH levels, $CaCO_3$ content, organic matter content, and silt + clay composition were treated with three levels of Cd (0, 20, and 200 mg Cd kg^{-1} soil) using a nitrate solution. These soil samples were incubated at field capacity for 1 day, 14 days (2 weeks), and 28 days (4 weeks), and the available Cd fraction was measured using 0.005 M DTPA extraction followed by atomic absorption spectrophotometry. The findings revealed a decline in Cd availability over time, indicating aging effects. This decline followed the order of 1 day >2 weeks >4 weeks for all soil types. However, the extent of decrease varied based on the contamination levels. In soils contaminated with 20 mg Cd kg^{-1} soil and aged for 2 weeks, the lowest Cd extraction was observed when 40% of the soil consisted of the silt + clay fraction. On the other hand, after 4 weeks of aging, the lowest extraction occurred in soils treated with 5% $CaCO_3$. When soil contamination reached 200 mg Cd kg^{-1}, the addition of 40% silt + clay resulted in the lowest Cd recovery at both 2 and 4 weeks of aging. These results shed light on the natural attenuation process of Cd contamination and provide insights into the capacity of different soils to act as sinks for Cd.

An experiment was proceeded by Dubey et al. (2022) to reveal the effect of rice residue biochar on cadmium and lead bioavailability in Indian mustard sown in loamy sand soil. The soil was spiked with varying levels of Cd (ranging from 0 to

40 mg kg^{-1}) and Pb (ranging from 0 to 200 mg kg^{-1}). In addition, the metal-treated soil pots were supplemented with different amounts (0%, 1%, 2%, and 4% w/w) of rice residue biochar. The Indian mustard crops were cultivated for a period of 60 days with recommended nitrogen and phosphorus doses. The incorporation of biochar up to a 2% rate led to an increase in the above-ground biomass of Indian mustard, but this trend reversed when biochar was applied at a 4% rate, in comparison to the unamended soil. Interestingly, the application of biochar resulted in a noticeable reduction in the accumulation of both Cd and Pb by the crop, with this effect becoming more pronounced as the biochar application rate increased. Notably, the efficacy of biochar in mitigating the impact of Pb was more pronounced compared to its effect on Cd. In conclusion, the findings suggest that rice residue biochar holds promise in diminishing the solubility, availability, and movement of Pb and Cd in Indian mustard plants.

The study investigates the effect of rice straw biochar on the bioavailability and accumulation of cadmium (Cd) by Indian mustard in sewage water irrigated soil artificially spiked with Cd (Dubey et al. 2020). The study reveals that rice straw biochar can be a cost-effective amendment for the remediation of soil contaminated with Cd by immobilizing it, thereby reducing its availability in soil and plants. The study suggests that rice straw biochar using surplus biomass available in large quantities in northwestern India could be a sustainable strategy to remediate Cd-contaminated soil and significantly contribute to reducing its toxicity in the food chain. However, the effectiveness of rice straw biochar in remediation of Cd and other heavy metal-contaminated soils should be tested under field conditions before its commercial use by the farmers.

The effect of rice residue biochar on the bioavailability and accumulation of cadmium (Cd) by Indian mustard in a loamy sand soil spiked with three levels each of Cd and lead (Pb) in all possible combinations by Kalyani et al. (2022). The study found that the addition of rice residue biochar decreased the mean DTPA-Cd in soil and its concentrations in plants, making it a potential remediation material for polluted soil. The biochar was produced from rice residues using an intermediate pyrolysis technique, and its properties were analyzed. The study concludes that further research is needed to determine the addition rate and frequency for the long-term aging process of biochar and the fate of sequestered contaminants under field conditions. Several remediation technologies are available for treating heavy metal-contaminated soils, and biochar has been found to be effective in reducing the mobility of heavy metals in soils.

7 Limitations

Although phytoremediation is cost-effective and efficient for low to moderate levels of metal pollution in soil, it does have certain limitations. The process of phytoremediation can be time consuming, and the low biomass and slow growth rate of some plants can lead to unsustainable plant growth. Additionally, proper disposal of the polluted biomass is crucial, as it may be considered dangerous waste. In climate-affected tropical and sub-tropical regions, some plants' accumulation capacity of metals may decrease due to pest attacks. Establishing and maintaining vegetation can be challenging in sites with high toxic metal levels. Moreover, there is a risk of accumulated metals entering the food chain if the biomass is mishandled. The success of phytoremediation largely depends on climatic and weather conditions. Many remediation studies are conducted in controlled indoor environments and pilot-scale demonstrations, particularly for conventional methods.

8 Conclusion

Surpassing metal thresholds in agricultural products presents a substantial peril to human well-being. This contamination has precipitated a decline in the agricultural economy and has incited instability in the socio-economic landscape. Therefore, the most pragmatic strategy for farmland resources enriched with metals is to undergo remediation while sustaining cultivation. To detoxify soil with elevated Cd levels and promote secure crop production, it is imperative to expeditiously amalgamate the capabilities of established techniques, including chemical elution, solidification, and stabilization, alongside phytoremediation and comprehensive field management. This all-encompassing approach will effectively combat metal contamination and ensure the sustainable use of agricultural land.

References

Abdu N, Abdullahi AA, Abdulkadir A (2017) Heavy metals and soil microbes. Environ Chem Lett 15:65–84

Alengebawy A, Abdelkhalek ST, Qureshi SR, Wang M-Q (2021) Heavy metals and pesticides toxicity in agricultural soil and plants: ecological risks and human health implications. Toxics 9(3):42

Ali H, Khan E (2019) Trophic transfer, bioaccumulation, and biomagnification of non-essential hazardous heavy metals and metalloids in food chains/webs—concepts and implications for wildlife and human health. Hum Ecol Risk Assess Int J 25(6):1353–1376

Alloway BJ (2012) Heavy metals in soils: trace metals and metalloids in soils and their bioavailability, vol 22. Springer

Asati A, Pichhode M, Nikhil K (2016) Effect of heavy metals on plants: an overview. Int J Appl Innov Eng Manag 5(3):56–66

Ayangbenro AS, Babalola OO (2017) A new strategy for heavy metal polluted environments: A review of microbial biosorbents. Int J Environ Res Public Health 14(1):94

Basta NT, McGowen SL (2004) Evaluation of chemical immobilization treatments for reducing heavy metal transport in a smelter-contaminated soil. Environ Pollut 127(1):73–82

Batáriová A, Spěváčková V, Beneš B, Čejchanová M, Šmíd J, Černá M (2006) Blood and urine levels of Pb, Cd and Hg in the general population of the Czech Republic and proposed reference values. Int J Hyg Environ Health 209(4):359–366

Bernard A (2008) Cadmium & its adverse effects on human health. Indian J Med Res 128(4): 557–564

Bernhard D, Rossmann A, Wick G (2005) Metals in cigarette smoke. IUBMB Life 57(12):805–809

Chaney RL, Angle JS, Broadhurst CL, Peters CA, Tappero RV, Sparks DL (2007) Improved understanding of hyperaccumulation yields commercial phytoextraction and phytomining technologies. J Environ Qual 36(5):1429–1443

Dubey S, Sikka R, Singh D (2020) Effect of rice straw biochar on the bioavailability of cadmium in soil to Indian Mustard (Brassica juncea L. czern.) in relation to cadmium spiking. Agric Res J 57(6):832–838. https://doi.org/10.5958/2395-146X.2020.00122.2

Dubey S, Sikka R, Dhaliwal J (2022) Bioavailability of Cd and Pb to Indian mustard (Brassica juncea (L.) Czern.) as influenced by Biochar. Agric Res J 59(5):864–869. https://doi.org/10.5958/2395-146X.2022.00121.1

Dudka S, Miller WP (1999) Accumulation of potentially toxic elements in plants and their transfer to human food chain. J Environ Sci Health B 34(4):681–708

Dusenge ME, Duarte AG, Way DA (2019) Plant carbon metabolism and climate change: elevated CO_2 and temperature impacts on photosynthesis, photorespiration and respiration. New Phytol 221(1):32–49

Edelstein M, Ben-Hur M (2018) Heavy metals and metalloids: sources, risks and strategies to reduce their accumulation in horticultural crops. Sci Hortic 234:431–444. https://doi.org/10.1016/j.scienta.2017.12.039

Ernstberger H, Zhang H, Tye A, Young S, Davison W (2005) Desorption kinetics of Cd, Zn, and Ni measured in soils by DGT. Environ Sci Technol 39(6):1591–1597

Fernández-Luqueño F, López-Valdez F, Medina-Pérez G (2023) Bio and nanoremediation of hazardous environmental pollutants. CRC Press

Geng H-X, Wang L (2019) Cadmium: toxic effects on placental and embryonic development. Environ Toxicol Pharmacol 67:102–107

Ghosh M, Singh SP (2005) A review on phytoremediation of heavy metals and utilization of it's by products. Asian J Energy Environ 6(4):18

Gill M (2014) Heavy metal stress in plants: a review. Int J Adv Res 2(6):1043–1055

Gwenzi W, Mangori L, Danha C, Chaukura N, Dunjana N, Sanganyado E (2018) Sources, behaviour, and environmental and human health risks of high-technology rare earth elements as emerging contaminants. Sci Total Environ 636:299–313

Han D, Wu X, Li R, Tang X, Xiao S, Scholz M (2021) Critical review of electro-kinetic remediation of contaminated soils and sediments: mechanisms, performances and technologies. Water Air Soil Pollut 232(8):335. https://doi.org/10.1007/s11270-021-05182-4

He ZL, Yang XE, Stoffella PJ (2005) Trace elements in agroecosystems and impacts on the environment. J Trace Elem Med Biol 19(2):125–140. https://doi.org/10.1016/j.jtemb.2005.02.010

Holmgren GGS, Meyer MW, Chaney RL, Daniels RB (1993) Cadmium, lead, zinc, copper, and nickel in agricultural soils of the United States of America. Wiley Online Library

Hussain M, Farooq M, Nawaz A, Al-Sadi AM, Solaiman ZM, Alghamdi SS, Ammara U, Ok YS, Siddique KH (2017) Biochar for crop production: potential benefits and risks. J Soils Sediments 17:685–716

Hynninen A (2010) Zinc, cadmium and lead resistance mechanisms in bacteria and their contribution to biosensing. University of Helsinki

Ibañez S, Talano M, Ontañon O, Suman J, Medina MI, Macek T, Agostini E (2016) Transgenic plants and hairy roots: exploiting the potential of plant species to remediate contaminants. New Biotechnol 33(5B):625–635. https://doi.org/10.1016/j.nbt.2015.11.008

Inaba T, Kobayashi E, Suwazono Y, Uetani M, Oishi M, Nakagawa H, Nogawa K (2005) Estimation of cumulative cadmium intake causing Itai–itai disease. Toxicol Lett 159(2): 192–201

Kalyani P, Sikka R, Mavi MS (2022) Effect of rice residue biochar on bioavailability of cadmium to Indian mustard under combined lead and cadmium spiking in a loamy sand soil. Agric Res J 59(5)

Kamalvand AA, Hosseini Sarghein S, Karamian R (2022) Impact of cadmium stress on growth and physiological responses of fenugreek (Trigonella foenum-graecum L.). J Plant Physiol Breed 12(1):51–65

Khan A, Khan S, Khan MA, Qamar Z, Waqas M (2015) The uptake and bioaccumulation of heavy metals by food plants, their effects on plants nutrients, and associated health risk: a review. Environ Sci Pollut Res 22:13772–13799

Khanna K, Kohli SK, Kaur R, Handa N, Bakshi P, Sharma P, Ohri P, Bhardwaj R (2022) Reconnoitering the efficacy of plant growth promoting rhizobacteria in expediting phytoremediation potential of heavy metals. J Plant Growth Regul 2022:1–29

Kollah B, Patra AK, Mohanty SR (2016) Aquatic microphylla Azolla: a perspective paradigm for sustainable agriculture, environment and global climate change. Environ Sci Pollut Res 23: 4358–4369

Li D (2004) Electrokinetics in microfluidics. Elsevier.

Liu L, Li W, Song W, Guo M (2018) Remediation techniques for heavy metal-contaminated soils: principles and applicability. Sci Total Environ 633:206–219

Mani D, Kumar C (2014) Biotechnological advances in bioremediation of heavy metals contaminated ecosystems: an overview with special reference to phytoremediation. Int J Environ Sci Technol 11:843–872

McClements DJ, Grossmann L (2022) Next-generation plant-based foods. Springer

Mezynska M, Brzóska MM (2018) Environmental exposure to cadmium—a risk for health of the general population in industrialized countries and preventive strategies. Environ Sci Pollut Res 25:3211–3232

Mobasser S, Firoozi AA (2016) Review of nanotechnology applications in science and engineering. J Civil Eng Urban 6(4):84–93

Mrayyan B, Hamdi MR (2006) Management approaches to integrated solid waste in industrialized zones in Jordan: a case of Zarqa City. Waste Manag 26(2):195–205

Mudgal V, Madaan N, Mudgal A, Singh RB, Mishra S (2010) Effect of toxic metals on human health. Open Nutraceuticals J 3(1)

Nambunmee K (2014) Osteotoxicity in the inhabitants of a cadmium polluted area. J Health Res 28(3):211–217

Ogunkunle CO, Odulaja DA, Akande FO, Varun M, Vishwakarma V, Fatoba PO (2020) Cadmium toxicity in cowpea plant: Effect of foliar intervention of nano-TiO2 on tissue Cd bioaccumulation, stress enzymes and potential dietary health risk. J Biotechnol 310:54–61

Olaolu T, Jibulu F, Damilare R (2023) Hepatotoxic effect of cadmium and available therapeutic options. Thai J Pharm Sci 46(6):631–637

Pandey G, Madhuri S (2014) Heavy metals causing toxicity in animals and fishes. Res J Anim Vet Fish Sci 2(2):17–23

Payne TE, Brendler V, Ochs M, Baeyens B, Brown PL, Davis JA, Ekberg C, Kulik DA, Lutzenkirchen J, Missana T (2013) Guidelines for thermodynamic sorption modelling in the context of radioactive waste disposal. Environ Model Softw 42:143–156

Qin X, Xia Y, Hu C, Yu M, Shabala S, Wu S, Tan Q, Xu S, Sun X (2021) Ionomics analysis provides new insights into the co-enrichment of cadmium and zinc in wheat grains. Ecotoxicol Environ Saf 223(112):623. https://doi.org/10.1016/j.ecoenv.2021.112623

Rajendran S, Priya AK, Kumar PS, Hoang TK, Sekar K, Chong KY, Khoo KS, Ng HS, Show PL (2022) A critical and recent developments on adsorption technique for removal of heavy metals from wastewater - a review. Chemosphere 303(135):146

Razo I, Carrizales L, Castro J, Díaz-Barriga F, Monroy M (2004) Arsenic and heavy metal pollution of soil, water and sediments in a semi-arid climate mining area in Mexico. Water Air Soil Pollut 152:129–152

Roy M, Giri AK, Dutta S, Mukherjee P (2015) Integrated phytobial remediation for sustainable management of arsenic in soil and water. Environ Int 75:180–198
Schaefer HR, Dennis S, Fitzpatrick S (2020) Cadmium: Mitigation strategies to reduce dietary exposure. J Food Sci 85(2):260–267
Shah K, Nahakpam S, Chaturvedi V, Singh P (2019) Cadmium-induced anatomical abnormalities in plants. In: Cadmium toxicity and tolerance in plants. Elsevier, pp 111–139
Shukla KP, Singh NK, Sharma S (2010) Bioremediation: developments, current practices and perspectives. IntechOpen
Sikdar S, Kundu M (2018) A review on detection and abatement of heavy metals. ChemBioEng Rev 5(1):18–29
Sikka R (2011) Risk assessment of cadmium contaminated soils varying in physico-chemical characteristics with reference to aging effect. J Res 48(3 and 4):112–118
Sikka R, Nayyar V (2012) Cadmium accumulation and its effects on uptake of micronutrients in Indian Mustard [Brassica juncea (L.) Czern.] grown in a loamy sand soil artificially contaminated with cadmium. Commun Soil Sci Plant Anal 43(4):672–688. https://doi.org/10.1080/00103624.2012.644007
Sikka R, Nayyar V, Sidhu SS (2009) Monitoring of Cd pollution in soils and plants irrigated with untreated sewage water in some industrialized cities of Punjab, India. Environ Monit Assess 154(1):53–64. https://doi.org/10.1007/s10661-008-0377-4
Singh SR, Bhadra A, Kademane A, Shrivastava R, Patil SJ (2023) Phytoremediation - a promising approach for pollution management. Phytoremediation. https://doi.org/10.48047/ecb/2023.12.si10.0045
Sun Y, Li H, Guo G, Semple KT, Jones KC (2019) Soil contamination in China: Current priorities, defining background levels and standards for heavy metals. J Environ Manag 251(109):512
Tang X, Li Q, Wu M, Lin L, Scholz M (2016) Review of remediation practices regarding cadmium-enriched farmland soil with particular reference to China. J Environ Manag 181:646–662
Trentacosta CJ, Mulligan DJ (2020) New directions in understanding the role of environmental contaminants in child development: four themes. New Dir Child Adolesc Dev 2020(172):39–51
Ullah S, Khan J, Hayat K, Abdelfattah Elateeq A, Salam U, Yu B, Ma Y, Wang H, Tang Z-H (2020) Comparative study of growth, cadmium accumulation and tolerance of three chickpea (Cicer arietinum L.) cultivars. Plants 9(3):310
Vig K, Megharaj M, Sethunathan N, Naidu R (2003) Bioavailability and toxicity of cadmium to microorganisms and their activities in soil: a review. Adv Environ Res 8(1):121–135
Violante A, Cozzolino V, Perelomov L, Caporale AG, Pigna M (2010) Mobility and bioavailability of heavy metals and metalloids in soil environments. J Soil Sci Plant Nutr 10(3):268–292
Wang Z-H, Li S-X, Malhi S (2008) Effects of fertilization and other agronomic measures on nutritional quality of crops. J Sci Food Agric 88(1):7–23. https://doi.org/10.1002/jsfa.3084
Wang P, Chen H, Kopittke PM, Zhao F-J (2019) Cadmium contamination in agricultural soils of China and the impact on food safety. Environ Pollut 249:1038–1048
Yang Q, Li Z, Lu X, Duan Q, Huang L, Bi J (2018) A review of soil heavy metal pollution from industrial and agricultural regions in China: pollution and risk assessment. Sci Total Environ 642:690–700
Ye C, Li S, Zhang Y, Zhang Q (2011) Assessing soil heavy metal pollution in the water-level-fluctuation zone of the Three Gorges Reservoir, China. J Hazard Mater 191(1–3):366–372
Yilmaz B, Terekeci H, Sandal S, Kelestimur F (2020) Endocrine disrupting chemicals: exposure, effects on human health, mechanism of action, models for testing and strategies for prevention. Rev Endocr Metab Disord 21:127–147
Zand AD, Tabrizi AM, Heir AV (2020) Co-application of biochar and titanium dioxide nanoparticles to promote remediation of antimony from soil by Sorghum bicolor: metal uptake and plant response. Heliyon 6(8)

Molecular Mechanism of Tolerance of Cadmium Toxicity in Plants

Dipti Srivastava and Neerja Srivastava

Abstract Environmental pollution is among most dangerous challenges for the wellbeing of organisms. Due to anthropological and agricultural practices, heavy metals are typically present in soil or are transferred into the environment. Cadmium-polluted soil emphasizes two key points: first, they affect the plant's life cycle by reducing crop yields, and second, they are absorbed as well as accumulated in plant tissues and ultimately enter inside the food chain to harm animals as well as humans. The major goal of plant biotechnology studies is to know how plants deal with cadmium toxicity and the aim of plant breeders is to generate plants that can withstand cadmium exposure.

Keywords Heavy metal stress · Cadmium stress · Stress proteins

1 Introduction

Cadmium (Cd) is a highly poisonous material for living species and human beings, affecting both terrestrial as well as aquatic life forms (Chellaiah 2018). Concentration of Cd increases in cultivated land because of agriculture and industrial growth (Bojórquez et al. 2016). Cd gets entry into environment through various human actions and environmental releases (Abbas et al. 2014). As Cd moves rapidly in polluted soil, it accumulates inside plants cultivated in soil contaminated with Cd and poses significant health risks for both animals and humans (Chen et al. 2016). Cd toxicity has its effects on several organs but is mainly stored in the kidney, causing significant damage such as renal tubular destruction, pulmonary emphysema, and kidney stones (Mahajan and Kaushal 2018). Calcium is replaced by cadmium in minerals due to same charge, ionic radius as well as chemistry (Kubier et al. 2019). As a result, it readily enters inside the body as well as accumulates in large amounts in several organs (Hajeb et al. 2014). Cd poisoning harms the liver as well as bones

D. Srivastava · N. Srivastava (✉)
Department of Biochemistry, SLSBT, CSJM University, Kanpur, Kanpur, Uttar Pradesh, India
e-mail: neerjasrivastava@csjmu.ac.in

A. K. Jha, N. Kumar (eds.), *Cadmium Toxicity Mitigation*,
https://doi.org/10.1007/978-3-031-47390-6_14

and might prevent calcium absorption inside the body (Lata et al. 2019). Reduced development and chlorosis are clearly recognized signs of plant Cd toxicity (Jali et al. 2016). Toxicity enhancement limits the development of the necrotic plant (Hermans et al. 2011). Plants are influenced by toxicity through limiting carbon fixation with reducing chlorophyll content as well as photosynthetic activity (Gallego et al. 2012). Relative water content of leaf, vascular conductance, and transpiration is reduced by cadmium causing osmotic stress and causes physiological damage in plants (Rizwan et al. 2016). Cd poisoning produces excessive amounts of ROS, which harms plant membranes as well as destroys cell biomolecules together with organs (Abbas et al. 2017). Cadmium also inhibits iron as well as zinc uptake by plants, which causes leaf chlorosis (Xu et al. 2017). Cadmium usually inhibits movement as well as absorption of calcium, phosphorous magnesium, potassium, and manganese (Nazar et al. 2012). Cadmium toxicity has been under active investigation, with several findings on the harmful effects of cadmium on the productivity of crops, reactive oxygen species production, and lipid peroxidation with possible remedial measures (Baruah et al. 2019; Bashir et al. 2018; Farooq et al. 2020; Hussain et al. 2021; Haider et al. 2021; Qin et al. 2020; Zhang et al. 2019).

Cadmium is usually dangerous for the majority of plant species, but metallophytes from cadmium-rich soils have developed Cd exclusion processes that limit its entrance as well as prevent its movement toward aboveground tissue. Though, a handful of supposed hyperaccumulating plants common on metal-enriched soils can store in their aboveground organs more Cd than is normally observed in plants without experiencing phytotoxicity (Baker and Brooks 1989). Many researchers are interested in the complex processes carried out by those plants that avert cadmium intake or facilitate extraction of metal from soil, handling, partitioning, and cellular detoxification. To stop cadmium from accessing the food chain, it is necessary to reveal the genetic as well as physiological processes in plants which have the potential to evade cadmium stress through prevention of its absorption or transfer. Also, knowledge of the process of cadmium storage in the vegetative portions of hyperaccumulating plants is vital for a potential method in which plants might be utilized for ecological remediation (phytoremediation) of environments polluted with cadmium (DalCorso et al. 2008).

2 Fundamental Chemical Properties of Cadmium

Cd is a periodic group IIB member with atomic number 48. It has the same chemical properties as other group IIB members, particularly Zn and Hg. Cd often occurs in natural geological conditions as Zn and Hg. Cd(II) is more stable and is found in this state in most natural water systems (Baes and Mesmer 1976). Capability of cadmium for binding to amines, halide ions, ammonia as well as cyanide shows that it shares resemblances to the majority of transition metal series ions. Cd is a relatively volatile white shiny and darkening material having 321 and 767 °C melting and boiling points respectively with 26.8 K cal per mol heat of vaporization (Cotton and

Wilkinson 1966). This last characteristic allows its entry into the surroundings, as an important constituent in the universal cadmium cycle (Laws 1993; Hasan et al. 2009)

3 Sources of Cadmium in the Environment

Cd contamination in groundwater as well as soil has been documented worldwide (Chellaiah 2018). Cd is highly persistent biologically, although it has toxicologic consequences and persists in the environment for several years after organisms consume it (Weggler et al. 2004). Mining, metal industry, transport, landfills, accidents, emissions, combustion as well as sewage discharge Cd to groundwater and soil (Bigalke et al. 2017). In addition, the application of artificial phosphate fertilizers having cadmium often causes high cadmium concentrations in groundwater as well as soil (Kubier et al. 2019). Groundwater intrusion mechanisms have been studied in Australia, the United Kingdom, Denmark, Canada, Germany, Finland, Norway, Sweden New Zealand, and the United States, among others (Taylor et al. 2016). According to investigations, adding artificial phosphate fertilizers with cadmium changes the chemistry of soil and affects soil biota as well as microbial population (Bigalke et al. 2017). Cd is found in refined petroleum products because it is present naturally (Buekers 2007). It can gain access to food chain as well as harm organisms (Grant 2011). The sources of transfer of cadmium could be both dispersed and localized. Localized sources such as mines, industrial sites, or abandoned landfills (Cloquet et al. 2006) and dispersed sources such as wastewater reuse, atmospheric functions as well as agricultural events are the primary sources of Cd distribution in soil and the surroundings (Knappe et al. 2008). Globally, primary Cd sources are landfills, Cd-Ni batteries and municipal solid waste (Khan et al. 2017). Cadmium is mainly used in Cd-Ni batteries, which have high capacity, less maintenance, longer service life, as well as excellent electrical with physical stress tolerance (Wuana and Okieimen 2011). In Europe, municipal solid waste comprises 0.3–12 mg per kg Cd, while landfill leachate possesses 0.5–3.4 g per kg Cd (UNEP 2010). Polyvinyl chloride stabilizers, compounds, pigments, coatings, and coatings are also Cd-containing products (Sprynskyy et al. 2011). According to Weggler et al. (2004), compost and sludge leachate account for 2–5% of Cd deposition in soil. Other sources of leachate are aerial deposition and farm manure, which account for 30–55 percent, 15–50 percent, and 10–25 percent Cd accumulation in soil, respectively (Belon et al. 2012). Cd coating gives excellent protection from corrosion to vehicles, particularly in extremely corrosive atmospheres like aviation along with marine (Zhang et al. 2009). Unlike anthropogenetic discharges and actions, the natural difference of minerals as well as rocks in the adjoining soils might increase Cd levels in the surroundings (Bigalke et al. 2017). Because cadmium is easily stimulated, soil must be considered an important transient reservoir of Cd that can rapidly affect groundwater concentrations at different time scales, including more than a decade (dry versus wet years) as well as annually (rainy versus dry season) (Sprynskyy et al. 2011). Elimination of Cd from the soil by harvesting is one likely explanation for the decrease in Cd concentration (Kubier et al. 2019). Considerable Cd enhancement (77 mg Cd per kg P_2O_5), was observed in the east Mediterranean (Azzi et al. 2017) as well as European nations, where values varied between 36 and 60 mg Cd/per kg P_2O_5 (Six and Smoulders 2014). Cd contamination in soil from Zn smelters contributed 74 mg/kg due to atmospheric deposition and 344 mg/kg due to solid waste leaching (Voglar and Leštan 2010). Contamination in groundwater occurs at the same time from multiple sources as well as happens in single location, which may limit the recognition of important cadmium sources, pathways, and geogenic processes (Zhu et al. 2013a, Haider et al. 2021).

4 Cadmium Stress in Plant

High concentrations of cadmium, nickel, zinc, as well as copper are harmful to plants, although it has been observed that lead causes phytotoxicity (Foy et al. 1978). Cadmium concentrations in unpolluted soil solutions vary between 0.04 and 0.32 millimolar, and concentrations between 0.32 and 1 mM are regarded as polluted (Di Toppi and Gabbrielli 1999; Hasan et al. 2009).

4.1 Cadmium Uptake and Transport in Plant

Cadmium can be taken up from soil or water by higher plants on the basis of its accessibility as well as quantity; however, a very small amount is received straight from surroundings (Clemens 2006). The bioavailability of cadmium and other heavy metals for plant uptake is influenced by soil pH, organic acids presence as well as rhizosphere (Benavides et al. 2005). For instance, the uptake of cadmium by maize is observed to be reduced in acidic and high organic material soils (Benavides et al. 2005). Cd absorption is also affected by the amount of other nutrients, e.g., calcium, zinc, and iron in the soil as adding calcium or zinc reduces the consumption of cadmium (Cosio et al. 2004). As more than 200 mV membrane potential inside root epidermal cells, a powerful driving force for assimilation of cation, toxic heavy metals compete with plant cells and enter through the same transport systems used for micronutrient assimilation: especially Cd ion assimilation. It happens via the same transmembrane transporters employed in iron, calcium, copper, and magnesium with zinc assimilation (Roth et al. 2006; Papoyan et al. 2007). Cadmium gets entered inside the roots quickly by the cortex and could move toward the xylem through apoplastic as well as symplastic pathways, making a complex by means of organic matter, phytochelatins, or acids (Salt et al. 1995). After internalization of Cd complexes into tracheal components, they spread throughout the plant. Accumulation of Cd in developing fruits is thought to occur through phloem-facilitated movement, which means complete uptake of heavy metals inside plants (Benavides et al. 2005; DalCorso et al. 2008).

Cadmium is quickly absorbed and moved to the aerial plant parts (Shanmugaraj et al. 2019). Cadmium absorption capacity differs between plant species as well as genotypes because of structural variability, plant physiological properties (Gupta et al. 2019; Yadav et al. 2018; Tiryakioglu et al. 2006) as well as phases of plant growth with age (Manousaki and Kalogerakis 2009). Features of root-like structure (Laghlimi et al. 2015; Fahr et al. 2013; Redjala et al. 2011), dimensions (Thakur et al. 2016) as well as area (Yadav et al. 2018; Zhi-bin et al. 2016; Thakur et al. 2016; Redjala et al. 2011) significantly affect metal uptake. Initially, Cd penetrates into the roots of plants and harms the root system as well as plant structure (Hasan et al. 2007). Cd uptake through root cells' plasma membrane is regulated through the difference in electrochemical potential in root apoplasts and cytosol (Benavides et al. 2005). Enough energy is provided by membrane potential for taking up cadmium even in quite less cadmium concentrations (Pinto et al. 2004). The Cd sorption energy in roots at low Cd activity shows biphasic characteristics of the saturable constituents of the sorption solution and a linear contribution at greater Cd activity (Nazar et al. 2012). In roots, Cd absorption can happen either in the form of inorganic complexes such as $CdCl^+$, $CdCl_2$, and $Cd^{2+}SO_4$, or in organic forms like plant metallophore complexes (Kubier et al. 2019). Soil acidification enhances cadmium bioavailability in plants as root fluid improves solubility. In soil solution, Cd generally occurs as Cd^{2+} as well as Cd chelates (Abedi and Mojiri 2020; Lux et al. 2011). Two types of adsorption pathways participating in metal ions uptake like cadmium are apoplastic and symplastic (Ismael et al. 2018; Haider et al. 2021).

In addition, root hairs have crucial role in cadmium uptake (Zheng et al. 2011), as plants having thin hairy roots have a greater level of metal uptake and accumulation (Gupta et al. 2019). Usually, micronutrients are absorbed in divalent state (Fontes et al. 2014; Gupta et al. 2016). Cd^{2+} must pass through the cell walls of roots through similar transporters as magnesium, calcium, copper, iron as well as zinc transporters (Ismael et al. 2018; Fontes et al. 2014; Villiers et al. 2012). Cd can move to the roots of plants via cell walls by diffusion from the soil solution (Redjala et al. 2011; Rog Young et al. 2015). In addition, Cd uses active transport for crossing the root cells' plasma membrane via non-specific transport proteins of membranes like ZIP zinc transporter, IRT iron transporter with metal pumping ATPase (Sebastian and Prasad 2018; Gallego et al. 2012; Yamaguchi et al. 2011; Wu et al. 2015). Several additional transporters like P-type ATPase, NRAMP family, ABC transporter, LCT transporter, CAX as well as CE family are found to be implicated in the translocation of cadmium (Gallego et al. 2012; Song et al. 2017). Apoplastic as well as symplastic pathways in plant cells are implicated to transfer cadmium into the xylem (Gallego et al. 2012; Thakur et al. 2016; Wu et al. 2015). Water moves along with metal ions in plants through the apoplastic route in free spaces which is not limited by means of membranes, and involves transport via cell walls as well as intercellular spaces (Hart et al. 2002). Water passes through the symplast in symplastic pathway, which possesses plasmodesmata and cytoplasm with slight contacts in neighboring cell cytoplasm (Lopez and Barclay 2017). In the plant root apoplast, cations are stored and controlled through cell wall exchange features which are primarily dependent on pH (i.e., metal ions adsorption through soil solution), while root cell wall functional groups like carboxylic are slowly deprotonated as pH of soil increases (DalCorso et al. 2008; Meychik and Yermakov 2001). Therefore, in the apoplast of root, there is electrostatic interaction of metal cations with carboxylate groups and thus first phase is fast as well as random showing that energy is not needed (after passive energy) (Yin et al. 2015). The symplastic route is an active mechanism based upon metabolic activity and is slower in comparison to apoplastic route (Ismael et al. 2018). However, because of the type as well as the metal and other ions concentration, every phase has a different meaning. Cadmium goes through an apoplastic route via the root cell membrane (Begum et al. 2019; Abedi and Mojiri 2020; Haider et al. 2021).

Cd was shown to accumulate inside the root apoplasts of barley (*Hordeum vulgare*) cultivated in Petri dishes on filter paper after 24 h recovery (Tamas et al. 2008). Rice seedlings cultivated on cadmium in a hydroponic test with developed apoplast barriers, were observed with fewer root tips/root surface area, resulting in reduced cadmium uptake as well as transfer (Huang et al. 2019). Mature apoplast barriers have been reported to reduce direct transport of Cd from the xylem of apoplast (Redjala et al. 2011). It was observed by Van Belleghem et al. (2007) inside the apoplast and symplast, that chelation of Cd with phosphorus as well as sulfur, respectively. The absorbed cadmium which is transferred from the aerial portions transported to the cytosol and pericycle cells, ultimately going into the xylem (Redjala et al. 2011; Wu et al. 2015, Thakur et al. 2016; El Rasafi et al. 2022).

5 Cadmium Accumulation and Toxicity in Plants

ABC transporters participating in the storage of cadmium in vacuoles or NRAMPs (natural resistance-associated macrophage proteins) implicated in divalent cation movement (Williams et al. 2000) were reported to be favorably activated in plants under 10 μM cadmium exposure, but less in plants under 100 μM cadmium exposure. 10 μM Cd also stimulates two metabolic enzymes alanine aminotransferase and hexokinase, which appears to have function in cellular responses to various abiotic stresses (De Sousa and Sodek 2003). The first one is a crucial enzyme in the synthesis of glutamate (Glu), the building block of glutamylcysteine as well as GSH. Existing data on the adverse influences of soil cadmium upon various physiological systems prompted the investigation on mechanisms of induction of cadmium resistance in plants. The cadmium binding to the sulfhydryl enzymes protein groups leads to detrimental effects of Cd on the function of plant membranes. Activity of H-ATPase is implicated in element assimilation through roots was found to be significantly lower in the papilla cell membrane in comparison to cadmium tolerant cucurbits (Obata et al. 1996). Cadmium has been shown to cause breaks in DNA strands, cross-linking of DNA protein, chromosomal aberrations, oxidative DNA damage, and gene expression disruption, all of which lead to increased proliferation, programmed cell death, and/or repair of DNA (Mouron et al. 2004). According to Ünyayar et al. (2010), cadmium causes both cytotoxicity and genotoxicity in *Vicia faba*, leading to sister chromosome exchange (SCE), a marker of chronic DNA damage. Antioxidant enzymes as well as antioxidants play a significant part in defending *V. faba* from continuous exposure to cadmium, neutralizing its genotoxic influences. Further Cd-induced ECS mechanisms investigation related to antioxidant system responses may provide novel understandings that in what way the mutagenic cadmium effects are altered through mechanisms of plant adaptation in Cd-polluted environments. Intracellular signalling as well as apoptotic routes are noticeably damaged after cadmium exposure.

Soil polluted with cadmium alters soil characteristics precisely near the roots, i.e., soil rhizosphere as well as affects key component uptake needed in normal plant growth as well as development. Influence of cadmium over intake of nutrients is decided through the level of soil pollution with differences in tolerances of plant species as well as organs to harmful heavy metals. Ueno et al. (2009) investigated the genotypic variation in cadmium level of shoot in 146 rice accessions of rice core-collection and observed a huge difference in the cadmium storage as well as cadmium tolerance in shoots. There is substantial variation in Cd resistance between species as well as varieties, but inconsistencies occur between the outcomes of different experimentations. These variations could be due to characteristic dissimilarities of various species as well as varieties in the storage and distribution of cadmium in roots as well as shoots, with their capacity to regulate cadmium in the roots. The distribution of Cd into various plant organs is very significant in the cadmium toxicity to plants. The concentration of cadmium accumulated in plants is restricted by various reasons, like (1) bioavailability of cadmium into the rhizosphere; (2) the cadmium movement rate inside the roots through apoplastic or symplastic routes; (3) fixed cadmium percentage in roots as cadmium–phytochelatin complexes and accumulated in vacuoles; and (4) rate of wood loading and Cd displacement. The root cell wall CEC (cation exchange capacity) might have a metal uptake function also. High CEC indicates strong metal adsorption inside the cell wall, increasing metal ions' availability for transmembrane transport while limiting metal outflow. Elevated CEC can also increase the metal level inside the cytoplasm, affecting metal resistance. A presumed chromatin remodelling factor, termed OXS3, was detected in testing for cadmium resistance of *B. juncea* cDNA library in *Schizosaccharomyces pombe* (Blanvillain et al. 2009). Over-induction of a cadmium hypersensitive OXS3 mutant enhanced cadmium resistance. Verbruggen et al. (2009) suggested that OXS3 can shield DNA or change its transcriptional selectivity. Although maximum cadmium is chelated prior to entering inside vacuole, cadmium/proton antagonists like CAX2, CAX4, as well as probably MHX, can transport Cd directly into the vacuole (Korenkov et al. 2007). In tobacco, overexpression of AtCAX2 or AtCAX4 increases cadmium as well as zinc movement inside root tonoplastic vesicles with increase in cadmium storage inside the roots of plants under cadmium exposure (Korenkov et al. 2007).

Heavy metals' interaction with soil properties has a significant impact on ecosystems because they reduce the heavy metals' bioavailability, which is beneficial to the environment. It is shown that some plants can alter their H^+ secretion in an active or passive manner under the influence of heavy metals. These rhizosphere-induced pH fluctuations play an important part in many pH-dependent nutrients bioavailability, as well as possibly hazardous metals with many trace metals (Hinsinger et al. 2005). Cd uptake by soil roots is affected by soil variables like Cd content, levels of pH and organic matter. Nutrient interactions with heavy metals, as well as soil can happen in the plant. Interaction of nutrients with metals happens in the soil through surface absorption, precipitation, and binding to organic compounds (Pantazis et al. 2007). The rhizome is a significant environmental link to connects plant roots to the soil. The effect of root exudates over bioavailability as well as toxicity of cadmium is due to changes in rhizosphere pH, redox capacity, rhizobacteria number and activity, and potential for cadmium ion chelation. Certain motile physiological modifications occur when plants are cultivated under heavy metal stress, which then stimulate physical and chemical reaction series in the rhizosphere affecting their metabolism in the plant and soil system, which might be helpful in reducing availability of metals and their absorption through plants. Thus, it is clear that the investigations in roots rhizosphere are amongst significant concerns regarding toxicity as well as resistance to metals. Mench and Martin (1991) found that the low MW organic acids discharged from plant roots play a major part in the solubility as well as availability of heavy metals, and cadmium availability will decrease if cadmium binds to the complex of cadmium chelate during root secretion. Certain plants, like wheat and buckwheat, produce organic acids like malic acid, oxalic acid, and citric acid, which may bind with cadmium and limit the entry of cadmium into the roots. Additionally, organic acid phosphates' interaction with cadmium ions will make cadmium phosphate complexes inaccessible to plants (Nazar et al. 2012).

6 Impact of Cadmium Toxicity on Plants

6.1 Seed Germination

Cadmium toxicity causes abnormalities and general growth retardation in several different plant species as it is not essential for plants (Zhang et al. 2019; Tran and Popova 2013). After long-term cadmium exposure, roots become necrotic, rotten as well as slimy, which limits root as well as shoot elongation and causes chlorosis and leaf rolling (Abbas et al. 2017). Cd poisoning inhibits root growth in the soil root zone. This is thought to be due to an abnormal expansion of the apical cortex and epidermal cell layer (Rascio and Navari-Izzo 2011). Cadmium toxicity reduces mitotic division in meristem cells, leading to a decrease in root length as well as dry biomass, while increasing root width (Seth et al. 2008; Gratão et al. 2009). Under cadmium stress, there is an increase in the size of cortical tissues and parenchymal cells that play a role in raising the resistance of plants to solute flow and water and can cause an increase in root diameter (DalCorso et al. 2008; Ismael et al. 2018). Alterations in leaf morphology under cadmium stress leads to changes in chloroplast superstructure with less chlorophyll content, chlorosis as well as reduced photosynthetic activity (Miyadate et al. 2011). Cd treatment suppresses root growth as well as morphological changes in rice (Rascio et al. 2008). Cd stress reduces root enlargement in various crops such as wheat, rice as well as tomato (Rizwan et al. 2012; Hussain et al. 2015; Abbas et al. 2017). Exposure to Cd causes chromosomal abnormalities in root tips of peas causing disturbances in meiosis as well as root elongation (Tran and Popova 2013). Under extreme Cd toxicity, reductions in length, tip area, and number of roots were linked with cadmium stress and showed less resource storage potential for nutrients and water in plants (Lu et al. 2013). Cadmium stress has been observed to decrease mineral intake as well as photosynthesis in plants, which reduces yield and crop quality (Rizwan et al. 2016). Cadmium poisoning causes fragmentation, premature defoliation, stickiness, bridging, and root retardation (Liu et al. 2010). Cadmium concentrations of 250 micromolar for only 24 h caused mitotic disruption of pea roots (Lee et al. 2010). Unexpected nuclear hypertrophy and spreading in pea roots have been reported as a result of cadmium poisoning (Fusconi et al. 2007). After 24 h, Cd-rich treatment induced inhibition and abnormalities in the mitotic, chromosomal, and micronucleus index of onion (Allium cepa L.) (Seth et al. 2008). DNA damage is also reported in root cells (Seth et al. 2008). Modest cadmium levels may also change plant metabolism (Younis et al. 2016). Cadmium stress causes a number of symptoms in plant leaves, like chlorosis, withering, growth retardation, with necrosis, and plants may display these deadly signs when cadmium concentrations exceed 3.30 milligrams per kilogram in tissues of plants (Ismael et al. 2018). The total leaf area as dry weight of some plant portions decreased significantly under the influence of Cd (Jinadasa et al. 2016). Plant growth retardation due to Cd stress might be related to decreased respiration, nutrient and water uptake, photosynthesis, nitrogen and carbon assimilation as well as antioxidant activity (Rizwan et al. 2017). It was observed that greater cadmium concentrations reduces growth of cell in whole chickpeas (*Cicer arietinum L.*), lentils (*Lens culinaris L.*), alfalfa (*Medicago sativa L.*), wheat, corn, spinach, and soybeans (Rizwan et al. 2012; Rehman et al. 2015; Younis et al. 2016; Abbas et al. 2017; Zhang et al. 2019).

6.2 Plant Growth and Development

Cadmium is not necessary for plants, and its stress produces widespread disruptions and development retardation in various plant species (Tran and Popova 2013; Zhang et al. 2019). After long-term Cd exposure, roots become necrotic, rotting, and slimy, leading to decreased root and shoot lengths, leaf rolling, as well as yellowing (Abbas et al. 2017). Cd stress limits accessory root production in the rhizosphere and causes primary roots to become twisted, rigid, and brown (Krantev et al. 2008). This is because of aberrant growth of cortical cell layers within the epidermal parietal area (Rascio and Navari-Izzo 2011). Cd stress inhibits meristem cell development, resulting in decreased dry biomass, root length, as well as increased root diameter (Seth et al. 2008; Gratão et al. 2009). Under cadmium, an enhancement in the size of the cortical and parenchymal cells, which increases plant resistance to solvent and water fluxes, may result in an enhancement in diameter of the root (DalCorso et al. 2008; Ismail et al. 2018). Cd stress causes alterations in chloroplast structure with less chlorophyll level, resulting in chlorosis as well as decreased photosynthetic activity (Miyadate et al. 2011).

Cd exposure decreased root growth as well as morphological changes in rice plants (Rascio et al. 2008). Cd exposure has been demonstrated in previous research to lower root length in numerous crops, like wheat, rice, as well as tomato (Rizwan et al. 2012; Hussain et al. 2015; Abbas et al. 2017). Cadmium exposure causes chromosomal abnormalities at the taproot tip, which results in meiosis and reduced root length (Tran and Popova 2013). Cd stress was associated with decreased root area, root length, as well as root tip number, indicating that the capacity to store resources (i.e., plant nutrients and water) is reduced under high Cd stress (Lu et al. 2013). Cd toxicity impairs mineral uptake as well as photosynthesis in plants, resulting in lower yield and crop quality (Rizwan et al. 2016). Cadmium poisoning commonly causes debris, premature shedding, adhesions, omissions, and root plaques (Liu et al. 2010). Only 24 h of exposure to high Cd concentrations (i.e., 250 M) causes mitotic disruption in pea roots (Lee et al. 2010). Due to Cd toxicity, abnormal proliferation and nuclear distribution have been detected in pea roots (Fusconi et al. 2007). After 24 h, large doses of Cd produced inhibition and abnormalities in mitotic, chromosomal, and micronucleus indices in onions (Allium cepa L.) (Seth et al. 2008). Root follicular cells were also found to have DNA damage (Seth et al. 2008). Plant metabolism can also be altered by relatively low Cd concentrations (Younis et al. 2016). Several indications of cadmium stress have been described in leaves, including yellowing, wilting, stunting, and necrosis, and plants may exhibit these toxic symptoms when cadmium concentrations are present in the tissues. 3–30 mg kg^{-1} for plants and animals (Ismael et al. 2018). Cadmium also dramatically reduced the total leaf area and dry weight of numerous plant components (Jinadasa et al. 2016). Plant development is reduced under Cd stress due to decreased nutrient and water absorption, respiration, photosynthesis, nitrogen and carbon assimilation, and antioxidant activity (Rizwan et al. 2017). High Cd concentrations have been shown to reduce cell as well as complete plant growth in chickpeas (*Cicer arietinum L.*), lentils (*Lens culinaris L.*), alfalfa (*Medicago sativa L.*), wheat, corn, spinach, and soybeans (Rizwan et al. 2012; Rehman et al. 2015; Younis et al. 2016; Abbas et al. 2017; Zhang et al. 2019).

6.3 Nutrient Uptake

Cd ions in soil can affect root uptake and subsequent plant nutrient distribution and transport (Rochayati et al. 2011). Cadmium has been demonstrated to affect plant absorption, storage, and utilization of a variety of elements (including Ca, Mg, P, and K), as well as water consumption (Tran and Popova 2013). Sugar beetroot roots exposed to Cd show Fe deficit (Chang et al. 2003). After Cd treatment, the absorption of P, K, S, Ca, Zn, Mn, and B in peas was dramatically reduced (Metwally et al.

2005). Treatment with 1.0 M Cd considerably reduced the amounts of P, K, Ca, Mg, Cu, Fe, Mn, Zn, Mo, and B in the roots of barley (*Hordeum vulgare* L.) without damaging the roots. Concentration of shoots (Guo et al. 2007). Cd toxicity reduces Ca and K absorption in salt (*Atriplex halimus* L.) (Kinay 2018). Cadmium poisoning affects the amount of nitrogen, calcium, magnesium, and phosphorus in lucerne and sprouts (Zhang et al. 2019). Cadmium can reduce nitrate uptake and transport from the roots to the aerial sections of *Cucubalus campionis* (*Silene cucubalus* L.) via modulating the enzymatic activity of nitrate reductase in shoots (Tran and Popova 2013). Cadmium treatment lowers nodulation in soybeans due to insufficient ammonia uptake and nitrogen fixation in the rhizosphere (Karina et al. 2003). Laccase activity is required for the production of lignin (Tran and Popova 2013). Soybean plants treated with cadmium tended to increase lactase activity, accelerate lignin production, and decrease root length during early root development (Yang et al. 2007). Competition for essential nutrients with Cd and other harmful metals affects macro and micronutrient transport in poplar (*Populus jacquemontiana L.*) (Solti et al. 2011). Cadmium has the ability to prevent iron chelation and absorption into the nucleus (Popova et al. 2012). Cd competes with Ca transporters in alkaline soil, limiting Ca transport in plant roots' woody matter (Choppala et al. 2014). Several phytonutrients influence Cd toxicity and plant availability in both direct and indirect ways (Rizwan et al. 2016). Direct strategies include reducing Cd solubility in soil via adsorption and precipitation (Matusik et al. 2008), intense competition for substances between phytonutrients and Cd ions, and Cd accumulation in vegetative organs to prevent sequestration in seeds and edible components (Zhao et al. 2005). Indirect methods include boosting plant yield and biomass and decreasing physiological stress to reduce Cd ion concentrations (Tran et al. 2011).

6.4 Oxidative Damage

Trace metals often affect plants directly or indirectly via ROS production (Ehsan et al. 2014; Nagajyoti et al. 2010). Cd stress prevents photooxidation of photoelectron system II (PSII) by limiting electron transport (Farooq et al. 2016). Thus, Cd may indirectly promote ROS generation by degrading leaf chloroplasts (Gallego et al. 2012). Cadmium poisoning is associated with enhanced generation of ROS in the electron transport chain of mitochondria (Heyno et al. 2008). Cadmium exposure induces plasma membrane-associated NADPH oxidase synthesis in peroxisomes as well as ROS production in rice and legumes (Hasan et al. 2009; Tran and Popova 2013). ROS synthesis in plant cells occurs rapidly; for example, pine (*Pinus sylvestris L.*) under 50 mM exposure to cadmium increased ROS generation within 6 h (Schützendübel et al. 2001).

Lucerne exposed to Cd for 6–24 h resulted in fast peroxide accumulation and lowered levels of homoglutathione as well as glutathione, causing redox imbalance (Gutsch et al. 2019). Cadmium stress kills tobacco cells by raising NADPH oxidase levels and H_2O_2 fatty acid synthesis (Gill and Tuteja 2010). ROS damage in plants includes protein and lipid peroxidation as well as DNA damage (Younis et al. 2016). Cadmium poisoning causes an increase in peroxidation of lipids in plants including peas, sunflower (*Helianthus annuus L.*), and common bean (*Phaseolus vulgaris L.*) (Chaoui et al. 1997). Cd-associated DNA damage harms cell membranes and nucleic acids, harms photosynthetic proteins, and affects protein synthesis, impairing overall organism development (Abbas et al. 2017). DNA damage causes growth retardation, premature cleavage, fragmentation, adhesion, and an increase in the frequency of double as well as single bridges (Kranner and Colville 2011). Proteins as well as amino acids behave like hydrogen peroxide and superoxide solvents. These are not restricted to non-enzymatic scavengers, namely tocopherol, ascorbic acid, carotenoids, and glutathione (GSH); with enzymatic scavengers namely, peroxidase (POD), glutathione reductase (GR), ascorbate peroxidase (APX), dehydroascorbate reductase (DHAR), monodehydroascorbate reductase (MDHAR), superoxide dismutase (SOD), and catalase (CAT) (Shahid et al. 2014). Catalase, POD, APX, GR, and SOD activities varied with Cd concentration and diversity. Cadmium has been demonstrated to increase SOD activity in soybeans, wheat, and peas (Milone et al. 2003). Cadmium poisoning lowers the activity of SOD as well as CAT enzymes in beans, sunflower, plus peas (Sandalio et al. 2001). Greater activity of CAT enhances resistance in cadmium-infected Greenland cultivars (*Groenlandia densa L.*) (Yılmaz and Parlak 2011).

Cadmium toxicity increases GPX as well as APX activity in fescue (*Ceratophyllum demersum L.*) and wheat while decreasing POD activity in canola (*Brassica juncea* L.) (Tran and Popova 2013). DHAR and MDHAR activities in canola treated with 10 M Cd were elevated (Markovska et al. 2009). On the other hand, Cd toxicity enhanced erythrocyte activity in canola, wheat, cotton (*Gossypium malvaceae L.*), with mung beans (*Vigna mung L.*) (Gill and Tuteja 2010). Cadmium toxicity had no effect on pea POD activity (Tran and Popova 2013), but raised radish POD value (El-Beltagi et al. 2010). Tobacco usage raises proline levels, lowering cadmium's inhibitory effect on cell development and proliferation (Islam et al. 2009).

6.5 *Effect on Photosynthetic System*

Cd is a strong photosynthetic inhibitor (Vassilev et al. 2005). In oilseeds, legumes, and cereals, a linear link inhibition of transpiration and photosynthesis was established (Zhang et al. 2019; Younis et al. 2016). Cadmium stress harms photosynthetic machinery, especially the photoelectron system as well as the light-collecting Complexes I and II (Hasan et al. 2009). Fe enhances chlorophyll concentration and is involved in the production of other pigments which are directly implicated in photosynthetic light capture (Akinola and Ekiyoyo 2006). Cd inhibits Fe-reducing enzyme (Fe^{3+}), resulting in deficit of iron (Fe^{2+}) and greatly disrupting the photosynthetic mechanism and machinery (Hasan et al. 2009).

Furthermore, cadmium causes closing of stomata as well as lowers photosynthesis in higher plants. Many crops, including gerbera (*Thlaspi caerulescens L.*) (Küpper et al. 2007), rapeseed (Ali et al. 2015), beans, barley, maize (Popova et al. 2012), wheat (Abbas et al. 2017), and chickpea (Wahid et al. 2008), are affected by Cd toxicity. Cadmium's primary targets include the photosynthetic apparatus, pigments, and the production of carotene as well as chlorophyll (Rafiq et al. 2014). The combination of cadmium and production of chlorophyll in seed crops like rapeseed causes chlorosis (Baryla et al. 2001). The decrease in chlorophyll content caused by cadmium was more pronounced at leaf surfaces (i.e., in stomatal guard cells) than in mesophyll cells (Rucińska-Sobkowiak 2016).

Cd poisoning affects cell size, stomatal conductance, and stomatal density in the epidermis of leaf, as well as causing cell division and chloroplast formation to be disrupted (Abbas et al. 2017). Cadmium ions have been demonstrated to impair chloroplast structure and function in plants such as sugar beetroot, spinach, wheat (Sersen and Kral'ova 2001; Younis et al. 2016), chickpeas, as well as common beans (Wahid et al. 2007). The enzymes RUBPCase as well as PEPCase are necessary to fix carbon and effective photosynthesis (Zhang et al. 2007). Cd ions inhibit RuBPCase activity as well as directly alter its structure through displacing magnesium ions, key cofactors in processes of carboxylation and regulate RuBPCase activity in oxygen-containing reactions (Tran and Popova 2013).

Cd irreversibly cleaves large as well as small RuBPCase subunits, inhibiting enzyme activity completely (Noor et al. 2018). Cd influences the PSII water oxidation complex (OEC) by substituting Ca in the Ca/Mn group that forms the oxygen evolution center (Dinakar et al. 2009) or through changing locations of the binding sites of Qb (Hayat et al. 2011). Cd lowers parameters of gas exchange, causing chloroplast structure as well as photosynthetic pigments to degrade (Yadav 2010; Noor et al. 2018).

6.6 *Impact on Amino Acids and Proteins*

Tremendous fluctuations in the environment can change gene expression, which changes cellular protein diversity (Kieffer et al. 2009). Protein transport changes in a high-stress environment could be a molecular indication of plant stress response (Nanjo et al. 2011). The proteomics approach is a valuable tool for assessing plant stress resistance. New evidence has emerged in recent years suggesting the protein stress response occurs rather quickly in stressed plants (Tran and Popova 2013). Heat shock proteins are proteins that aid eukaryotes in their stress response (Jali et al. 2016). After cadmium treatment, a heat shock protein (70 kDa phosphoprotein) was produced (Popova et al. 2012).

Low-temperature pretreatment prior to Cd exposure protected Peruvian tomatoes (*Lycopersicon peruvianum* L.) from cell membrane damage (Yabuta 2016). Pea cells have been demonstrated to be protected against cadmium toxicity by HSP71 and the pathogen-associated protein PrP4A (Rodríguez-Serrano et al. 2009). Wheat plants exposed to $CdCl_2$ (50 μM) after 48 h developed a 51 kDa soluble protein known as Cd toxicity-related protein in root tissues under the plasma membrane as well as at the outer margin of the tonoplast (Mittra et al. 2008). It was observed that after 56 or 14 days of cadmium treatment, pathogenesis-related proteins as well as heat shock proteins are elevated in European poplar (*Populus tremula L.*) leaves (Sergeant et al. 2014). Cadmium has been shown to impact the production of 36 rice proteins (Lee et al. 2010). Tomatoes treated with a low dosage of cadmium (10 μM) transformed 36 polypeptides, whereas tomatoes treated with a high Cd (100 μM) transformed 41 polypeptides (Semane et al. 2010).

6.7 Plant–Water Relations

The major tasks of plant roots are the intake of inorganic nutrients and water, supporting, and defending the plant stem into soil, accumulating nutrients as well as promoting vegetative reproduction (Feleafel and Mirdad 2013; Rucińska-Sobkowiak 2016). They can be the first points of contact with dangerous metals like cadmium ions as well as tend to retain greater amounts of the metal as compared to terrestrial plants (Burkhead et al. 2009). The storage of cadmium ions in tissues can affect absorption of soil water and reduce the amount of water in plant roots (Chen et al. 2004). Cadmium exposure causes water deficit in plants like beans, sunflower, lettuce, barley, and pea (Schat et al. 1997; Rucińska-Sobkowiak 2016). Interaction of cadmium ions directly with guard cells or initial effects of cadmium storage in plant parts such as stems and roots triggers stomatal closure (Bazihizina et al. 2014). The effects of cadmium over plant–soil water interactions are considered different from its impact on soil water availability, root development, decreased water absorption, as well as other adverse effects (Burkhead et al. 2009). When the soil is enriched with Cd, the permeability of the soil solution can be less as compared to root cell sap (Małecka et al. 2008). The soil solution strongly limits the amount of water that plants can absorb under these conditions, causing osmotic pressure (Rucińska-Sobkowiak 2016). When the salinity reaches 10 3 M, it is considered that osmotic pressure exists (Levitt 1972). Plant water uptake is more likely regulated indirectly via changes in endogenous properties like changes in root structure as well as morphology (Hayat et al. 2012). The total area of goat willow (*Salix caprea L.*) roots was significantly reduced after exposure to Cd (Vaculı et al. 2012). In cadmium-stressed plants abnormalities such as reduced hair surface area, elongation of primary root (Gallego et al. 2012), rapid death, and stunting of secondary roots (Lux et al. 2011), affect the water-to-water ratio. Structural changes caused by cadmium weaken the root–soil connection and weaken the ability of plants to absorb groundwater (Feleafel and Mirdad 2013). Exposure to cadmium causes various water-associated modifications throughout the plant. Cadmium stress reduces water absorption as well as inhibits water movement for small distances in pathways of the root apoplast and symplast (Kudoyarova et al. 2015). In addition, cell wall thickening caused by cadmium ions or other cell wall deposition increases apoplastic resistance to water flow (Hashem 2014). The ability to diffuse water across the membrane is possibly impaired because of variations in aquaporin function (Le et al. 2015). These changes modify water transport through vascular system as well as also reduce root resin secretion (Kaznina et al. 2014). Reluctance to transport water over long distances leads to reduced leaf water and subsequent leaf desiccation (Osakabe et al. 2014). A quick decrease in root vacuole, permeability, as well as intercalation in stem and leaf tissues can be considered as those events which increase plant water retention (Chmielowska-Bąk et al. 2014; Haider et al. 2021).

7 Mechanism of Plant Responses to Cd

Plants, like every other species, have developed a complicated arrangement of homeostatic systems to reduce the effects of unnecessary metal ions exposure. To evade cadmium toxicity, terrestrial plants have developed active as well as passive mechanisms for removing heavy metals from the cellular environment. Plant secretions like malate or citrate combine with metals of soil as a primary line of defense against cadmium stress (Delhaize and Ryan 1995). Second, the cell wall by its pectin sites and histidyl groups with extracellular carbohydrates like scar tissue, mucus may have an important function in trapping harmful ions and stopping their contents from entering into cytoplasm (Di Toppi and Gabbrielli 1999). However, when concentrations of toxic elements exceed the physical adsorption limits of these barriers, an active metabolism occurs that produces chelating substances like phytochelatin and in some cases metallothionein, which participate in the detoxifying heavy metals and prevent them in certain parts of the cell. In addition, Cd resistance, like resistance to other abiotic stimuli, includes stress-related protein synthesis and signalling molecules like salicylic and abscisic acids, ethylene as well as heat shock proteins (Di Toppi and Gabbrielli 1999). Signalling pathways are distinguished by a complex genes interaction, in which transcription factors have an important function, as control of their expression has been shown to significantly alter the plant stress response (Uno et al. 2000). With the use of genetic engineering, many putative genes of Cd stress response can be recognized: for example, cDNA amplicon length polymorphism (AFLP) identified various genes which are activated in *Brassica juncea* after cadmium exposure (Fusco et al. 2005). Affymetrix DNA arrays were used to investigate modifications in the *Arabidopsis* transcriptome under Cd and Pb exposure (Kovalchuk et al. 2005).

7.1 Cadmium Induces Modulation of Gene Expression

Heavy metal stress responses are based upon a complex signalling route inside the cell, which starts with the detection of the heavy metal or heavy metal-linked symptoms and culminates in the transcriptional control of metal-induced genes

(Singh et al. 2002). Not much is known about the molecular machinery of metal-responsive signalling, as well as transcription factors under heavy metal stress are distinguished by differential expression analysis (Fusco et al. 2005). Heavy metal-responsive TFs, like other stress-related TFs, share similar signalling pathways and are thus induced by abiotic stimuli like cold, desiccation, salicylic acid as well as H_2O_2 (Singh et al. 2002). In addition, there is an interaction of cadmium resistance with pathogen protection signals (Suzuki et al. 2001). Cd influences the ERF protein expressions of the APETALA2 (AP2)/ethylene-responsive factor-binding protein (EREBP) family. These transcription factors' family members can combine with many pathogenesis-linked promoters as well as dehydration response factors (DRE modules) (Singh et al. 2002). ERF1 and ERF2 genes were increased in *A. thaliana* roots after 2 h exposure to Cd (Weber et al. 2006). In addition, Cd was shown to activate DREB2A, which interacts specially with DRE motif in the rd29A promoter area and stimulates its transcription in plants under cadmium exposure. Cold stress, salt, and desiccation have previously been shown to be beneficial for Rd29A (Suzuki et al. 2001). A bZIP-like DNA-binding protein OBF5, produced by Cd, was observed to bind in the promoter area in the gene encoding glutathione S-transferase, implicated in scavenging ROS as well as detoxifying xenobiotics (Suzuki et al. 2001). In addition, Fusco et al. (2005) found that short exposure to Cd induced the BjCdR15, a bZIP protein expression in *B. juncea*. This transcription factor regulates the many metal transporters expression and is implicated in the long-space transport of cadmium from roots to shoots, and its overexpression improves cadmium resistance and storage in *A. thaliana* as well as tobacco shoots (Farinati et al. 2010). A transcription factor WRKY53 of WRKY family, was found to be differentially expressed in cadmium exposed *Thlaspi caerulescens* plants. It is also regulated by different environmental stimuli, like salinity, cold, drought as well as salicylic acid, and appears to be involved in stress-linked signalling and affects TF activity beyond directly activating genes (Wei et al. 2008). MYeloblastosis proteins (MYBs), especially R2R3 MYB subtype, also react to heavy metal stress. After cadmium and zinc treatment in *A. thaliana*, MYB4 is induced more strongly (Van de Mortel et al. 2008) and in roots, proteins MYB43, MYB48, and MYB124 are specially induced by cadmium (Weber et al. 2006). MYB28 is strongly induced in *T. caerulescens* in Zn deficiency and greater amount of cadmium and is involved in the Glucosinolate (GSL) synthesis regulation (Van de Mortel et al. 2008). GSL is a significant form of sulfur accumulation and its production is affected by modification in biotic/abiotic stress factors and nutritional status (Hirai et al. 2007). Cadmium inhibits food and sulfur metabolism. Modulation of different groups of TFs can reveal the complexity of plant response to cadmium, by signal detection to the intracellular transduction cascade that activates genes involved in Cd uptake, transport, transformation, and detoxification.

7.2 *Phytochelatins*

Cadmium may stimulate the formation of phytochelatins (PC), small metal-bound peptides. The general structure of PC is (-Glu-Cys)n-X, where n can be between 2 and 11 and X can be any amino acid like Ala, Gly, Glu or Gln and Ser (Cobbett and Goldsbrough 2002). PC binds to Cd and forms many complexes with molecular weights of approximately 2500-3600 Da due to Cys-thiol groups that shield the cytosol with free cadmium ions (Cobbett 2000). The building block of PCS synthesis Glutathione is catalyzed by PCS synthase of cytosol. Heavy metals have been shown to constitutively express PCS and activate it post-translationally (Cobbett and Goldsbrough 2002). Phytochelatins are thought to function in cellular homeostasis as well as transport essential nutrients like copper and zinc due to their affinity for metal ions (Thumann et al. 1991). They are needed for the dissolution of toxic metal toxicity, especially Cd, such as in *Arabidopsis* and *Schizosaccharomyces pombe*, confirmed through the cadmium-responsive cad1 mutants phenotype lacking activity of PCS (Ha et al. 1999). However, PC's extreme consumption did not induce hyper-resistance; in fact, although increased PC synthesis appears to enhance heavy metal accumulation in transgenic plants (Pomponi et al. 2006), and additional AtPCS genes determine sensitivity to Cd exposure (Lee et al. 2003). After production, PCs bind to heavy metal ions as well as enhance their movement as complexes to vacuoles, where it subsequently produce high molecular weight complexes (Clemens 2006). Different investigations have shown that the movement of HMW complexes through the tonoplast in *Arabidopsis* is facilitated by ABC (ATP-binding cassette transporters) (Cobbett and Goldsbrough 2002). It has been established that PC has function in the cadmium movement from the root to the embryo, as well as PC-dependent "efflux protection mechanism" helps keep Cd accumulation low in roots, thus increasing the probability of Cd accumulation to bring the Cd to the shoot (Gong et al. 2003).

7.3 *Metallothioneins*

Metallothionein (MT) is another cysteine-rich low molecular weight peptide that may bind metal ions through mercaptile bonds. Unlike PC, metallothionein is an mRNA translation product that is activated under heavy metal exposure (Cobbett and Goldsbrough 2002). In *E. coli* pea metallothionein (PsMTa) when produced may bind cadmium, zinc, and copper (Tommey et al. 1991). Additionally, Arabidopsis metallothioneins restored copper resistance in yeast strains which is metallothionein deficient (Zhou and Goldsbrough 1994). Though the importance of plant metallothionein in cadmium resistance is still widely understood, data indicate that they play a role in copper homeostasis (Cobbett and Goldsbrough 2002). In addition, mouse metallothionein is overexpressed in tobacco which increased Cd resistance in vitro (Pan et al. 1994), and *Brassica juncea* MT2, when expressed ectopically in

Arabidopsis thaliana, produced higher Cd and Cu tolerance (Zhigang et al. 2006). Several plant metallothionein genes are induced in all tissues in greater amounts in terms of transcript levels. *Arabidopsis* MT1a as well as MT2a appear to store in trichomes where they participate in the compartmentalization of heavy metal ions (Salt et al. 1995). *Arabidopsis* metallothionein expressed in phloem components, is involved in movement of metal ions (Garcia-Hernandez et al. 1998). Lastly, metallothionein genes are induced at different phases of plant development as well as in response to various environmental situations (Rauser 1999). Copper, cadmium, and aluminum and abiotic stressors like high temperature as well as nutrient deficiencies can activate the MT gene in wheat and rice (Cobbett and Goldsbrough 2002). Several MT genes have been recognized in mature fruit that are believed to participate in normal development (Rauser 1999).

7.4 Metal Ion Transporters

Terrestrial plants have very efficient metal ion absorption systems that allow plant roots to uptake metal ions as well as other inorganic compounds from the soil. Thus, metal transporters located in the tonoplast and plasma membrane have a significant function in maintaining metal homeostasis under physiological limits. Cadmium resistance is actually related to its exclusion or intracellular sequestration, facilitated through the stimulation of particular transport mechanisms.

In general, metal transporters defined for ion entry have limited selectivity. For instance, AtIRT1 located on the plasma membrane of root cells is the main root iron intake mechanism of *Arabidopsis*, but it may also transfer a substantial amount of cadmium (Korshunova et al. 1999). In contrast, intracellular transporters which transport metal ions from cytoplasm into the vacuole or outside the cell are very choosy. Like, the tonoplast transporters AtMTP1 as well as AtMTP3 take up only zinc in the vacuole (Kramer et al. 2007).

The ZIP family of transport proteins of plasma membrane having ZRT, IRT-like protein activated by Zn limitation in Arabidopsis roots and shoots is an important group of metal transporters. ZIP members occur in various plant species, with fungi, bacteria, and animals, and are suggested to be implicated in the transmembrane movement of divalent cations (López-Millán et al. 2004). They are believed to be involved in the assimilation of cadmium from soil inside root cells and the transfer of Cd from roots to shoots during the tree cycle (Kramer et al. 2007). Increased uptake of metals in roots facilitated via ZIP transporters appears to be an essential but insufficient factor for hyperaccumulating in *Arabidopsis halleri* and *T. caerulescens* (Kramer et al. 2007), and storage potential varied depending on the expression of these proteins in these plants. For example, ZIP9 possesses a high level of expression in roots in Zn-rich environment and is elevated in Zn-deficient shoots of *A. halleri* (Kramer et al. 2007). In contrast, ZIP9 was increased by Zn scarcity in both roots as well as shoots of *A. thaliana*. Similarly, under Zn deficiency, ZIP6 is highly expressed in hyperaccumulators but not in *A. thaliana* (Becher et al. 2004; Filatov et al. 2006).

Another important class of transmembrane proteins involved in metal transport and homeostasis is the naturally occurring macrophage metal ion transporter (NRAMP) family of proteins associated with drug resistance. Due to their capacity for transferring manganese, zinc, copper, iron$^+$, cadmium, nickel, as well as cobalt, these transporters are called "common metal ion transporters" (Nevo and Nelson 2006). NRAMP family members, such as ZIP transporters, share significant protein sequence homology between plants, yeasts, and mammals (Nevo and Nelson 2006). According to cDNA microarray data (Chiang et al. 2006), NRAMP gene expression is higher in hyperaccumulating species. They are induced in both roots as well as shoots and participate in the movement of metal cations through the plasma membrane inside the cytoplasm or through tonoplast (Kramer et al. 2007). These metal transporters are mainly involved in iron homeostasis in *A. thaliana*. AtNRAMP1, AtNRAMP3, and AtNRAMP4 of *Arabidopsis* NRAMP family, can facilitate iron, manganese, and cadmium uptake in heterologous systems (Curie et al. 2000). Overexpression of AtNramp3 caused root growth and increased Fe accumulation in Cd-sensitive *Arabidopsis* (Thomine et al. 2000). These results suggest that NRAMP metal transporters can transfer both iron and cadmium in plants (Thomine et al. 2000). P1B-ATPase (HMA) metal transporters hydrolyze ATP to transport metal ions outside the cytoplasm means outside the plasma membrane or vacuole. Transporters of exported metals have more options than transporters of imported metals: e.g., HMA members like HMA2, HMA3, and HMA4 transport only zinc and cadmium (Kramer et al. 2007). When produced in yeast, AhHMA4, AhHMA3, and TcHMA4 of the hyperaccumulator species *A. halleri* and *T. caerulescens,* respectively can grant zinc or cadmium tolerance (Bernard et al. 2004; Papoyan and Kochian 2004). Thus, it is proposed that TcHMA4, AhHMA4 as well as perhaps AtHMA4, its *A. thaliana* equivalent, can play a part in cadmium and zinc homeostasis by displacing metal ions from the cellular part (Kramer et al. 2007). In addition, these transporters are mostly induced in roots and shoot vascular system indicates that they may be involved in metal transport from root to shoot (Verret et al. 2004). ABC transporters are involved in multiple functions, including lipid catabolism, polar auxin transport, stomatal function, xenobiotics, disease resistance as well as metal detoxification (Kim et al. 2006; Rea 2007). An example is the ABC family of Arabidopsis mitochondria (AtATM), AtATM3 was increased in plant roots under cadmium and lead exposure. Furthermore, plants overexpressing AtATM3 were resistant to Cd, while AtATM3 mutants were more sensitive. In *Schizosaccharomyces pombe*, a AtATM3 homolog HMT1 is tonoplast transporter that transports cadmium–phytochelatin complexes. Likewise, it is hypothesized, but not proven, that AtATM3 is involved in the extrusion of the Cd–GSH complex inside the mitochondria as well as sensitivity of the mutation is due to oxidative damage because of cadmium storage (Kim et al. 2006). Another ABC transporter AtPDR8 is involved in metal homeostasis in *A. thaliana* and has been shown to be involved in cadmium as well as pathogen resistance (Kobae et al. 2006; Stein et al. 2006). Cadmium not only increases its induction, but its over-induction decreases cadmium accumulation in roots and shoots. It is primarily found in hairs and cuticle membranes of roots (Kim et al. 2007). AtPDR8 is thought to induce cadmium resistance by driving it from the plasma membrane into the apoplast (Kim et al. 2007). Lastly, "cation diffusion aid" (CDF) transporter members appear to facilitate the vacuolar compartmentalization, accumulation, and movement of metal ions from the cytoplasm to the outer compartment (Kramer et al. 2007). CDF transporters are identified in prokaryotes as well as eukaryotes and can transfer divalent metal cations like zinc cadmium, cobalt, iron, nickel, or manganese across membranes (Montanini et al. 2007).

7.5 Enzyme

Heavy metal toxicity, as mentioned earlier, affects the activity as well as accumulation of various enzymes (Prasad 1995). For instance, cadmium changes carbon-absorbing enzyme activity such as Rubisco, possibly by reacting with the protein's sulfhydryl groups and meddling with folding or activity (Prasad 1995). In addition, cadmium exposure stimulates DAG kinase and increases Mg-dependent ATPase activity in *B. juncea* roots, proposing that cadmium might induce lipid signalling pathways (Lang et al. 2005). Cadmium was observed to alter protein kinase expression in *Arabidopsis* (Suzuki et al. 2001) and is implicated in cadmium signalling pathways in rice as well as alfalfa (Romero-Puertas et al. 2007). Cadmium toxicity reduces the activity of enzymes implicated in primary nitrogen assimilation as well as nitrogen mobilization (Chaffei et al. 2004). Activity of Glutamine synthase was shown to be reduced. In shoots, the activity and expression of the chloroplast isoform of GS decreased, while transcription of the cytosolic gene increased. Cytosolic GS isoform mRNA increased in roots. When Cd suppresses plastid GS activity, plants activate cellular isoforms to compensate as well as regulate glutamine synthesis (Chaffei et al. 2004). On the other hand, a response mechanism to overcome heavy metal stress is the production of PC by PCS. PCS is activated, both in vivo and in vitro, by a wide range of metals and metalloids, such as cadmium, silver, lead, copper, mercury, zinc, arsenic, and gold (Schat et al. 2002). The activation process is still not known and PCS is thought to detect heavy metals by direct binding to metal ions, however, this binding is shown not to be necessary for catalytic activation (Vatamaniuk et al. 2000). Cd-induced PCS gene expression in *A. thaliana* and *B. juncea* has been investigated, with often conflicting results. Cazale and Clemens (2001) determined that the AtPCS1 as well as AtPCS2 genes were continuously induced and not transcriptionally controlled by cadmium, while Lee et al. (2002) found that the AtPCS1 transcript level is sensitive to Cd, but not the protein level. Furthermore, after long-term cadmium exposure, the amount of PCS protein increased in leaves but not in roots of *B. juncea* (Heiss et al. 2003), which showed that the effects of cadmium on the expression of PCS can differ depending on the organ as well as plant species. Cd promotes oxidation by encouraging ROS production as well as preventing antioxidant system scavenging (Schützendübel et al. 2001; Romero-Puertas et al. 2007). Catalase (CAT) is an important enzyme in protective reactions under oxidative stress. It is found just in peroxisomes which catalyzes the breakdown of H_2O_2 (Buchanan et al. 2000). Four different CAT sequences were cloned in *B. juncea* and cadmium exposure induced an enhancement in CAT3 transcript. This stimulation may be important to restrict excessive concentration of H_2O_2 and protect cells against oxidative stress (Lang et al. 2005). Cadmium stimulated CAT protein oxidation in pea plants to reduce CAT activity and protein levels. In response to Cd, CAT transcription increases as a compensatory mechanism (Romero-Puertas et al. 2007). For scavenging ROS, ascorbic acid-glutathione cycle begins. Cd regulates key enzymes in these processes by increasing ascorbate peroxidase activity which is the first enzyme of the cycle in *Phaseolus vulgaris* as well as *Pisum sativum* (Romero-Puertas et al. 2007). Additionally, one other enzyme glutathione reductase (GR) of the cycle is differentially activated in roots as well as leaves of cadmium-exposed legumes (Yannarelli et al. 2007). The enzyme superoxide dismutase protects cells against the accumulation of ROS. Long-term Cd treatment increases SOD activity in tomato seedlings (Dong et al. 2006). In addition, a prominent enhancement in activity was observed in wheat leaves when exposed to excess amount of cadmium, probably because of more superoxide generation (Lin et al. 2007). However, previous work displayed that activity of superoxide dismutase decreases with toxicity of cadmium in peas (Romero-Puertas et al. 2007; DalCorso et al. 2008).

7.6 Cadmium Signalling and Gene Regulation

The ability of plants to detect dramatic alterations in abiotic stress is found to be essential to adaptation or tolerance. Heavy metal tolerance in plants has been described as the capacity for persisting at metal concentrations that are harmful to different plant species which likely results from genotype–environment interactions. Tolerance mechanisms require synchronization of many complicated physiological as well as biochemical mechanisms like changes in overall gene expression, protein modifications, and modification in the composition of primary as well as secondary metabolites. Plant cells should be capable of recognizing stress signals as well as change them in the specific response, giving plants the capacity to survive in adverse environments. Functional genomic techniques have partly elucidated the complicated processes that control stress sensing with transmission and the expression as well as regulation of genes involved in plant stress responses (Urano et al. 2010).

Cd affects physiological signalling mechanisms, which is amongst causes of Cd toxicity. These effects are likely to enhance cellular responses for other adverse impacts of Cd, like direct disruption of thiol oxidation, enzymes as well as transporters, or elevated ROS concentrations. In the majority of cases, Cd avoids cell surface receptors and causes relatively everlasting modifications at another level of communication that can be non-physiologically increased or decreased, affecting cell function, transcription, and regulatory genes, which can lead to death of cell and/or stress-activated adaptation as well as survival (Thévenod 2009). For the detection of heavy metals, cell wall is an excellent site (Blinda et al. 1997). Cadmium-sensitive plants have less cadmium levels in the cell wall as well as higher cadmium levels in the vacuoles as compared to non-Cd-sensitive plants according to cell fractionation experiments (Uraguchi et al. 2009). When heavy metals are detected, plant cells activate special genes that resist stress stimuli. Thus, differential gene regulation is the result of a signaling cascade.

7.7 MAP Kinase, Ca-Calmodulin, and Hormones

The cadmium-activated enhancement in production of ROS might function as a cellular signal that initiates the response to stress. MAPKs (Mitogen-activated protein kinases) appear to be implicated in Cd-induced transcriptional responses, which may be activated by ROS in the presence of high Cd concentrations (Jonak et al. 2004). MAPK pathways in eukaryotes signify a signalling process consisting of three consecutively acting protein kinases: MAPK kinase kinase (MAPKKK), MAPK kinase (MAPKK), and MAPK (Schaeffer and Weber 1999). Those Ser/Thr protein kinases that are phosphorylated by MAPKKs are called MAPKKKs. Like this, information could be sent from downstream kinases to downstream targets through the phosphorylation cascade. MAPKK is accountable for the MAPK phosphorylation at Thr and Tyr after their activation through phosphorylation. Thus, the phosphorylation cascade is believed to be implicated in cadmium signalling to the nucleus as well as activation of the enzymes (Jonak et al. 2004). MAPKs involvement in cadmium signalling is widely observed in many mammalian cell types (Thévenod 2009 et al.; Liu and Kapron 2010). However, nothing is known about cadmium-activated MAPKs in cells of plants (Jonak et al. 2004; Yeh et al. 2004; Liu and Kapron 2010). In *Oryza sativa*, cadmium induces MBP kinase of 42 kilodaltons with properties of a mitogen-activated MAPK protein having activity of OsMPK3 and OsMPK6 also (Yeh et al. 2007). In addition, NADPH oxidase inhibitor diphenyleneiodonium (DPI) inhibited Cd-induced MAP kinase activity as well as cadmium-induced NADPH oxidase-like activity, proposing that ROS might be implicated in cadmium-activated MAP kinase activation (Yeh et al. 2007).

According to Liu and Kapron (2010), Cd stimulates MPK3 and MPK6 two *Arabidopsis* MAPKs, and ROS accumulation may be related to the activation of metal-activated kinase. Another effect of Cd-induced ROS accumulation in cellular compartments is altered signalling mediated by H_2O_2 and other reactive oxygen species. H_2O_2 is known for its importance as a signalling molecule that initiates defense mechanisms against biotic and abiotic stressors (Van Breusegem et al. 2008). Cadmium ions may inhibit Zn finger transcription factors activity by displacing zinc ions thus disrupting the transcription machinery (Di Toppi and Gabbrielli 1999). Cadmium replaced calcium ions in calmodulin proteins by mechanisms similar to those reported for Zn finger transcription factors, causing fluctuations in concentration of intracellular calcium as well as changing Ca-dependent signalling pathways (Perfus-Barbeoch et al. 2002). Ca as well as calmodulin ions are identified as external stimuli transmitters and their involvement in signalling of heavy metals is also proposed. Intracellular calcium levels enhanced dramatically under Cd exposure (DalCorso et al. 2008), which induces calmodulin-like proteins for interacting with calcium ions. Calmodulin proteins regulate many systems, like gene regulation, ion transport, metabolism, as well as stress tolerance, all of which synchronize in response to cadmium just partially (Yang and Poovaiah 2003). In radish, Cd^{2+} competes with Ca^{2+} to bind with specific ionic binding sites for inhibiting calmodulin-dependent phosphodiesterase activity (Rivetta et al. 1997). An Abc1 family member AtOS1, found in chloroplasts, is another potential component of Cd-induced oxidative stress signalling that involves phosphorylation. AtOS1 does not move cadmium, but appears important for cadmium resistance, perhaps by kinase activity (Jasinski et al. 2008). Several plant hormones as well as growth regulators may be involved in the responses of plants against cadmium toxicity, as observed for other abiotic stresses. Cd exposure can increase the levels of abscisic acid (ABA), jasmonate (JA), ethylene (ET), as well as salicylic acid (SA) (DalCorso et al. 2008). Cd has been shown to stimulate ET biosynthesis in several species of plants (Di Toppi and Gabbrielli 1999), but the molecular connection of ET biosynthesis with cadmium toxicity remains unknown (Di Toppi and Gabbrielli 1999). Synthesis of jasmonic acid and ethylene increases in peas under cadmium exposure (Rodríguez-Serrano et al. 2009). Salicylic acid increases cadmium-activated oxidative stress in *Arabidopsis*, as chlorophyll level substantially decreased with significant increase in lipid peroxidation in the wild-type seedlings, however, unchanged in SA-deficient lines after 5 days of cadmium treatment (Zawoznik et al. 2007). Exposure to cadmium enhanced the concentration of free salicylic acid in barley (*Hordeum vulgare*) roots (Metwally et al. 2003). It was concluded by scientists that salicylic acid attenuated cadmium toxicity via affecting other Cd detoxification pathways that are not yet fully understood. The importance of lipid-based signalling in plant defense against pathogens, mechanical stress, as well as herbivores is widely recognized. Oxylipins, especially those which belong to the family of jasmonate, are important signalling molecules for abiotic as well as biotic stress responses (Mithöfer et al. 2004). The interaction of jasmonic acid as well as oxylipin with heavy metal-derived signals has also been observed.

Despite the fact that pathogen or herbivore attacks or heavy metal exposure are very dissimilar stresses, plants have a consistent response: production as well as storage of secondary chemosensory substances (Mithöfer et al. 2004). Fatty acid hydroperoxide synthesis is mediated by ROS or catalyzed by enzymes. The presence of these molecules is important and serves as a starting point for the spontaneous or enzymatic synthesis of many oxylipins. Different biological responses induced by oxylipin can be facilitated through a clear structure–activity relationship (Farmer et al. 2003). Therefore, we may assume that the uncontrolled production of heavy metal oxylipin can trigger the secondary metabolism of plants through the formation of structurally similar compounds. If the architecture is not identical, it participates in defense responses against biological challenges (Mithöfer et al. 2004).

7.8 Transcription Factors and miRNAs

Using differential expression analysis, Fusco et al. (2005) identified many transcription factors (TFs) that may be sensitive to heavy metals. In pea (Romero-Puertas et al. 2007), *Arabidopsis* (Weber et al. 2006), and barley (Tamas et al. 2008) many different TFs have been reported to be induced, of which certain are constitutively upregulated in *A. helleri* or *T. caerulescens* (Weber et al. 2006; Van de Mortel et al. 2008). But, no information about their specific tasks is available. Heavy metal-responsive TFs, like other stress-related TFs, share the same signalling pathways and thus are activated via abiotic stimuli like cold, desiccation, SA, or hydrogen peroxide (Singh et al. 2002). Cd influences the ERF proteins expression member of the APETALA2 (AP2)/ethylene-responsive factor-binding protein (EREBP) family. They have been shown to bind to various promoters associated with pathogenesis as well as dehydration response factors (DRE motifs) (Singh et al. 2002; Weber et al. 2006). MYB4 (MYeloBlastosis protein) was also reported to be significantly induced under cadmium exposure in *A. thaliana* (Van de Mortel et al. 2008) but MYB48, MYB43, as well as MYB124 proteins were selectively stimulated through cadmium inside roots (Weber et al. 2006). In addition, MYB72 as well as bHLH100 TFs (both coiled-coil TFs) showed modified expression under Cd exposure, signifying their involvement in homeostasis of metal (Van de Mortel et al. 2008). MYB28 is highly inducted in *T. caerulescens* in Zn deficiency as well as greater cadmium levels and is implicated in the control of glucosinolate (GSL) production (Van de Mortel et al. 2008). GSL is a prominent form of sulfur accumulation and its production reacts to modifications in nutrient status as well as biotic and abiotic challenges (Hirai et al. 2007). OBF5, a bZIP family TF, joins to the glutathione S-transferase promoter to regulate its induction in a cadmium-dependent method (Suzuki et al. 2001). The TGA3 protein might also have function in regulating gene expression in response to cadmium exposure and biotic stress (Johnson et al. 2003). BjCdR15, an ortholog of TGA3, was recognized in *B. juncea* under short exposure to Cd (Fusco et al. 2005). This transcription factor regulates the induction of many metal transporters and is implicated in the long-distance transport of cadmium from root to shoot. Its over-induction in *A. thaliana* as well as tobacco improved tolerance as well as cadmium storage in shoots (Farinati et al. 2010). Another TF WRKY53 of WRKY family, is differentially expressed in Cd-exposed *T. caerulescens* plants. This gene is also altered by other environmental stresses like salinity or drought, and perhaps participate in the stress related signal transduction pathway regulating the activity of other TFs, rather than directly activating gene expression (Wei et al. 2008). Transcriptional as well as post-transcriptional regulation of gene is essential in response to exposure or deficiency of metal. On the other hand, the regulatory network of metal homeostasis has remained mostly unexplored. Recently, growing evidence has revealed that miRNAs, a large family of endogenous small RNAs (Carrington and Ambros 2003; Bartel 2004), have crucial roles in the modulation of gene expression, and they also function as key regulators in the alleviation of plant metal stresses (Sunkar and Zhu 2004; Lu et al. 2005). MiR393 and miR171 have been shown to be involved in alleviating Cd stress through the downregulation of target genes in *Brassica napus* and *Medicago truncatula* (Xie et al. 2007; Huang et al. 2010; Zhou et al. 2008). In addition, using standard sequencing methods, Huang et al. (2009) identified 19 new putative miRNAs in response to Cd. Ding et al. (2011) identified 19 Cd-responsive miRNAs in rice based on a miRNA microarray analysis, further validating six of them by quantitative real-time PCR (qPCR) and postulating the potential involvement of certain miRNAs in Cd tolerance in rice. Sub-miR604 regulates Cd tolerance by controlling lipid transfer protein (LTP) mRNA decay (Xie et al. 2007; Zhou et al. 2008) and has been assumed to target a WAK-like protein (Li et al. 2008). The study of miRNAs as well as their targets can give new information on plant responses to heavy metal stress (Gallego et al. 2012).

8 Hyperaccumulator Plants: A New Frontier of Plant Biotechnology

Hyperaccumulators are a type of plant that can store large quantities of hazardous metals in their tissues (Reeves and Baker 2000). Hyperaccumulation is a mechanism based on an internal super-tolerant process that prevents lethal amounts of accumulated metals and an efficient scavenging mechanism that allows impurities to be effectively absorbed (Salt 2006). About 400 metal hyperaccumulators have been reported so far (Eapen and D'Souza 2005). Majority of them are nickel and/or zinc hyperaccumulators and only some of known types are cadmium hyperaccumulators. The most common are *Thlaspi praecox*, *A. helleri, T. caerulescens*, and *Sedum alfredii* (Van de Mortel et al. 2008). *Thlaspi* species are polymetallic hyperaccumulators that store higher concentrations of zinc, cadmium, nickel, and lead (Mari et al. 2006), but *A. halleri* tolerates zinc, cadmium, and lead, and only excessively accumulates zinc and cadmium (Van Rossum et al. 2004). *S. alfredii* was previously found to be a Zn hyperaccumulator and shown to be a Cd hyperaccumulator as well (Zhou and Qiu 2005). Usually, the heavy metals accumulated in the roots of plants do not accumulate much, but hyperaccumulators are capable of transporting majority of the absorbed toxic elements to the shoots (Lasat et al. 1998). Therefore, metal transfer from root to shoot via xylem is an important factor in the hyperaccumulating phenotype. In this regard, it has recently been shown that root-to-shoot transport requires metal transporter HMA4. HMA4 is expressed in greater amounts in hyperaccumulating *A. halleri* than nontolerant *A. thaliana* (Hanikenne et al. 2008). At the molecular level, amino acids as well as organic acids are suggested to have function of heavy metal hyperaccumulation or tolerance (Sharma and Dietz 2006), however, there is no clear process of long-distance movement of metal linked to hyperaccumulation (Mari et al. 2006). Phytoremediation is a novel, cheap, and ecofriendly in situ technique that utilizes the capacity of plants to absorb heavy metals into their atmospheric tissues to restore damaged habitats (Alkorta et al. 2004). In this regard, hyperaccumulators act either as a direct pollutant or indirectly as a genetic resource to improve non-accumulative plants. In fact, ideal cultivars should have characteristics such as high biomass, fast growth rate, extensive and well-branched roots, and ease of harvesting. On the other hand, natural hyperaccumulators have less biomass and moderate growth rates. This limitation can be overpowered by moving the genetic potential accountable for overflow from hyperaccumulators to plants with pollution improving traits, thus creating accumulation potential and greater resistance to impurities. Poplar has recently emerged as a "genomics to sequence" model system and a promising alternative to solving plant pollution. It is known that a transgenic yellow poplar expressing a bacterial mercury-reducing enzyme has been engineered to increase the plant's ability to handle mercury (Rugh et al. 1998). Additionally, Indian mustard (*B. juncea*) is a promising target species due to its production of high biomass comparatively high metal storage, and well-established processing methods. Metal ions chelation movement of the metal or its complexes, and successive distribution into vacuoles are mechanisms by which biotechnology may help plants improve their metal processing capacity. contamination of these plants. For example, transfer of a single metal transporter, like HMA4, from *A. halleri* to *A. thaliana* stimulates metal rod loading in this non-accumulator (Hanikenne et al. 2008). The creation of transgenic yellow poplars expressing microbial mercury-reducing enzymes to improve mercury processing in plants is known (Rugh et al. 1998). Additionally, Indian mustard (*B. juncea*) is a viable target species due to its enormous biomass production, comparatively high metal accumulation, and well-established transformation. In general, biotechnology can help plants improve their ability to deal with pollutants by chelating metal ions, transporting the metal or its complexes, and then distributing them into the interstitial space. For example, one metal transporter, such as HMA4, transferred from *A. halleri* to *A. thaliana* enhanced the metal load in its unharvested shoots (Hanikenne et al. 2008). Another study showed that AtPCS1 gene expression improved cadmium as well as arsenic tolerance as well as storage in *B. juncea* (Gasic and Korban 2007) and tobacco (Pomponi et al. 2006). A BZIP transcription factor was recently shown to be differentially expressed in response to Cd treatment in *B. juncea* (Fusco et al. 2005), which increases cadmium storage and resistance in transgenic *Arabidopsis* as well as tobacco plants (Farinati et al. 2010). Furthermore, comparisons between hyperaccumulating and non-accumulating sister species (e.g., *A. helleri* and *A. thaliana*) indicate that hyperaccumulation may be related to gene copy number sequence mutations, and/or several expression levels of proteins that promote metal resistance (Plaza et al. 2007; Hanikenne et al. 2008). These results highlight that perhaps part of the genetic capacity for metal detoxification already exists in majority of plant species and these minor sequence modifications which affect both metal sensing as well as activation of suitable responses make the difference (DalCorso et al. 2008).

9 Conclusions

Plants have several mechanisms of action to deal with heavy metal (especially cadmium) toxicity. Studying different responses reveals unique mechanisms that help combat high-dose pollution. These systems (based on protein chelation, general metabolic reactions, or vacuole/cell wall sequestration) may or may not be present in economically important plant species and may affect plant products or plant waste. Various investigations have shown modified cadmium storage in plants engineered to express genes for heavy metal assimilation, translocation, or chelation. These results show the complexity of the biotechnology option to improve heavy metal tolerance in plants because plants overexpressing certain genes can express certain characteristics. a negative/unexpected point related to the host response to the transgene. The presence of "cross-homeostasis" was found to be an important determinant of the phenotype of plants with altered metal homeostasis gene expression. This cross-homeostasis is mainly because of cross-talk between the homeostatic processes of the various metals. Biotechnical methods are the most effective way to accumulate heavy metals in plants. Three components are particularly important for the development of effective engineering methods for the assimilation/accumulation of heavy metals in plants: (1) the use of certain stimulants, such as tissue-specific stimulants, which limit the transformation capacity of a specific tissue or organ; (2) integrated studies aimed at elucidating plant responses to a specific metal at the transcriptional, protein and metabolic levels; and (3) selecting a host for expression of a particular transgene. This last property is important because it is necessary to promote rapid growth and high biomass productivity. Further investigation of how different early and late plant species adapt and respond to cadmium will provide a specific process or perhaps a combination of molecular pathways that can be exploited to improve the reactivity of agricultural plants in polluted soil. In this regard, the combination of omics (transcriptomics, proteomics, and metabolomics) and advanced imaging techniques (especially mass spectrometry such as secondary ion mass spectrometry (SIMS) can provide a wealth of information covering many levels of biological complexity.

References

Abbas SZ, Rafatullah M, Ismail N, Lalung J (2014) Isolation, identification, and characterization of cadmium resistant Pseudomonas sp. M3 from industrial wastewater. J Waste Manag 2014:1–6. https://doi.org/10.1155/2014/160398

Abbas T, Rizwan M, Ali S, Adrees M, Zia-ur-Rehman M, Qayyum MF, Ok YS, Murtaza G (2017) Effect of biochar on alleviation of cadmium toxicity in wheat (Triticum aestivum L.) grown on Cd-contaminated saline soil. Environ Sci Pollut Res 25:25668–25680. https://doi.org/10.1007/s11356-017-8987-4

Abedi T, Mojiri A (2020) Cadmium uptake by wheat (Triticum aestivum L.): an overview. Plants 9: 500. https://doi.org/10.3390/plants9040500

Akinola MO, Ekiyoyo TA (2006) Accumulation of lead, cadmium and chromium in some plants cultivated along the bank of river Ribila at Odo-nla area of Ikorodu, Lagos state, Nigeria. J Environ Biol 27(3):597–599
Ali B, Gill RA, Yang S, Gill MB, Farooq MA, Liu D et al (2015) Regulation of cadmium-induced proteomic and metabolic changes by 5-aminolevulinic acid in leaves of Brassica napus L. PLoS One 10(4):e0123328
Alkorta I, Hernandez-Allica J, Becerril JM, Amezaga I, Albizu I, Garbisu C (2004) Recent findings on the phytoremediation of soils contaminated with environmentally toxic heavy metals and metalloids such as zinc, cadmium, lead, and arsenic. Rev Environ Sci Biotechnol 3:71–90
Azzi V, Kazpard V, Lartiges B, Kobeissi A, Kanso A, El Samrani AG (2017) Trace metals in phosphate fertilizers used in Eastern Mediterranean countries. Clean Soil Air Water 45:45
Baes CF, Mesmer RE (1976) The hydrolysis of cations. Krieger Publishing Company, Malabar, FL, p 489
Baker AJM, Brooks RR (1989) Terrestrial higher plants which hyperaccumulate metallic elements: a review of their distribution, ecology and phytochemistry. Biorecovery 1:81–126
Bartel DP (2004) MicroRNAs: genomics, biogenesis, mechanism, and function. Cell 116:281–297
Baruah N, Subham CM, Farooq M, Gogoi N (2019) Influence of heavy metals on seed germination and seedling growth of wheat, pea, and tomato. Water Air Soil Pollut 230:273–288. https://doi.org/10.1007/s11270-019-4329-0
Baryla A, Carrier P, Franck F, Coulomb C, Sahut C, Havaux M (2001) Leaf chlorosis in oilseed rape plants (Brassica napus) grown on cadmium-polluted soil: causes and consequences for photosynthesis and growth. Planta 212:696–709
Bashir S, Zhu J, Fu Q, Hu H (2018) Cadmium mobility, uptake and anti-oxidative response of water spinach (Ipomoea aquatic) under rice straw biochar, zeolite and rock phosphate as amendments. Chemosphere 194:579–587
Bazihizina N, Taiti C, Marti L, Rodrigo-Moreno A, Spinelli F, Giordano C et al (2014) Zn2+-induced changes at the root level account for the increased tolerance of acclimated tobacco plants. J Exp Bot 65(17):4931–4942
Becher M, Talke IN, Krall L, Kramer U (2004) Cross-species microarray transcript profiling reveals high constitutive expression of metal homeostasis genes in shoots of the zinc hyperaccumulator Arabidopsis halleri. Plant J 37:251–226
Begum N, Hu Z, Cai Q, Lou L (2019) Influence of PGPB inoculation on HSP70 and HMA3 gene expression in switchgrass under cadmium stress. Plants 8(11):504
Belon E, Boisson M, Deportes IZ, Eglin TK, Feix I, Bispo AO et al (2012) An inventory of trace elements inputs to French agricultural soils. Sci Total Environ 439:87–95
Benavides MP, Gallego SM, Tomaro ML (2005) Cadmium toxicity in plants. Braz J Plant Physiol. 17:21–34
Bernard C, Roosens N, Czernic P, Lebrun M, Verbruggen N (2004) A novel CPx-ATPase from the cadmium hyperaccumulator Thlaspi caerulescens. FEBS Lett. 569:140–148
Bigalke M, Ulrich A, Rehmus A, Keller A (2017) Accumulation of cadmium and uranium in arable soils in Switzerland. Environ Pollut 221:85–93
Blanvillain R, Kim JH, Wu S, Lima A, Ow DW (2009) Oxidative stress 3 is a chromatin-associated factor involved in tolerance to heavy metals and oxidative stress. Plant J 57(3):654–665. https://doi.org/10.1111/j.1365-313X.2008.03717.x
Blinda A, Koch B, Ramanjulu S, Dietz KJ (1997) De novo synthesis and accumulation of apoplastic proteins in leaves of heavy metal exposed barley seedlings. Plant Cell Environ 20:969–981
Bojórquez C, Frías Espericueta MG, Voltolina D (2016) Removal of cadmium and lead by adapted strains of Pseudomonas aeruginosa and Enterobacter cloacae. Rev Int Contam Ambient 32(4): 407–412
Buchanan BB, Gruissem W, Jones RL (2000) Biochemistry and molecular biology of plants. American Society of Plant Physiologists, Rockville, Maryland
Buekers J (2007) Fixation of cadmium, copper, nickel and zinc in soil: kinetics, mechanisms and its effect on metal bioavailability

Burkhead JL, Gogolin-Reynolds KA, Abdel-Ghany SE, Cohu CM, Pilon M (2009) Copper homeostasis. New Phytol 182:799–816. https://doi.org/10.1111/j.1469-8137.2009.02846.x
Carrington JC, Ambros V (2003) Role of microRNAs in plant and animal development. Science 301:336
Cazale AC, Clemens S (2001) Arabidopsis thaliana expresses a second functional phytochelatin synthase. FEBS Lett 507:215–219
Chaffei C, Pageau K, Suzuki A, Gouia H, Ghorbel HM, MascalauxDaubresse C (2004) Cadmium toxicity induced changes in nitrogen management in Lycopersicon esculentum leading to a metabolic safeguard through an amino acid storage strategy. Plant Cell Physiol. 45:1681–1693
Chang YC, Zouari M, Gogorcena Y, Lucena JJ, Abadía J (2003) Effects of cadmium and lead on ferric chelate reductase activities in sugar beet roots. Plant Physiol Biochem 41(11–12): 999–1005
Chaoui A, Mazhoudi S, Ghorbal MH, El Ferjani E (1997) Cadmium and zinc induction of lipid peroxidation and effects on antioxidant enzyme activities in bean (Phaseolus vulgaris L.). Plant Sci 127(2):139–147
Chellaiah ER (2018) Cadmium (heavy metals) bioremediation by Pseudomonas aeruginosa: a mini-review. Appl Water Sci 8:154. https://doi.org/10.1007/s13201-018-0796-5
Chen C-T, Chen T-H, Lo K-F, Chiu C-Y (2004) Effects of proline on copper transport in rice seedlings under excess copper stress. Plant Sci 166:103–111. https://doi.org/10.1016/j.plantsci.2003.08.015
Chen B, Stein AF, Castell N, Gonzalez-Castanedo Y, De La Campa AS, De La Rosa JD (2016) Modeling and evaluation of urban pollution events of atmospheric heavy metals from a large Cu-smelter. Sci Total Environ 539:17–25
Chiang HC, Lo JC, Yeh QC (2006) Genes associated with heavy metal tolerance and accumulation in Zn/Cd hyperaccumulator Arabidopsis halleri: a genomic survey with cDNA microarray. Environ Sci Technol 40:6792–6798
Chmielowska-Bąk J, Gzyl J, Rucińska-Sobkowiak R, Arasimowicz-Jelonek M, Deckert J (2014) The new insights into cadmium sensing. Front Plant Sci 5:245
Choppala G, Saifullah BN, Bibi S, Iqbal M, Rengel Z, Ok YS (2014) Cellular mechanisms in higher plants governing tolerance to cadmium toxicity cellular mechanisms in higher plants governing tolerance. Crit Rev Plant Sci 33:1–18
Clemens S (2006) Evolution and function of phytochelatin synthases. J Plant Physiol 163:319–332
Cloquet C, Carignan J, Libourel G, Sterckeman T, Perdrix E (2006) Tracing source pollution in soils using cadmium and lead isotopes. Environ Sci Technol 40(8):2525–2530
Cobbett CS (2000) Phytochelatins and their roles in heavy metal detoxification. Plant Physiol. 123: 825–832
Cobbett CS, Goldsbrough P (2002) Phytochelatins and metallothioneins: roles in heavy metal detoxification and homeostasis. Annu Rev Plant Physiol Plant Mol Biol 53:159–182
Cosio C, Martinoia E, Keller C (2004) Hyperaccumulation of cadmium and zinc in Thlaspi caerulescens and Arabidopsis halleri at the leaf cellular level. Plant Physiol. 134:716–725
Cotton FA, Wilkinson G (1966) Advanced inorganic chemistry: a comprehensive text, 2nd edn. John Wilen Publishers, New York, p 1135
Curie C, Alonso JM, Le Jean M, Ecker JR, Briat JF (2000) Involvement of NRAMPl from Arabidopsis thaliana in iron transport. Biochem J 347:749–755
DalCorso G, Farinati S, Maistri S, Furini A (2008) How plants cope with cadmium: staking all on metabolism and gene expression. J Integr Plant Biol. 50:1268–1280. https://doi.org/10.1111/j.17447909.2008.00737.x
De Sousa CAF, Sodek L (2003) Alanine metabolism and alanine aminotransferase activity in soybean (Glycine max) during hypoxia of the root system and subsequent return to normoxia. Environ Exp Bot 50(1):1–8. https://doi.org/10.1016/S0098-8472(02)00108-9
Delhaize EP, Ryan R (1995) Aluminium toxicity and tolerance in plants. Plant Physiol 107:315–321

Di Toppi LS, Gabbrielli R (1999) Response to cadmium in higher plants. Environ Exp Bot 41(2): 105–130

Dinakar N, Nagajyothi PC, Suresh S, Damodharam T, Suresh C (2009) Cadmium induced changes on proline, antioxidant enzymes, nitrate and nitrite reductases in Arachis hypogaea L. J Environ Biol 30(2):289–294

Ding YF, Chen Z, Zhu C (2011) Microarray-based analysis of cadmium responsive microRNAs in rice (Oryza sativa). J Exp Bot. https://doi.org/10.1093/jxb/err046

Dong J, Wu F, Zhang G (2006) Influence of cadmium on antioxidant capacity and four microelement concentrations in tomato seedlings (Lycopersicon esculentum). Chemosphere 64:1659–1666

Eapen S, D'Souza SF (2005) Prospects of genetic engineering of plants for phytoremediation of toxic metals. Biotechnol Adv 23:97–114

Ehsan S, Ali S, Noureen S, Mahmood K, Farid M, Ishaque W et al (2014) Citric acid assisted phytoremediation of cadmium by Brassica napus L. Ecotoxicol Environ Saf 106:164–172

El Rasafi T, Oukarroum A, Haddioui A, Song H, Kwon EE, Bolan N, Tack FMG, Sebastian A, Prasad MNV, Rinklebe J (2022) Cadmium stress in plants: a critical review of the effects, mechanisms, and tolerance strategies. Crit Rev Environ Sci Technol 52(5):675–726

El-Beltagi HS, Mohamed AA, Rashed MM (2010) Response of antioxidative enzymes to cadmium stress in leaves and roots of radish (Raphanus sativus L.). Not Sci Biol 2(4):76–82

Fahr M, Laplaze L, Bendaou N, Hocher V, Mzibri M, El Bogusz D, Smouni A (2013) Effect of lead on root growth. Front Plant Sci 4:175. https://doi.org/10.3389/fpls.2013.00175

Farinati S, DalCorso G, Varotto S, Furini A (2010) The Brassica juncea BjCdR15, an ortholog of Arabidopsis TGA3, is a regulator of cadmium uptake, transport and accumulation in shoots and confers cadmium tolerance in transgenic plants. New Phytol 185:964–978

Farmer EE, Alméras E, Krishnamurthy V (2003) Jasmonates and related oxylipins in plant responses to pathogenesis and herbivory. Curr Opin Plant Biol 6:372–378

Farooq MA, Ali S, Hameed A, Bharwana SA, Rizwan M, Ishaque W et al (2016) Cadmium stress in cotton seedlings: physiological, photosynthesis and oxidative damages alleviated by glycine betaine. South Afr J Bot 104:61–68

Farooq M, Ullah A, Usman M, Siddique KH (2020) Application of zinc and biochar help to mitigate cadmium stress in bread wheat raised from seeds with high intrinsic zinc. Chemosphere 260: 127652

Feleafel MN, Mirdad ZM (2013) Hazard and effects of pollution by lead on vegetable crops. J Agric Environ Ethics 26:547–567. https://doi.org/10.1007/s10806-012-9403-1

Filatov V, Dowdle J, Smirnoff N, Ford-Lloyd B, Newbury HJ, Macnair MR (2006) Comparison of gene expression in segregating families identifies genes and genomic regions involved in a novel adaptation, zinc hyperaccumulation. Mol Ecol 15:3045–3059

Fontes RLF, Pereira JMN, Neves JCL (2014) Uptake and translocation of Cd and Zn in two lettuce cultivars. An Acad Bras Cienc 86(2):907–922. https://doi.org/10.1590/0001-37652014117912

Foy CD, Chaney RL, White MG (1978) The physiology of metal toxicity in plants. Ann Rev Plant Physiol 29:511–566

Fusco N, Micheletto L, Dal Corso G, Borgato L, Furini A (2005) Identification of cadmium-regulated genes by cDNA-AFLP in the heavy metal accumulator Brassica juncea L. J Exp Bot 56:3017–3027

Fusconi A, Gallo C, Camusso W (2007) Effects of cadmium on root apical meristems of Pisum sativum L.: cell viability, cell proliferation and microtubule pattern as suitable markers for assessment of stress pollution. Mutat Res Genet Toxicol Environ Mutagen 632(1–2):9–19

Gallego SM, Pena LB, Barcia RA, Azpilicueta CE, Iannone MF, Rosales EP et al (2012) Unravelling cadmium toxicity and tolerance in plants: insight into regulatory mechanisms. Environ Exp Bot 83:33–46

Garcia-Hernandez M, Murphy A, Taiz L (1998) Metallothioneins 1 and 2 have distinct but overlapping expression patterns in Arabidopsis. Plant Physiol. 118:387–397

Gasic K, Korban SS (2007) Transgenic Indian mustard (Brassica juncea) plants expressing an Arabidopsis phytochelatin synthase (AtPCS1) exhibit enhanced As and Cd tolerance. Plant Mol Biol 64:361–369

Gill SS, Tuteja N (2010) Reactive oxygen species and antioxidant machinery in abiotic stress tolerance in crop plants. Plant Physiol Biochem 48(12):909–930

Gong JM, Lee DA, Schroeder JI (2003) Long-distance root-to-shoot transport of phytochelatins and cadmium in Arabidopsis. Proc Natl Acad Sci USA 100:10118–10123

Grant CA (2011) Influence of phosphate fertilizer on cadmium in agricultural soils and crops. Pedologist 3:143–155

Gratão PL, Monteiro CC, Rossi ML, Martinelli AP, Peres LE, Medici LO et al (2009) Differential ultrastructural changes in tomato hormonal mutants exposed to cadmium. Environ Exp Bot 67(2):387–394

Guo T-R, Zhang G-P, Zhou M-X, Wu F-B, Chen J-X (2007) Influence of aluminum and cadmium stresses on mineral nutrition and root exudates in two barley cultivars. Pedosphere 17(4): 505–512

Gupta N, Ram H, Kumar B (2016) Mechanism of Zinc absorption in plants: uptake, transport, translocation and accumulation. Rev Environ Sci Biotechnol 15(1):89–109. https://doi.org/10.1007/s11157-016-9390-1

Gupta N, Yadav KK, Kumar V, Kumar S, Chadd RP, Kumar A (2019) Trace elements in soil-vegetables interface: Translocation, bioaccumulation, toxicity and amelioration—a review. Sci Total Environ 651(Pt 2):2927–2942. https://doi.org/10.1016/j.scitotenv.2018.10.047

Gutsch A, Sergeant K, Keunen E, Prinsen E, Guerriero G, Renaut J, Hausman JF, Cuypers A (2019) Does long-term cadmium exposure influence the composition of pectic polysaccharides in the cell wall of Medicago sativa stems? BMC Plant Biol 19:271. https://doi.org/10.1186/s12870-019-1859-y

Ha SB, Smith AP, Howden R, Dietrich WM, Bugg S, O'Connell MJ et al (1999) Phytochelatin synthase genes from Arabidopsis and the yeast Schizosaccharomyces pombe. Plant Cell 11: 1153–1164

Haider FU, Liqun C, Coulter JA, Cheema SA, Wu J, Zhang R et al (2021) Cadmium toxicity in plants: Impacts and remediation strategies. Ecotoxicol Environ Saf 211:111887

Hajeb P, Sloth JJ, Shakibazadeh SH, Mahyudin NA, Afsah-Hejri L (2014) Toxic elements in food: occurrence, binding, and reduction approaches. Compr Rev Food Sci Food Saf 13(4):457–472

Hanikenne M, Talke IN, Haydon MJ, Lanz C, Nolte A, Motte P et al (2008) Evolution of metal hyperaccumulation required cis-regulatory changes and triplication of HMA4. Nature 453:391–395

Hart JJ, Welch RM, Norvell WA, Kochian LV (2002) Transport interactions between cadmium and zinc in roots of bread and durum wheat seedlings. Physiol Plant 116:73–78

Hasan SA, Ali B, Hayat S, Ahmad A (2007) Cadmium induced changes in the growth and carbonic anhydrase activity of chickpea. Turk J Biol 31:137–140

Hasan SA, Fariduddin Q, Ali B, Hayat S, Ahmad A (2009) Cadmium: toxicity and tolerance in plants. J Environ Biol 30(2):165–174

Hashem HA (2014) Cadmium toxicity induces lipid peroxidation and alters cytokinin content and antioxidant enzyme activity. Botany 92:1–7. https://doi.org/10.1139/cjb-2013-0164

Hayat S, Hasan SA, Ahmad A (2011) Growth, nitrate reductase activity and antioxidant system in cadmium stressed tomato (Lycopersicon esculentum Mill.) cultivars. BASE 15

Hayat S, Hayat Q, Alyemeni MN, Wani AS, Pichtel J, Ahmad A (2012) Role of proline under changing environments: a review. Plant Signal Behav 7:1456–1466. https://doi.org/10.4161/psb.21949

Heiss S, Wachter A, Bogs J, Cobbett C, Rausch T (2003) Phytochelatin synthase (PCS) protein is induced in Brassica juncea leaves after prolonged Cd exposure. J Exp Bot 54:1833–1839

Hermans C, Chen J, Coppens F, Inzé D, Verbruggen N (2011) Low magnesium status in plants enhances tolerance to cadmium exposure. New Phytol 192(2):428–436

Heyno E, Klose C, Krieger-Liszkay A (2008) Origin of cadmium-induced reactive oxygen species production: mitochondrial electron transfer versus plasma membrane NADPH oxidase. New Phytol 179(3):687–699

Hinsinger P, Plassard C, Jaillard B (2005) Rhizosphere: a new frontier for soil biogeochemistry. J Geochem Explor 88(1-3):210–213. https://doi.org/10.1016/j.gexplo.08.041

Hirai MY, Sugiyama K, Sawada Y, Tohge T, Obayashi T, Suzuki A, Araki R, Sakurai N, Suzuki H, Aoki K, Goda H (2007) Omics-based identification of Arabidopsis Myb transcription factors regulating aliphatic glucosinolate biosynthesis. Proc Natl Acad Sci 104(15):6478–6483

Huang SQ, Peng J, Qiu CX, Yang ZM (2009) Heavy metal-regulated new microRNAs from rice. J Inorg Biochem 103(2):282–287

Huang SQ, Xiang AL, Che LL, Chen S, Li H, Song JB, Yang ZM (2010) A set of miRNAs from Brassica napus in response to sulphate deficiency and cadmium stress. Plant Biotechnol J 8: 887–899

Huang L, Li WC, Tam NFY, Ye Z (2019) Effects of root morphology and anatomy on cadmium uptake and translocation in rice (Oryza sativa L.). J Environ Sci 75:296–306. https://doi.org/10.1016/j.jes.2018.04.005

Hussain MM, Saeed A, Khan AA, Javid S, Fatima B (2015) Differential responses of one hundred tomato genotypes grown under cadmium stress. Genet Mol Res 14:13162–13171

Hussain B, Ashraf MN, Abbas A, Li J, Farooq M (2021) Cadmium stress in paddy fields: effects of soil conditions and remediation strategies. Sci Total Environ 754:142188

Islam MM, Hoque MA, Okuma E, Banu MNA, Shimoishi Y, Nakamura Y, Murata Y (2009) Exogenous proline and glycine betaine increase antioxidant enzyme activities and confer tolerance to cadmium stress in cultured tobacco cells. J Plant Physiol 166(15):1587–1597

Ismael MA, Elyamine AM, Moussa MG, Cai M, Zhao X, Hu C (2018) Cadmium in plants: uptake, toxicity, and its interactions with selenium fertilizers. Metallomics 11(2):255–277

Jali P, Pradhan C, Das AB (2016) Effects of cadmium toxicity in plants: a review article. Scholars Acad J Biosci 4:1074–1081. https://doi.org/10.21276/sajb.2016.4.12.3

Jasinski M, Sudre D, Schansker G, Schellenberg M, Constant S, Martinoia E, Bovet L (2008) AtOSA1, a member of the Abc1-like family, as a new factor in cadmium and oxidative stress response. Plant Physiol 147:719–731

Jinadasa N, Collins D, Holford P, Milham PJ, Conroy JP (2016) Reactions to cadmium stress in a cadmium-tolerant variety of cabbage (Brassica oleracea L.): is cadmium tolerance necessarily desirable in food crops? Environ Sci Pollut Res 23:5296–5306

Johnson C, Boden E, Arias J (2003) Salicylic acid and NPR1 induce the recruitment of trans-activating TGA factors to a defense gene promoter in Arabidopsis. Plant Cell 15:1846–1858

Jonak C, Nakagami H, Hirt H (2004) Heavy metal stress. Activation of distinct mitogen-activated protein kinase pathways by copper and cadmium. Plant Physiol 136:3276–3283

Karina BB, Benavides MP, Gallego SM, Tomaro ML (2003) Effect of cadmium stress on nitrogen metabolism in nodules and roots of soybean plants. Funct Plant Biol 30:57–64

Kaznina NM, Titov AF, Batova YV, Laidinen GF (2014) The resistance of plants Setaria veridis (L.) Beauv. to the influence of cadmium. Biol Bull 41:428–433. https://doi.org/10.1134/s1062359014050045

Khan MA, Khan S, Khan A, Alam M (2017) Soil contamination with cadmium, consequences and remediation using organic amendments. Sci Total Environ 601:1591–1605

Kieffer P, Schröder P, Dommes J, Hoffmann L, Renaut J, Hausman JF (2009) Proteomic and enzymatic response of poplar to cadmium stress. J Proteomics 72(3):379–396

Kim DY, Bovet L, Kushnir S, Noh EW, Martinoia E, Lee Y (2006) AtATM3 is involved in heavy metal resistance in Arabidopsis. Plant Physiol. 140:922–932

Kim DY, Bovet L, Maeshima M, Martinoia E, Lee Y (2007) The ABC transporter AtPDR8 is a cadmium extrusion pump conferring heavy metal resistance. Plant J. 50:207–218

Kinay A (2018) Effects of cadmium on nicotine, reducing sugars and phenolic contents of Basma tobacco variety. Fresenius Environ Bull 27:9195–9202

Knappe F, Möhler S, Ostermayer A, Lazar S, Kaufmann C (2008) Comparative evaluation of substance inputs into soils via different input pathways. Umweltforschungsplan des Bundesministeriums für Umwelt, Naturschutz und Reaktorsicherheit

Kobae Y, Sekino T, Yoshioka H, Nakagawa T, Martinoia E, Maeshima M (2006) Loss of AtPDR8, a plasma membrane ABC transporter of Arabidopsis thaliana, causes hypersensitive cell death upon pathogen infection. Plant Cell Physiol 47:309–318

Korenkov V, Hirschi K, Weld JDC, Wagner GJ (2007) Enhancing tonoplast Cd/H antiport activity increases Cd, Zn, and Mn tolerance, and impacts root/shoot Cd partitioning in *Nicotiana tabacum L.* Planta 226(6):1379–1387. https://doi.org/10.1007/s00425-007-0577-0

Korshunova YO, Eide D, Clark WG, Guerinot ML, Pakrasi HB (1999) The IRT1 protein from Arabidopsis thaliana is a metal transporter with a broad substrate range. Plant Mol Biol 40:37–44

Kovalchuk I, Titov V, Hohn B, Kovalchuk O (2005) Transcriptome profiling reveals similarities and differences in plant responses to cadmium and lead. Mutat Res 570:149–161

Kramer U, Talke IN, Hanikenne M (2007) Transition metal transport. FEBS Lett 581:2263–2272

Kranner I, Colville L (2011) Metals and seeds: biochemical and molecular implications and their significance for seed germination. Environ Exp Bot 72(1):93–105

Krantev A, Yordanova R, Janda T, Szalai G, Popova L (2008) Treatment with salicylic acid decreases the effect of cadmium on photosynthesis in maize plants. J Plant Physiol. 165:920–931

Kubier A, Wilkin RT, Pichler T (2019) Cadmium in soils and groundwater: a review. Appl Geochem 108:104388. https://doi.org/10.1016/j.apgeochem.2019.104388

Kudoyarova GR, Dodd IC, Veselov DS, Rothwell SA, Veselov YS (2015) Common and specific responses to availability of mineral nutrients and water. J Exp Bot 66(8):2133–2144

Küpper H, Parameswaran A, Leitenmaier B, Trtílek M, Šetlík I (2007) Cadmium-induced inhibition of photosynthesis and long-term acclimation to cadmium stress in the hyperaccumulator Thlaspi caerulescens. New Phytol 175(4):655–674

Laghlimi M, Baghdad B, Hadi HE, Bouabdli A (2015) Phytoremediation mechanisms of heavy metal contaminated soils: a review. Open J Ecol 05(08):375–388. https://doi.org/10.4236/oje.2015.58031

Lang M, Zhang Y, Chai T (2005) Identification of genes up-regulated in response to Cd exposure in Brassica juncea L. Gene 363:151–158

Lasat MM, Baker A, Kochian LV (1998) Altered Zn compartmentation in the root symplasm and stimulated Zn absorption into the leaf as mechanisms involved in Zn hyperaccumulation in Thlaspi caerulescens. Plant Physiol. 118:875–883

Lata S, Kaur HP, Mishra T (2019) Cadmium bioremediation: a review. Int J Pharm Sci Res 10: 4120–4128

Laws EA (1993) Aquatic pollution: An introductory text, 2nd edn. John Wilen Publishers, New York, p 11

Le GH, Philippe F, Domon J-M, Gillet F, Pelloux J, Rayon C (2015) Cell wall metabolism in response to abiotic stress. Plants 4:112–166

Lee S, Moon JS, Domier LL, Korban SS (2002) Molecular characterization of phytochelatin synthase expression in transgenic Arabidopsis. Plant Physiol Biochem 40:727–773

Lee S, Moon JS, Ko TS, Petros D, Goldsbrough PB, Korban SS (2003) Overexpression of Arabidopsis phytochelatin synthase paradoxically leads to hypersensitivity to cadmium stress. Plant Physiol 131:656–663

Lee K, Bae DW, Kim SH, Han HJ, Liu X, Park HC et al (2010) Comparative proteomic analysis of the short-term responses of rice roots and leaves to cadmium. J Plant Physiol 167(3):161–168

Levitt J (1972) Salts and other stress. In: Responses of plants to environmental stresses. Academic Press, pp 489–530

Li H, Zhou SY, Zhao WS, Su SC, Peng YL (2008) A novel wall-associated receptor-like protein kinase gene, OsWAK1, plays important roles in rice blast disease resistance. Plant Mol Biol 69: 337–346

Lin R, Wang X, Luo Y, Du W, Guo H, Yin D (2007) Effects of soil cadmium on growth, oxidative stress and antioxidant system in wheat seedlings (Triticum aestivum L.). Chemosphere 69:89–98

Liu J, Kapron CM (2010) Differential induction of MAP kinase signalling pathways by cadmium in primary cultures of mouse embryo limb bud cells. Reprod Toxicol 29:286–291

Liu J, Cao C, Wong M, Zhang Z, Chai Y (2010) Variations between rice cultivars in iron and manganese plaque on roots and the relation with plant cadmium uptake. J Environ Sci 22(7): 1067–1072

Lopez FB, Barclay GF (2017) Plant anatomy and physiology. In: Badal S, Delgoda R (eds) Pharmacognosy. Academic Press, Boston, pp 45–60. https://doi.org/10.1016/B978-0-12-802104-0.00004-4

López-Millán AF, Ellis DR, Grusak MA (2004) Identification and characterization of several new members of the ZIP family of metal ion transporters in Medicago truncatula. Plant Mol Biol 54: 583–596

Lu S, Sun YH, Shi R, Clark C, Li L, Chiang VL (2005) Novel and mechanical stress-responsive microRNAs in Populus trichocarpa that are absent from Arabidopsis. Plant Cell 17:2186–2203

Lu Z, Zhang Z, Su Y, Liu C, Shi G (2013) Cultivar variation in morphological response of peanut roots to cadmium stress and its relation to cadmium accumulation. Ecotoxicol Environ Saf 91: 147–155

Lux A, Martinka M, Vaculık M, White PJ (2011) Root responses to cadmium in the rhizosphere: a review. J Exp Bot 62(1):21–37. https://doi.org/10.1093/jxb/erq281

Mahajan P, Kaushal J (2018) Role of phytoremediation in reducing cadmium toxicity in soil and water. J Toxicol 2018:1–16

Małecka A, Piechalak A, Morkunas I, Tomaszewska B (2008) Accumulation of lead in root cells of Pisum sativum. Acta Physiol Plant 30:629–637. https://doi.org/10.1007/s11738-008-0159-1

Manousaki E, Kalogerakis N (2009) Phytoextraction of Pb and Cd by the Mediterranean saltbush (AtripLex halimus L.): metal uptake in relation to salinity. Environ Sci Pollut Res Int 16(7): 844–854. https://doi.org/10.1007/s11356-009-0224-3

Mari S, Gendre D, Pianelli K, Ouerdane L, Lobinski R, Briat JF et al (2006) Root-to-shoot long-distance circulation of nicotianamine and nicotianamine-nickel chelates in the metal hyperaccumulator Thlaspi caerulescens. J Exp Bot 57:4111–4122

Markovska YK, Gorinova NI, Nedkovska MP, Miteva KM (2009) Cadmium-induced oxidative damage and antioxidant responses in Brassica juncea plants. Biol Plant 53:151–154

Matusik J, Bajda T, Manecki M (2008) Immobilization of aqueous cadmium by addition of phosphates. J Hazard Mater 152(3):1332–1339

Mench M, Martin E (1991) Mobilization of cadmium and other metals from two soils by root exudates of Zea mays L., Nicotiana tabacum L and Nicotiana rustica L. Plant Soil 132:187–196

Metwally A, Finkemeier I, Georgi M, Dietz KJ (2003) Salicylic acid alleviates the cadmium toxicity in barley seedling. Plant Physiol 132:272–281

Metwally A, Safronova VI, Belimov AA, Dietz KJ (2005) Genotypic variation of the response to cadmium toxicity in Pisum sativum L. J Exp Bot 56:167–178

Meychik N, Yermakov I (2001) Ion exchange properties of plant root cell walls. Plant Soil 234: 181–193

Milone MT, Sgherri C, Clijsters H, Navari-Izzo F (2003) Antioxidative responses of wheat treated with realistic concentration of cadmium. Environ Exp Bot 50(3):265–276

Mithöfer A, Schulze B, Boland W (2004) Biotic and heavy metal stress response in plants: evidence for common signals. FEBS Lett 566:1–5

Mittra B, Sharma S, Das AB, Henry SL, Das TK, Ghosh P et al (2008) A novel cadmium induced protein in wheat: characterization and localization in root tissue. Biol Plant 52:343–346

Miyadate H, Adachi S, Hiraizumi A, Tezuka K, Nakazawa N, Kawamoto T et al (2011) OsHMA3, a P1B-type of ATPase affects root-to-shoot cadmium translocation in rice by mediating efflux into vacuoles. New Phytol 189(1):190–199

Montanini B, Blaudez D, Jeandroz S, Sanders D, Chalot M (2007) Phylogenetic and functional analysis of the cation diffusion facilitator (CDF) family: improved signature and prediction of substrate specificity. BMC Genomics 8:107

Mouron SA, Grillo CA, Dulout FN, Golijow CD (2004) A comparative investigation of DNA strand breaks, sister chromatid exchanges and K-Ras Gene mutations induced by cadmium salts in cultured human cells. Mutat Res 568(1):221–231. https://doi.org/10.1016/j.mrfmmm.2004.08.012

Nagajyoti PC, Lee KD, Sreekanth TVM (2010) Heavy metals, occurrence and toxicity for plants: a review. Environ Chem Lett 8:199–216

Nanjo Y, Nouri MZ, Komatsu S (2011) Quantitative proteomic analyses of crop seedlings subjected to stress conditions; a commentary. Phytochemistry 72(10):1263–1272

Nazar R, Iqbal N, Masood A, Khan MIR, Syeed S, Khan NA (2012) Cadmium toxicity in plants and role of mineral nutrients in its alleviation. Am J Plant Sci 3:1476–1489. https://doi.org/10.4236/ajps.2012.310178

Nevo Y, Nelson N (2006) The NRAMP family of metal-ion transporters. Biochim Biophys Acta 1763:609–620

Noor W, Umar S, Mir MY, Shah D, Majeed G, Hafeez S et al (2018) Effect of cadmium on growth, photosynthesis and nitrogen metabolism of crop plants. IBM J Res Dev 18:100–106

Obata H, Inoue N, Umebayashi M (1996) Effect of Cd on plasma membrane ATPase from plant roots differing in tolerance to Cd. Soil Sci Plant Nutr 42(2):361–366

Osakabe Y, Osakabe K, Shinozaki K, Tran LSP (2014) Response of plants to water stress. Front Plant Sci 5. https://doi.org/10.3389/fpls.2014.00086

Pan A, Yang M, Tie F, Li L, Chen Z, Ru B (1994) Expression of mouse metallothionein-I gene confers cadmium resistance in transgenic tobacco plants. Plant Mol Biol 24:341–351

Pantazis V, Kalavrouziotis I, Deligiannakis I (2007) Cu-Zn accumulation on soil plant system irrigated with wastewater. In: Proceedings in IWA facing sludge diversities: challenges, risks and opportunities, Antalya, 28–30 March 2007

Papoyan A, Kochian LV (2004) Identification of Thlaspi caerulescens genes that may be involved in heavy metal hyperaccumulation and tolerance. Characterization of a novel heavy metal transporting ATPase. Plant Physiol 136:3814–3823

Papoyan A, Pineros M, Kochian LV (2007) Plant Cd2+ and Zn2+ status effects on root and shoot heavy metal accumulation in Thlaspi caerulescens. New Phytol 175:51–58

Perfus-Barbeoch L, Leonhardt N, Vavaddeur A, Forestier C (2002) Heavy metal toxicity: cadmium permeates through calcium channels and disturbs the plant water status. Plant J 32:539–548

Pinto AP, Mota AD, De Varennes A, Pinto FC (2004) Influence of organic matter on the uptake of cadmium, zinc, copper and iron by sorghum plants. Sci Total Environ 326(1–3):239–247

Plaza S, Tearall KL, Zhao FJ, Buchner P, McGrath SP, Hawkesford MJ (2007) Expression and functional analysis of metal transporter genes in two contrasting ecotypes of the hyperaccumulator Thlaspi caerulescens. J Exp Bot 58:1717–1172

Pomponi M, Censi V, Di Girolamo V, De Paolis A, Sanita di Toppi L, Aromolo R et al (2006) Overexpression of Arabidopsis phytochelatin synthase in tobacco plants enhances Cd2+ tolerance and accumulation but not translocation to the shoot. Planta 223:180–190

Popova LP, Maslenkova LT, Ivanova AP, Stoinova Z (2012) Role of salicylic acid in alleviating heavy metal stress. In: Ahmad P, Prasad MNV (eds) Environmental adaptations and stress tolerance of plants in the era of climate change. Springer, New York, pp 441–466

Prasad MNV (1995) Cadmium toxicity and tolerance in vascular plants. Environ Exp Bot 35:525–545

Qin S, Liu H, Nie Z, Rengel Z, Gao W, Li C, Peng Z (2020) Toxicity of cadmium and its competition with mineral nutrients for uptake by plants: a review. Pedosphere 30:168–180

Rafiq MT, Aziz R, Yang X, Xiao W, Stoffella PJ, Saghir A et al (2014) Phytoavailability of cadmium (Cd) to Pak choi (Brassica chinensis L.) grown in Chinese soils: a model to evaluate the impact of soil Cd pollution on potential dietary toxicity. PLoS One 9(11):e111461

Rascio N, Navari-Izzo F (2011) Heavy metal hyperaccumulating plants: how and why do they do it? And what makes them so interesting? Plant Sci 180(2):169–181

Rascio N, Dalla Vecchia F, La Rocca N, Barbato R, Pagliano C, Raviolo M et al (2008) Metal accumulation and damage in rice (cv. Vialone nano) seedlings exposed to cadmium. Environ Exp Bot 62(3):267–278

Rauser WE (1999) Structure and function of metal chelators produced by plants: the case for organic acids, amino acids, phytin and metallothioneins. Cell Biochem Biophys 31:19–48

Rea PA (2007) Plant ATP-binding cassette transporters. Annu Rev Plant Biol 58:347–375

Redjala T, Zelko I, Sterckeman T, Legue V, Lux A (2011) Relationship between root structure and root cadmium uptake in maize. Environ Exp Bot 71(2):241–248. https://doi.org/10.1016/j.envexpbot.2010.12.010

Reeves RD, Baker AJM (2000) Metal-accumulating plants. In: Raskin I, Ensley BD (eds) Phytoremediation of toxic metals: using plants to clean up the environment. Wiley, New York, pp 193–229

Rehman MZU, Rizwan M, Ghafoor A, Naeem A, Ali S, Sabir M, Qayyum MF (2015) Effect of inorganic amendments for in situ stabilization of cadmium in contaminated soils and its phyto-availability to wheat and rice under rotation. Environ Sci Pollut Res 22:16897–16906

Rivetta A, Negrini N, Cocucci M (1997) Involvement of Ca2+-calmodulin in Cd2+ toxicity during the early phases of radish (Raphanus sativus L.) seed germination. Plant Cell Environ 20:600–608

Rizwan M, Meunier JD, Miche H, Keller C (2012) Effect of silicon on reducing cadmium toxicity in durum wheat (Triticum turgidum L. cv. Claudio W.) grown in a soil with aged contamination. J Hazard Mater 209:326–334

Rizwan M, Ali S, Abbas T, Zia-ur-Rehman M, Hannan F, Keller C et al (2016) Cadmium minimization in wheat: a critical review. Ecotoxicol Environ Saf 130:43–53

Rizwan M, Ali S, Adrees M, Ibrahim M, Tsang DCW, Zia-Ur-Rehman M, Zahir ZA, Rinklebe J, Tack FMG, Ok YS (2017) A critical review on effects, tolerance mechanisms and management of cadmium in vegetables. Chemosphere 182:90–105

Rochayati S, Du Laing G, Rinklebe J, Meissner R, Verloo M (2011) Use of reactive phosphate rocks as fertilizer on acid upland soils in Indonesia: accumulation of cadmium and zinc in soils and shoots of maize plants. J Plant Nutr Soil Sci 174(2):186–194

Rodríguez-Serrano M, Romero-Puertas MC, Pazmino DM, Testillano PS, Risueño MC, Del Río LA, Sandalio LM (2009) Cellular response of pea plants to cadmium toxicity: cross talk between reactive oxygen species, nitric oxide, and calcium. Plant Physiol 150(1):229–243

Rog Young K, Yoon JK, Kim TS, Yang JE, Owens G, Kim KR (2015) Bioavailability of heavy metals in soils: Definitions and practical implementation - a critical review. Environ Geochem Health 37(6):1041–1061. https://doi.org/10.1007/s10653-015-9695-y

Romero-Puertas MC, Corpas FJ, Rodríguez-Serrano M, Gómez M, del Río LA, Sandalio LM (2007) Differential expression and regulation of antioxidative enzymes by Cd in pea plants. J Plant Physiol 164:1346–1357

Roth U, Von Roepenack-Lahaye E, Clemens S (2006) Proteome changes in Arabidopsis thaliana roots upon exposure to Cd2+. J Exp Bot 57:4003–4013

Rucińska-Sobkowiak R (2016) Water relations in plants subjected to heavy metal stresses. Acta Physiol Plant 38:257. https://doi.org/10.1007/s11738-016-2277-5

Rugh CL, Senecoff JE, Meagher RB, Merkle SA (1998) Development of transgenic yellow poplar for mercury phytoremediation. Nat Biotechnol 16:925–928

Salt DE (2006) An extreme plant lifestyle: metal hyperaccumulation. In: Taiz L, Zeiger E (eds) Companion to plant physiology. Sinauer Associates, Sunderland, MA

Salt DE, Prince RC, Pickering IJ, Raskin I (1995) Mechanisms of cadmium mobility and accumulation in Indian mustard. Plant Physiol 109:1427–1433

Sandalio LM, Dalurzo HC, Gomez M, Romero-Puertas MC, Del Rio LA (2001) Cadmium-induced changes in the growth and oxidative metabolism of pea plants. J Exp Bot 52(364):2115–2126

Schaeffer HJ, Weber MJ (1999) Mitogen-activated protein kinases: specific messages from ubiquitous messengers. Mol Cell Biol 19:2435–2444

Schat H, Sharma SS, Vooijs R (1997) Heavy metal-induced accumulation of free proline in a metal-tolerant and a nontolerant ecotype of Silene vulgaris. Physiol Plant 101:477–482. https://doi.org/10.1111/j.1399-3054.1997.tb01026.x

Schat H, Llugany M, Vooijs R, Hartley-Whitaker J, Bleeker PM (2002) The role of phytochelatins in constitutive and adaptive heavy metal tolerances in hyperaccumulator and non-hyperaccumulator metallophytes. J. Exp Bot 53:2381–2392

Schützendübel A, Schwanz P, Teichmann T, Gross K, Langenfeld-Heyser R, Godbold DL et al (2001) Cadmium-induced changes in antioxidative systems, hydrogen peroxide content, and differentiation in Scots pine roots. Plant Physiol 75:887–898

Sebastian A, Prasad MNV (2018) Exogenous citrate and malate alleviate cadmium stress in Oryza sativa L.: probing role of cadmium localization and iron nutrition. Ecotoxicol Environ Saf 166: 215–222. https://doi.org/10.1016/j.ecoenv.2018.09.084

Semane B, Dupae J, Cuypers A, Noben JP, Tuomainen M, Tervahauta A et al (2010) Leaf proteome responses of Arabidopsis thaliana exposed to mild cadmium stress. J Plant Physiol 167(4): 247–254

Sergeant K, Kieffer P, Dommes J, Hausman J-F, Renaut J (2014) Proteomic changes in leaves of poplar exposed to both cadmium and low-temperature. Environ Exp Bot 106:112–123. https://doi.org/10.1016/j.envexpbot.2014.01.007

Sersen F, Kral'ova K (2001) New facts about CdCl2 action on the photosynthetic apparatus of spinach chloroplasts and its comparison with HgCl2 action. Photosynthetica 39:575–580

Seth CS, Misra V, Chauhan LKS, Singh RR (2008) Genotoxicity of cadmium on root meristem cells of Allium cepa: cytogenetic and Comet assay approach. Ecotoxicol Environ Saf 71(3): 711–716

Shahid M, Pourrut B, Dumat C, Nadeem M, Aslam M, Pinelli E (2014) Heavy metal-induced reactive oxygen species: phytotoxicity and physicochemical changes in plants. In: Whitacre D (ed) Reviews of Environmental Contamination and Toxicology Volume 232. Reviews of Environmental Contamination and Toxicology (Continuation of Residue Reviews), vol 232. Springer, Cham. https://doi.org/10.1007/978-3-319-06746-9_1

Shanmugaraj BM, Malla A, Ramalingam S (2019) Cadmium stress and toxicity in plants: an overview. In: Hasanuzzaman M, Prasad MNV, Nahar K (eds) Cadmium toxicity and tolerance in plants. Elsevier, pp 1–17. https://doi.org/10.1016/b978-0-12-814864-8.00001-2

Sharma SS, Dietz KJ (2006) The significance of amino acids and amino acid-derived molecules in plant responses and adaptation to heavy metal stress. J Exp Bot 57:711–726

Singh BK, Foley RC, Onate-Sanchez L (2002) Transcription factors in plant defense and stress responses. Curr Opin Plant Biol 5:430–436

Six L, Smolders E (2014) Future trends in soil cadmium concentration under current cadmium fluxes to European agricultural soils. Sci Total Environ 485:319–328

Solti Á, Sárvári É, Tóth B, Basa B, Lévai L, Fodor F (2011) Cd affects the translocation of some metals either Fe-like or Ca-like way in poplar. Plant Physiol Biochem 49(5):494–498

Song Y, Jin L, Wang X (2017) Cadmium absorption and transportation pathways in plants. Int J Phytoremediation 19:133–141

Sprynskyy M, Kowalkowski T, Tutu H, Cozmuta LM, Cukrowska EM, Buszewski B (2011) The adsorption properties of agricultural and forest soils towards heavy metal ions (Ni, Cu, Zn, and Cd). Soil Sediment Contam 20(1):12–29

Stein M, Dittgen J, Sanchez-Rodriguez C, Hou BH, Molina A, Schulze Lefert P et al (2006) Arabidopsis PEN3/PDR8, an ATP binding cassette transporter, contributes to nonhost resistance to inappropriate pathogens that enter by direct penetration. Plant Cell 18:731–746

Sunkar R, Zhu JK (2004) Novel and stress-regulated microRNAs and other small RNAs from Arabidopsis. Plant Cell 16:2001–2019

Suzuki N, Koizumi N, Sano H (2001) Screening of cadmium-responsive genes in Arabidopsis thaliana. Plant Cell Environ 24:1177–1188

Tamas L, Dudikova J, Durcekova K, Haluskova L, Huttova J, Mistrık I, Olle M (2008) Alterations of the gene expression, lipid peroxidation, proline and thiol content along the barley root exposed to cadmium. J Plant Physiol 165:1193–1203

Taylor M, Kim N, Smidt G, Busby C, McNally S, Robinson B et al (2016) Trace element contaminants and radioactivity from phosphate fertiliser. In: Phosphorus in agriculture: 100% zero. Springer, pp 231–266

Thakur S, Singh L, Wahid ZA, Siddiqui MF, Atnaw SM, Din MFM (2016) Plant-driven removal of heavy metals from soil: Uptake, translocation, tolerance mechanism, challenges, and future perspectives. Environ Monit Assess 188(4):204. https://doi.org/10.1007/s10661-016-5211-9

Thévenod F (2009) Cadmium and cellular signaling cascades: to be or not to be? Toxicol Appl Pharmacol 238:221–239

Thomine S, Wang R, Ward JM, Crawford NM, Schroeder JI (2000) Cadmium and iron transport by members of a plant metal transporters family in Arabidopsis with homology to Nramp genes. Proc Natl Acad Sci USA 97:4991–4996

Thumann J, Grill E, Winnacker EL, Zenk MH (1991) Reactivation of metal-requiring apoenzymes by phytochelatin-metal complexes. FEBS Lett 284:66–69

Tiryakioglu M, Eker S, Ozkutlu F, Husted S, Cakmak I (2006) Antioxidant defense system and cadmium uptake in barley genotypes differing in cadmium tolerance. J Trace Elem Med Biol 20(3):181–189. https://doi.org/10.1016/j.jtemb.2005.12.004

Tommey AM, Shi J, Lindsay WP, Urwin PE, Robinson NJ (1991) Expression of the pea gene PsMTa in *E. coli*—metal binding properties of the expressed protein. FEBS Lett 292:48–52

Tran TA, Popova LP (2013) Functions and toxicity of cadmium in plants: recent advances and future prospects. Turk J Bot 37(1):1–13

Tran TA, Paunova S, Nedeva D, Popova L (2011) Nitric oxide alleviates cadmium toxicity on photosynthesis in pea plants. Compte Rendus l' Acad Bulg Sci 64:1137–1142

Ueno D, Koyama E, Kono I, Ando T, Yano M, Ma JF (2009) Identification of a novel major quantitative trait locus controlling distribution of cd between roots and shoots in rice. Plant Cell Physiol 50(12):2223–2233. https://doi.org/10.1093/pcp/pcp160

UNEP (2010) Final review of scientific information on cadmium. United Nations Environment Programme, p 201

Uno Y, Furihata T, Abe H, Yoshida R, Shinozaki K, Yamaguchi-Shinozaki K (2000) Arabidopsis basic leucine zipper transcription factors involved in an abscissic acid dependent signal transduction pathway under drought and high-salinity. Proc Natl Acad Sci U S A 97:11632–11637

Ünyayar S, Değer AG, Çelik A, Çekiç FÖ, Çevik S (2010) Cadmium-induced antioxidant status and sisterchromatid exchanges in Vicia faba L. Turk J Biol 34(4):413–422

Uraguchi S, Kiyono M, Sakamoto T, Watanabe I, Kuno K (2009) Contributions of apoplasmic cadmium accumulation, antioxidative enzymes and induction of phytochelatins in cadmium tolerance of the cadmium-accumulating cultivar of black oat (Avena strigosa Schreb). Planta 230:267–276

Urano K, Kurihara Y, Seki M, Shinozaki K (2010) 'Omics' analyses of regulatory networks in plant abiotic stress responses. Curr Opin Plant Biol 13:1–7

Vaculı KM, Konlechner C, Langer I, Adlassnig W, Puschenreiter M, Lux A, Hauser MT (2012) Root anatomy and element distribution vary between two Salix caprea isolates with different Cd accumulation capacities. Environ Pollut 163:117–126. https://doi.org/10.1016/j.envpol.2011.12.031

Van Belleghem F, Cuypers A, Semane B, Smeets K, Vangronsveld J, d'Haen J, Valcke R (2007) Subcellular localization of cadmium in roots and leaves of Arabidopsis thaliana. New Phytol 173(3):495–508. https://doi.org/10.1111/j.1469-8137.2006.01940.x

Van Breusegem F, Bailey-Serres J, Mittler R (2008) Unraveling the tapestry of networks involving reactive oxygen species in plants. Plant Physiol 147:978–984

Van de Mortel JE, Schat H, Moerland PD, Loren V, van Themaat E, Van Der Ent S, Blankestijn H et al (2008) Expression differences for genes involved in lignin, glutathione and sulphate

metabolism in response to cadmium in Arabidopsis thaliana and the related Zn/Cd-hyperaccumulator Thlaspi caerulescens. Plant Cell Environ 31:301–324
Van Rossum F, Bonnin I, Fenart S, Pauwels M, Petit D, Saumitou-Laprade P (2004) Spatial genetic structure within a metallicolous population of Arabidopsis halleri, a clonal, self-incompatible and heavy-metal-tolerant species. Mol Ecol 13:2959–2967
Vassilev A, Perez-Sanz A, Semane B, Carleer R, Vangronsveld J (2005) Cadmium accumulation and tolerance of two Salix genotypes hydroponically grown in presence of cadmium. J Plant Nutr 28(12):2159–2177
Vatamaniuk OK, Mari S, Lu Y-P, Rea PA (2000) Mechanism of heavy metal ion activation of phytochelatin (PC) synthase-blocked thiols are sufficient for PC synthase-catalyzed transpeptidation of glutathione and related thiol peptides. J Biol Chem 275:31451–31459
Verbruggen N, Hermans C, Schat H (2009) Molecular mechanisms of metal hyperaccumulation in plants. New Phytol 181(4):759–776
Verret F, Gravot A, Auroy P, Leonhardt N, David P, Nussaume L et al (2004) Overexpression of AtHMA4 enhances root-to-shoot translocation of zinc and cadmium and plant metal tolerance. FEBS Lett 576:306–312
Villiers F, Jourdain A, Bastien O, Leonhardt N, Fujioka S, Tichtincky G, Parcy F, Bourguignon J, Hugouvieux V (2012) Evidence for functional interaction between brassinosteroids and cadmium response in Arabidopsis thaliana. J Exp Bot 63(3):1185–1200. https://doi.org/10.1093/jxb/err335
Voglar GE, Leštan D (2010) Solidification/stabilisation of metals contaminated industrial soil from former Zn smelter in Celje, Slovenia, using cement as a hydraulic binder. J Hazard Mater 178(1–3):926–933
Wahid A, Ghani A, Ali I, Ashraf MY (2007) Effects of cadmium on carbon and nitrogen assimilation in shoots of mungbean [Vigna radiata (L.) Wilczek] seedlings. J Agron Crop Sci 193(5):357–365
Wahid A, Ghani A, Javed F (2008) Effect of cadmium on photosynthesis, nutrition and growth of mungbean. Agron Sustain Dev 28:273–280
Weber M, Trampczynska A, Clemens S (2006) Comparative transcriptome analysis of toxic metal responses in Arabidopsis thaliana and the Cd-hypertolerant facultative metallophyte Arabidopsis halleri. Plant Cell Environ 29:950–963
Weggler K, McLaughlin MJ, Graham RD (2004) Effect of chloride in soil solution on the plant availability of biosolid-borne cadmium. J Environ Qual 33:496–504
Wei W, Zhang Y, Han L, Guan Z, Chai T (2008) A novel WRKY transcriptional factor from Thlaspi caerulescens negatively regulates the osmotic stress tolerance of transgenic tobacco. Plant Cell Rep 27:795–803
Williams LE, Pittman JK, Hall JL (2000) Emerging mechanisms for heavy metal transport in plants. Biochim Biophys Acta 1465(1-2):104–126. https://doi.org/10.1016/S0005-2736(00)00133-4
Wu Z, Zhao X, Sun X, Tan Q, Tang Y, Nie Z, Hu C (2015) Xylem transport and gene expression play decisive roles in cadmium accumulation in shoots of two oilseed rape cultivars (Brassica napus). Chemosphere 119:1217–1223. https://doi.org/10.1016/j.chemosphere.2014.09.099
Wuana RA, Okieimen FE (2011) Heavy metals in contaminated soils: a review of sources, chemistry, risks and best available strategies for remediation. Int Sch Res Notices 2011:402647
Xie FL, Huang SQ, Guo K, Xiang AL, Zhu YY, Nie L, Yang ZM (2007) Computational identification of novel microRNAs and targets in Brassica napus. FEBS Lett 581:1464–1474
Xu ZM, Li QS, Yang P, Ye HJ, Chen ZS, Guo SH, Wang LL, He BY, Zeng EY (2017) Impact of osmoregulation on the differences in Cd accumulation between two contrasting edible amaranth cultivars grown on Cd-polluted saline soils. Environ Pollut 224:89–97. https://doi.org/10.1016/j.envpol.2016.12.067
Yabuta Y (2016) Functions of heat shock transcription factors involved in response to photooxidative stresses in Arabidopsis. Biosci Biotechnol Biochem 80:1254–1263. https://doi.org/10.1080/09168451.2016.1176515

Yadav SK (2010) Heavy metals toxicity in plants: an overview on the role of glutathione and phytochelatins in heavy metal stress tolerance of plants. South Afr J Bot 76(2):167–179
Yadav KK, Gupta N, Kumar A, Reece LM, Singh N, Rezania S, Ahmad Khan S (2018) Mechanistic understanding and holistic approach of phytoremediation: a review on application and future prospects. Ecol Eng 120:274–298. https://doi.org/10.1016/j.ecoleng.2018.05.039
Yamaguchi N, Mori S, Baba K, Kaburagi-Yada S, Arao T, Kitajima N, Hokura A, Terada Y (2011) Cadmium distribution in the root tissues of solanaceous plants with contrasting root-to-shoot Cd translocation efficiencies. Environ Exp Bot 71(2):198–206. https://doi.org/10.1016/j.envexpbot.2010.12.002
Yang T, Poovaiah BW (2003) Calcium/calmodulin-mediated signal network in plants. Trends Plant Sci 8:505–512
Yang YJ, Cheng LM, Liu ZH (2007) Rapid effect of cadmium on lignin biosynthesis in soybean roots. Plant Sci 172(3):632–639
Yannarelli GG, Fernandez-Alvarez AJ, Santa-Cruz DM, Tomaro ML (2007) Glutathione reductase activity and isoforms in leaves and roots of wheat plants subjected to cadmium stress. Phytochemistry 68:505–512
Yeh CM, Hsiao LJ, Huang HJ (2004) Cadmium activates a mitogen-activated protein kinase gene and MBP kinases in rice. Plant Cell Physiol 45:1306–1312
Yeh CM, Chien PS, Huang HJ (2007) Distinct signalling pathways for induction of MAP kinase activities by cadmium and copper in rice roots. J Exp Bot 58:659–671
Yılmaz DD, Parlak KU (2011) Changes in proline accumulation and antioxidative enzyme activities in Groenlandia densa under cadmium stress. Ecol Indic 11(2):417–423
Yin Y, Wang Y, Liu Y, Zeng G, Hu X, Hu X et al (2015) Cadmium accumulation and apoplastic and symplastic transport in Boehmeria nivea (L.) Gaudich on cadmium-contaminated soil with the addition of EDTA or NTA. RSC Adv 5(59):47584–47591
Younis U, Malik SA, Rizwan M, Qayyum MF, Ok YS, Shah MHR, Rehman RA, Ahmad N (2016) Biochar enhances the cadmium tolerance in spinach (Spinacia oleracea) through modification of Cd uptake and physiological and biochemical attributes. Environ Sci Pollut Res 23:21385–21394. https://doi.org/10.1007/s11356-016-7344-
Zawoznik MS, Groppa MD, Tomaro ML, Benavides MP (2007) Endogenous salicylic acid potentiates cadmium-induced oxidative stress in Arabidopsis thaliana. Plant Sci 173:190–197
Zhang C-J, Chen L, Shi D-W, Chen G-X, Lu C-G, Wang P, Sun L (2007) Characteristics of ribulose-1,5-bisphosphate carboxylase and C4 pathway key enzymes in flag leaves of a super-high-yield hybrid rice and its parents during the reproductive stage. South Afr J Bot 73:22–28. https://doi.org/10.1016/j.sajb.2006.05.002
Zhang H, Dang Z, Zheng LC, Yi XY (2009) Remediation of soil co-contaminated with pyrene and cadmium by growing maize (Zea mays L.). Int J Environ Sci Technol 6:249–258
Zhang F, Liu M, Li Y, Che Y, Xiao Y (2019) Effects of arbuscular mycorrhizal fungi, biochar and cadmium on the yield and element uptake of Medicago sativa. Sci Total Environ 655:1150–1158
Zhao Z, Zhu YG, Cai YL (2005) Effects of zinc on cadmium uptake by spring wheat (Triticum aestivum, L.): long-time hydroponic study and short-time109 Cd tracing study. J Zhejiang Univ Sci 6A:643–648
Zheng R, Li H, Jiang R, Romheld V, Zhang F, Zhao FJ (2011) The role of root hairs in cadmium acquisition by barley. Environ Pollut 159(2):408–415. https://doi.org/10.1016/j.envpol.2010.10.034
Zhi-bin L, He J, Polle A, Rennenberg H (2016) Heavy metal accumulation and signal transduction in herbaceous and woody plants: paving the way for enhancing phytoremediation efficiency. Biotechnol Adv 34(6):1131–1148. https://doi.org/10.1016/j.biotechadv.2016.07.003
Zhigang A, Cuijie L, Yuangang Z, Yejie D, Wachter A, Gromes R et al (2006) Expression of BjMT2, a metallothionein 2 from Brassica juncea, increases copper and cadmium tolerance in Escherichia coli and Arabidopsis thaliana, but inhibits root elongation in Arabidopsis thaliana seedlings. J Exp Bot 57:3575–3582

Zhou J, Goldsbrough PB (1994) Functional homologs of animal and fungal metallothionein genes from Arabidopsis. Plant Cell 6:875–884

Zhou W, Qiu B (2005) Effects of cadmium hyperaccumulation on physiological characteristics of Sedum alfredii Hance (Crassulaceae). Plant Sci 169:737–745

Zhou ZS, Huang SQ, Yang ZM (2008) Bioinformatic identification and expression analysis of new microRNAs from Medicago truncatula. Biochem Biophys Res Commun 374:538–542

Zhu CW, Wen HJ, Zhang YX, Fan HF, Fu SH, Xu J, Qin TR (2013a) Characteristics of Cd isotopic compositions and their genetic significance in the lead zinc deposits of SW China. Sci China Earth Sci 56:2056–2065

Oxidative Stress in Cadmium Toxicity in Animals and Its Amelioration

Chanchal Singh, Raghubir Singh, and Apoorva Shekhar

Abstract The present chapter discusses the intricate relationship between cadmium toxicity and oxidative stress in animals. Cadmium, a highly toxic heavy metal, is prevalent in various environmental settings and is known to accumulate in animal tissues, leading to detrimental health effects. The chapter highlights how exposure to cadmium triggers an imbalance between the production of reactive oxygen species (ROS) and the organism's antioxidant defense mechanisms, resulting in oxidative stress. This oxidative stress can induce cellular damage, disrupt physiological processes, and contribute to the development of various diseases in animals. The chapter also explores potential strategies to mitigate the adverse effects of cadmium-induced oxidative stress. These strategies encompass a range of interventions, including the use of antioxidants, chelating agents, dietary modifications, and various pharmacological approaches. By ameliorating oxidative stress, these interventions aim to minimize the harmful impacts of cadmium toxicity on animal health and overall well-being.

Keywords Oxidative stress · Cadmium toxicity · Animals · Amelioration · Antioxidants

1 Introduction

In recent decades, the rapid industrialization and urbanization of the modern world have led to an unprecedented rise in environmental pollutants, posing significant threats to both human health and ecosystem stability. Among these pollutants, heavy

C. Singh (✉) · A. Shekhar
Department of Veterinary Physiology and Biochemistry, College of Veterinary Science, Guru Angad Dev Veterinary and Animal Sciences University, Ludhiana, Punjab, India

R. Singh
Department of Veterinary Public Health, College of Veterinary Science and Animal Husbandry, Jalukie, Nagaland, India

Central Agricultural University Imphal, Imphal, Manipur, India

A. K. Jha, N. Kumar (eds.), *Cadmium Toxicity Mitigation*,
https://doi.org/10.1007/978-3-031-47390-6_15

metals have emerged as one of the most concerning due to their persistent nature and potential to accumulate in various environmental compartments. Cadmium, a non-essential heavy metal, has gained particular attention due to its widespread presence and toxic effects on living organisms; particularly animals. Cadmium is the seventh most toxic heavy metal as per ATSDR ranking. It is highly toxic to humans and animals (Roman et al. 2002; Patra et al. 2006; Swarup et al. 2007) and not essential to physiological and biochemical functions. It is a by-product of zinc production which humans or animals may get exposed to at work or in the environment. Once this metal gets absorbed by humans, it will accumulate inside the body throughout life. Cadmium is toxic to virtually every system in the animal body. Cadmium toxicity is a multi-faceted issue that affects a wide range of biological systems and processes within animals (Jayamurali et al. 2021). One of the underlying mechanisms through which cadmium exerts its toxic effects is oxidative stress. Oxidative stress occurs when there is an imbalance between the production of reactive oxygen species (ROS) and the cellular antioxidant defense mechanisms, leading to cellular damage and dysfunction. Cadmium-induced oxidative stress has been implicated in various pathological conditions, including DNA damage, lipid peroxidation, protein oxidation, and disruption of cellular signalling pathways. The molecular mechanisms through which cadmium triggers ROS production and interferes with antioxidant defense systems are examined in detail. Special attention is given to the intricate interplay between oxidative stress and inflammatory responses, as well as the potential for cadmium-induced oxidative damage to contribute to chronic diseases in animals. Amid the alarming consequences of cadmium toxicity, there is growing interest in identifying strategies to ameliorate its harmful effects. This chapter provides insights into potential amelioration approaches aimed at reducing cadmium-induced oxidative stress. These approaches encompass dietary interventions, administration of antioxidants, and the exploration of natural compounds with chelating properties to mitigate cadmium accumulation and its subsequent impact on oxidative stress-related damage.

2 Sources and Routes of Cadmium Exposure in Animals

Animals can be exposed to cadmium through various sources and routes, both in the natural environment and due to human activities. Some of the primary sources and routes of cadmium exposure in animals include:

A. **Industrial Emissions:**

Animals living in proximity to industrial areas may be exposed to cadmium through the inhalation of air contaminated with cadmium emissions. Lui (2003) describes the concurrent poisoning of lead (Pb) and Cd in sheep and horses near a non-ferrous metal smelter in China. Affected horse's mean blood Cd concentrations were 170 μg/l compared with control horse's blood concentrations of 30 μg/kg. While mean blood Cd concentrations in the affected sheep were

370 μg/kg compared with 20 μg/ kg in control animals. Research conducted in an industrialized area of North West Spain, Galicia reported blood concentrations of Cd in 6–10 month old calves and cows (Lopez Alonso et al. 2000). The mean blood Cd concentration was 0.373 and 0.449 μg/l in calves and cows, respectively.

Despite the similar blood concentrations, kidney, liver, and muscle concentrations were higher in the calves located in the industrial area compared with those located in a rural area; suggestive that although exposed to higher Cd, this is not always associated with raised blood concentrations. This Spanish study concludes that trace element status of the calves was affected, with almost half the calves in the industrialized zone demonstrated to have lowered tissue Cu concentrations, underlying the importance of heavy metals' effects on trace element status. Compared to these Spanish studies, research conducted in India, reports substantially higher concentrations of blood Cd, using similar methodologies. Patra et al. (2005) determined blood and milk Cd concentrations in 210 lactating cows reared and kept within a 2-km radius of a number of different industrial units or in a non-polluted area to serve as controls. Their results are suggestive that cows reared and kept near a steel manufacturing plant had higher blood Cd concentrations (mean 232 μg/l; ranging from 90 to 410 μg/l) compared with cows kept near other industrial sites or in a non-polluted area (mean 28 μg/l; ranging from 0 to 50 μg/l). Interestingly, further research by the same group suggests that whole blood Fe was lower in the cows near the steel processing plant compared with those in the non-polluted areas (Patra et al. 2006).

B. **Environmental Contamination:**

(a) Soil: Cadmium is released into the soil through natural weathering processes, volcanic activity, and human activities like industrial emissions, waste disposal, and use of cadmium-containing fertilizers and sewage sludge.
(b) Water: Cadmium can contaminate water bodies through industrial discharges, agricultural runoff, and leaching from soil.
(c) Air: Cadmium-containing particulate matter can be released into the air from industrial processes, waste incineration, and vehicle emissions, eventually settling into soil and water bodies.

C. **Food Chain:**

(a) Grazing Animals: Herbivorous animals may ingest cadmium-contaminated vegetation while grazing, leading to its accumulation in their bodies.
(b) Predatory Animals: Carnivorous and omnivorous animals can be exposed to cadmium by consuming other animals that have accumulated the metal in their tissues.

D. **Drinking Water:**
Contaminated groundwater and surface water sources can be a significant route of cadmium exposure for animals, especially in areas with high industrial or agricultural activities.

E. **Agricultural Practices:**
   (a) Use of Cadmium-Containing Pesticides: In some regions, cadmium-containing pesticides have been used in agriculture, and animals may be exposed through ingestion of treated crops or contaminated water.
   (b) Fertilizers and Soil Amendments: Cadmium present in certain fertilizers and soil amendments can lead to its accumulation in crops and, consequently, in animals that consume these crops.

F. **Waste and Landfills:**
Animals scavenging at landfills or living near waste disposal sites can be exposed to cadmium from discarded items and industrial waste.

G. **Mining and Smelting:**
Animals living near mining and smelting sites are at risk of cadmium exposure due to the release of the metal during extraction and processing activities.

H. **Contaminated Wildlife Habitats:**
Cadmium can accumulate in the tissues of certain plants and animals, and those higher up in the food chain may experience increased exposure.

3 Cadmium Toxicity and Mechanisms of Oxidative Stress

3.1 Molecular Mechanisms of Cadmium-Induced Oxidative Stress

Cadmium-induced oxidative stress in cells involves a series of interconnected molecular mechanisms that lead to the generation of reactive oxygen species (ROS) and subsequent damage to cellular components (Valko et al. 2005; Das and Al-Naemi 2019). These mechanisms include:

3.1.1 ROS Production

Cadmium can stimulate the production of ROS within cells. It can undergo redox cycling, where it cycles between different oxidation states, leading to the generation of superoxide radicals (O_2^-) and hydrogen peroxide (H_2O_2).

Mitochondrial Dysfunction: Cadmium can accumulate in mitochondria, the cellular powerhouses responsible for energy production. Increased cadmium levels disrupt the electron transport chain, leading to electron leakage and increased ROS production within mitochondria.

Metallothionein Depletion: Cadmium has a high affinity for metallothionein, a protein involved in cellular metal homeostasis and detoxification. When cadmium binds to metallothionein, it sequesters the protein, reducing its availability to scavenge other metal ions, such as zinc and copper. This disruption in metallothionein function can result in the accumulation of free metal ions, which can contribute to ROS production through Fenton-like reactions.

Activation of NADPH Oxidase: Cadmium exposure can lead to the activation of NADPH oxidase, an enzyme complex responsible for the production of ROS as part of the cell's immune response to pathogens. However, in the presence of cadmium, NADPH oxidase can become overactive, leading to excessive ROS production and oxidative stress.

Inhibition of Antioxidant Enzymes: Cadmium can interfere with the activity of antioxidant enzymes, such as superoxide dismutase (SOD), catalase (CAT), and glutathione peroxidase (GPx). These enzymes are crucial in neutralizing ROS and maintaining cellular redox balance. By inhibiting their function, cadmium reduces the cell's ability to counteract ROS, leading to an accumulation of oxidative stress.

Induction of Inflammatory Pathways: Cadmium exposure can activate pro-inflammatory pathways, leading to the release of cytokines and chemokines. Inflammatory responses can, in turn, stimulate the production of ROS by immune cells as part of the host defense mechanism, further contributing to oxidative stress.

3.1.2 Protein and DNA Damage

Protein Damage

(a) Protein Oxidation: Cadmium exposure leads to the generation of reactive oxygen species (ROS), such as superoxide radicals and hydrogen peroxide. These ROS can directly interact with amino acid residues in proteins, leading to the oxidation of specific side chains, such as cysteine, methionine, and histidine. Protein oxidation can alter protein structure and function, rendering them less effective or even non-functional.

(b) Disruption of Enzyme Function: Many enzymes are susceptible to cadmium-induced oxidative damage. Inhibition or alteration of enzyme activity can disrupt vital cellular processes and metabolic pathways.

(c) Formation of Protein–Cadmium Complexes: Cadmium can bind to various proteins, forming protein–cadmium complexes. This interaction can result in changes to the protein's conformation and function, leading to cellular dysfunction.

(d) Proteasomal Dysfunction: Cadmium exposure can impair the function of proteasomes, cellular complexes responsible for degrading damaged and misfolded proteins. Inhibition of proteasomes can lead to the accumulation of damaged proteins within cells, further contributing to cellular dysfunction.

DNA Damage

(a) Generation of Reactive Oxygen Species (ROS): As mentioned earlier, cadmium exposure results in ROS production. ROS can attack DNA, causing oxidative damage to DNA bases (Lin et al. 2007). One of the most common forms of oxidative DNA damage caused by ROS is 8-hydroxyguanine (8-OHdG), which can lead to mutations if not repaired properly.
(b) Induction of DNA Strand Breaks: Cadmium exposure can directly induce single-strand and double-strand breaks in DNA. These breaks can disrupt DNA replication and repair processes, leading to genomic instability.
(c) Inhibition of DNA Repair Mechanisms: Cadmium can interfere with DNA repair mechanisms, compromising the cell's ability to fix DNA damage and leading to the accumulation of mutations.
(d) Formation of Cadmium–DNA Adducts: Cadmium can form adducts with DNA bases, resulting in structural alterations that may interfere with DNA replication and transcription.

The cumulative effect of protein and DNA damage induced by cadmium leads to cellular dysfunction, increased susceptibility to diseases, and may contribute to the development of cancer. These processes are particularly concerning in organs with high rates of cellular turnover, such as the kidneys and lungs, where cadmium-induced damage can have severe consequences for organ function and overall health.

3.1.3 Disruption of Calcium Homeostasis

Cadmium can interfere with cellular calcium signalling, leading to disturbances in intracellular calcium levels. Dysregulated calcium levels contribute to ROS production and cellular damage.

The cumulative effect of these molecular mechanisms results in an imbalance between ROS production and the cell's antioxidant defense system, leading to oxidative stress. This oxidative stress can have deleterious effects on cell function, contributing to various pathological conditions, including organ damage, apoptosis, and cellular dysfunction. Understanding these molecular mechanisms is essential for developing effective strategies to mitigate cadmium-induced oxidative stress and its associated toxic effects in animals and humans. Antioxidant supplementation and other protective measures can be employed to counteract oxidative stress and reduce the adverse consequences of cadmium exposure.

3.2 Enzymatic Markers of Oxidative Stress in Cadmium Toxicity

Enzymatic markers of oxidative stress play a crucial role in assessing the extent of oxidative damage caused by cadmium toxicity. These enzymes are involved in the body's defense against reactive oxygen species (ROS) and help maintain redox balance. However, under conditions of cadmium exposure, these enzymatic markers can be affected, leading to alterations in their activity levels. Some of the key enzymatic markers of oxidative stress in cadmium toxicity include:

Superoxide Dismutase (SOD): SOD is an essential antioxidant enzyme that catalyzes the dismutation of superoxide radicals (O_2^-) into hydrogen peroxide (H_2O_2) and molecular oxygen. There are three isoforms of SOD in mammals: cytosolic copper/zinc SOD (Cu/Zn-SOD), mitochondrial manganese SOD (Mn-SOD), and extracellular copper/zinc SOD (Ec-SOD). Cadmium exposure can lead to a reduction in SOD activity, contributing to the accumulation of superoxide radicals and oxidative stress.

Catalase (CAT): Catalase is an enzyme that converts hydrogen peroxide (H_2O_2) into water (H_2O) and molecular oxygen (O_2). It serves as a crucial defense mechanism against H_2O_2 accumulation. Cadmium exposure can lead to decreased CAT activity, resulting in elevated levels of hydrogen peroxide and increased oxidative stress.

Glutathione Peroxidase (GPx): GPx is an enzyme that uses reduced glutathione (GSH) to detoxify hydrogen peroxide (H_2O_2) and lipid hydroperoxides. It is a key player in maintaining cellular redox balance. Cadmium exposure can inhibit GPx activity, reducing the cell's ability to scavenge hydrogen peroxide and lipid peroxides, leading to increased oxidative damage.

Glutathione S-Transferase (GST): GSTs are a family of enzymes involved in the detoxification of various electrophilic compounds. They play a role in protecting cells against oxidative damage caused by reactive metabolites of cadmium. However, chronic cadmium exposure may lead to decreased GST activity due to metal-induced inhibition.

Thioredoxin (Trx) and Thioredoxin Reductase (TrxR): Thioredoxin and thioredoxin reductase are essential components of the cellular antioxidant defense system. They help maintain the reduced state of proteins by reducing disulfide bonds. Cadmium exposure can interfere with the Trx/TrxR system, leading to increased oxidative stress.

Glutathione Reductase (GR): GR is an enzyme that maintains the reduced form of glutathione (GSH) in cells. GSH is a vital intracellular antioxidant that scavenges ROS. Cadmium exposure can inhibit GR activity, compromising the cell's ability to regenerate GSH and counteract oxidative stress. Measurement of these enzymatic markers in biological samples, such as blood, tissues, or urine, can serve as valuable indicators of oxidative stress and the impact of cadmium toxicity on the antioxidant defense system. Assessing the changes in enzymatic activities can aid in understanding the severity of cadmium-induced oxidative stress and guide the development of therapeutic approaches to mitigate its harmful effects.

4 Effects of Cadmium Toxicity in Animals

A review of cadmium toxicity in animals provides a comprehensive overview of the adverse effects of cadmium exposure on various biological systems and the potential mechanisms underlying these effects. Cadmium, a heavy metal, is widely distributed in the environment due to industrial activities, agricultural practices, and natural processes. Animals can be exposed to cadmium through contaminated water, soil, and food sources, making it a significant concern for both wildlife and domestic animals (Genchi et al. 2020).

4.1 *Physiological Effects*

Cadmium toxicity affects numerous physiological processes within animals. It tends to accumulate in certain organs, primarily the liver, kidneys, and lungs, where it disrupts cellular functions. The metal interferes with essential metals like zinc, calcium, and iron, disrupting their roles in enzymatic activities and cellular processes. This leads to cellular dysfunction and contributes to a range of health issues, including impaired growth, reproductive problems, and compromised immune responses (Sharma and Agrawal 2005).

4.2 *Oxidative Stress*

One of the prominent mechanisms through which cadmium exerts its toxicity is oxidative stress. Cadmium stimulates the generation of reactive oxygen species (ROS) within cells while impairing the antioxidant defense mechanisms. This imbalance results in oxidative damage to lipids, proteins, and DNA. Oxidative stress not only damages cellular components but also triggers inflammation and contributes to the development of various diseases, including cancer, cardiovascular diseases, and neurodegenerative disorders.

4.3 *Genotoxicity and Carcinogenicity*

Cadmium is known for its genotoxic and carcinogenic properties. It can directly damage DNA, leading to mutations and the potential development of cancer. The accumulation of DNA damage over time increases the risk of malignant transformation. Animal studies have demonstrated an association between cadmium exposure and the development of tumors in various organs, further highlighting its carcinogenic potential. Cadmium effectively induces cancer at multiple sites and by various routes.

4.4 Impact on Reproduction and Development

Cadmium can adversely affect reproductive outcomes in animals. It accumulates in the ovaries and testes, disrupting reproductive hormone balance and impairing fertility. Maternal exposure to cadmium during pregnancy can lead to developmental abnormalities in offspring, affecting both physical and cognitive aspects.

5 Organ-Specific Effects of Cadmium

Cadmium is a toxic heavy metal that can accumulate in various organs and tissues, leading to organ-specific effects in animals and humans. The distribution and severity of these effects depend on the duration and level of cadmium exposure, as well as individual factors such as age, gender, and overall health (Sarkar et al. 2013). Many studies worldwide have reported the concentrations of Cd in meat, liver, and kidneys of meat producing animals. Regulatory limits of maximum Cd levels in muscle, liver, and kidneys of cattle for human consumption within the European Union are set by the Commission Regulation No. 1881/2006 (amended by No. 629/2008) at 0.05, 0.5, and 1.0 mg/kg wet weight, respectively. Research work conducted by the same laboratory suggests that samples obtained from calves in industrialized areas have higher concentrations of liver and kidney concentrations of Cd (Cd: liver 0.030 kidney 0.161 mg/kg) compared with those in calves from rural areas (Cd: liver 0.023, kidney 0.096 mg/kg, Miranda et al. 2005). Jamaica is an island known for its Cd-enriched soils (Lalor et al. 1998). Nriagu et al. 2009 reported higher concentrations of Cd in bovine livers (geometric mean 0.378 mg/kg) and kidneys (geometric mean 1.48 mg/kg) compared with other studies. Age was highly associated with kidney Cd concentrations and older cows had much higher concentrations of Cd compared with younger animals.

Here are some of the organ-specific effects of cadmium toxicity:

5.1 Kidneys

Cadmium has a high affinity for the kidneys, where it accumulates over time. This organ is particularly vulnerable to cadmium toxicity (Shi et al. 2017).

Cadmium-induced oxidative stress can damage the renal tubular cells and impair their function, leading to tubular dysfunction and proteinuria (Yan and Allen 2021). Chronic cadmium exposure can cause renal tubular necrosis and renal failure, a condition known as "Itai-Itai disease." The kidney is the critical organ in humans and mammals exposed for long periods to relatively small amounts of cadmium that might occur in foods (Satarug and Moore 2004). Long-term exposure to cadmium leads to several morpho-pathological changes in the kidneys. Cd accumulates in the kidney as a result of its preferential uptake by receptor-mediated endocytosis of freely filtered and metallothionein-bound Cd (Cd-MT) in the renal proximal tubule. Internalized Cd-MT is degraded in endosomes and lysosomes, releasing free Cd_2 into the cytosol, where it can generate reactive oxygen species (ROS) and activate cell death pathways. Studies in cattle suggest that females accumulate increased Cd in kidneys compared with males (Lopez et al. 2000). The form of Cd administered may affect the degree of nephrotoxicity. A single injection of Cd bound to MT at doses as low as 0.2 mg/kg was nephrotoxic in mice, whereas administration of Cd chloride up to 3 mg/kg did not affect renal function (Dorian et al. 1995).

5.2 Reproductive Organs

Cadmium can accumulate in the testes and ovaries. In males, cadmium exposure can lead to testicular damage, reduced sperm quality, and impaired fertility. In females, cadmium toxicity can cause ovarian dysfunction and adverse effects on reproductive health.

Reproductive and Developmental Consequences of Cadmium Toxicity
Cadmium toxicity can have significant reproductive and developmental consequences in both males and females. Exposure to cadmium during critical periods of reproduction and development can lead to adverse effects on fertility, pregnancy outcomes, and the health of offspring. Some of the reproductive and developmental consequences of cadmium toxicity are as follows:

Reproductive Consequences
Male Reproductive System:

Reduced Sperm Quality: Cadmium exposure can lead to decreased sperm count, motility, and viability, impairing male fertility.

Testicular Damage: Cadmium can cause structural and functional damage to the testes, leading to testicular inflammation and degeneration.

Hormonal Disruption: Cadmium can interfere with hormone regulation in males, affecting testosterone production and other reproductive hormones (Xiong et al. 2021).

Female Reproductive System:

Ovarian Dysfunction: Cadmium exposure can disrupt ovarian function, leading to irregular menstrual cycles and decreased fertility (Tribowo et al. 2014).

Adverse Pregnancy Outcomes: Cadmium exposure during pregnancy may increase the risk of miscarriage and preterm birth.

Developmental Consequences
Embryonic Development:

Cadmium exposure during early pregnancy can lead to developmental defects in the embryo, affecting organogenesis and leading to congenital malformations. Cadmium exposure during pregnancy can lead to reduced fetal growth, resulting in low birth weight and small-for-gestational-age infants.

Neuro-developmental Effects:

Cadmium can cross the placental barrier and affect fetal brain development. Prenatal exposure to cadmium is associated with cognitive impairments, learning disabilities, and behavioral problems in children.

5.3 Immune System

Cadmium exposure can suppress the immune system, making individuals more susceptible to infections and diseases. The severity of organ-specific effects depends on the dose and duration of cadmium exposure. Long-term exposure to high levels of cadmium can result in chronic toxicity, whereas acute exposure to high concentrations can cause acute toxicity with severe symptoms. Reducing cadmium exposure through environmental regulations and adopting protective measures is crucial to preventing and mitigating the organ-specific effects of cadmium toxicity.

Impaired Immune Function
Prenatal and postnatal exposure to cadmium can weaken the immune system in infants and children, increasing their susceptibility to infections and illnesses. Immunotoxicity and oxidative stress are closely linked in the context of cadmium toxicity. Cadmium exposure can have profound effects on the immune system, leading to immunotoxicity, and one of the key mechanisms underlying this effect is the induction of oxidative stress. Here is how cadmium-induced oxidative stress contributes to immunotoxicity:

Generation of Reactive Oxygen Species (ROS): Cadmium exposure leads to the production of reactive oxygen species (ROS) within immune cells, such as macrophages and lymphocytes. ROS, including superoxide radicals (O_2^-), hydrogen peroxide (H_2O_2), and hydroxyl radicals (OH^-), are highly reactive molecules that can cause damage to cellular components, including lipids, proteins, and DNA.

Oxidative Damage to Immune Cells: The ROS generated in response to cadmium exposure can attack immune cells, causing oxidative damage to their cell membranes, proteins, and genetic material. This damage can impair immune cell function, reduce their lifespan, and compromise their ability to mount an effective immune response.

5.4 Other Organs

The liver is another target organ for cadmium accumulation. Cadmium can induce oxidative stress in liver cells, leading to lipid peroxidation and hepatotoxicity. It can also interfere with liver enzymes and disrupt hepatic metabolic processes. Inhalation of cadmium-containing dust and fumes can lead to pulmonary toxicity. Cadmium can cause irritation and inflammation in the respiratory system, leading to respiratory symptoms such as coughing and difficulty breathing. Cadmium can replace calcium in bones, leading to skeletal demineralization. Chronic cadmium exposure is associated with decreased bone density and increased risk of fractures. Cadmium exposure has been linked to cardiovascular diseases, including hypertension and atherosclerosis. It can induce oxidative stress in vascular endothelial cells and impair vascular function. Cadmium can cross the blood–brain barrier and accumulate in the brain. It can cause neurotoxic effects, leading to cognitive impairment, memory deficits, and behavioral changes. Ingestion of cadmium-contaminated food and water can lead to gastrointestinal disturbances, including nausea, vomiting, and abdominal pain. Cadmium can interfere with hormone regulation and disrupt endocrine function.

6 Amelioration Strategies in Cadmium Toxicity

Cadmium is a toxic heavy metal that poses significant risks to both human health and the environment. Its widespread presence in various sources, including contaminated water, soil, and food, highlights the urgent need for effective amelioration strategies. These strategies are essential to mitigate the harmful effects of cadmium toxicity and promote the well-being of both animals and humans. Here is why amelioration strategies are crucial:

Health Protection
Cadmium toxicity can lead to a range of health issues, including kidney damage, cardiovascular problems, and increased cancer risk. Implementing amelioration strategies can help reduce the risk of these health problems and improve the overall quality of life for exposed individuals.

Environmental Preservation
Cadmium contamination can have detrimental effects on ecosystems and wildlife. Implementing amelioration strategies helps prevent the accumulation of cadmium in the environment, preserving biodiversity and ecosystem functionality.

Food Safety
Cadmium can accumulate in food crops, posing a direct risk to human health through the consumption of contaminated food products. By reducing cadmium levels in soil and water, amelioration strategies contribute to safer food production and consumption.

6.1 Antioxidant Supplementation in Amelioration of Cadmium Toxicity

Antioxidant supplementation has shown promise in ameliorating cadmium toxicity by mitigating the harmful effects of oxidative stress. Antioxidants neutralize reactive oxygen species (ROS) and prevent oxidative damage to cellular components, thereby protecting cells and tissues from the adverse effects of cadmium exposure (Kara et al. 2007; Shaikh et al. 1999; Unsal et al. 2020). Here are some commonly used antioxidants that have been studied for their potential to counteract cadmium toxicity:

6.1.1 Vitamins (e.g., Vitamin C and Vitamin E) in Cadmium Toxicity

Vitamins, such as Vitamin C (ascorbic acid) and Vitamin E (tocopherol), play essential roles in mitigating the adverse effects of cadmium toxicity, particularly by their antioxidant properties. These vitamins act as free radical scavengers and help protect cells and tissues from oxidative damage induced by cadmium exposure. Here is how Vitamin C and Vitamin E are involved in countering cadmium toxicity:

Vitamin C (Ascorbic Acid)
Antioxidant Defense: Vitamin C is a water-soluble antioxidant that readily donates electrons to neutralize free radicals, including ROS produced during cadmium toxicity. By doing so, Vitamin C helps prevent oxidative damage to proteins, lipids, and DNA.

Regenerating other Antioxidants: Vitamin C can regenerate other antioxidants, such as Vitamin E, by reducing the oxidized forms back to their active states. This regenerating property enhances the overall antioxidant capacity of the cell.

Metal Chelation: Vitamin C can also act as a metal chelator, binding to cadmium ions and reducing their bioavailability and toxicity.

Vitamin E (Tocopherol)
Lipid Peroxidation Inhibition: Vitamin E is a fat-soluble antioxidant that specifically protects cell membranes and lipids from oxidative damage caused by cadmium-induced ROS. It prevents the propagation of lipid peroxidation chain reactions in cell membranes, preserving their integrity and function (Al-Attar 2011).

Synergistic Effects with Vitamin C: Vitamin E works synergistically with Vitamin C to scavenge free radicals and maintain the antioxidant defense system.

Studies have demonstrated the protective effects of Vitamin C and Vitamin E against cadmium-induced oxidative stress and associated toxicities in various organs, including the liver, kidneys, lungs, and brain (Nemmiche et al. 2007). These vitamins have been shown to reduce lipid peroxidation, increase antioxidant enzyme activities, and preserve cellular function in the presence of cadmium exposure.

Glutathione (GSH): Glutathione is a crucial endogenous antioxidant that plays a central role in the detoxification of ROS and other harmful molecules. It acts as a free radical scavenger and helps regenerate other antioxidants, such as Vitamins C and E. Supplementation with GSH or its precursors, such as N-acetylcysteine (NAC), has been shown to reduce oxidative stress and protect against cadmium-induced toxicity in various tissues (Goncalves et al. 2010).

It is essential to ensure an adequate intake of these vitamins through a balanced diet rich in fruits, vegetables, nuts, and seeds, as they contribute to overall health and can help counteract the damaging effects of environmental toxicants like cadmium. However, excessive supplementation of these vitamins should be avoided, as excessive doses may not provide additional benefits and may even have adverse effects. Always consult a healthcare professional before starting any supplementation regimen. Furthermore, while Vitamins C and E can help ameliorate oxidative stress induced by cadmium, reducing exposure to cadmium at the source remains the most effective strategy to prevent its toxic effects on health. Public health measures and environmental regulations aimed at minimizing cadmium exposure are crucial to safeguarding the well-being of individuals and populations.

6.1.2 Minerals (e.g., Selenium and Zinc) in Amelioration of Cadmium Toxicity

Minerals such as selenium and zinc have been studied for their potential in ameliorating cadmium toxicity. These minerals play essential roles in the body's antioxidant defense system and can counteract the harmful effects of cadmium-induced oxidative stress (Messaoudi et al. 2009). Indeed, in rats dietary Zn and Se seem to exert a cooperative effect in the protection of Cd-induced hepatic damage, but not renal damage (El Heni et al. 2008). Here is how selenium and zinc contribute to reducing the toxicity of cadmium:

Selenium

Glutathione Peroxidase (GPx) Activation: Selenium is an essential component of the antioxidant enzyme GPx. GPx helps detoxify hydrogen peroxide and lipid hydroperoxides, which are produced as part of the cellular response to cadmium exposure. Selenium supplementation can enhance GPx activity, thereby reducing oxidative stress and protecting cells from damage caused by cadmium-induced ROS.

Antioxidant Synergy: Selenium works in synergy with other antioxidants, such as Vitamin C and Vitamin E, to scavenge free radicals and neutralize ROS. The combined action of these antioxidants helps to maintain cellular redox balance and prevent oxidative damage induced by cadmium (Poli et al. 2022).

Metal Binding: Selenium can bind to cadmium and other heavy metals, forming less toxic complexes that can be excreted from the body. By reducing the bioavailability of cadmium, selenium helps limit its toxic effects on tissues and organs.

Zinc

Superoxide Dismutase (SOD) Activity: Zinc is an essential cofactor for the antioxidant enzyme copper/zinc superoxide dismutase (Cu/Zn-SOD). SOD catalyzes the dismutation of superoxide radicals (O_2^-) into hydrogen peroxide (H_2O_2) and molecular oxygen, thus reducing the levels of the highly reactive superoxide radicals. Zinc supplementation can enhance SOD activity, leading to decreased oxidative stress caused by cadmium exposure.

Metallothionein Induction: Zinc can stimulate the synthesis of metallothionein, a metal-binding protein that plays a crucial role in cellular metal detoxification. Metallothionein can bind to cadmium and sequester it, reducing its availability to cause damage in the cell.

Anti-Inflammatory Effects: Zinc has anti-inflammatory properties and can help mitigate the inflammatory responses triggered by cadmium exposure. By reducing inflammation, zinc indirectly contributes to a decrease in oxidative stress.

Both selenium and zinc are essential trace elements for human health, and their supplementation may offer protective benefits against cadmium toxicity. However, as with any supplementation, it is important to avoid excessive intake, as high levels of these minerals can also have adverse effects. A balanced and varied diet that includes selenium-rich foods (e.g., Brazil nuts, fish, and poultry) and zinc-rich foods (e.g., meat, legumes, and nuts) can provide adequate amounts of these minerals to support the body's antioxidant defenses and potentially ameliorate the toxic effects of cadmium exposure.

6.1.3 Polyphenols and Flavonoids

Polyphenols and flavonoids are bioactive compounds found in various plant-based foods, known for their potential health benefits and protective effects against various stressors, including heavy metal toxicity like cadmium, for example, β-cryptoxanthin was found to have preventive effects against cadmium-induced oxidative stress in the rat testis (Liu et al. 2016). Cadmium is a toxic heavy metal that can be harmful to both humans and animals. It can enter the body through contaminated water, food, and air, and can accumulate in various organs, particularly the kidneys. Exposure to cadmium has been associated with oxidative stress, inflammation, and various health issues. Polyphenols and flavonoids are known for their antioxidant and anti-inflammatory properties. They can potentially mitigate the adverse effects of cadmium toxicity through several mechanisms:

1. **Antioxidant Activity:** Cadmium exposure leads to the production of reactive oxygen species (ROS), causing oxidative stress and cellular damage. Polyphenols and flavonoids have strong antioxidant properties, which can help neutralize ROS and protect cells from damage.
2. **Metal Chelation:** Some polyphenols have the ability to chelate or bind to heavy metals like cadmium. This can reduce the absorption and distribution of cadmium in the body, limiting its toxic effects.
3. **Anti-inflammatory Effects:** Cadmium exposure triggers inflammation in tissues. Polyphenols and flavonoids have anti-inflammatory properties that can help mitigate the inflammatory response and reduce tissue damage.
4. **Detoxification Enzyme Activation:** Certain polyphenols and flavonoids can stimulate the body's detoxification enzymes, enhancing the removal of heavy metals from the body.
5. **Cellular Protection:** Polyphenols and flavonoids may help protect cellular structures from cadmium-induced damage, preserving cell function and integrity.

It is essential to note that while antioxidant supplementation can be beneficial in mitigating cadmium toxicity, it should be used with caution and under proper guidance. Excessive antioxidant supplementation may have adverse effects, and the optimal dosage and duration should be determined based on individual needs and specific health conditions. Moreover, the most effective approach to counteract cadmium toxicity involves reducing exposure to cadmium at its source. Implementing environmental regulations, adopting safe waste management practices, and promoting sustainable agricultural and industrial practices can help minimize cadmium exposure and reduce the need for extensive antioxidant supplementation to combat its toxic effects.

6.2 Chelation Therapy

Chelation therapy is a medical treatment used to remove heavy metals, including cadmium, from the body by administering chelating agents. Chelating agents are compounds that can bind to heavy metals and form complexes that are more easily excreted through urine. Chelation therapy is primarily used in cases of severe heavy metal poisoning, but its effectiveness and safety can vary depending on the specific circumstances and the metal involved (Kim et al. 2019).

In the context of cadmium toxicity, chelation therapy has been explored as a potential treatment option (Flora et al. 2007, 2008). However, it is important to note that chelation therapy for cadmium toxicity is not as widely accepted or commonly used as it is for other heavy metals like lead or mercury. This is because cadmium behaves differently in the body, and its toxic effects are not solely due to its presence but also its impact on various cellular processes.

Key considerations regarding chelation therapy for cadmium toxicity include:

Chelating Agents: Common chelating agents used for heavy metal removal, such as EDTA (ethylene diaminetetra acetic acid), DMSA (dimercapto succinic acid), and DMPS (dimercapto propane sulfonate), have been studied for their potential to remove cadmium. However, their effectiveness in significantly reducing cadmium levels and improving health outcomes is still debated (Sompamit et al. 2010; Tandon et al. 2003).

Limited Efficacy: Cadmium tends to accumulate in specific tissues like the kidneys over a long period, and it forms stable complexes that are not as easily removed by chelation therapy as other metals. Chelation therapy might not effectively target cadmium stored in these tissues.

Potential Risks: Chelation therapy carries risks, including potential side effects and complications. Chelating agents can bind to essential minerals and trace elements, leading to imbalances and deficiencies.

Individualized Approach: The decision to use chelation therapy should be made on a case-by-case basis, considering the severity of cadmium toxicity, the overall health of the individual (or animal), and the potential benefits and risks of the treatment.

6.3 Phytochemical Interventions

Phytochemical interventions involve using bioactive compounds derived from plants to mitigate the adverse effects of cadmium toxicity. Phytochemicals, including polyphenols, flavonoids, antioxidants, and other compounds found in various plant-based foods, have been studied for their potential to counteract the harmful effects of cadmium exposure. For example, olive oil and colocynth oil were found to be protective against cadmium-induced oxidative stress in the liver of Wistar rats (Amamou et al. 2015). *Tinospora cordifolia* extract prevented cadmium-induced oxidative stress and hepatotoxicity in experimental rats (Baskaran et al. 2018). Protective effect of the artichoke (Cynara scolymus) leaf extract against cadmium toxicity-induced oxidative stress, hepato-renal damage, and immunosuppressive and hematological disorders in rats (El-Boshy et al. 2017). Here are some ways phytochemicals which may offer interventions for cadmium toxicity:

1. **Antioxidant Activity:** Cadmium exposure leads to the generation of reactive oxygen species (ROS) and oxidative stress, causing cellular damage. Many phytochemicals, such as flavonoids, anthocyanins, and resveratrol, possess strong antioxidant properties. They can neutralize ROS, reducing oxidative damage and protecting cells from cadmium-induced harm (Saleh 2014).
2. **Metal Chelation:** Certain phytochemicals have the ability to bind to heavy metals like cadmium. By forming complexes with cadmium ions, these compounds can reduce the absorption and distribution of cadmium in the body, potentially limiting its toxic effects.

3. **Anti-inflammatory Effects:** Cadmium exposure triggers inflammation in tissues, contributing to its toxic effects. Phytochemicals with anti-inflammatory properties, such as curcumin and quercetin, can help mitigate the inflammatory response and decrease tissue damage (Kanter et al. 2016; Park et al. 2021).
4. **Enhancement of Detoxification Enzymes:** Some phytochemicals can stimulate the body's detoxification enzymes, which are responsible for metabolizing and eliminating toxic compounds like cadmium. Inducing these enzymes can enhance the removal of cadmium from the body.
5. **Cellular Protection:** Phytochemicals may protect cellular structures and organelles from cadmium-induced damage, helping to maintain cell function and integrity.
6. **Kidney Protection:** Cadmium has a strong affinity for the kidneys, where it can cause significant damage. Certain phytochemicals have been studied for their potential to protect kidney tissues from cadmium-induced nephrotoxicity. Examples of plant-based foods rich in phytochemicals that might offer interventions against cadmium toxicity include berries (blueberries, strawberries), cruciferous vegetables (broccoli, cauliflower), green tea, turmeric, garlic, ferulic acid, and citrus fruits (Sanjeev et al. 2019).
7. **Cardio and Hepatic Protection**: Aframomum melegueta extract was reported to have ameliorative potential in cadmium-induced hepatic damage and oxidative stress in male Wistar rats (Oyinloye et al. 2016a, b). Studies conducted on Wistar rats also revealed Cardioprotective and antioxidant influence of aqueous extracts from *Sesamum indicum* seeds on oxidative stress (Oyinloye et al. 2016b). While phytochemicals show promise in mitigating cadmium toxicity, it is essential to consider factors such as the specific compounds, dosages, and individual health conditions. The effectiveness of phytochemical interventions can vary based on these factors. As with any health-related interventions, consulting a healthcare professional or veterinarian is crucial before making any changes to your or your pet's diet or treatment plan.

7 Conclusion

In this chapter, we have delved into the intricate relationship between cadmium toxicity and oxidative stress in animals, along with promising strategies for ameliorating these adverse effects. Cadmium, a ubiquitous environmental pollutant, poses a significant threat to both animal health and ecosystem integrity. Cadmium-induced oxidative stress arises from a complex interplay of mechanisms, including the generation of reactive oxygen species (ROS) and impaired antioxidant defense systems. This oxidative assault inflicts damage upon cellular components, leading to a cascade of detrimental effects on various physiological systems. From disruption of cellular membranes to DNA damage and protein oxidation, the repercussions of cadmium-induced oxidative stress are far-reaching. Researchers have tirelessly pursued interventions aimed at countering cadmium toxicity. Antioxidant supplementation, chelating agents, and dietary modifications have emerged as potential tools to alleviate the burden of oxidative stress. These strategies, although not without their limitations, have shown promise in reducing oxidative damage and mitigating the toxic impact of cadmium on animals. Future investigations should seek to elucidate the precise mechanisms of cadmium-induced oxidative stress in different animal species, explore novel amelioration approaches, and assess the long-term sustainability of these interventions.

References

Al-Attar AM (2011) Antioxidant effect of vitamin E treatment on some heavy metals-induced renal and testicular injuries in male mice. Saudi J Biol Sci 18:63–72

Amamou F, Nemmiche S, Kaouthar Meziane R, Didi A, Yazit SM, Chabane-Sari D (2015) Protective effect of olive oil and colocynth oil against cadmium-induced oxidative stress in the liver of Wistar rats. Food Chem Toxicol 78:177–184

Baskaran R, Priya LB, Kumar VS, Padma VV (2018) Tinospora cordifolia extract prevents cadmium-induced oxidative stress and hepatotoxicity in experimental rats. J Ayurveda Integr Med 9(4):252–257

Das SC, Al-Naemi HA (2019) Cadmium toxicity: oxidative stress, inflammation and tissue injury, vol 7. Academic Publisher, p 144

Dorian C, Gattone VH, Klaassen CD (1995) Discrepancy between the nephrotoxic potencies of cadmium-metallothionein and cadmium chloride and the renal concentration of cadmium in the proximal convoluted tubules. Toxicol Appl Pharmacol 130:161–168

El Heni J, Messaoudi I, Hamouda F, Abdelhamid K (2008) Protective effects of selenium (Se) and zinc (Zn) on Cd (Cd) toxicity in the liver and kidney of the rat: histology and Cd accumulation. Food Chem Toxicol 46:3522–3527

El-Boshy M, Ashshi A, Gaith M, Qusty N, Bokhary T, AlTaweel N, Abdelhady M (2017) Studies on the protective effect of the artichoke (Cynarascolymus) leaf extract against cadmium toxicity-induced oxidative stress, hepatorenal damage, and immunosuppressive and hematological disorders in rats. Environ Sci Pollut Res 24:12372–12383

Eybl V, Kotyzová D, Lešetický L, Bludovská M, Koutenský J (2006) The influence of curcumin and manganese complex of curcumin on cadmium-induced oxidative damage and trace elements status in tissues of mice. J Appl Toxicol 26(3):207–212

Flora SJS, Flora G, Saxena G, Mishra M (2007) Arsenic and lead induced free radical generation and their reversibility following chelation. Cell Mol Biol 53(1):26–47

Flora SJS, Mittal M, Mehta A (2008) Heavy metal induced oxidative stress & its possible reversal by chelation therapy. Indian J Med Res 128(4):501–523

Genchi G, Sinicropi MS, Lauria G, Carocci A, Catalano A (2020) The effects of cadmium toxicity. Int J Environ Res Public Health 17(11):3782

Goncalves JF, Fiorenza AM, Spanevello RM, Mazzanti CM, Bochi GV, Antes FG et al (2010) N-acetylcysteine prevents memory deficits, the decrease in acetylcholinesterase activity and oxidative stress in rats exposed to cadmium. Chem Biol Interact 186(1):53–60

Jayamurali D, Varier KM, Liu W, Raman J, Ben-David Y, Shen X, Gajendran B (2021) An overview of heavy metal toxicity. In: Metal, metal oxides and metal sulphides for biomedical applications. Springer, pp 323–342

Kanter M, Aktoz T, Aktas C, Ozen F, Yarali O, Kanter B (2016) Role of quercetin in cadmium-induced oxidative stress, testicular damage, and apoptosis in rats. Anal Quant Cytopathol Histpathol 38(1):45–51

Kara H, Cevik A, Konar V, Dayangac A, Yilmaz M (2007) Protective effects of antioxidants against cadmium-induced oxidative damage in rat testes. Biol Trace Elem Res 120:205–211

Kim JJ, Kim YS, Kumar V (2019) Heavy metal toxicity: an update of chelating therapeutic strategies. J Trace Elem Med Biol 54:226–231

Lalor GC, Rattray R, Simpson P, Vutchkov MK (1998) Heavy metals in Jamaica soils. Part 3: the distribution of Cd in Jamaican soils. La Revista Internacional de Contaminación Ambiental 14: 7–12

Lin AJ, Zhang XH, Chen MM, Qing CAO (2007) Oxidative stress and DNA damages induced by cadmium accumulation. J Environ Sci 19(5):596–602

Liu XR, Wang YY, Fan HR, Wu CJ, Kumar A, Yang LG (2016) Preventive effects of β-cryptoxanthin against cadmium-induced oxidative stress in the rat testis. Asian J Androl 18(6):920

Lopez Alonso M, Benedito JL, Miranda M, Castillo C, Hernández J, Shore RF (2000) Arsenic, cadmium, lead, copper and zinc in cattle from Galicia, NW Spain. Sci Total Environ 246:237–248

Lopez A, Benedito M, Miranda JL, Castillo M, Hernández C, Shore RF (2000) Arsenic, cadmium, lead, copper and zinc in cattle from Galicia, NW Spain. Sci Total Environ 246:237–248

Lui ZP (2003) Lead poisoning combined with cadmium in sheep and horses in the vicinity of non-ferrous metal smelters. Sci Total Environ 309(1–3):117–126

Messaoudi I, El Heni J, Hammouda F, Saïd K, Kerkeni A (2009) Protective effects of selenium, zinc, or their combination on cadmium-induced oxidative stress in rat kidney. Biol Trace Elem Res 130:152–161

Miranda M, Lopez-Alonso M, Castillo C, Hernandez J, Benedito JL (2005) Effects of moderate pollution on toxic and trace metal levels in calves from a polluted area of northern Spain. Environ Int J 31:543–548

Nemmiche S, Chabane-Sari D, Guiraud P (2007) Role of α-tocopherol in cadmium-induced oxidative stress in Wistar rat's blood, liver and brain. Chem Biol Interact 170(3):221–230

Oyinloye BE, Ajiboye BO, Ojo OA, Musa HM, Onikanni SA, Ojo AA (2016a) Ameliorative potential of Aframomum melegueta extract in cadmium-induced hepatic damage and oxidative stress in male Wistar rats. J Appl Pharm Sci 7(6):94–99

Oyinloye BE, Ajiboye BO, Ojo OA, Nwozo SO, Kappo AP (2016b) Cardioprotective and antioxidant influence of aqueous extracts from Sesamum indicum seeds on oxidative stress induced by cadmium in Wistar rats. Pharmacogn Mag 12(Suppl 2):S170

Park JH, Lee BM, Kim HS (2021) Potential protective roles of curcumin against cadmium-induced toxicity and oxidative stress. J Toxicol Environ Health B 24(3):95–118

Patra RC, Swarup D, Naresh R, Kumar P, Shekhar P, Ranjan R (2005) Cadmium level in blood and milk from animals reared around different polluting sources in India. Bull Environ Contam Toxicol 74:1092–1097

Patra RC, Swarup D, Sharma MC, Naresh R (2006) Trace mineral profile in blood and hair from cattle environmentally exposed to lead and cadmium around different industrial units. J Vet Med 53:511–517

Poli V, Madduru R, Aparna Y, Kandukuri V, Motireddy SR (2022) Amelioration of cadmium-induced oxidative damage in Wistar rats by vitamin C, zinc and N-acetylcysteine. Med Sci 10(1):7

Roman TRN, Lima EG, Azoubel R, Batigalia F (2002) Cadmium toxicity in man. HB Cientifica 9: 43–48

Saleh AAS (2014) Potential effects of some antioxidants against lead toxicity. Int J Adv Res 2(5): 824–834

Sanjeev S, Bidanchi RM, Murthy MK, Gurusubramanian G, Roy VK (2019) Influence of ferulic acid consumption in ameliorating the cadmium-induced liver and renal oxidative damage in rats. Environ Sci Pollut Res 26(20):20631–20653

Sarkar A, Ravindran G, Krishnamurthy V (2013) A brief review on the effect of cadmium toxicity: from cellular to organ level. Int J Biotechnol Res 3(1):17–36

Satarug S, Moore MR (2004) Adverse health effects of chronic exposure to low-level cadmium in foodstuffs and cigarette smoke. Environ Health Perspect 112:1099–1103

Shaikh ZA, Vu TT, Zaman K (1999) Oxidative stress as a mechanism of chronic cadmium-induced hepatotoxicity and renal toxicity and protection by antioxidants. Toxicol Appl Pharmacol 154(3):256–263

Sharma RK, Agrawal M (2005) Biological effects of heavy metals: an overview. J Environ Biol 26(2):301–313

Shi L, Cao H, Luo J, Liu P, Wang T, Hu G, Zhang C (2017) Effects of molybdenum and cadmium on the oxidative damage and kidney apoptosis in duck. Ecotoxicol Environ Saf 145:24–31

Sompamit K, Kukongviriyapan U, Donpunha W, Nakmareong S, Kukongviriyapan V (2010) Reversal of cadmium-induced vascular dysfunction and oxidative stress by meso-2,3-dimercaptosuccinic acid in mice. Toxicol Lett 198(1):77–82

Swarup D, Naresh R, Varshney VP, Balagangatharathilagar M, Kumar P, Nandi D, Patra RC (2007) Changes in plasma hormones profile and liver function in cows naturally exposed to lead and cadmium around different industrial areas. Res Vet Sci 82:16–21

Tandon SK, Singh S, Prasad S, Khandekar K, Dwivedi VK, Chatterjee M, Mathur N (2003) Reversal of cadmium induced oxidative stress by chelating agent, antioxidant or their combination in rat. Toxicol Lett 145(3):211–217

Tribowo JA, Arizal MH, Nashrullah M, Aditama AR, Utama DG (2014) Oxidative stress of cadmium-induced ovarian rat toxicity. Int J Chem Eng Appl 5(3):254

Unsal V, Dalkıran T, Çiçek M, Kölükçü E (2020) The role of natural antioxidants against reactive oxygen species produced by cadmium toxicity: a review. Adv Pharm Bull 10(2):184

Valko MMHCM, Morris H, Cronin MTD (2005) Metals, toxicity and oxidative stress. Curr Med Chem 12(10):1161–1208

Xiong L, Zhou B, Liu H, Cai L (2021) Comprehensive review of cadmium toxicity mechanisms in male reproduction and therapeutic strategies. Rev Environ Contam Toxicol 258:151–193

Yan LJ, Allen DC (2021) Cadmium-induced kidney injury: oxidative damage as a unifying mechanism. Biomol Ther 11(11):1575

SPRINGER NATURE

GPSR Compliance

The European Union's (EU) General Product Safety Regulation (GPSR) is a set of rules that requires consumer products to be safe and our obligations to ensure this.

If you have any concerns about our products, you can contact us on ProductSafety@springernature.com

In case Publisher is established outside the EU, the EU authorized representative is:

Springer Nature Customer Service Center GmbH
Europaplatz 3
69115 Heidelberg, Germany

The manufacturer's authorised representative in the EU is Springer Nature Customer Service Centre GmbH, Europaplatz 3, 69115 Heidelberg, Germany. If you have any concerns regarding our products, please contact ProductSafety@springernature.com

Printed and bound by CPI Group (UK) Ltd, Croydon, CR0 4YY
28/09/2026
02224056-0006